S.N. Singh • R.D. Tripathi
Environmental Bioremediation Technologies

Shree N. Singh
Rudra D. Tripathi
(Eds.)

# Environmental Bioremediation Technologies

With 58 Figures, 1 in colour

Springer

**Editors**

**Dr. Shree N. Singh**
Deputy Director and Head
Ecotoxicology & Bioremediation
National Botanical Research Institute
Lucknow 226001, (India)

**Dr. Rudra D. Tripathi**
Scientist E-II & Group Leader
Ecotoxicology & Bioremediation
National Botanical Research Institute
Lucknow 226001
India

ISBN 10 3-540-34790-9 **Springer Berlin Heidelberg New York**
ISBN 13 978-3-540-34790-3 **Springer Berlin Heidelberg New York**

Library of Congress Control Number: 2006927433



**Springer is a part of Springer Science+Business Media**
springeronline.com


Cover design: E. Kirchner, Heidelberg
Production: A. Oelschläger
Typesetting: Camera-ready by the Editors

Printed on acid-free paper 30/2132/AO 54321

# Foreword

Environmental contamination from both natural and anthropogenic sources is, today, a major environmental concern due to pervasiveness and persistence of many toxicants. It is considered as an inevitable evil of our progress and modernization. To decontaminate the soils, sediments and waters, polluted by anthropogenic activities, the scientists and technologists have evolved different technologies over the years. Although we have to pay high cost for physical and chemical environmental technologies, but they are not eco-friendly and safe. Hence, it was deeply realized to develop viable technologies employing microbes and plants to remediate not only metallic residues and radionuclides, but also the xenobiotic compounds like PCBs, PAHs, PCPs, petroleum sludge and the military wastes. No doubt, the scientists have also got some success in this endeavour and as the result, many companies are in place today to promote the sale of plant or microbe-based technologies to deal with specific environmental contamination challenges. Besides, these technologies are self-driven and do not disturb the sites in cleaning process.

In order to give a boost to this technology, I would like to appreciate the sincere efforts of my colleagues, Dr. S.N. Singh and Dr. R.D. Tripathi, both senior scientists of Ecotoxicology and Bioremediation Group of our institute, to publish this volume which contains latest information on the various aspects of bioremediation to deal with specific environmental contaminants. I hope this book will serve as a ready reckoner to the new researchers and also help the scientists working in this area in identifying the gaps for research. I consider this book a value addition to the scientific knowledge on bioremediation – an emerging and promising technology of today.

Rakesh Tuli
Director
NBRI, Lucknow
India

# Preface

Environmental bioremediation is an emerging technology because conventional methods to clean up the environment are cost-intensive and eco-unfriendly. In this technology, we employ from micro-organisms to higher plants to treat hazardous organic and metallic residues or by-products which enter into soils and sediments from various processes associated with domestic, municipal, agricultural, industrial and military activities. Hazardous materials may render harm to humans, livestock, wildlife, crops or native plants through handling, ingestion, application to land or other distributions of the contaminated materials into the environment.

No doubt, naturally occurring micro-organisms degrade the hazardous organic wastes including xenobiotic compounds, such as pesticides, polycyclic aromatic hydrocarbons (PAHs) and polychlorinated biphenyls (PCBs) in due course of time. However, metallic residues can not be degraded in composting, but may be converted into organic combinations that have less bioavailability than mineral combinations of the heavy metals. In addition, microbes can transform the oxidation states of several toxic metals and increase their bioavailability in the rhizosphere to be taken up by metal hyperaccumulating plants. This technology is termed as phytoremediation and has received a lot of attention in recent years due to its cost effectiveness solar driven and high efficiency. In addition, biotechnology provides us tools to accelerate the phytoremediation process through either over expression of genes responsible for the sequestration of metals in plants or gene transfer from low biomass accumulating metal hyperaccumulator plants to high biomass yielding non-accumulating plants.

To address this problem, we present before you an edited volume which focuses on different aspects of environmental bioremediation, such as (i) Accumulation, detoxification and bioremediation of heavy metals and radionuclides by plants and microbes (ii) Biotechnological approaches to enhance phytoremediation efficiency (iii) Bioremediation of petroleum sludge and polycyclic aromatic hydrocarbons (PAHs) (vi) Fungal-based treatment of textile wastewater and PCP-contaminated soil (v) Use of aquatic macrophytes in metal and nutrient removal (vi) Application of biofilms in porus media: mathematical modeling and numerical simulation (vii) Phytomonitoring and phytoremediation of air pollutants and (viii) Nanotechnology for bioremediation of heavy metals. These aspects have been dealt with in 21 chapters contributed by the leading workers, drawn from world over, in their own fields.

In this endeavour, we, the editors were not alone, but assisted by many people. We thank Director, NBRI, Dr. Rakesh Tuli, for his kind support and encouragement to this task. Besides, we would like to acknowledge all the contributors who responded to our request and contributed their chapters enthusiastically, containing the latest information on the relevant aspects. We also record our appreciation to all those, more particularly Dr. Todd R. Sandrin, USA, who helped us in editing the some of the manuscripts for value addition. The services rendered by our own research workers, Dr. Amitosh Verma, Dr. Sanjay Dwivedi, Dr. Larisha Tyagi, Dr. Vinay Singh Baghel, Mrs. Seema Mishra, Mr. Sudhakar Srivastava, Ms. Ragini Singh, Mrs. Babita Kumari, Mrs. Sudha Dwivedi, Ms. Sadhana Tiwari, Mr. Rishabh Kr. Tripathi and Mr. Deepak Pandey were remarkable and appreciable. Mr. Dilip Kumar Chakraborty deserves special thanks for his relentless efforts for computer work to prepare the manuscript on camera ready format.

Lastly, the editors acknowledge their family members for their inspiration, endurance and moral support during this period.

S.N. Singh
R.D. Tripathi

# Contents

# Contributors

**Banat, I.M.,** Biotechnology Research Group, School of Biomedical Sciences, University of Ulster, Coleraine – BT52 1SA, Northern Ireland, UK

**Banu, Sara Parwin,** Centre for Plant Molecular Biology, Tamil Nadu Agricultural University, Coimbatore 641 003, Tamil Nadu, INDIA

**Bezerra, Rui M.,** CETAV-Department of Biological and Environmental Engineering, University of Trás-os-Montes e Alto Douro, Apartado 1013, 5001-911 Vila Real, PORTUGAL

**Bruschi, Mireille,** Unité de Bioénergétique et Ingénierie des Protéines, UPR 9036, Institut de Biologie Structurale et Microbiologie, IFR88, C.N.R.S., 31 chemin Joseph Aiguier, 13402 Marseille Cedex 20, FRANCE, Email: bruschi@ibsm.cnrs-mrs.fr

**Chen-Charpentier Benito M.,** Department of Mathematics, University of Wyoming, Laramie, WY 82071-3036, USA, bchen@uwyo.edu

**D'Souza, S.F.,** Nuclear Agriculture & Biotechnology Division, Bhabha Atomic Research Centre, Mumbai 400 085, INDIA

**Dhania, Geeta,** Department of Bioscience, M.D. University, Rohtak 124 001, INDIA

**Dias, Albino A.,** CETAV-Department of Biological and Environmental Engineering, University of Trás-os-Montes e Alto Douro, Apartado 1013, 5001-911 Vila Real, PORTUGAL

**Duncan S.D.,** Department of Biological Sciences, University of Waikato, Private Bag 3105, Hamilton, NEW ZEALAND

**Dwivedi, U.N.,** Department of Biochemistry, University of Lucknow, Lucknw-226007, INDIA

**Eapen, Susan,** Nuclear Agriculture & Biotechnology Division, Bhabha Atomic Research Centre, Mumbai 400 085, INDIA, Email: eapenhome@yahoo.com

**Farrell, R.L.,** Department of Biological Sciences, University of Waikato, Private Bag 3105, Hamilton, NEW ZEALAND, Email: r.farrell@waikato.ac.nz

**Giovannetti, Luciana,** Dipartimento di Biotecnologie Agrarie – Sez. Microbiologia, University of Florence, ITALY, Email: luciana.giovannetti@unifi.it

**Goulhen, Florence,** Unité de Bioénergétique et Ingénierie des Protéines, UPR 9036, Institut de Biologie Structurale et Microbiologie, IFR88, C.N.R.S., 31 chemin Joseph Aiguier, 13402 Marseille Cedex 20, FRANCE

**Greenway, Margaret,** School of Environmental Engineering, Griffith University, Nathan, Queensland 4111, AUSTRALIA, Email: m.greenway@griffith.edu.au

**Grill, Erwin,** Technical University of Munich, Centre of Scienctific Research Weihenstephan, Department of Bioscience, Institute of Botany, Am Hochanger-4, 85350 Weihenstephan, GERMANY

**Gunasekaran, P.,** Department of Microbial Technology, Centre for Excellence in Genomic Sciences, School of Biological Sciences, Madurai Kamaraj University, Madurai 625 021, INDIA, Email: pguna@eth.net

**Heart Sunil,** School of Environmental Engineering, Griffith University, Nathan Campus, Brisbane, Queensland – 4111, AUSTRALIA

**Hoffman, Douglas R.,** Department of Biology and Microbiology, University of Wisconsin Oshkosh, Oshkosh, Wisconsin, USA

**Jain, R.K.,** Institute of Microbial Technology, Sector 39-A, Chandigarh-160 036, INDIA, Email: rkj@imtech.res.in

**Jaiwal, Pawan K.,** Department of Bioscience, M.D. University, Rohtak 124001, INDIA

**Jha, V.K.,** Institute of Life Sciences, Nalco Square, Bhubaneswar 751 023, Orissa, INDIA

**Kamaludeen,** Centre for Plant Molecular Biology, Tamil Nadu Agricultural University, Coimbatore 641 003, Tamil Nadu, INDIA

**Kapur, Manisha,** Institute of Microbial Technology, Sector 39-A, Chandigarh-160 036, INDIA

**Kobayashi, Nobuyuki,** Wind Engineering Research Center, Tokyo Polytechnic University, 1583, Iiyama, Atsugi, Kanagawa 243-0297, JAPAN

**Kojouharov Hristo V.,** Department of Mathematics, University of Texas at Arlington, Arlington, TX 76019-0408, USA

**Labana, Sumeet,** Institute of Microbial Technology, Sector 39-A, Chandigarh-160 036, INDIA

**Lamar R.T.,** Earthfax Development Corporation, Logan, UTAH

**Lord, R.,** Clean Environment Management Centre, School of Science and Technology, University of Teesside, Middlesbrough – TS13BA, Tees Valley, UK

**Malik, Deepak K.,** Institute of Microbial Technology, Sector 39-A, Chandigarh-160 036, INDIA

**Mishra, Seema,** Ecotoxicology and Bioremediation Group, National Botanical Research Institute, Rana Pratap Marg, Lucknow 226 001, INDIA

**Murooka, Yoshikatsu,** Department of Biotechnology, Graduate School of Engineering, Osaka University, 2-1 Yamada-oka, Suita-shi, Osaka 565-0871, JAPAN, Email: murooka@bio.eng.osaka-u.ac.jp

**Prakash, Dhan,** Institute of Microbial Technology, Sector 39-A, Chandigarh-160 036, INDIA

**Prasad, M.N.V.,** Department of Plant Sciences, University of Hyderabad, Hyderabad-500 046, INDIA, Email: mnvsl@uohyd.ernet.in

**Rahman, K.S.M.,** Clean Environment Management Centre, School of Science and Technology, University of Teesside, Middlesbrough – TS13BA, Tees Valley, UK, Email: P.Rahman@tees.ac.uk

**Rahman, T.J.,** Biotechnology Research Group, School of Biomedical Sciences, University of Ulster, Coleraine – BT52 1SA, Northern Ireland, UK

**Rajendran, P.,** Department of Zoology, Vivekananda College, Tiruvedakam West, Madurai 625 217, INDIA

**Ramasamy, K.,** Centre for Plant Molecular Biology, Tamil Nadu Agricultural University, Coimbatore 641 003, Tamil Nadu, INDIA, Email: ramasamytnau@yahoo.com

**Sahu, B.B.,** Institute of Life Sciences, Nalco Square, Bhubaneswar 751 023, Orissa, INDIA

**Sampaio, Ana,** CETAV-Department of Biological and Environmental Engineering, University of Trás-os-Montes e Alto Douro, Apartado 1013, 5001-911 Vila Real, PORTUGAL

**Sandrin, Todd R.,** Department of Biology and Microbiology, University of Wisconsin Oshkosh, Oshkosh, Wisconsin, USA, Email: sandrin@uwosh.edu

**Sharma, Asha,** Department of Bioscience, M.D. University, Rohtak 124 001, INDIA

**Shaw, B.P.,** Institute of Life Sciences, Nalco Square, Bhubaneswar 751 023, Orissa, INDIA, Email: b_p_shaw@yahoo.com

**Singh, Poorinima,** Department of Biochemistry, University of Lucknow, Lucknw-226007, INDIA

**Singh, Rana P.,** Department of Bioscience, M.D. University, Rohtak 124 001, INDIA, Email: rana_psingh@rediffmail.com

**Singh, S.N.,** Environmental Science Division, National Botanical Research Institute, Lucknow 226 001, INDIA, Email: drsn_s@rediffmail.com

**Singh, Shraddha,** Nuclear Agriculture & Biotechnology Division, Bhabha Atomic Research Centre, Mumbai 400 085, INDIA

**Sinha Rajiv K.**[1]**,** School of Environmental Engineering, Griffith University, Nathan Campus, Brisbane, Queensland – 4111, AUSTRALIA

**Sriprang, Rutchadaporn,** BIOTEC Center for Genetic Engineering and Biotechnology, National Science and Technology Development Agency, 113 Phaholyothin Rd., Klong 1, Klong Luang, Pathumthani, 12120, THAILAND

**Srivastava, Sudhakar,** Ecotoxicology and Bioremediation Group, National Botanical Research Institute, Rana Pratap Marg, Lucknow 226 001, INDIA

**Street G.,** Clean Environment Management Centre, School of Science and Technology, University of Teesside, Middlesbrough – TS13BA, Tees Valley, UK

**Tandon, P.K.,** Department of Botany, Lucknow University, Lucknow, INDIA

**Thwaites J.M.,** Department of Biological Sciences, University of Waikato, Private Bag 3105, Hamilton, NEW ZEALAND

**Tripathi, R.D.,** Ecotoxicology and Bioremediation Group, National Botanical Research Institute, Rana Pratap Marg, Lucknow 226 001, INDIA, Email: tripathi_rd@rediffmail.com

**Upadhyay, Jeetendra K.,** Wind Engineering Research Center, Tokyo Polytechnic University, 1583, Iiyama, Atsugi, Kanagawa 243-0297, JAPAN, Email: upadhyay@arch.t-kougei.ac.jp

**Verma, Amitosh,** Environmental Science Division, National Botanical Research Institute, Lucknow 226 001, INDIA

**Viti, Carlo,** Dipartimento di Biotecnologie Agrarie – Sez. Microbiologia, University of Florence, ITALY

**White R.B.,** Earthfax Development Corporation, Logan, UTAH

# 1

# Bioremediation of Organic and Metal Co-contaminated Environments: Effects of Metal Toxicity, Speciation, and Bioavailability on Biodegradation

Todd R. Sandrin and Douglas R. Hoffman

Department of Biology and Microbiology, University of Wisconsin Oshkosh, Oshkosh, Wisconsin, USA, Email: sandrin@uwosh.edu

## 1. Introduction

Forty percent of the hazardous waste sites on the U. S. Environmental Protection Agency's National Priority List (NPL) are co-contaminated with metal and organic pollutants (Sandrin et al. 2000). Metals most frequently found at Superfund sites include arsenic, barium, cadmium, chromium, lead, mercury, nickel and zinc. Common organic co-contaminants include petroleum, chlorinated solvents, pesticides and herbicides. Conventional approaches to removing the organic pollutants at these sites, such as pump and treat, are costly and often ineffective (NRC 1994). Bioremediation is a viable alternative to conventional technologies, but metal toxicity at co-contaminated sites may limit its utility. Many studies report that metals inhibit general microbial activity (e.g., litter decomposition, methanogenesis, acidogenesis, nitrogen transformation), but a few have specifically investigated the impact of metals on organic pollutant biodegradation. The fact, that metals affect a myriad of microbial activities suggests that metals have the potential to affect the biodegradation of organics in co-contaminated environments. In some studies, metals have no impact or have a stimulatory effect on microbial activity. Thus, the effect of metals on organic pollutant biodegradation remains poorly characterized. This review discusses: 1) the toxicity of metals to microorganisms, 2) the roles metal speciation and bioavailability play in governing the extent to which metals affect organic pollutant biodegradation, 3) reported effects of metals on aerobic and anaerobic biodegradation, 4) patterns in which metals affect biodegradation, and 5) approaches to increasing organic biodegradation in co-contaminated systems.

## 2. Metal Toxicity to Microorganisms

An understanding of mechanisms of metal toxicity is essential in anticipating to what extent, metals will inhibit pollutant biodegradation by a particular population of microorganisms. A lucid and comprehensive understanding of modes of metal toxicity may lead to the development of novel technologies to mitigate metal toxicity in metal and organic co-contaminated environments. Mechanisms of metal toxicity to microorganisms have been studied extensively, and several excellent reviews are now available (Nies 1992; Rouch et al. 1995a; Ji and Silver 1995; Silver and Phung 1996; Rosen 1996; Silver 1996; Nies 1999). Despite this sizable body of work, the precise mechanisms of the toxicity of many metals remain unclear. Hence Nies so astutely observed in his review of microbial metal toxicity and resistance, "We are just beginning to understand the metabolism of heavy metals" (Nies 1999).

### 2.1 Metal Chemistry

Incompletely filled d-orbitals allow metals to form complex compounds with organic ligands, such as the proteins (Nies 1999), nucleic acids, and cell wall materials of microorganisms (Toth and Tomasovicova 1989). This binding is beneficial in the case of some metals such as calcium, magnesium, manganese, copper, and zinc. These metals serve as enzyme co-factors in complex biochemical processes; however, at high concentrations, the same essential metals can form non-specific complexes with organic ligands. This leads to toxicity. In addition, some metals, such as mercury, cadmium, and silver, form such strong complexes with organic ligands that they are rarely used in biochemical process (Nies and Silver 1995). For example, only one enzyme, carbonic anhydrase utilized by a marine diatom, is known to use cadmium as a cofactor (Lane and Morel 2000; Lane et al. 2005).

Metals bind to functional groups of biological molecules with varying affinities and can be classified as either hard or soft. Hard metals (e.g., sodium, potassium, magnesium, calcium, manganese and iron) are small cations that are not readily polarizable, while soft metals (e.g., copper, lead, cadmium, mercury, and silver) are larger cations that are very polarizable due to their large number of electrons (Hughes and Poole 1991). Hard metals prefer to bind to ligands containing oxygen, such as carboxylic acid, sulfate, and phosphate functional groups. In contrast, soft metals preferentially bind to ligands containing sulfur, such as the sulfhydryl ($-SH_2$) groups found in proteins.

### 2.2 Heavy Metal Uptake

Of course, for a metal to bind to an essential protein, nucleic acid or membrane component, the metal must first be taken up by the cell. Differentiating between toxic and non-toxic metals is a complex cellular process. The structures of many metals, toxic and non-toxic, are remarkably similar. For instance, manganese, iron, cobalt, nickel, copper and zinc have ionic diameters which vary by less than 14% (from 138-160 pm) (CRC 1991). In addition, each of these cations is divalent. Serving as further disguise, some metals can coordinate with oxygen in such a way as to resemble common innocuous molecules. Arsenate ($AsO_4^{3-}$) resembles phosphate ($PO_4^{3-}$), while chromate ($CrO_4^{2-}$) is remarkably similar to sulfate ($SO_4^{2-}$). Evolution has endowed microorganisms with effective mechanisms to distinguish between toxic and non-toxic metals. Two general types of uptake mechanisms have been described: 1) selective, substrate-specific uptake systems that are slow and require considerable energy (ATP) and 2) substrate-non-specific, fast systems that transport metals using a chemiosmotic gradient rather than ATP (Nies and Silver 1995). Fast, nonspecific uptake systems are constitutively expressed, while slower, specific, energy-consuming uptake systems are inducible (Nies and Silver 1995).

An example of a fast, non-specific uptake system is the magnesium uptake system, CorA, found in Gram negative bacteria, archaea and baker's yeast. This system is responsible for the uptake of a variety of cations in addition to magnesium, including nickel, cobalt, zinc, and manganese. Two common fast transport systems that heavy metals often exploit to enter cells are Pit (phosphate inorganic transport) and the sulfate transport system. Arsenate is able to enter via Pit, while chromate can infiltrate cells via the sulfate transport system (Nies 1999). Slow, specific metal uptake systems include the P-type ATPases that transport zinc, manganese, cadmium, magnesium, calcium, potassium, copper, lead and silver (Fagan and Saier 1994).

## 2.3 Interaction of Heavy Metals with Cellular Components

Even highly evolved, substrate-specific uptake mechanisms may not prevent entry of a toxic metal into a cell. Once inside, metal cations can interact with various cellular components including cell membranes, proteins, and nucleic acids. Interactions of metals with these cellular components have been linked to toxicity (Toth and Tomasovicova 1989). Baath (1989) reported that copper and zinc disrupt the cell membrane. Furthermore, an early step in metal uptake may be binding of the metal to the cell surface. The outer membrane of Gram negative bacteria effectively complexes metals including sodium, calcium, magnesium, strontium, nickel, manganese, lead, and iron. In addition, the thin layer of peptidoglycan of Gram negative bacteria can bind metals, albeit not nearly as effectively as the thick layer of peptidoglycan of Gram positive bacteria which contain teichoic acid, a potent metal chelator (Beveridge and Doyle 1989).

The ability of cell surfaces to complex metals lies in their net negative charge at normal growth pH. In Gram negative bacteria, the net negative charge of the cell surface results from the phosphate and carboxyl groups of lipopolysaccharide molecules (Goldberg et al. 1983; Volesky 1990), while the negative charge in Gram positive bacteria results largely from teichoic acid. A more negative cell surface charge may more effectively attract and bind toxic metal cations, thus rendering the cell more susceptible to the toxic effects of the metal (Rai et al. 1996).

Interactions of metals with cellular proteins are more commonly implicated in causing toxicity than interactions of metals with membranes. Toxic metals readily bind to sulfhydryl groups of proteins. As mentioned above, soft cations, such as cadmium and lead, preferentially bind sulfur-containing ligands over oxygen-containing ones. This binding affects the structure and function of the protein. Interestingly, the dissociation constants of soft metals complexed to sulfhydryl groups correlate well with the minimum inhibitory concentration (MIC) of the same metals. This illustrates the importance of the ability of a metal to bind to proteins in determining its toxicity (Nies 1999).

### 2.4 Substitution for Essential Metabolites

If both hard and soft cations are present, soft cations will replace hard cations on ligands. This can lead to substitution of an essential metabolite by a toxic metal. The resemblance of some deleterious heavy metals to essential metals not only allows them to enter the cell, but also to exert their toxic effects via substitution. For example, chromate is often mistakenly used as sulfate, arsenate is mistaken for phosphate, cadmium is used as an enzyme co-factor instead of zinc or calcium, nickel and cobalt replace iron, and zinc is commonly mistaken for magnesium. All of these mistaken identities result in the construction of an unstable, inhibited, or non-functional enzyme or other biological molecule (Nies and Silver 1995; Nies 1999).

### 2.5 Heavy Metal Induced Oxidative Stress

The toxicity of heavy metals to Gram negative bacteria is due, in part, to oxidative stress (Kachur et al. 1998). Metal cations may bind two glutathione molecules, forming a bis-glutathione molecule that reacts with diatomic oxygen to yield oxidized bis-glutathione, the metal cation, and hydrogen peroxide. The oxidized bis-glutathione must be reduced using NADPH; however, the metal cation released in the process is once again free to re-initiate this process and continue imposing considerable oxidative stress on the cell (Nies 1999).

## 3. Metal Speciation and Bioavailability

Despite the substantial information concerning mechanisms of metal toxicity, meaningful quantitative data on responses of pollutant-degrading microorganisms to metals is still lacking. This is largely due to the fact that making comparisons between concentrations of metals that inhibit biodegradation reported by different studies is exceedingly difficult. For example, five orders of magnitude separate literature reports of concentrations of zinc that inhibit biodegradation (Table 1). While it should be noted that not all studies attempted to pinpoint the lowest concentration that inhibits biodegradation, many disparities likely result from variations in metal bioavailability between studies.

Most commonly, metal inhibition of biodegradation has been related to the total metal concentration in a system. This may not be the most appropriate predictor of metal toxicity, as suggested by the wide range of total metal concentrations reported to inhibit biodegradation (Table 1). The concentration of the most bioavailable form (i.e., species) of the metal (commonly held to be the free, ionic, solution-phase metal species) is likely a better indicator of the extent to which a metal will inhibit biodegradation. In media commonly used to study metal toxicity, metals exist in a number of different species in addition to the free, ionic species. Depending on medium characteristics described below, metals can exist as free ions (possibly with different oxidation states), hydroxo-complexes, or be complexed to organic or inorganic ligands (Hughes and Poole 1991; Twiss et al. 2001). The distribution of these different metal forms is referred to as metal speciation.

### 3.1 Factors Affecting Metal Speciation and Toxicity

It is well-established that different metal species vary in their biological reactivity (Hughes and Poole 1991; Traina and Laperche 1999; Twiss et al. 2001; Behra et al. 2002). Certain metal species are more likely than others to associate with biochemically active sites (e.g., enzymes) and initiate biological responses. In this review, we define bioavailability as the ability of a metal species to access these sites. In the case of organic-degrading microbes, interactions of metals with enzymes results in the inactivation of enzymes necessary for biodegradation (e.g., monoxygenases, dioxygenases) or of enzymes used in the general metabolism (Nies 1999; Baldrian et al. 2000; Sandrin and Maier 2003). There is still some debate as to which metal species are most bioavailable. Currently, though, there is a considerable amount of evidence suggesting that free, ionic, solution-phase metal species are most bioavailable (Angle and Chaney 1989; Traina and Laperche 1999; Behra, et al. 2002). Despite being highly bioavailable, the free ionic metal concentration may represent only a small fraction of the total metal species distribution in a solid or aqueous medium. For these reasons, it is of paramount importance to understand what properties of metal toxicity test systems impact metal speciation and metal bioavailability. Two of the most important of these properties are medium chemical composition and pH.

**Table 1.** Reported metal concentrations that inhibit aerobic (A) and anaerobic (B) biodegradation and/or transformation of organic pollutants

**A.**

| Metal | Organic | Lowest metal conc. reported to reduce biodegradation | Microbe(s) Studied | Environment | pH | Reference |
|---|---|---|---|---|---|---|
| $Cd^{2+}$ | 2,4-D | 0.060 mg/g[a] | *Alcaligenes eutrophus* JMP134 | soil microcosms | 8.2 | Roane et al. (2001) |
| $Cd^{2+}$ | 2,4-D | 0.060 mg/g[a] | *Alcaligenes eutrophus* JMP134 | field-scale bioreactors | 8.2 | Roane et al. (2001) |
| $Cd^{2+}$ | 2,4-DME | 0.100 mg/l[a] | indigenous community | sediment (microcosm) | 6.5 | Said and Lewis (1991) |
| $Cd^{2+}$ | 2,4-DME | 0.629 mg/l[a] | indigenous community | aufwuchs[e] (microcosm) | 5.6 | Said and Lewis (1991) |
| $Cd^{2+}$ | PHEN | 1 mg/l[b] | indigenous community | soil microcosms | 7.6 | Maslin and Maier (2000) |
| $Cd^{2+}$ | NAPH | 1 mg/l[b] | *Burkholderia* sp. | dilute mineral salts medium containing 1.4 mM phosphate | 6.5 | Sandrin et al. (2000) |
| $Cd^{2+}$ | diesel fuel | 1.1 mg/l[a] | Enrichment culture | MES-buffered mineral salts medium containing 0.33 mM phosphate | 6.8 | Riis et al. (2002) |
| $Cd^{2+}$ | 2,4-D | >3 mg/l[a] | *Alcaligenes eutrophus* JMP134 | mineral salts medium | 6.0 | Roane et al. (2001) |
| $Cd^{2+}$ | 2,4-D | 24 mg/l[a] | *Alcaligenes eutrophus* JMP134 | mineral salts medium containing cadmium resistant isolate | 6.0 | Roane et al. (2001) |

| | | | | | | |
|---|---|---|---|---|---|---|
| $Cd^{2+}$ | 4CP, 3CB, 2,4D, XYL, IPB, NAPH, BP | <25.3 - 50.6 mg/l[a,c] | *Alcaligenes* spp., *Pseudomonas* spp., *Moraxella* sp. | Tris-buffered minimal medium plates | 7.0 | Springael et al. (1993) |
| $Cd^{2+}$ | TOL | 37 mg/l[a] | *Bacillus* sp. | mineral salts medium containing 36 mM phosphate | 5.9 | Amor et al. (2001) |
| $Cd^{2+}$ | EDTA | 562 mg/l[d] | Enrichment culture | MOPS-buffered minimal medium | 7.0 | Thomas et al. (1998) |
| $Co^{2+}$ | 4CP,3CB,2,4D, XYL, IPB, NAPH, BP | <13.3 - 1,330 mg/l[a, c] | *Alcaligenes* spp., *Pseudomonas* spp., *Moraxella* sp. | Tris-buffered minimal medium plates | 7.0 | Springael et al. (1993) |
| $Co^{2+}$ | NTA | 116.9 mg/l[d] | enrichment culture | PIPES-buffered mineral salts medium | 7.0 | White and Knowles (2003) |
| $Co^{2+}$ | EDTA | 292 mg/l[d] | Enrichment culture | MOPS-buffered minimal medium | 7.0 | Thomas et al. (1998) |
| $Cr^{3+}$ | 2,4-DME | 0.177 mg/l[a] | indigenous | aufwuchs[e] (microcosm) | 6.1 | Said and Lewis (1991) |
| $Cr^{6+}$ | diesel fuel | 2.32 mg/l[a] | Enrichment culture | MES-buffered mineral salts medium containing 0.33 mM phosphate | 6.8 | Riis et al. (2002) |
| $Cr^{6+}$ | 4CP, 3CB, 2,4D, XYL, IPB, NAPH, BP | <131 mg/l[a c] | *Alcaligenes* spp., *Pseudomonas* spp., *Moraxella* sp. | Tris-buffered minimal medium plates | 7.0 | Springael et al. (1993) |
| $Cu^{2+}$ | PH | 0.01 mg/l[a] | *Acinetobacter calcoaceticus* AH strain | bioreactor medium containing 0.15 mM phosphate | 7.8 | Nakamura and Sawada (2000) |
| $Cu^{2+}$ | 2,4-DME | 0.027 mg/l[a] | indigenous community | aufwuchs[e] (microcosm) | 5.0 | Said and Lewis (1991) |
| $Cu^{2+}$ | 2,4-DME | 0.076 mg/l[a] | indigenous community | sediment (microcosm) | 6.1 | Said and Lewis (1991) |

| | | | | | | |
|---|---|---|---|---|---|---|
| $Cu^{2+}$ | NTA | 3.18 mg/l[d] | *Chelatobacter heintzii* ATCC 29600 | PIPES-buffered mineral salts medium | 7.0 | White and Knowles (2000) |
| $Cu^{2+}$ | crude oil | 6.30 mg/l[a] | *Pseudomonas* sp. | mineral salts medium containing 31 mM phosphate | 7.2 | Benka-Coker and Ekundayo (1998) |
| $Cu^{2+}$ | diesel fuel | 6.35 mg/l[a] | Enrichment culture | MES-buffered mineral salts medium containing 0.33 mM phosphate | 6.8 | Riis et al. (2002) |
| $Cu^{2+}$ | PHB | 8 mg/l[b] | *Acidovorax delafieldii* | agar plates containing 4.70 mM phosphate | 6.9 | Birch and Brandl (1996) |
| $Cu^{2+}$ | crude oil | 11.25 mg/l[a] | *Micrococcus* sp. | mineral salts medium containing 31 mM phosphate | 7.2 | Benka-Coker and Ekundayo (1998) |
| $Cu^{2+}$ | 4CP, 3CB, 2,4-D, XYL, IPB, NAPH, BP | <14.3 -71.6 mg/l[a,c] | *Alcaligenes* sp., *Pseudomonas* spp., *Moraxella* sp. | Tris-buffered minimal medium plates | 7.0 | Springael et al. (1993) |
| $Cu^{2+}$ | NTA | 127.1 mg/l[d] | enrichment culture | PIPES-buffered mineral salts medium | 7.0 | White and Knowles (2003) |
| $Cu^{2+}$ | NTA | 127.1 mg/l[d] | *Mesorhizobium* sp. NCIMB 13524 | PIPES and phosphate-buffered mineral salts media | 7.0 | White and Knowles (2003) |
| $Cu^{2+}$ | EDTA | 318 mg/l[d] | Enrichment culture | MOPS-buffered minimal medium | 7.0 | Thomas et al. (1998) |
| $Hg^{2+}$ | 2,4-DME | 0.002 mg/l[a] | indigenous community | aufwuchs[e] (microcosm) | 6.8 | Said and Lewis (1991) |
| $Hg^{2+}$ | diesel fuel | 4 mg/l[a] | Enrichment culture | MES-buffered mineral salts medium containing 0.33 mM phosphate | 6.8 | Riis et al. (2002) |
| $Hg^{2+}$ | 4 CP, 3 CB, 2,4-D, XYL, IPB, NAPH, BP | <45.2 - 226 mg/l[a,c] | *Alcaligenes* sp., *Pseudomonas* spp., *Moraxella* sp. | Tris-buffered minimal medium plates | 7.0 | Springael et al. (1993) |

| | | | | | | |
|---|---|---|---|---|---|---|
| $Mn^{2+}$ | crude oil | 28.2 mg/l[a] | *Micrococcus* sp. | mineral salts medium containing 31 mM phosphate | 7.2 | Benka-Coker and Ekundayo (1998) |
| $Mn^{2+}$ | crude oil | 317.0 mg/l[a] | *Pseudomonas* sp. | mineral salts medium containing 31 mM phosphate | 7.2 | Benka-Coker and Ekundayo (1998) |
| $Ni^{2+}$ | 4 CP, 3 CB, 2,4-D, XYL, IPB, NAPH, BP | 5.18 - 10.3 mg/l[a,c] | *Alcaligenes* sp., *Pseudomonas* spp., *Moraxella* sp. | Tris-buffered minimal medium plates | 7.0 | Springael et al. (1993) |
| $Ni^{2+}$ | diesel fuel | 5.9 mg/l[a] | Enrichment culture | MES-buffered mineral salts medium containing 0.33 mM phosphate | 6.8 | Riis et al. (2002) |
| $Ni^{2+}$ | TOL | 20 mg/l[a] | *Bacillus* sp. | mineral salts medium containing 36 mM phosphate | 5.9 | Amor et al. (2001) |
| $Ni^{2+}$ | NTA | 117.4 mg/l[d] | *Mesorhizobium* sp. NCIMB 13524 | PIPES and phosphate-buffered mineral salts media | 7.0 | White and Knowles (2003) |
| $Ni^{2+}$ | EDTA | 293 mg/l[d] | Enrichment culture | MOPS-buffered minimal medium | 7.0 | Thomas et al. (1998) |
| $Pb^{2+}$ | crude oil | 1.41 mg/l[a] | *Micrococcus* sp. | mineral salts medium containing 31 mM phosphate | 7.2 | Benka-Coker and Ekundayo (1998) |
| $Pb^{2+}$ | crude oil | 2.80 mg/l[a] | *Pseudomonas* sp. | mineral salts medium containing 31 mM phosphate | 7.2 | Benka-Coker and Ekundayo (1998) |
| $Pb^{2+}$ | diesel fuel | 41.4 mg/l[a] | Enrichment culture | MES-buffered mineral salts medium containing 0.33 mM phosphate | 6.8 | Riis et al. (2002) |
| $Zn^{2+}$ | 2,4-DME | 0.006 mg/l[a] | indigenous community | sediment (microcosm) | 6.4 | Said and Lewis (1991) |
| $Zn^{2+}$ | 2,4-DME | 0.041 mg/l[a] | indigenous community | aufwuchs[e] (microcosm) | 5.6 | Said and Lewis (1991) |

| | | | | | | |
|---|---|---|---|---|---|---|
| $Zn^{2+}$ | crude oil | 0.43 mg/l[a] | *Pseudomonas* sp. | mineral salts medium containing 31 mM phosphate | 7.2 | Benka-Coker and Ekundayo (1998) |
| $Zn^{2+}$ | crude oil | 0.46 mg/l[a] | *Micrococcus* sp. | mineral salts medium containing 31 mM phosphate | 7.2 | Benka-Coker and Ekundayo (1998) |
| $Zn^{2+}$ | TOL | 2.8 mg/l[a] | *Bacillus* sp. | mineral salts medium containing 36 mM phosphate | 5.9 | Amor et al. (2001) |
| $Zn^{2+}$ | pH | 10 mg/l[a] | *Acinetobacter calcoaceticus* AH strain | bioreactor medium containing 0.15 mM phosphate | 7.8 | Nakamura and Sawada (2000) |
| $Zn^{2+}$ | 4 CP, 3 CB, 2,4-D, XYL, IPB, NAPH, BP | <29.5 - 736 mg/l[a,c] | *Alcaligenes* sp., *Pseudomonas* spp., *Moraxella* sp. | Tris-buffered minimal medium plates | 7.0 | Springael et al. (1993) |
| $Zn^{2+}$ | diesel fuel | 65.4 mg/l[a] | Enrichment culture | MES-buffered mineral salts medium containing 0.33 mM phosphate | 6.8 | Riis et al. (2002) |
| $Zn^{2+}$ | NTA | 130.8 mg/l[d] | *Mesorhizobium* sp. NCIMB 13524 | PIPES-buffered mineral salts media | 7.0 | White and Knowles (2003) |

**Abbreviations:**
2,4D, 2,4-dichlorophenoxy acetic acid; 2,4-DME, 2,4-dichlorophenoxy acetic acid methyl ester; BP, biphenyl; CB, chlorobenzoate; CP, chlorophenol; EDTA, ethylenediaminetetraacetic acid; IPB, isopropylbenzene; MES, morpholinoethane sulfonic acid; MOPS, 3-(N-morpholino)propanesulfonic acid; NAPH, naphthalene; NTA, nitrilotriacetic acid; PH, phenol; PHB, poly (3-hydroxybutyrate); PHEN, phenanthrene; PIPES, Piperazine-N,N'-bis(2-ethanesulfonic acid); TOL, toluene; XYL, xylene

[a] value represents total metal added to system

[b] value represents solution phase concentration of metal present in system

[c] value represents Minimum Inhibitory Concentration (MIC) calculated by multiplying Maximum Tolerated Concentration (MTC) by a factor of 2.25. MIC = MTC*2.25.

[d] metal was complexed to a biodegradable organic (NTA or EDTA)

[e] floating algal mats

**B.**

| Metal | Organic | Lowest metal conc. reported to reduce biodegradation | Microbe(s) Studied | Environment | pH | Reference |
|---|---|---|---|---|---|---|
| $Cd^{2+}$ | HCB | 0.001 mg/g[a] | indigenous community | microcosms containing contaminated sediment | NR | Jackson and Pardue (1998) |
| $Cd^{2+}$ | TCA | 0.01 mg/l[b] | indigenous community | laboratory soil microcosms containing rice paddy and bottomland hardwood soils | 6.9-7.4 | Pardue et al. (1996) |
| $Cd^{2+}$ | TCA | 0.2 mg/l[b] | indigenous community | laboratory soil microcosms containing organic matter-rich soil | 6.8 | Pardue et al. (1996) |
| $Cd^{2+}$ | 2CP, PH, BEN, 3CB | 0.5-1.0 mg/l[a] | indigenous community | aqueous sediment enrichment in anaerobic growth medium | 7.0 | Kuo and Genthner (1996) |
| $Cd^{2+}$ | TCE | 5 mg/l[a] | *Burkholderia picketti* PK01 | mineral salts medium containing 44 mM phosphate; denitrifying conditions | NR | Degraffenreid and Shreve (1998) |
| $Cd^{2+}$ | 2CP, 3CP | 20 mg/l[a] | indigenous community | sediment slurry | 7.0 | Kong (1998) |
| $Cr^{6+}$ | 2CP, PH, BEN, 3CB | 0.01-0.5 mg/l[a] | indigenous community | aqueous sediment enrichment in anaerobic growth medium | 7.0 | Kuo and Genthner (1996) |
| $Cr^{6+}$ | 2CP, 3CP | 20 mg/l[a] | indigenous community | sediment slurry | 7.0 | Kong (1998) |
| $Cr^{c}$ | OD | 5,000 µg/g[a] | indigenous community | clay-containing sediment slurry | 6.5 | DeLaune et al. (1998) |
| $Cu^{2+}$ | 2CP, PH, BEN, 3CB | 0.1-1.0 mg/l[a] | indigenous community | aqueous sediment enrichment in anaerobic growth medium | 7.0 | Kuo and Genthner (1996) |
| $Cu^{2+}$ | 2,4-DANT, RDX | 4 mg/g[a] | indigenous community | soil slurry containing 50 mM phosphate buffer | 6.5 | Roberts et al. (1998) |

| | | | | | | |
|---|---|---|---|---|---|---|
| $Cu^{2+}$ | 4-ADNT | 8 mg/g[a] | indigenous community | soil slurry containing 50 mM phosphate buffer | 6.5 | Roberts et al. (1998) |
| $Cu^{2+}$ | 2CP, 3CP | 20 mg/l[a] | indigenous community | sediment slurry | 7.0 | Kong (1998) |
| $Hg^{2+}$ | 2CP, PH, BEN, 3CB | 0.1-1.0 mg/l[a] | indigenous community | aqueous sediment enrichment in anaerobic growth medium | 7.0 | Kuo and Genthner (1996) |
| $Pb^{2+}$ | HCB | 0.001 mg/g[a] | indigenous community | microcosms containing contaminated sediment | NR | Jackson and Pardue (1998) |
| $Pb^{2+}$ | 2,4-DANT, RDX | >1 mg/g[a] | indigenous community | soil slurry containing 50 mM phosphate buffer | 6.5 | Roberts et al. (1998) |
| $Zn^{2+}$ | 2,4-DANT | 1.5 mg/g[a] | indigenous community | soil slurry containing 50 mM phosphate buffer | 6.5 | Roberts et al. (1998) |
| $Zn^{2+}$ | PCP | 2 mg/l[a] | indigenous community | anaerobic digester sludge in a liquid medium containing 0.6 mM phosphate | NR | Jin and Bhattacharya (1996) |
| $Zn^{2+}$ | PCP | 8.6 mg/l[a] | indigenous community | anaerobic enrichment cultures in serum bottles | NR | Majumdar et al. (1999) |
| $Zn^{2+}$ | NB | 10 mg/l[a] | indigenous community | anaerobic enrichment cultures in serum bottles | NR | Majumdar et al. (1999) |

**Abbreviations:**
2,4-DANT, 2,4-diamino-6-nitrotoluene; 4-ADNT, 4-amino-2,6-dinitrotoluene; BEN, benzoate; CB, chlorobenzoate; CP, chlorophenol; HCB, hexachlorobenzene; NB, nitrobenzene; NR, not reported; OD, octadecane; PCP, pentachlorophenol; PH, phenol; RDX, hexahydro-1,3,5-trinitro-1,3,5-triazine; TCA, trichloroaniline; TCE, trichloroethylene.

[a] value represents total metal added to system

[b] value represents solution phase concentration of metal present in system

[c] oxidation state not specified

### 3.1.1 Chemical Composition

To accurately characterize metal speciation, the chemical composition of the medium must be known. This requires use of a chemically defined medium to ensure that all components capable of interacting with metals are taken into consideration (Hughes and Poole 1991; Twiss et al. 2001). Many complex microbiological media contain extracts (e.g., yeast extract and beef extract) that vary in their precise chemical composition. Common to many studies investigating organic pollutant biodegradation, a minimal medium is often used. Minimal media typically consist of a solution of mineral salts amended with an organic pollutant targeted for degradation as the sole source of carbon (Springael et al. 1993; Benka-Coker and Ekundayo 1998; Amor et al. 2001; Roane et al. 2001; Sandrin and Maier 2002). Some studies have also used sediment or soil slurries taken directly from the environment to monitor the biodegradation of an added organic, while others have used a combination of these approaches by placing a defined amount of sediment or soil into a minimal medium containing an organic pollutant (Said and Lewis 1991; Pardue et al. 1996; Delaune et al. 1998; Kong 1998; Roberts et al. 1998; Maslin and Maier 2000). Regardless of the type of medium, the buffering system has a dramatic impact on metal speciation and bioavailability. Because buffers are often present at higher concentrations than other medium components and may contain agents that reduce metal bioavailability, their impact on metal speciation and bioavailability must be considered (Hughes and Poole 1991; Teresa et al. 2000; Vasconcelos and Leal 2002).

A variety of buffers have been used in studies examining effects of metals on biodegradation. Phosphate buffers, probably among the most common buffers used in microbiology, have been used in the majority of studies (Birch and Brandl 1996; Benka-Coker and Ekundayo 1998; Amor et al. 2001; Nakamura and Sawada 2000). Phosphate readily sequesters metals and reduces their bioavailability via the formation of insoluble metal-phosphate species. In fact, phosphate is so efficient at metal sequestration that it has been used as a metal-complexing agent in a few studies to reduce free ionic metal concentrations (Ruby et al. 1994; White and Knowles 2000). The remarkable ability of phosphate to reduce bioavailable metal concentrations is illustrated in Figure 1A, which shows predicted concentrations of free ionic metals as a function of phosphate concentration in a medium commonly used in biodegradation studies, Bushnell Haas medium (Difco™, Sparks, MD). With a relatively low phosphate concentration of 2.27 mM, 44% less free ionic cadmium exists in the medium containing phosphate than in the same medium not containing phosphate. Some metals are more sensitive to phosphate precipitation than others. As shown in Figure 1A, cobalt bioavailability is predicted to remain high (95% remains in the free, ionic form) as the phosphate concentration is raised to 15 mM, but the concentration of free, ionic nickel is predicted to fall to 21% of its concentration in the medium free of phosphate. Metal-phosphate species are quite insoluble, even at neutral to mildly acidic pH values.

The ability of phosphate buffers to precipitate metals has been taken for granted in several metal toxicity studies. In their review of metal speciation, Hughes and Poole (1991) describe the difficulty of detecting metal precipitates in a turbid culture. Metal-phosphate precipitates can present many problems, especially if culture turbidity is used as the measure of growth and biodegradation. Precipitates can easily be misinterpreted as cell biomass, making growth measurements misleading and inaccurate. In their study of nitrilotriacetic acid (NTA) biodegradation, White and Knowles avoided this problem by acidifying their samples prior to measuring culture turbidity (White and Knowles 2000; 2003). Lowering the pH dissolved any metal-phosphate precipitates present in the samples. Other techniques have been developed to overcome problems with phosphate precipitation. For example, Malakul *et al.* (1998) replaced phosphate with glycerophosphate. In this form, phosphate will not readily bind metals and cause precipitation. Glycerophosphate, though, can potentially serve as a carbon source for organic-degrading microbes, thus decreasing the effectiveness of pollutant biodegradation and confounding interpretation of biodegradation data based solely on biomass measurements. Metal-phosphate precipitation can also be reduced by decreasing the phosphate concentration. This allows higher metal levels to be tested while reducing precipitation and increasing bioavailability. Though, caution should be exercised as the buffering capacity of the medium will be compromised as the phosphate concentration is reduced.

Metals tend to remain more bioavailable in the presence of zwitterionic buffers (such as HEPES (4-(2-hydroxyethyl)-1-piperazineethanesulfonic acid), MES (morpholinoethanesulfonic acid), MOPS (3-(N-morpholino)propanesulfonic acid), and PIPES (1,4-piperazinebis(ethanesulfonic acid)) than in the presence of phosphate buffers. This is due to the fact that these buffers do not interact with metals as strongly as phosphate buffers. At pH 7.2, Mash *et al.* (2003) reported that MES and MOPS buffers (each at 50 mM) did not complex copper, while HEPES (35 mM) showed some copper complexation. PIPES buffer (0.8 mM) did not complex copper (Vasconcelos et al. 1998). Despite its frequent use in metal toxicity studies, little metal complexation data is available for Tris-base (2-amino-2-(hydroxymethyl)-1,3-propanediol). Available data, however, suggest that Tris-base is capable of complexing many metals, though to what extent is not clear (Twiss et al. 2001). Because of their limited interaction with metals, many have recommended the use of MES, MOPS, and PIPES in metal toxicity studies, presuming studies are conducted in the operational pH range of the buffers (6.1-7.5) (Twiss et al. 2001; Mash et al. 2003).

While some buffers do not complex metals, many inorganic ligands, such as $Cl^-$, $NO_3^-$, $OH^-$, $SO_3^-$ and $SO_4^{2-}$, have strong metal-complexing capabilities and high affinities for many metals. Metals complexed with these ligands usually remain soluble; however, their bioavailability is thought to be lower than free, ionic metals (Reed and Nonavinakere 1992; Janos 1993; Bianchini and Bowles 2002).

### 3.1.2 pH

Metal speciation and bioavailability are also dependent on pH. In general, metals are more bioavailable at acidic pH values (Hughes and Poole 1991; Twiss et al. 2001). Under acidic conditions, free ionic metal species are thought to be more prevalent, likely from the saturation of metal binding sites with protons ($H^+$). This saturation limits interactions between metals and potential metal-complexing ligands. Also, under basic conditions, metals tend to form hydroxy-metal complexes. Figure 1B illustrates the predicted pH-dependent loss of free ionic metal species in Bushnell Haas medium amended with a total concentration of 100 μM of one of several metals. Depending on the particular metal, hydroxo-metal complexes may be soluble (e.g., $CdOH^+$, $NiOH^+$, $ZnOH^+$) or insoluble (e.g., $Cr(OH)_3$, $Fe(OH)_3$). The dependence of metal bioavailability on pH varies between different metals. For example, at pH 7, 68 μM cobalt is predicted to exist in the free, ionic form, whereas only 4.1 μM nickel remains in the same form. Free, ionic concentrations of lead, copper, cadmium, and zinc are predicted to be considerably lower.

Medium pH acts in conjunction with phosphate content to profoundly impact free ionic metal levels. Figure 1C shows predicted free ionic concentrations of cadmium in Bushnell-Haas medium initially amended with 100 μM total cadmium, adjusted to different pH values, and containing variable amounts of phosphate. Small changes in pH or phosphate concentration can have large effects on free ionic metal concentrations. For example, $Cd^{2+}$ levels decreased dramatically as pH and phosphate concentration increased. At pH 7 in the presence of 0, 0.15, 1.5, 15, 30, and 50 mM phosphate, $Cd^{2+}$ levels were predicted to be 66, 64, 17, 1.5, 0.93, and 0.65 μM, respectively. Of the studies summarized in Table 1, nine used a medium containing a mean phosphate concentration of ~19 mM and were adjusted to a mean pH of 6.8. Thus, bioavailable concentrations of metals in studies cited in Table 1 are likely much lower than the reported total metal concentrations.

Because pH strongly influences free ionic levels of metals and their bioavailability, maintaining pH throughout the duration of an experiment is necessary. This requires selection of an appropriate buffering system. Biodegradation studies are typically conducted at neutral to mildly acidic pH values. Use of buffers whose operational pH range lies in this region is recommended to avoid dramatic changes in pH. The operational pH range of a buffer is dependent on the pKa of the weak acid(s) used to buffer the medium. A buffer's pKa value represents the pH at which one-half of the buffering agent is protonated. Using a buffer at a pH significantly higher or lower than its pKa will result in a poorly buffered medium. Excretion of acidic metabolic end products by microbes can reduce the pH of marginally buffered media and can result in unanticipated metal speciation events (Hughes and Poole 1991; Twiss et al. 2001).

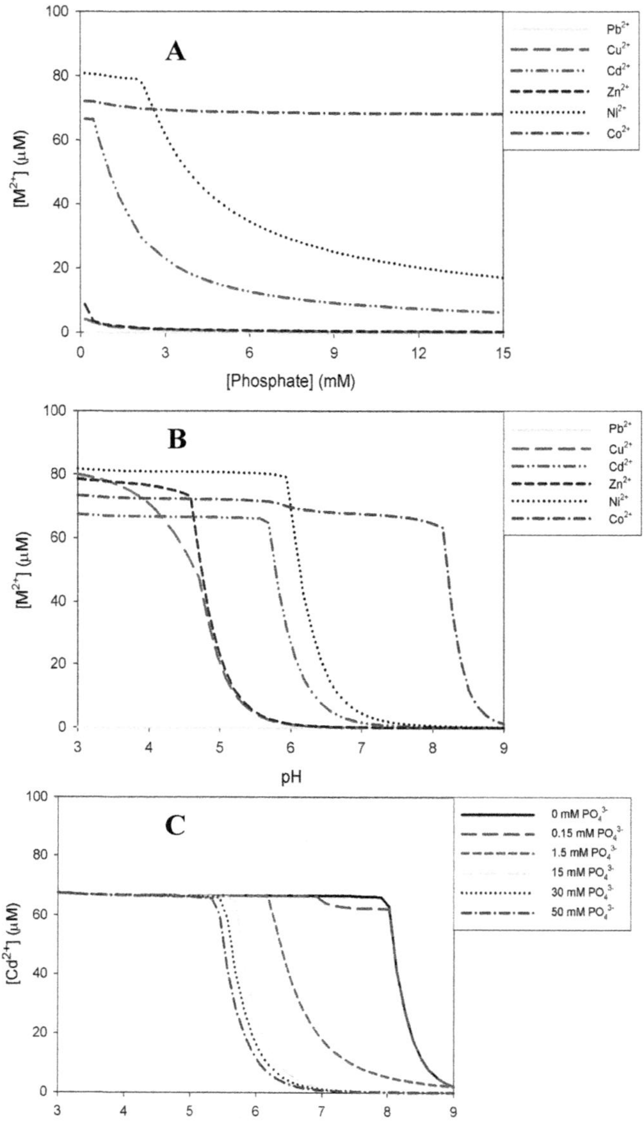


**Fig. 1.** Effect of phosphate concentration (A), pH (B), and interactions of phosphate and pH (C) on solution-phase, ionic metal ([$M^{2+}$]) and cadmium ([$Cd^{2+}$]) concentrations as predicted by MINEQL+ geochemical modeling software (Environmental Research Software, Hallowell, ME, USA) in Bushnell-Haas broth (Difco, Sparks, MD) amended with 100 μM total lead, copper, cadmium, zinc, nickel, or cobalt. When prepared according to the manufacturer's specifications, Bushnell-Haas broth contains 15 mM pH and has a pH of 7.0 ±0.2. The pH of the medium in (A) was set at 6.5

### 3.2 Metal Speciation and Bioavailability to affect Biodegradation in Soil

Metal bioavailability is often low in soil systems. This is due largely to the composition and pH of many soils studied. For example, in soil systems used to investigate effects of cadmium on phenanthrene biodegradation, 394 mg total cadmium/kg were added, but only 3 mg cadmium/L were actually bioavailable (Maslin and Maier 2000). Similarly, only 1% of the total zinc used in the work of Majumdar et al. (1999) was in the aqueous phase. Kong (1998) found that soluble metal concentrations in treatments initially amended with 20 mg total metal/L were below detection limits of 0.03-0.04 mg/L. At 100 mg total metal/L, only 1 mg cadmium/L and less than 0.12 mg copper and chromium/L were found in the aqueous phase.

In the soil environment, organic matter and clay mineral content are important factors that can reduce metal bioavailability. Thus, as increasing amounts of metal are added, toxicity is observed only after binding sites on organic matter become saturated with metal cations. For instance, Pardue *et al.* (1996) found that only 0.01 mg solution phase cadmium/L was required to inhibit trichloroaniline dechlorination in a mineral dominated soil, while 0.2 mg solution phase cadmium/L was required for inhibition in an organic matter dominated soil. This increase in tolerance to cadmium was correlated to saturation of metal binding sites on the organic matter. Similarly, only bioavailable cadmium has been reported to inhibit dehalogenation in microcosms containing cadmium-contaminated sediment (Jackson and Pardue 1998). Furthermore, Said and Lewis (1991) reported that biodegradation of a common herbicide, 2,4-dichlorophenoxy acetic acid methyl ester (2,4-DME), was much more sensitive to metal inhibition in aufwuchs (floating algal mats) than in sediments. The authors suggested that this was due to higher metal binding by sediments than by aufwuchs. Roberts et al.. (1998) observed inhibition of 2,4-diamino-6-nitrotoluene biodegradation at an undetectable concentration of soluble lead (below 1 mg/L) in treatments initially containing 10,000 mg total lead/kg. The phosphate buffer in this study may have caused this large reduction in lead bioavailability. Clay minerals have also been shown to reduce metal bioavailability. Clays with high cation exchange capacities, such as montmorillonite, appear to reduce metal bioavailability and toxicity most (Babich and Stotzky 1977). In fact, the profound impacts of clays on the bioavailability of toxic metals have prompted investigations into the use of clays to reduce metal toxicity as described later in this review.

### 3.3 Measurement of Bioavailable Metal

Reporting of bioavailable metal concentrations is a vital step towards standardizing experiments to determine effects of metals on organic pollutant biodegradation. Bioavailable metal concentrations can be estimated from solution phase metal concentrations using tools such as ion selective electrodes, which measure only ionic solution phase metals. A number of promising tools are in development that use biological systems to quantify solution phase and bioavailable metal concentrations. One of the most attractive aspects of these tools is that they can be used in complex systems, such as microbiological media and soil. The first such tool is the immunoassay which can detect solution phase metal concentrations in low µg/L range. Immunoassays have been developed for cadmium, lead, cobalt, nickel, and zinc. An immunoassay for mercury is commercially available (Blake et al. 1998; Khosraviani et al. 1998). A second tool is the use of bioreporters. These are whole cells that produce a protein with measurable activity (e.g., LacZ) or light in response to bioavailable metal. Bioreporters for detection of mercury have been created using both the *lacZ* system (Rouch et al. 1995b) and the luminescent *lux* system (Selifonova et al. 1993; Corbisier et al. 1999). While a bioreporter measures bioavailable metal, it should be emphasized that depending on the metal resistance mechanisms of the bioreporter system used, measurement of bioavailable metal can vary. A review of applications, advantages and limitations of immunoassays and bioreporters for metal detection is available (Neilson and Maier 2001).

In addition to biological-based approaches, geochemical modeling software (e.g., MINTEQA2, MINEQL+) can be used to predict metal speciation as a function of pH and ionic strength (Pardue et al. 1996). At least three computational models have been developed to predict the impact of metals on organic biodegradation (Jin and Bhattacharya 1996; Nakamura and Sawada 2000; Amor et al. 2001). These models account for metal inhibition by adding metal inhibition constants (e.g., $K_i$) to conventional microbial growth and/or degradation equations. For instance, Amor *et al.* (2001) used a form of the Andrew's equation (often used to describe microbial growth with inhibition) to model effects of cadmium, zinc, and nickel on rates of alkylbenzene biodegradation:

$$\mu = \mu_{max} S/(K_s + S + S^2/K_i), \tag{1.1}$$

Where $\mu$ is the alkylbenzene biodegradation rate
$\mu_{max}$ is the maximum alkylbenzene biodegradation rate
S is the alkylbenzene concentration
$K_s$ is the alkylbenzene concentration that yields $\frac{1}{2}\mu_{max}$
$K_i$ is the metal inhibition constant.

None of these models incorporates metal speciation and bioavailability. Thus, data generated by these models may only be meaningful for the medium or soil that was used to develop the model. For example, the medium used by Nakamura and Sawada (2000) was adjusted to a pH of 7.8 and contained 0.147 mM phosphate. Likewise, the medium used by Amor et al. (2001) was adjusted to a pH of 5.9 and contained 36 mM phosphate. In both media, much of the metal may precipitate. Thus, these models are likely to underpredict metal toxicity in systems that have a lower pH and/or less phosphate.

# 4. Metal Inhibition of Biodegradation

The impacts of metals on many general microbial activities including litter decomposition, methanogenesis and acidogenesis, nitrogen transformation, biomass generation, and enzymatic (e.g., dehydrogenase) activity have been studied extensively (Mosey 1976; Doelman and Haanstra 1979a; Doelman and Haanstra 1979b; Capone et al. 1983; Pankhania and Robinson 1984; Babich and Stotzky 1985; Rogers and Li 1985; Kouzelikatsiri et al. 1988; Baath 1989; Hickey et al. 1989; Nandan et al. 1990; Burkhardt et al. 1993; Lin 1993; Bardgett and Saggar 1994; Masakazu and Itaya 1995; Knight et al. 1997). Metals including copper, zinc, cadmium, chromium (III and VI), nickel, mercury, and lead have been reported to inhibit each of these processes. In contrast, some metals have been observed to stimulate activity. For example, Baath (1989) noted that both inhibitory and stimulatory effects of lead on carbon mineralization have been observed. Equally perplexing, the addition of some metals including mercury, lead, nickel, cadmium, and copper, stimulated methanogenesis in anoxic salt sediments (Capone et al. 1983) and nickel (< 300 mg total nickel/L) stimulated acidogenesis (Lin 1993). As illustrated below, available data on the effect of metals on organic pollutant biodegradation is not extensive, but demonstrates that metals have the potential to inhibit pollutant biodegradation under both aerobic and anaerobic conditions.

## 4.1 Effects of Metals on Aerobic Biodegradation

Metals have been shown to inhibit the aerobic biodegradation of a variety of organic pollutants (Table 1A). For example, copper, cadmium, mercury, zinc and chromium (III) were found to inhibit the aerobic biodegradation of 2,4-DME in lakewater samples inoculated with either a sediment or an aufwuch (floating algal mat) sample (Said and Lewis 1991). Zinc, with a minimum inhibitory concentration (MIC) of 0.006 mg total Zn/L, was most toxic in sediment samples; however, in aufwuch samples, mercury was most toxic with an MIC of 0.002 mg total Hg/L. A pure culture study using a naphthalene-degrading *Burkholderia* sp. reported an MIC of 1 mg bioavailable cadmium/L (Sandrin et al. 2000). This MIC was in the same range as the MICs reported by Said and Lewis (1991) for cadmium (0.1 mg total cadmium/L for sediment samples and 0.629 mg/L for aufwuch samples). The fact, that different microorganisms were used in each study likely accounts for differences between the reported MICs.

Springael et al. (1993) also showed that metals inhibited biodegradation of a variety of organic contaminants by several bacterial genera in pure culture. Reported MICs were 2 to 4 orders of magnitude higher than those reported by Said and Lewis (1991) (see Table 1A). The large discrepancies between MICs reported by these two studies are likely due to differences in the test system used in each study. Springael *et al.* (1993) quantified metal toxicity on solid agar media, while Said and Lewis quantified metal toxicity in liquid culture. Colony growth, that occurs on solid media, may have aided in protection against metal toxicity and resulted in higher MICs.

Metal inhibition has also been observed in soil systems. For example, 60 mg total cadmium/kg, inhibited the biodegradation of the herbicide 2,4-dichlorophenoxyacetic acid (2,4-D) in a soil system that was inoculated with the 2,4-D-degrader *Alcaligenes eutrophus* JMP134 (Roane et al. 2001). This study was performed both in small-scale microcosms and larger 5-gallon mesocosms showing similar metal sensitivity. Experiments have also been performed investigating the impact of metals on biodegradation by the indigenous soil community (Maslin and Maier 2000). In this case, the impact of cadmium on phenanthrene degradation in two desert soils was measured over a nine-day period. Results showed a 5-day increase in lag period for phenanthrene degradation in the presence of 1 and 2 mg bioavailable cadmium/L and complete inhibition at 3 mg bioavailable cadmium/L.

Effects of metal toxicity on biodegradation are not only limited to aromatic contaminants. The impact of copper toxicity on biodegradation of a polymer commonly used for medical, agricultural, and industrial purposes, polyhydroxybutyrate (PHB), has also been investigated (Birch and Brandl 1996). The polymer is used in agriculture both as a film mulch and as a long-term delivery device for fertilizers. In both applications, the material is expected to biodegrade after it has served its purpose; however, treatment of agricultural fields with sewage sludge (which is often rich in copper) can increase the soil metal content. To determine the impact of copper toxicity on PHB biodegradation, a PHB-containing agar overlay was placed on media containing a concentration gradient of copper. Plates were inoculated with a PHB-degrading strain of *Acidovorax delafieldii*. The concentration of copper along the gradient was determined by measuring copper in filter paper that was in contact with the gradient. Using this novel approach, the authors found that 8 to 15 mg bioavailable copper/L were required to inhibit PHB biodegradation.

Not all studies have investigated the impact of single metals on biodegradation of only a single, pure organic. Benka-Coker and Ekundayo (1998) investigated the impact of zinc, lead, copper and manganese on crude oil biodegradation by a *Micrococcus* sp. and a *Pseudomonas* sp. Biodegradation was reduced most by zinc (concentrations as low as 0.43 mg total zinc/L) and least by manganese (concentrations as low as 28.2 mg total manganese/L). Interestingly, combinations of metals were reported to be less toxic than some single metals. For instance, toxicity of 0.5 mg total zinc/L was mitigated by addition of 0.5 mg total copper, lead, and manganese/L. Most recently, Riis *et al.* (2002) reported inhibition of diesel fuel biodegradation in liquid cultures by combinations of metals, including copper, nickel, and zinc.

Some readily biodegradable organic pollutants, such as ethylenediamine-tetraacetic acid (EDTA) and NTA, interact strongly with metals. Despite the ubiquity of these compounds in wastewater, there is a paucity of information in the literature regarding the biodegradability of metal-organic complexes. Biodegradation of several EDTA-metal complexes, including complexes containing cadmium, nickel, cobalt, and copper, has been reported to be much slower than biodegradation of EDTA alone (Thomas et al. 1998). Similarly, *Chelatobacter heintzii* ATCC 29600 readily degraded free NTA, but was unable to degrade NTA complexed by copper, nickel, or cobalt (White and Knowles 2000). Complexation of NTA by the same metals reduced NTA biodegradation by *Mesorhizobium* sp. NCIMB 13524 (White and Knowles 2003). Additional organic pollutants capable of complexing and interacting with metals do exist. For this reason and the fact that the bioavailability of metals complexed to various organic ligands has not been well-characterized, more research in this area is warranted.

### 4.2 Effects of Metals on Anaerobic Biodegradation

Anaerobic catabolic pathways often represent the sole process for biodegradation of highly halogenated organics such as trichloroethene (TCE) and perchloroethene (PCE) (Alexander 1999). Many of these solvents have been discarded with metals. For this reason, several studies have addressed the effects of metal toxicity on the biodegradation of organic pollutants by anaerobic bacterial consortia (Table 1B).

Only 5 mg total cadmium/L has been reported to reduce TCE biodegradation (Degraffenreid and Shreve 1998). Representative of additional solution studies, Kuo and Genthner (1996) investigated the impact of cadmium, copper, chromium, and mercury on dechlorination and biodegradation by an anaerobic bacterial consortium isolated from an aquatic sediment. The consortium was capable of completely degrading 2-chlorophenol (2CP), 3-chlorobenzoate, phenol and benzoate. Results showed that different activities (e.g., dehalogenation, biodegradation, and methanogenesis) were affected differently by each metal. For example, biodegradation of 3-chlorobenzoate was inhibited most by cadmium and chromium, biodegradation of benzoate was most sensitive to copper, and phenol biodegradation was most reduced by mercury. In general, the addition of low levels of metals (0.1-2.0 mg total metal/L) lengthened acclimation periods and decreased dechlorination and biodegradation rates. Concentrations from 0.5-5 mg total metal/L completely inhibited either dechlorination or biodegradation. Similar results have been reported elsewhere. Kamashwaran and Crawford (2001) found that cadmium reduced pentachlorophenol biodegradation rates. Kuo and Genthner (1996) point out that their results suggest that, in addition to adversely affecting degraders in a consortium, metals may affect non-degrading consortium members that play a vital but indirect role in the degradation process. For instance, members of the consortium that produce reducing equivalents for reductive dehalogenation or remove dechlorinated products from the system to allow further dehalogenation may be deleteriously impacted, thus reducing the overall rate and extent of biodegradation.

Such an indirect mode of toxicity has also been implicated in the mechanism by which metals inhibited the anaerobic biodegradation of trinitrotoluene (TNT) metabolites (Roberts et al. 1998). Copper, zinc, and lead did not affect establishment of anaerobic conditions in a bioreactor, nor did these metals impact loss of the parent TNT compound; however, subsequent removal of TNT degradation intermediates was reduced by each of the metals. For instance, lead (total concentrations > 1000 mg/kg) delayed degradation of a TNT biodegradation intermediate (2,4-diamino-6-nitrotoluene) by as many as nine days. Zinc (1500 mg total zinc/kg) delayed degradation of the same intermediate by eight days. Copper (4000 and 8000 mg total copper/kg) completely inhibited removal of this intermediate. Thus, it is important to consider the effects of metals on populations of microorganisms other than those biodegrading the parent compound.

Soil type affects the extent to which metals inhibit biodegradation. For example, Pardue et al. (1996) examined the impact of cadmium on reductive dehalogenation of trichloroaniline in different soils. As described above, in microcosms containing two mineral-dominated soils, only 0.01 mg solution phase cadmium/L was required to inhibit reductive dehalogenation. In microcosms containing an organic matter-dominated soil, more than an order of magnitude higher cadmium concentration (0.2 mg solution phase cadmium/L) was required to inhibit dehalogenation. Furthermore, results showed that the dehalogenation pathway expressed in soil exposed to cadmium was different than the pathway expressed in cadmium-stressed soil. This suggests that cadmium stress selected for a different, dominant dehalogenating population than was found in the cadmium-free soil. Sediments have also been shown to mediate metal toxicity. The impact of metals on reductive dehalogenation of hexachlorobenzene in a waste lagoon sediment co-contaminated with cadmium and lead has been investigated (Jackson and Pardue 1998). In this study, cadmium and lead inhibited reductive dehalogenation, but only when not bound to sediment material.

### 4.3 Relationships between Metal Concentration and Inhibition of Biodegradation

It should be noted that the literature contains reports that metals do not inhibit some biodegradative processes. For example, cadmium (≤500 mg total cadmium/L) and mercury (≤100 mg total mercury/L) did not affect biodegradation of a variety of polycyclic aromatic hydrocarbons (PAHs) by the fungus *Pleurotus ostreatus* in soil (Baldrian et al. 2000). Similarly, Delaune et al. (1998) investigated the effects of chromium and lead on crude oil biodegradation. Those metals did not affect overall total hydrocarbon biodegradation, chromium (5,000 μg total chromium/g) reduced biodegradation of a constituent hydrocarbon of the oil, octadecane. This reduction occurred only under reducing conditions. Similarly, a suite of metals (copper, nickel, and zinc, at 31.8, 29.3, and 32.7 mg total metal/L, respectively) had no effect on diesel fuel biodegradation in soil slurries by an indigenous community of degraders (Riis et al. 2002); however, the same metals at 25-fold lower concentrations inhibited diesel fuel degradation in liquid culture by a community of degraders extracted from the soil. As with several other studies described throughout this review (Said and Lewis 1991; Pardue et al. 1996), the low bioavailability of metals in these studies may account for the fact that inhibitory effects were not observed. Furthermore, metal toxicity in the study conducted by Baldrian *et al.* (2000) may have been ameliorated by the acidity of the soil in which the experiments were conducted, as has been described previously (Franklin et al. 2000; Sandrin and Maier 2002).

When metals inhibit biodegradation, their effects are not always dose-dependent. The data presented thus far suggest that there is a direct, dose-dependent relationship between the amount of toxic metal in a co-contaminated environment and the extent of metal inhibition of organic biodegradation (Fig. 2A); however, there is an evidence for two semi-dose dependent patterns of metal effects on organic biodegradation.

#### 4.3.1 Semi-Dose Dependent Pattern 1

The results of several studies suggest that metals stimulate activity until a maximum level of stimulation is reached. Thereafter, metal toxicity increases with increasing metal concentration (Fig. 2B). All of these studies used consortia, not single isolates. For this reason, it is likely that this pattern results from differential toxicity effects, wherein one population that is sensitive to metal stress competes in some way with another, metal-tolerant population expressing the activity of interest (e.g., biodegradation). Inhibition of the more sensitive population reduces competition for resources needed by the metal tolerant population expressing the activity of interest. Capone *et al.* (1983) provide an evidence supporting this view point. Methanogenesis was stimulated by the addition of some metals. As the authors suggested, this may have resulted from differential inhibition of the methane and non-methane producing microorganisms. Metals may have selected for a metal-resistant, methanogenic population and reduced competition from a metal-sensitive, non-methanogenic population. Similarly, Kuo and Genther (1996) reported that low concentrations of metals stimulated biodegradation. Hexavalent chromium (0.01 mg total chromium/L) increased the biodegradation rate of phenol by 177% and that of benzoate by 169% over controls containing no metals. Other metals exhibited similar effects. Copper and cadmium (both at 0.01 mg total metal/L) increased the benzoate biodegradation rate 185% and the 2-chlorophenol biodegradation rate by 168%. Mercury (1-2 mg total mercury/L) increased the biodegradation rates of 2-chlorophenol and 3-chlorophenol by 133-154%.

Other studies report similar results with various consortia (Sterritt and Lester 1980; Hughes and Poole 1989). These groups suggested the stimulatory effect may be due to metals reducing competition for reducing equivalents or nutrients between metal-resistant degraders and metal-sensitive non-degraders. As in the work of Capone et al. (1983), Kuo and Genthner (1996), and Roberts et al. (1998), the impact of metals on microbially mediated processes in these studies may be mainly due to effects of metals on a population other than the one carrying out the process of interest, the existence of this semi-dose dependent pattern of metal effects underscores the importance of considering not only the physiological impact of a toxic metal on a degrading population of interest, but also the ecological impact of the toxic metal.

### *4.3.2 Semi-Dose Dependent Pattern 2*

The second semi-dose dependent pattern is one in which low concentrations of metals increasingly inhibit activity until a maximum level of inhibition is reached and, thereafter, metal toxicity decreases with increasing metal concentration (Fig. 2C). The data of Said and Lewis (1991) generally shows that 2,4-DME biodegradation decreased in a dose-dependent fashion; however, a closer examination of these data reveals that the maximal degradation rate ($V_{max}$) of 2,4-DME was less in the presence of 10 μM cadmium (0.61 ± 0.03 μg 2,4-DME/L/min) than in the presence of 100 μM cadmium (0.74 ± 0.00 μg 2,4-DME/L/min). In a subsequent study, a similar pattern of inhibition was observed as populations of 2,4-D degraders in a cadmium contaminated soil were more resistant to cadmium toxicity at a higher concentration of cadmium (40 mg total cadmium/L) than at a lower concentration of cadmium (20 mg total cadmium/L) (Roane and Pepper 1997). Pattern 2 responses to metals might be explained by microbial community dynamics. High metal concentrations may create selective pressure for metal-resistant, organic-degrading microorganisms that reduced competition from metal-sensitive non-degrading microorganisms, thus increasing biodegradation at higher metal concentrations. At the level of single cells, though, it is possible that high metal concentrations may more rapidly induce a metal resistance mechanism important in cadmium detoxification (e.g., an efflux pump) than low metal concentrations.

In summary, the existence of semi-dose dependent patterns of metal effects on biodegradation complicates understanding and predicting metal toxicity in the environment. As demonstrated by the patterns described above, metals may impact both the physiology and ecology of pollutant degrading microorganisms. For this reason, models designed to predict the impact of metals on biodegradation may fail to do so accurately unless they include both physiological and ecological effects of metals on organic-degrading microorganisms.

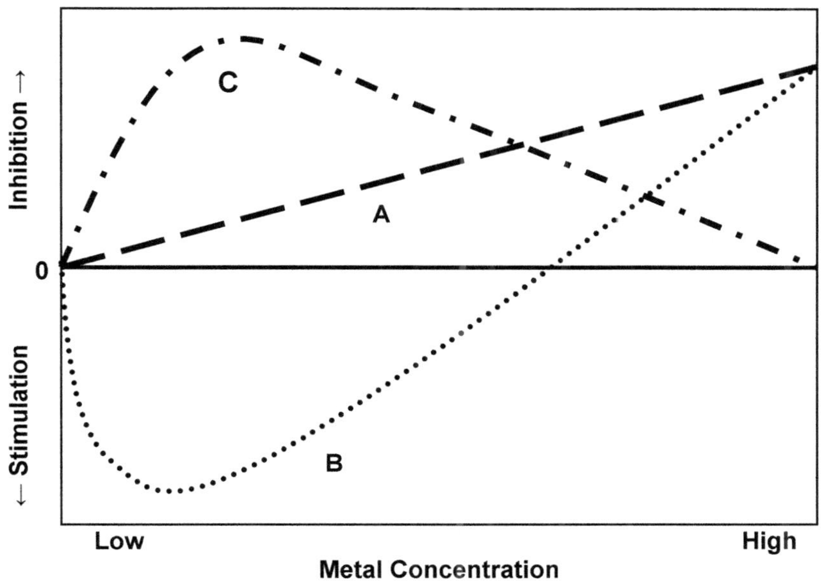


**Fig. 2.** Reported patterns in which metals affect organic pollutant biodegradation: the most commonly reported, dose-dependent pattern (A), semi-dose-dependent pattern 1 (B), and semi-dose-dependent pattern 2 (C)

## 5. Strategies to Enhance Biodegradation in Co-contaminated Environments

Several approaches have been described to reduce the extent to which metals inhibit organic biodegradation. Specifically, each approach has involved lowering metal bioavailability and/or increasing metal resistance to facilitate biodegradation. Approaches include inoculation with metal resistant microorganisms and the addition of materials that can reduce metal bioavailability including calcium carbonate, phosphate, clay minerals, and surfactants.

### 5.1 Metal Resistant Bacteria

Microorganisms employ a variety of mechanisms to cope with toxic metals. These have been reviewed thoroughly elsewhere (Nies 1992; Ji and Silver 1995; Nies and Silver 1995; Rosen 1996; Silver 1996; Silver and Phung 1996; Nies 1999). Resistance mechanisms include intracellular and extracellular metal sequestration, metal reduction, metal efflux pumps, and production of metal chelators such as metallothioneins and biosurfactants. Despite the ubiquity and efficacy of microbial metal resistance mechanisms, a few studies have attempted to exploit them to increase pollutant biodegradation in co-contaminated systems.

Introduction of metal-resistance mechanisms into pollutant-degrading bacteria may represent a viable strategy to mitigate metal-inhibition of organic pollution biodegradation. Springael *et al.* (1993) showed that metal inhibition of biodegradation could be reduced by the introduction of metal resistance genes into biodegrading microorganisms. For example, strains containing metal resistance genes degraded both polychlorinated biphenyls (PCBs) and 2,4-D in the presence of either 1 mM nickel or 2 mM zinc. Biodegradation of these compounds by organisms without introduced resistance genes was inhibited at the same metal concentrations.

A single study has investigated inoculation of metal-contaminated soil with metal-resistant bacteria to enhance organic contaminant biodegradation (Roane et al. 2001). In this study, soil microcosms were co-contaminated with 2,4-D (500 mg/kg) and cadmium (60 mg total cadmium/kg). Inoculation with *Alcaligenes eutrophus* JMP134, a 2,4-D degrader, was required because this soil did not contain an active 2,4-D-degrading population. JMP134, though, was sensitive to cadmium. To achieve rapid degradation of the 2,4-D, it was necessary to inoculate the metal-contaminated soil with both JMP134 and a cadmium resistant isolate, *Pseudomonas* H1, which accumulates cadmium intracellularly. These results suggest that inoculation with metal-sequestering microorganisms will foster increased biodegradation in the presence of a toxic metal.

## 5.2 Treatment Additives

Treatment additives, such as calcium carbonate, phosphate, cement, manganese oxide and magnesium hydroxide can reduce metal bioavailability and mobility in metal-contaminated sites (Ruby, et al. 1994; Traina and Laperche 1999; Hettiarachchi et al. 2000). In spite of this, only a single study has examined the impact of such reductions on metal toxicity to soil microorganisms. Jonioh et al. (1999) examined the effect of calcium carbonate on the toxicity of lead to microorganisms isolated from a military rifle range soil contaminated with lead and other heavy metals. Calcium carbonate was found to reduce lead toxicity when added at 1, 2.5, 5, and 10% (w/w). Toxicity was determined using the Microtox® assay (which uses a luminescence assay to determine viability). The effective concentration of contaminated soil required for a 50% reduction in loss of luminescence (EC50) increased from 14% in the absence of calcium carbonate to 75% in the presence of 10% calcium carbonate. Calcium carbonate decreased lead leachability and raised the soil pH. Because lead bioavailability typically decreases as pH increases, the additive likely reduced lead toxicity by reducing its bioavailability. Such promising results suggest that treatment additives may play key roles in future viable approaches to remediating metal and organic co-contaminated sites.

### 5.3 Clay Minerals

Clay minerals can reduce metal bioavailability and toxicity. The addition of kaolinite (1 to 20%) or montmorillonite (1 to 5%) to an agar medium containing cadmium reduced the toxicity of cadmium to several fungi including *Aspergillus niger* and *Trichoderma viride*, to bacteria including *Bacillus megaterium, Agrobacter tumefaciens,* and to an actinomycete, *Nocardia corallina* (Babich and Stotzky 1977). Similarly, in solution studies, bentonite and vermiculite (at 3% each) reduced the toxicity of 150 mg total cadmium/L to *Streptomyces bottropensis* (Kamel 1986). Kaolinite also reduced cadmium toxicity, but more was required (6% vs. 3%) and less protection was afforded than with the other clays. In general, protection increased with clay concentration. The protective ability of each clay correlated well with its cation exchange capacity (CEC). For example, the most effective clay, vermiculite, had a CEC of 108.7 meq/g, while the least effective clay, kaolinite, had a CEC of only 4.8 meq/g.

The effect of clay addition on metal toxicity was less pronounced in soil than in the plate and solution studies described above. Babich and Stotzky (1977) found that 3 to 12% montmorillonite was required to reduce cadmium toxicity to various fungi in soil; however, kaolinite failed to reduce toxicity. The low CEC of kaolinite appeared to explain its failure to reduce metal bioavailability and hence toxicity, as in the results of plate studies.

### 5.4 Chelating Agents

Chelating agents have been used to mitigate metal toxicity to organic-degrading microorganisms. EDTA has been shown to reduce the toxicity of cadmium to *Chlorella* sp. of nickel to algae (Spencer and Nichols 1983) and an actinomycete (Babich et al. 1983), and of copper to bacteria and algae (Sunda and Guillard 1976); however, the toxicity of EDTA to many microorganisms and its limited biodegradability may reduce its suitability for application to the bioremediation of co-contaminated environments (Braide 1984; Ibim et al. 1992; Borgmann and Norwood 1995; Ogundele 1999). In addition, biodegradation of metal-EDTA complexes may be slow (Thomas et al. 1998). Thus, the use of other chelating agents to reduce metal toxicity is of interest.

A commercially available chelating resin (Chelex 100) and surfactant-modified clays reduced cadmium toxicity during biodegradation of naphthalene (Malakul et al. 1998). Clays were modified by adsorbing a cationic surfactant to the clay surface to which various metal-binding ligands (e.g. palmitic acid) were attached via hydrophobic interactions. Naphthalene biodegradation occurred at higher cadmium concentrations in the presence of the modified clays than in controls containing either no clay or unmodified clay. The abilities of the resin and the modified clays to reduce cadmium toxicity were quantitatively related to the metal adsorption characteristics of the two chelating agents.

Biosurfactants (i.e., microbially produced surfactants) show promise for enhancing organic biodegradation in metal and organic co-contaminated environments. Sandrin et al. (2000) showed that a rhamnolipid biosurfactant produced by *Pseudomonas aeruginosa* reduced cadmium toxicity during naphthalene biodegradation by a *Burkholderia* sp. in solution studies. The mechanism by which the biosurfactant reduced cadmium toxicity appeared to involve a combination of rhamnolipid complexation of cadmium and rhamnolipid-induced lipopolysaccharide release from the outer membrane of the degrader (Leive 1965; Goldberg et al. 1983; Al-Tahhan et al. 2000). Later, Maslin and Maier (2000) used the same biosurfactant to reduce cadmium toxicity during biodegradation of phenanthrene by indigenous populations in two soils co-contaminated with phenanthrene and cadmium. Serial additions of rhamnolipid increased phenanthrene mineralization from 7.5 to 35% in one soil and from 10 to 58% in the second soil. Serial applications were necessary due to biodegradation of rhamnolipid which occurred in two to three weeks. The possibility for *in situ* biosurfactant production is being investigated to make this approach more cost-effective.

## 6. Conclusions and Future Directions

The current body of knowledge concerning metal effects on biodegradation is still in its infancy, yet the timely and cost-effective remediation of metal and organic co-contaminated sites will require a lucid understanding of factors important in determining the extent to which toxic metals inhibit organic biodegradation. Past attempts to measure impacts of metals on biodegradation are difficult to interpret, because they have generally been based on total metal rather than solution phase or bioavailable metal concentrations. This has resulted in reported inhibitory concentrations of metals that vary by as many as 5 orders of magnitude. A critical first step will be to consistently report solution phase or bioavailable metal concentrations so that legitimate comparisons of biodegradation behaviors in co-contaminated sites can be made. Currently, a useful approximation is to measure and use solution phase metal data; however, new methods of defining and determining bioavailable metal are rapidly being developed. Despite the enormous variance among reported inhibitory concentrations of metals, it remains clear that metals have the potential to inhibit organic biodegradation in both aerobic and anaerobic systems. The mechanisms and patterns by which metals inhibit biodegradation vary with the composition and complexity of each system and include both physiological and ecological components. A more thorough understanding of these systems, taking into account various levels of complexity, is needed to develop new approaches to bioremediate co-contaminated sites. Nevertheless, there already exist several approaches including addition of metal resistant microorganisms and additives that reduce metal bioavailability. Field trials are needed to validate these approaches.

Acknowledgements. This work was supported by the U.S. Environmental Protection Agency's Science to Achieve Results (STAR) Fellowship Program, the University of Wisconsin Oshkosh Faculty Development Program, and the UW Oshkosh Vander Putten International Fund.

## References

CRC handbook of chemistry and physics. (1991). 72nd ed. CRC Press, Cleveland, OH

Al-Tahhan RA, Sandrin TR, Bodour AA, Maier RM (2000) Rhamnolipid-induced removal of lipopolysaccharide from *Pseudomonas aeruginosa*: effect on cell surface properties and interaction with hydrophobic substrates. Appl Environ Microb 66:3262-3268

Alexander M (1999) Biodegradation and bioremediation. Academic Press, San Diego, CA

Amor L, Kennes C, Veiga MC (2001) Kinetics of inhibition in the biodegradation of monoaromatic hydrocarbons in presence of heavy metals. Bioresource Technol 78:181-185

Angle JS, Chaney RL (1989) Cadmium resistance screening in nitrilotriacetate-buffered minimal media. Appl Environ Microb 55:2101-2104

Baath E (1989) Effects of heavy-metals in soil on microbial processes and populations (a review). Water Air Soil Poll 47:335-379

Babich H, Schiffenbauer M, Stotzky G (1983) Further studies on environmental factors that modify the toxicity of nickel to microbes. Regul Toxicol Pharmacol 3:82-99

Babich H, Stotzky G (1977) Sensitivity of various bacteria, including actinomycetes, and fungi to cadmium and the influence of pH on sensitivity. Appl Environ Microb 33:681-695

Babich H, Stotzky G (1985) Heavy-metal toxicity to microbe-mediated ecologic processes - a review and potential application to regulatory policies. Environ Res 36:111-137

Baldrian P, Der Wiesche CI, Gabriel J, Nerud F, Zadrazil F (2000) Influence of cadmium and mercury on activities of ligninolytic enzymes and degradation of polycyclic aromatic hydrocarbons by Pleurotus ostreatus in soil. Appl Environ Microb 66:2471-2478

Bardgett RD, Saggar S (1994) Effects of heavy metal contamination on the short-term decomposition of [14C] glucose in a pasture soil. Soil Biol Biochem 26:727-733

Behra R, Landwehrjohann R, Vogal L, Wagner B, Sigg L (2002) Copper and zinc content of periphyton from two rivers as a function of dissolved metal concentration. Aquat Sci 64:300-306

Benka-Coker MO, Ekundayo JA (1998) Effects of heavy metals on growth of species of Micrococcus and Pseudomonas in a crude oil/mineral salts medium. Bioresource Technol 66:241-245

Beveridge TJ, Doyle RJ (1989) Metal ions and bacteria. John Wiley and Sons, New York

Bianchini A, Bowles KC (2002) Metal sulfides in oxygenated aquatic systems: implications for the biotic ligand model. Comp Biochem Phys C 133:51-64

Birch L, Brandl H (1996) A rapid method for the determination of metal toxicity to the biodegradation of water insoluble polymers. Fresen J Anal Chem 354:760-762

Blake DA, Blake RC, Khosraviani M, Pavlov AR (1998) Immunoassays for metal ions. Anal Chim Acta 376:13-19

Borgmann U, Norwood WP (1995) edta toxicity and background concentrations of copper and zinc in Hyalella azteca. Can. J Fish Aquat Sci 52:875-881

Braide VB (1984) Calcium EDTA toxicity - renal excretion of endogenous trace- metals and the effect of repletion on collagen degradation in the rat. Gen Pharmacol 15:37-41

Burkhardt C, Insam H, Hutchinson TC, Reber HH (1993) Impact of heavy-metals on the degradative capabilities of soil bacterial communities. Biol Fert Soils 16:54-156

Capone DG, Reese DD, Kiene RP (1983) Effects of metals on methanogenesis, sulfate reduction, carbon-dioxide evolution, and microbial biomass in anoxic salt-marsh sediments. Appl Environ Microb 45:1586-1591

Corbisier P, Van Der Lelie D, Borremans B, Provoost A, De Lorenzo V, Brown NL, Lloyd JR, Hobman JL, Csoregi E, Johansson G, Mattiasson B (1999) Whole cell- and protein-based biosensors for the detection of bioavailable heavy metals in environmental samples. Anal Chim Acta 387:235-244

Degraffenreid N, Shreve GS (1998) The effect of cadmium on the kinetics of trichloroethylene biodegradation by Pseudomonas (Burkolderia) picketti pk01 under denitrifying conditions. Water Res 32:3398-3402

Delaune RD, Mulbah C, Devai I, Lindau CW (1998) Effect of chromium and lead on degradation of south Louisiana crude oil in sediment. J Environ Sci Heal A 33:527-546

Doelman P and Haanstra L (1979) Effect of lead on soil respiration and dehydrogenase-activity. Soil Biol. Biochem. 11:475-479

Doelman P, Haanstra L (1979) Effects of lead on the decomposition of organic-matter. Soil Biol Biochem 11:481-485

Fagan MJ and Saier MHJ (1994) P-type atpases of eukaryotes and bacteria: sequence comparisons and construction of phylogenetic trees. J Mol Evol 38:57-99

Franklin NM, Stauber JL, Markich SJ, Lim RP (2000) pH-dependent toxicity of copper and uranium to a tropical freshwater alga (Chlorella sp.). Aquat Toxicol 48:275-289

Goldberg SS, Cordeiro MN, Silva Pereira AA, Mares-Guia ML (1983) Release of lipopolysaccharide (LPS) from cell surface of Trypanosoma cruzi by EDTA. Int J Parasitol 13:11-18

Hettiarachchi GM, Pierzynski GM, Ransom MD (2000) In situ stabilization of soil lead using phosphorus and manganese oxide. Environ Sci Technol 34:4614-4619

Hickey RF, Vanderwielen J, Switzenbaum MS (1989) The effect of heavy-metals on methane production and hydrogen and carbon-monoxide levels during batch anaerobic sludge-digestion. Water Res 23:207-218

Hughes MN, Poole RK (1989) Metal Toxicity. In: Hughes MN, Poole RK (eds) Metals and Microorganisms, Chapman and Hall, New York, pp 252-302

Hughes MN, Poole RK (1991) Metal speciation and microbial growth - the hard (and soft) facts. J Gen Microbiol 137:725-734

Ibim SEM, Trotman J, Musey PI, Semafuko WEB (1992) Depletion of essential elements by calcium disodium EDTA treatment in the dog. Toxicology 73:229-237

Jackson WA, Pardue JH (1998) Assessment of metal inhibition of reductive dechlorination of hexachlorobenzene at a superfund site. Environ Toxicol Chem 17:1441-1446

Janos P (1993) Separation of some metals as their anionic oxalate complexes by reversed-phase ion-interaction chromatography. J Chromatogr 635:257-263

Ji GY, Silver S (1995) Bacterial-resistance mechanisms for heavy-metals of environmental concern. J Ind Microbiol 14:61-75

Jin PK, Bhattacharya SK (1996) Anaerobic removal of pentachlorophenol in presence of zinc. J Environ Eng-Asce 122:590-598

Jonioh V, Obbard JP, Stanforth RR (1999) Impact of treatment additives used to reduce lead solubility and microbial toxicity in contaminated soils. In: Leeson A, Alleman BC (eds) Bioremediation of Metals and Inorganic Compounds, Battelle Press, Columbus, OH, pp 7-12

Kachur AV, Koch CJ, Biaglow JE (1998) Mechanisms of copper-catalyzed oxidation of glutathione. Free Radical Res 28:259-269

Kamashwaran SR, Crawford DL (2001) Anaerobic biodegradation of pentachlorophenol in mixtures containing cadmium by two physiologically distinct microbial enrichment cultures. J Ind Microbiol Biot 27:11-17

Kamel Z (1986) Toxicity of cadmium to two Streptomyces species as affected by clay minerals. Plant Soil 93:193-205

Khosraviani M, Pavlov AR, Flowers GC, Blake DA (1998) Detection of heavy metals by immunoassay: optimization and validation of a rapid, portable assay for ionic cadmium. Environ Sci Technol 32:137-142

Knight BP, Mcgrath SP, Chaudri AM (1997) Biomass carbon measurements and substrate utilization patterns of microbial populations from soils amended with cadmium, copper, or zinc. Appl Environ Microb 63:39-43

Kong IC (1998) Metal toxicity on the dechlorination of monochlorophenols in fresh and acclimated anaerobic sediment slurries. Water Sci Technol 38:143-150

Kouzelikatsiri A, Kartsonas N, Priftis A (1988) Assessment of the toxicity of heavy-metals to the anaerobic- digestion of sewage-sludge. Environ Technol Lett 9:261-270

Kuo CW, Genthner BRS (1996) Effect of added heavy metal ions on biotransformation and biodegradation of 2-chlorophenol and 3-chlorobenzoate in anaerobic bacterial consortia. Appl Environ Microb 62:2317-2323

Lane TW, Morel FMM (2000) A biological function for cadmium in marine diatoms. P Natl Acad Sci USA 97:4627-4631

Lane TW, Saito MA, George GN, Pickering IJ, Prince RC, Morel FMM (2005) A cadmium enzyme from a marine diatom. Nature 435:42

Leive L (1965) Release of lipopolysaccharide by EDTA treatment of E. coli. Biochem Biophys Res Commun 21:290-296

Lin CY (1993) Effect of heavy-metals on acidogenesis in anaerobic-digestion. Water Res 27:147-152

Malakul P, Srinivasan KR, Wang HY (1998) Metal toxicity reduction in naphthalene biodegradation by use of metal-chelating adsorbents. Appl Environ Microb 64:4610-4613

Masakazu A, Itaya S (1995) Effects of copper on the metabolism of 14C-labeled glucose in soil in relation to amendment with organic materials. Soil Sci Plant Nutr 41:245-252

Mash HE, Chin YP, Sigg L, Hari R, Xue HB (2003) Complexation of copper by zwitterionic aminosulfonic (good) buffers. Anal Chem 75:671-677

Maslin P, Maier RM (2000) Rhamnolipid enhanced mineralization of phenanthrene in organic-metal co-contaminated soils. Bioremed J 4:295-308

Mosey FE (1976) Assessment of the maximum concentration of heavy metals in crude sewage which will not inhibit the anaerobic digestion of sludge. Water Pollut Control 10-19

Nakamura Y, Sawada T (2000) Biodegradation of phenol in the presence of heavy metals. J Chem Technol Biot 75:137-142

Nandan R, Tondwalkar V, Ray PK (1990) Biomethanation of spent wash - heavy-metal inhibition of methanogenesis in synthetic medium. J Ferment Bioeng 69:276-281

National Research Council (1994) Alternatives for groundwater cleanup. In: Kavanaugh MC (ed) National Academy Press, Washington DC

Neilson JW, Maier RM (2001) Biological techniques for measuring organic and metal contaminants in environmental samples. In: Clapp CE, Hayes MHB, Senesi N, Bloom PR, Jardine PM (eds) Humic Substances and Chemical Contaminants, Soil Science Society of America, Madison, WI, pp 255-273

Nies DH (1992) Resistance to cadmium, cobalt, zinc, and nickel in microbes. Plasmid 27:17-28

Nies DH (1999) Microbial heavy-metal resistance. Appl Microbiol Biot 51:730-750

Nies DH, Silver S (1995) Ion efflux systems involved in bacterial metal resistances. J Ind Microbiol 14:186-199

Ogundele MO (1999) Cytotoxicity of EDTA used in biological samples: effect on some human breast-milk studies. J Appl Toxicol 19:395-400

Pankhania IP, Robinson JP (1984) Heavy-metal inhibition of methanogenesis by Methanospirillum hungatei gp1. Fems Microbiol Lett 22:277-281

Pardue JH, Kongara S, Jones WJ (1996) Effect of cadmium on reductive dechlorination of trichloroaniline. Environ Toxicol Chem 15:1083-1088

Rai LC, Rai PK, Mallick N (1996) Regulation of heavy metal toxicity in acid-tolerant Chlorella: physiological and biochemical approaches. Environ Exp Bot 36:99-109

Reed BE, Nonavinakere SK (1992) Metal adsorption by activated carbon - effect of complexing ligands, competing adsorbates, ionic-strength, and background electrolyte. Separ Sci Technol 27:1985-2000

Riis V, Babel W, Pucci OH (2002) Influence of heavy metals on the microbial degradation of diesel fuel. Chemosphere 49:559-568

Roane TM, Josephson KL, Pepper IL (2001) Dual-bioaugmentation strategy to enhance remediation of cocontaminated soil. Appl Environ Microb 67:3208-3215

Roane TM, Pepper IL (1997) Microbial Remediation of Soils Co-contaminated with 2,4-Dichlorophenoxy Acetic Acid and Cadmium. In: Proceedings of the 12th Annual Conference on Hazardous Waste Research: Building Partnerships for Innovative Technologies. Manhattan, KS: Great Plains/Rocky Mountain Hazardous Substance Research Center, pp 343-356

Roberts DJ, Venkataraman N, Pendharkar S (1998) The effect of metals on biological remediation of munitions-contaminated soil. Environ Eng Sci 15:265-277

Rogers JE, Li SW (1985) Effect of metals and other inorganic-ions on soil microbial activity - soil dehydrogenase assay as a simple toxicity test. B Environ Contam Tox 34:858-865

Rosen BP (1996) Bacterial resistance to heavy metals and metalloids. J. Biol. Inorg. Chem 1:273-277

Rouch, D.A., Lee, B.T.O., and Morby, A.P. (1995) Understanding cellular-responses to toxic agents - a model for mechanism-choice in bacterial metal resistance. J Ind Microbiol 14:132-141

Rouch DA, Parkhill J, Brown NL (1995) Induction of bacterial mercury-responsive and copper-responsive promoters - functional differences between inducible systems and implications for their use in gene-fusions for in vivo metal biosensors. J Ind Microbiol 14:349-353

Ruby MV, Davis A, Nicholson A (1994) In situ formation of lead phosphates in soils as a method to immobilize lead. Environ Sci Technol 28:646-654

Said WA, Lewis DL (1991) Quantitative assessment of the effects of metals on microbial- degradation of organic-chemicals. Appl Environ Microb 57:1498-1503

Sandrin TR, Chech AM, Maier RM (2000) A rhamnolipid biosurfactant reduces cadmium toxicity during naphthalene biodegradation. Appl Environ Microb 66:4585-4588

Sandrin TR, Maier RM (2002) Effect of pH on cadmium toxicity, speciation, and accumulation during naphthalene biodegradation. Environ Toxicol Chem 21:2075-2079

Sandrin TR, Maier RM (2003) Impact of metals on the biodegradation of organic pollutants. Environ Health Persp 111:1093-1101

Selifonova O, Burlage R, Barkay T (1993) Bioluminescent sensors for detection of bioavailable Hg(II) in the environment. Appl Environ Microb 59:3083-3090

Silver S (1996) Bacterial resistances to toxic metal ions - a review. Gene 179:9-19

Silver S, Phung LT (1996) Bacterial heavy metal resistance: new surprises. Annu Rev Microbiol 50:753-789

Spencer DF, Nichols LH (1983) Free nickel inhibits growth of two species of green algae. Environ Pollut Ser 31:97-104

Springael D, Diels L, Hooyberghs L, Krepsk S, Mergeay M (1993) Construction and characterization of heavy metal resistant haloaromatic-degrading Alcaligenes eutrophus strains. Appl Environ Microb 59:334-39

Sterritt RM, Lester JN (1980) Interactions of heavy metals with bacteria. Sci Total Environ 14:5-17

Sunda W, Guillard RRC (1976) The relationship between cupric ion activity and the toxicity of copper to phytoplankton. J Mar Res 34:511-529

Teresa M, Vasconcelos SD, Almeida CMR, Lage OM, Sansonetty F (2000) Influence of zwitterionic pH buffers on the bioavailability and toxicity of copper to the alga amphidinium carterae. Environ Toxicol Chem 19:2542-2550

Thomas RAP, Lawlor K, Bailey M, Macaskie LE (1998) Biodegradation of metal-EDTA complexes by an enriched microbial population. Appl Environ Microb 64:1319-1322

Toth D, Tomasovicova D (1989) Effect of pollutants on microbial metabolism. In: Microbial interactions with chemical water pollution. Halstead Press, John Wiley and Sons, New York, pp 12-22

Traina SJ, Laperche V (1999) Contaminant bioavailability in soils, sediments, and aquatic environments. Proc Natl Acad Sci USA 96:3365-3371

Twiss MR, Errecalde O, Fortin C, Campbell PGC, Jumarie C, Denizeau F, Berkelaar E, Hale B, Van Rees K (2001) Coupling the use of computer chemical speciation models and culture techniques in laboratory investigations of trace metal toxicity. Chem Spec Bioavailab 13:9-24

Vasconcelos Mtsd, Azenha Mago, Almeida CMR (1998) Copper(II) complexation properties and surfactant activity of 3-[n,n-bis(2-hydroxyethyl)amino]-2-hydroxypropanesulfonic acid and n-(2-hydroxyethyl)piperazine-n '-2-hydroxypropanesulfonic acid pH buffers which may affect trace metal speciation in *in vitro* studies. Anal Biochem 265:193-201

Vasconcelos Mtsd, Leal MFC (2002) Influence of n-2-hydroxyethylpiperazine-n '-2-ethanesulfonic acid pH buffer on the biological response of marine algae. Environ Toxicol Chem 21:404-412

Volesky B (1990) Biosorption of Heavy Metals. In: Volesky B (ed) CRC Press, Boston, MA

White VE, Knowles CJ (2000) Effect of metal complexation on the bioavailability of nitrilotriacetic acid to Chelatobacter heintzii ATCC 29600. Arch Microbiol 173:373-382

White VE, Knowles CJ (2003) Degradation of copper-NTA by Mesorhizobium sp NCIMB 13524. Inter Biodeter Biodegr 52:143-150

# 2

# New Bioremediation Technologies to Remove Heavy Metals and Radionuclides using Fe(III)-, Sulfate- and Sulfur- Reducing Bacteria

Mireille Bruschi and Florence Goulhen

Unité de Bioénergétique et Ingénierie des Protéines, UPR 9036, Institut de Biologie Structurale et Microbiologie, IFR88, C.N.R.S., 31 chemin Joseph Aiguier, 13402 Marseille Cedex 20, FRANCE, Email: bruschi@ibsm.cnrs-mrs.fr

## 1. Introduction

Microbial mineral formation and dissolution converged to produce a new field of research on bacterial-metal interaction developed within the last decade, called geomicrobiology. This new field tries to elucidate the role that microbes play or have played in specific geological processes and gives information about the earliest geochemical signals of life on earth. Furthermore, an understanding of bacterial-metal interactions provides the basis of improved models of metal cycling and the environmental impact of such transformations.With the need for new and low-cost technologies to remove heavy metals and radionuclides polluting the environment, the knowledge of the mechanisms, by which microorganisms interact with metals, has been recently developed (Lloyd et al. 2002; Barton et al. 2003; Lloyd 2003).

Iron and manganese are the two most abundant reactive metals in the earth's-crust, and the origin of life is initially connected to the ability of iron to readily cycle between Fe(III) and Fe(II) states. Some of the earliest geochemical signals of life on earth are the conversion of Fe(II) dissolved in the archaeon seas to Fe(III) oxides deposits. This conversion is possibly a result of the Fe(II) oxidizing microorganisms.

Today, Fe(III) is very abundant at the earth's surface, but is very insoluble at neutral pH and so microorganisms, which require iron to support growth, have developed siderophores which are the evolutionary response to the appearance of $O_2$ in the atmosphere and responsible the concomitant oxidation of Fe(II) to Fe(III).

A wide variety of microorganisms is capable of dissimilatory Fe(III) reduction which is the early form of microbial respiration. These bacteria use molecular hydrogen, lactate, pyruvate or acetate as their electron donor and

Fe(III) as electron acceptor. Many of them are also able to use Mn(IV) as electron acceptor, reducing it to Mn(II).

In this first group of bacteria, the growth is coupled to the reduction of Fe(III) and Mn(IV). In the second group, some metals like selenium and arsenic, can be used by some bacteria to support growth, but the other heavy metals are toxic and lethal for the bacteria, hence they have developed detoxification strategies in which the reduction of the metal gives a less toxic element (Most toxic heavy metals are less soluble and toxic when reduced than oxidized).

The need to remediate extensive metal contamination of groundwater and soils from heavy metals and radionuclides has stimulated an increased interest to find new metalresistant microorganisms and new bioremediation processes. Indeed, laboratory microorganisms, such as *Escherichia coli*, are not good candidates to be used in bioremediation processes, as they are not adapted to heavy metals contamined environments.

The aim of this article is to provide an overview of the development of technologies, using the activity of Fe(III)-, sulfate- and sulfur- anaerobic bacteria to remove heavy metals and metalloids from ground waters and soils.

## 2. Microbial Reduction of Metals by Fe(III)-reducing Bacteria

Fe(III) reduction has been highly conserved during evolution (Lonergan et al. 1996). A wide diversity of microorganisms are able to reduce Fe(III) or Mn(IV) (Lloyd 2003). Nevertheless, the present chapter will focus on the *Geobacter* sp. and *Desulfuromonas* sp., included in the Geobacteraceae group. Indeed, the Geobacteraceae group is divided in two sub-groups: the *Geobacter* cluster and the *Desulforomonas* cluster (Lonergan et al. 1996).

### 2.1 *Geobacter*

*Geobacter* species are microorganisms able to colonize habitats with elevated metal concentrations. Dissimilatory Fe(III) reduction is a well-known environmental process in various environment, such as sediments, shallow aquifers and in the deep surface. A recent study (Cummings et al. 2003) has clearly shown that various phylotypes of *Geobacter* sp. could be isolated from pristine and metal-contaminated sites. The persistence of *Geobacter* species is highly important, since it provides a glimpse of its use in the bioremediation processes of heavy metal-contaminated sites. Moreover, Childers et al. (2002) demonstrated that some *Geobacter* sp. accesses Fe(III) oxides by chemotaxis. These findings pinpoint the reason why among the Fe(III)-reducing bacteria, *Geobacter* sp. are the most abundant community in sediment environments, suggesting that they could be considered a kind of natural environmental clean-up bacteria and new tools for bioremediation processes.

Various species of *Geobacter* have been isolated and characterized. In the early 1990's, Lovley and co-workers (1993a) characterized *Geobacter metallireducens*, a strict anaerobic bacterium, able to reduce various metals such as Mn(IV) or U(VI). *Desulfuromonas acetoxidans* is the closest relative of *G. metallireducens*. On the other side, *Geobacter sulfurreducens*, isolated from an hydrocarbon contaminated ditch, by Caccavo et al. (1994), was the first bacterium described that is able to couple the oxidation of hydrogen (or acetate) to Fe(III) reduction. Various heavy metals, such as Cr(VI) and more particularly Tc(VII), are reduced by *G. sulfurreducens* and *G. metallireducens* (Lloyd et al. 2000; Liu et al. 2002). More recently, *Geobacter hydrogenophilus*, *Geobacter chapellei* and *Geobacter grbiciae* (Coates et al. 2001) and *Geobacter bremensis* sp. *nov.* and *Geobacter pelophilus* sp. *nov.* (Straub and Buchholz-Cleven 2001) were also isolated.

Mechanisms of the reduction of Fe(III) and Mn(IV) have been extensively studied, using *Geobacter* species as a model (Lloyd 2003). Cytochromes are heme enzymes involved in the electron transfer chain coupled to metal reduction (Fe(III) for example). Metals, such as gold, silver, mercury and chromate, considered as electron acceptors, were reduced by *G. metallireducens* c-type cytochromes (Lovley et al. 1993a; Coates et al. 1996). Moreover, Lloyd (2003) has showed that c-type cytochromes of *G. metallireducens* transfer electrons to soluble Au(III) (Lovley et al. 1993a).

The first study, reporting the purification and characterization of a c-type cytochrome from *G. sulfurreducens*, indicated that a small molecular weight periplasmic protein (9.6 kDa) functions as an Fe(III)-reductase (Seeliger et al. 1998). However, another Fe(III)-reductase, described by Gaspard et al. (1998), is a molecular weight c-type cytochrome associated with peripheral outer membrane. Investigations by the Lovley's group (Lloyd et al. 1999c) demonstrated that the periplasmic 9.6 kDa cytochrome c was not an electron shuttle to Fe(III). The 9.6 kDa cytochrome closest relative was the three-hemic cytochrome $c_7$ from *Desulforomonas acetoxidans* (Seeliger et al. 1998) which is a multihemic, low potential cytochrome c homologous to the cytochrome $c_3$ isolated from sulfate reducing bacteria. This cytochrome was cloned and expressed in *Escherichia coli* and is able to reduce metals *in vitro* (Londer et al. 2002). Its structure was elucidated at 1.45Å (Pokkuluri et al. 2004). Other c-type cytochromes were also characterized (Magnuson et al. 2000 and 2001; Leang et al. 2003). Up to date, more than 100 c-type cytochromes could be found in the *G. sulfurreducens* genome (Methe et al. 2003). Lloyd et al. (2003) have reported the biochemical and genetic analysis of one small periplasmic c-type designated as PpcA.

Interestingly, a $c_7$ cytochrome of another species of *Geobacter* (*G. metallireducens*) which is highly homologous to *G. sulfurreducens*, was also purified and characterized, with an apparent molecular weight of 9.5 kDa and triheme per molecule, homologous with *D. acetoxidans* cytochrome $c_7$ (Afkar and Fukumori 1999).

Other proteins, such as hydrogenases, may be involved in the reduction of Tc(VII). A direct enzymatic reduction or a Fe(II)-mediated reduction of Tc(VII) by Fe(III)-reducing bacteria has been highlighted (*G. sulfurreducens*) by Lloyd et al. (2000).

While *Geobacter* is able to reduce metals and radionuclides, there have been a few reports that pinpoint their potential contributions to a bioremediation process (Lovley 1995; Lovley 1997). The scientific community is just beginning to decipher the physiology and metabolism of *Geobacter* species, and we are at the discovery stage of their potent use in the bioremediation process. Several reports indicate that the adding of electron donors *in situ* stimulate the microbial reduction by *Geobacter* community. Microbial reduction of U(VI) to insoluble U(IV) of uranium-contaminated sub-surface sediments was assayed by Finneran et al. (2002). It was found that the nitrate content of the sediments had a negative impact on the reduction of Fe(III) to Fe(II) and U(VI) to U(IV) by *G. metallireducens*, since nitrate has to be reduced first (Finneran et al. 2002). At the same time, a reduction of uranium in samples from U(VI)-contaminated aquifer sediments (Holmes et al. 2002) and from the aquifer itself (Anderson et al. 2003) amended with acetate, was clearly associated with a reduction of Fe(III) by the *Geobacter* community.

More recently, a U(VI)- and Tc(VII)-contaminated aquifer was *in situ* reduced while *Geobacter* was stimulated with electron donors, even if the site was highly nitrate concentrated (Istok et al. 2004). Members of Geobacteraceae are not able to grow at high salinities, nevertheless, a high U(VI) concentration in a saline aquifer sediments could be reduced in water by the addition of acetate (Nevin et al. 2003). The groundwater geochemistry of contaminated aquifers, amended with electron donors, was monitored using bio-markers: microbial biofilms including *Geobacter* and nitrate-reducing microorganisms (Peacock et al. 2004).

A genomic approach could be useful in the bioremediation process, since *Geobacter sulfurreducens* has been sequenced (Methe et al. 2003). A genetic system has been recently developed (Coppi et al. 2001) in which *Geobacter* could be replaced by *Ralstonia eutropha*, able to neutralize Cadmium *via* the expression of a mouse-metallo-fusion protein (Lovley and Lloyd 2000). Recombinant indigeneous soil microorganisms, expressing metallothioneins (cysteine rich proteins able to bind heavy metals), could be used in the polluted soils (Valls et al. 2000). Indeed, they described a recombinant *Ralstonia eutropha* strain able to support and adsorb high $Cd^{2+}$ concentration in soils.

Although the important role of *Geobacter* in the geochemistry of the sub-surface environment has been clearly described (Lovley 1997), but their potential uses in *in situ* or *ex situ* bioreaction configurations have not been yet developed. Therefore, biochemical (molecular biology and genomics) and ecological approaches, leading to improving methods for using *Geobacter* as bioremediation agent, will undoubtedy make an impact in the future of the environmental biotechnology.

### 2.2 *Desulfuromonas*

Bacteria that are able to grow by linking the oxidation of acetate to the reduction of elemental sulfur have been known, since Pfennig and Biebl (1976) described the isolation of *D. acetoxidans*. Recently two other species, *D. palmitatis* and *D. thiophila* have also been described (Coates et al. 1995; Finster et al. 1997).

*Desulfuromonas*, a sulfur reducing bacterium, is strictly anaerobic, gram negative, flagellated and rod-shaped. It acquires its energy from sulfur respiration and completely oxidizes acetate with S to carbon dioxide via the citric acid cycle (Widdel and Pfennig 1991). Reduction of S produces hydrogen sulfide ($H_2S$) which can react with heavy metal ions to form less toxic insoluble metal sulfides (Kim et al. 2001). Furthermore, these bacteria are also able to enzymatically reduce and precipitate these heavy metals (Aubert et al. 1998a,b). Several studies, focused on the bioenergetic metabolism of sulfur reducing bacteria, have led to the characterization of various metalloproteins and in particular, multiheme low potential cytochromes (Bruschi 1994; Bruschi et al. 1997), the most abundant being the cytochrome $c_7$. The biological function of cytochrome $c_7$ is not clearly established, but it has been proposed to have a role as an electron transfer protein in the sulfur metabolism of this bacterium, acting as a terminal reductase in the metabolic pathway by directly reducing elemental sulfur to sulfide; it has also been suggested that cytochrome $c_7$ could be involved in the dissimilatory reduction of Fe(III) and Mn(IV) to obtain energy growth by these bacteria (Roden and Lovley 1993).

The three dimensional structure of this triheme cytochrome determined by nuclear magnetic resonance shows that the orientation of the three heme groups is similar to that of the tetrahemic cytochrome with the heme 2 lacking (Banci et al. 1996).

The three heme groups have negative redox potentials ranging from –102 to –177 mV. Electrochemistry experiments have demonstrated the direct reduction of Fe(III), Mn(IV), V(V) (Lojou et al. 1998b; Lojou and Bianco 1999) by multihemic cytochrome whereas mitochondrial c-type cytochromes did not exhibit any activity (Lojou et al. 1998a).

The interaction between Cr(VI) and cytochrome $c_7$ was chosen as a model for the reduction of metals by $c_3$-type cytochromes, as the three dimensional structure of the oxidized and the reduced states of this cytochrome have been solved using NMR studies (Banci et al. 1996). $^1$H NMR experiments have been performed using reduced cytochrome $c_7$ (by a catalytic amount of hydrogenase representing the smallest amount necessary to reduce its physiological partner (Brugna et al. 1998): the $c_3$-type cytochrome) and various amounts of Cr(VI). Figure 1 shows a single binding site near heme IV, the heme with the highest reduction potential (Dolla et al. 1991; Assfalg et al. 2002). An electron flow involving the three hemes and the protein chain is proposed to explain the reaction which is potentially important for the construction of biosensors.

Moreover, several multihemic cytochrome c of higher molecular weight (50, 65 and 250 kDa) have been characterized by Bruschi et al. (1997), and Pereira et al. (1997) exhibiting several domains and high thermal stability (Giudici et al. 2003). The genome sequence of the bacteria, presently under study, reveals a very high number of multihemic cytochromes as observed in *Geobacter sulfurreducens* genome. Considering these similarities with the presence of multihemic domains and low potential redox, these cytochromes could be related to cytochrome $c_7$ and show also a metal reduction activity and could be used to select high performance metal-reductase bacteria or to develop biosensors.

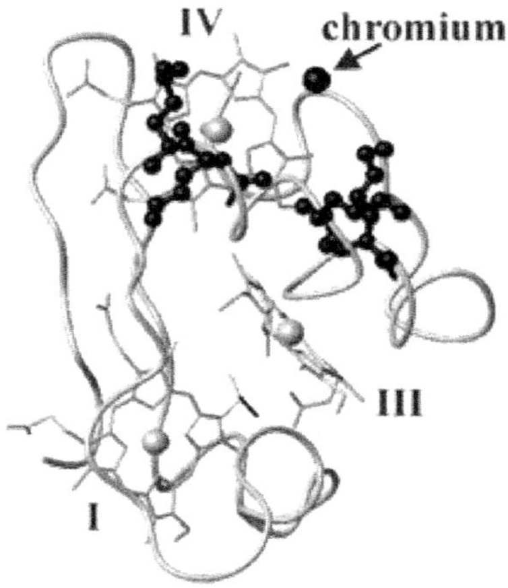


**Fig. 1.** The Cr(III) binding site on cytochrome $c_7$ from *Desulfuromonas acetoxidans*. The Cr(III) ion is shown as a black sphere, and the hemes are labeled by roman numbers (Assfalg et al. 2002)

## 3. Microbial Interaction with Toxic Metals by Sulfate-reducing Bacteria

In contrast to the first group of bacteria in which the metal is used as a terminal acceptor in the metabolism, sulfate-reducing bacteria (SRB) are not able to use the metal to support growth. SRB are strict anaerobic bacteria, requiring a redox potential of less than -200 mV (Postgate 1984) and are naturally present in waters and soils. These microorganisms are found in various sites contaminated with metals, metalloids and pollutants, which are lethal to other bacteria. The first isolated SRB was *Spirullum desulfuricans* (reclassified as *Desulfovibrio* included in Desulfovibrionaceae group) in 1895 (Beyerinck 1895). At the end of the 1980's, the role of SRB on the bioremediation of technetium was highlighted (Pignolet et al. 1989). Now-a-days, SRB are of increasing economic and industrial importance, since the European criteria, regarding the heavy metal rejected in the environment, are more drastic. The ability to reduce metals to a less toxic form, associated with its precipitation, is a potentially useful process for bioremediation.

SRB are able to couple the oxidation of organic compounds or $H_2$ with the reduction of sulfate as electron acceptor. During this process, the dissimilatory sulfate reduction, leads to the production of $H_2S$ which is dissimilated into the environment and can reduce heavy metals. SRB, in addition to the chemical indirect reduction due to the production of $H_2S$, can also reduce metals *via* enzymatic pathway involving $c_3$-type cytochromes (Lovley and Phillips 1992; Lovley et al. 1993a,b). *Desulfovibrio desulfuricans* can not only reduce the soluble toxic form of U(VI) to insoluble U(IV) (Lovley and Phillips 1992; Tucker et al. 1996), but also Cr(VI), Mo(VI), Se(VI) (Tucker et al. 1998), Pd (Lloyd et al. 1998) and Tc(VII) (Lloyd et al. 1999a,b). The metal-reductase activity of the $c_3$ cytochrome has been decribed in the case of several heavy metals, such as U(VI), Cr(VI), Fe(III) (Lovley et al. 1993b; Lovley and Phillips 1994; Lojou et al. 1998a,b; Michel et al. 2001; Elias et al. 2004), Pd (Lloyd et al. 1998) and Tc (Lloyd et al. 1999a,b). All of these recent studies emphazised a wide metal reduction activity among SRB associated with cytochrome $c_3$, which are periplasmic proteins. When exposed to heavy metal ions, bacteria grown in the presence of high Cr(VI) concentration accumulate precipitates of trivalent chromium at its cell surface (Fig. 2) (Goulhen et al. 2005). These findings are consistent with a direct electron transfer to the metal by cytochromes and hydrogenases, which are periplasmic proteins.

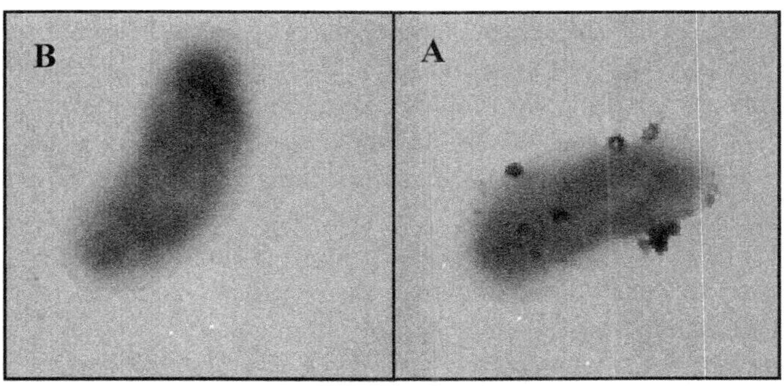


**Fig. 2.** Electron micrographs of unstained preparations of *D. vulgaris* Hildenborough grown in the absence of Cr(VI) (panel A) or in the presence of 250 μM Cr(VI) (panel B)

In order to develop potentially new bioremediation processes, it is required to select the most efficient and heavy metal-resistant strains. Thus, it is highly important to select strains from various contaminated sites and to evaluate their potent activity. Nevertheless, the SRB community behaviour (regarding adaptation) in contaminated sites is poorly documented and has to be evaluated to decipher the response of the SRB in changing environment. A study has recently focused on the high diversity and characterization of SRB in a groundwater uranium contaminated site (Chang et al. 2001). *Desulfotomaculum* sp. was predominant in this site and *Desulfovibrio* sp. was isolated from a parcel exhibiting lower uranium concentrations (Chang et al. 2001). Models, able to forecast the activity of SRB regarding heavy metals, are to be developed, since, for example, copper has more inhibitory effects than zinc on SRB (Utgikar et al. 2001; Utgikar et al. 2003). Interestingly, the cultivation of *D. vulgaris* in the presence of Cu(II) and Hg(II) increases the lag phase and final biomass yield. Toxic metal adaptation appeared to be an ATP-dependent mechanism (Chang et al. 2004).

A *Desulfovibrio* strain highly resistant to copper (about 10 fold normal level) was recently isolated. The *pco* gene, well-known to play an important role in copper resistance, was present on a plasmid of that particular strain which could be used as an interesting bioremediation tool (Karnachuk et al. 2003).

Various SRB, including *Desulfovibrio* and *Desulfomicrobium* species, were evaluated regarding their enzymatic reduction of Cr(VI). Intact cells of *D. norvegicum* showed the best Cr(VI)- reducing activity (up to 500 μM Cr(VI)) compared to *D. escambiense*, D. vulgaris Hildenborough, *D. gigas*, *D. desulfuricans*, a strain named BRGM isolated from a gold mine (France) and new strains isolated from black smokers (Pacific ocean) (Michel et al. 2001; Michel et al. 2003a). The Cr(VI) acts as a stressing agent at high concentrations, leading to an increasing bacterial cells fragility, since bacteria become long filament (default in cell division) and c-type cytochromes could be found in culture supernatant (Michel et al. 2001; Bruschi et al. 2003). The effects of Cr(VI) on bioenergetic metabolism were monitored using isothermal microcolorimetry (Chardin et al. 2002). An extension of the lag growth phase and deep changes in the bacterial metabolism of lactate were observed in the presence of high Cr(VI) concentration. The growth was inhibited with a concomitant energy production, which suggests that lactate is catabolized for lowering the redox potential to maintain survival conditions for sulphate-reducing bacteria. Indeed, Cr(VI) reduction is a protective escape to keep the bacterial environment favourable (Chardin et al. 2002; Bruschi et al. 2003). Microcalorimetry could be a potent criterion to evaluate the effects of the metal concentration on bacteria and to choose the best strain needed to decontaminate a polluted environment (Bruschi et al. 2003).

As metal reduction can also be achieved enzymatically, the metal reductase activity of purified cytochromes $c_3$ and hydrogenases have been studied. On the basis of amino-acid sequence and three dimensional comparisons of multihemic cytochromes, characterized by bishistidinyl axial iron coordination and low redox potentials, we have proposed that all these cytochromes belong to the cytochrome $c_3$ superfamily and that they have a common ancestral origin (Bruschi et al. 1992; Bruschi et al. 1994; Bruschi 1994). As we have demonstrated that the all the cytochrome $c_3$ tested and the cytochrome $c_7$ have a metal reductase activity, we could propose that the other multihemic cytochromes c described in sulfate and sulfur reducing bacteria, as they possess the common tetraheme motif as building block, could have also a similar metal reductase activity (Czjzek et al. 1996; Czjzek et al. 2002). In order to pinpoint the SRB strain demonstrating the highest metal reductase activity, Michel *et al.* (2001) compared the chromate reductase activity of various c-type cytochromes, concluding that $c_3$-cytochrome from *D. norvegicum* presented the highest activity (Table 1). The monohemic cytochrome $c_{553}$ from *D. vulgaris,* characterized by a higher redox potential and mitochondrial cytochrome c, was also tested and showed no metal reductase activity. This result suggests that a negative redox potential is necessary for enzymatic reduction. The Cr(VI) reductase activity of site directed mutagenesis mutants of cytochrome $c_{3,}$ named respectively H22M and H35M (the histidine residue of the sixth axial ligand heme 1 and 2 respectively has been replaced by a methionine residue), have been tested. A decrease of 15% in the chromate reductase activity is observed for mutant H22M suggesting that heme 1 is crucial for chromate reduction.

**Table 1.** Cr(VI) reductase activity of wild type and mutated purified cytochromes (Michel et al. 2001)

| Enzymes | Cr(VI) reduction rate (µmol Cr(VI)/min/ µmol enzyme) |
|---|---|
| Cytochrome $c_3$ (*Desulfovibrio vulgaris* Hildenborough) | $5.08 \pm 0.23$ |
| Cytochrome $c_3$ (*Desulfomicrobium norvegicum*) | $9.60 \pm 0.76$ |
| Cytochrome $c_7$ (*Desulfuromonas acetoxidans*) | $5.07 \pm 0.23$ |
| Cytochrome $c_{553}$ (*Desulfovibrio vulgaris* Hildenborough) | No activity |
| Cytochrome H35M (*Desulfovibrio. vulgaris* Hildenborough) | $5.20 \pm 0.25$ |
| Cytochrome H22M (*Desulfovibrio vulgaris* Hildenborough) | $4.43 \pm 0. 3$ |

To test the involvement of the cytochrome $c_3$ in the *in vivo* U(VI) reduction, a cytochrome $c_3$ mutant has shown one half of the rate of reduction (Payne et al. 2002). In addition to polyhemic cytochromes, low redox proteins, present in the periplasmic space of SRB, exhibited a metal reductase activity. Lloyd and coworkers (1999a,b) indicated that the Tc(VII)-reductase activity of *D. desulfuricans* was associated with a periplasmic hydrogenase. More recently, it was reported that Tc(VII) could be reduced by the [NiFe] hydrogenase alone or acting with $c_3$-type cytochrome of *D. fructosovorans* (De Luca et al. 2001). Cr(VI) could be reduced by [Fe], [NiFe], and [NiFeSe] hydrogenases (Michel et al. 2001; Chardin et al. 2003). The highest Cr(VI) rate was observed for purified [Fe] hydrogenase from DvH compared to [Ni-Fe-Se] hydrogenase from *D. norvegicum* (Michel et al. 2001). Moreover, a chromate or oxidative stress applied on DvH cells leads to an overexpression of the periplasmic [Fe] hydrogenase (Fournier et al. 2004).

To summurize, as listed in Table 2, the most frequently reported proteins involved in metal reduction are cytochromes and hydrogenases isolated from Fe(III)-, sulfur- and sulfate- reducing bacteria.

**Table 2.** c-cytochromes and hydrogenases from Fe(III)-, sulfur- and sulfate- reducing bacteria involved in metal reduction

| Organism | Protein | Metal | Reference |
|---|---|---|---|
| *G. metallireducens* GS-15 | c-cytochromes | Fe(III), Au (III), Ag (I), Hg (II), Cr (VI) | Lovley et al. (1993a) |
| *G. metallireducens* H-2 | c-cytochromes | Fe(III), Cr(VI), Au(III), Ag(I),Hg(II), W(VI), U(VI), V(V), Mo(VI) | Coates et al. (1996) |
| *G. metallireducens* 172 | c-cytochromes | Fe(III), Cr(VI), Au(III), Ag(I),Hg(II), V(V) | Coates et al. (1996) |
| *G. sulfurreducens* | c-cytochrome<br>c-cytochrome (FerA)<br>c-cytochrome (OmcB)<br>cytochrome $c_7$ (PpcA)<br>hydrogenase | Fe(III)<br>Fe(III)<br>Fe(III)<br>Fe(III), U(VI)<br>Tc(VII) | Gaspard et al. (1998)<br>Magnuson et al. (2000; 2001)<br>Leang et al. (2003)<br>Lloyd et al. (2003)<br>Lloyd et al. (2000) |
| *D. vulgaris Hildenborough* | cytochrome $c_3$<br>cytochrome $c_3$<br>cytochrome $c_3$<br>[Fe] hydrogenase | Cr(VI)<br>Fe(III)<br>U(VI)<br>Cr(VI) | Lovley and Philips (1994), Michel et al. (2001)<br>Lojou et al. (1998a,b)<br>Lovley et al. (1993b)<br>Michel et al. (2001) |
| *D. acetoxidans* | cytochrome $c_7$<br>c-cytochrome<br>cytochrome $c_7$ | Fe(III), Cr(VI), Mn(IV), V(V)<br>Mn(IV), Fe(III)<br>Cr(VI) | Lojou et al. (1998a,b)<br>Roden and Lovley (1993)<br>Michel et al. (2001) |
| *D. fructosovorans* | [Fe] hydrogenase<br>cytochrome $c_3$<br>[NiFe] hydrogenase | Tc(VII)<br>Tc(VII)<br>Cr(VI) | De Luca et al. (2001)<br>De Luca et al. (2001)<br>Bruschi (unpublished data) |
| *D. desulfuricans* | Hydrogenase<br>Hydrogenase | Tc(VII)<br>Pd(II) | Lloyd et al. (1999a)<br>Lloyd et al. (1998) |
| *D. gigas* | cytochrome $c_3$ | Fe(III) | Lojou et al. (1998b) |
| *D. norvegicum* | [NiFeSe] hydrogenase<br>cytochrome $c_3$<br>cytochrome $c_3$ | Cr(VI)<br>Cr(VI)<br>Fe(III) | Michel et al. (2001)<br>Michel et al. (2001)<br>Lojou et al. (1998a,b) |

These redox proteins are not acting as terminal electron acceptors for heavy metals. For instance, *D. vulgaris* Hildenborough can reduce Cr(VI) using several enzymes involved in the electron chain transfer, but the reduction of this metal does not support growth (Chardin et al. 2002). Researches, elucidating the mechanisms of bacterial metal reduction, are all the most important, as they will improve the bacterial use conditions during bioremediation processes. Indeed, novel bioremediation approaches could emerge to treat contaminated environments.

## 4. Development of Biosensors

Over the last one decade, biosensors have been developed to be used in wide applications. Biosensors offer the potential to measure quickly, cheaply and accurately the contamination degree of environmental sites. There are two major markets for biosensors. The first one is concerned with clinical and health care fields and needs miniaturization and the second one is for environmental purposes for detection and control. Analysis methods could be largely refined with the development of biocaptors, since this kind of approach allows the detection and the direct quantification of a chemical compound in complex media.

Various studies have reported the construction of biosensors, using genetically engineered bacteria (reviewed in D'Souza 2001). More specifically, biosensors for the detection of heavy metals, have been developed (Verma and Singh 2005). These biosensors have used two distinct methods to detect heavy metals ions: (i) proteins (antibodies, enzymes) or (ii) whole cells (genetically modified or not). Various sensors were designed to evaluate the heavy metals concentration, for example, nickel and copper (Forzani et al. 2005), mercury (Hobman et al. 2000), cadmium (Blake et al. 2001), arsenic, iron and lead (Radhika et al. 2005).

We would like to present more specifically biosensors using proteins that exhibit a metal reductase activity. As we have previously demonstrated that the all the cytochrome $c_3$ tested and the cytochrome $c_7$ have a metal reductase activity, we could propose that the other multihemic cytochromes c described in sulfate and sulfur reducing bacteria, could have also a similar metal reductase activity. It is to be noticed that all these cytochromes have distinct and low redox potentials hemes and show remarkable properties of thermostability until 125°C for some of them (Florens et al. 1995). Recent studies have demonstrated that hydrogenases and other redox proteins with negative redox potentials (like ferredoxins) can also reduce metals. However, hydrogenases are proteins that are usually sensitive to oxygen and are produced in low amount by bacteria. On the contrary, cytochrome $c_3$ is stable towards many physico-chemical factors, such as pH, temperature, oxygen, salt, ageing (Bianco et al. 1986; Florens et al. 1995) and are still stable and active once immobilised at the electrode using various immobilisation techniques. These enzymes are naturally produced at high levels by sulfate reducing bacteria and are also overproduced in specific hosts. Indeed, cytochrome $c_3$ are better candidates for the construction of biosensors.

Different procedures used for constructing protein/enzyme-modified electrodes have been developed, in particular, adsorption on an electrode surface, covalent attachment, imprisonment in a layer by layer assembly and entrapment within cast films or a dialysis membrane. The performances of such modified electrodes with electroactive proteins or enzymes, attached to their active surface, have been compared (Bianco 2002).

A first approach to test the ability of the cytochromes $c_3$ family to achieve the remediation of metal contaminated water has been attempted in the case of iron-containing solution (Lojou et al. 1998b). In this study, the ability of poly ester-sulfonic acid ionomer (Eastman AQ-29D), cast on the electrode surface, is able to immobilize the cytochrome.A catalytic current is detected from cyclic voltammetry experiments where Fe(III) serves as the substrate to oxidize the cytochrome.

The membrane electrode technology offers an alternative to the entrapment of cytochromes within a polymer film. This technology has been extensively used in the case of other metals well known for their high toxicity, in particular Cr(VI), V(V) and U(VI) (Lojou et al. 1998a; Lojou and Bianco 1999). A rapid survey of important parameters, such as pH, metal concentration, nature and concentration of the supporting electrolyte, provides significant advantages. Most of the metals, in sediments and soils, are in the form of various insoluble oxides. In this approach, metal oxide and cytochrome $c_7$ were deposited and entrapped "within" the membrane electrode (Lojou and Bianco 1999).

An amperometric cytochrome $c_3$-based biosensor was constructed for chromate determination (Michel et al. 2003b). Several processes of enzyme mobilisation have been tested and the best results were obtained with dialysis membranes which allowed the determination of Cr(VI) from 0.2 to 6.84 mg/L with a small amount of cytochrome $c_3$ (372 ng of protein) required to construct the biosensor.

## 5. Development of Bioreactors

A number of bio-processes, based on the activity of sulfate and metal reducing bacteria to prevent heavy metal and metalloid pollution from ground waters and soils, have been developed. The objectives of these studies are to obtain improved biological tools and to develop low-cost effective and reliable technological alternatives for bioremediation.

Chemical treatments for the removal of heavy metals from contaminated materials are chemical extraction with acids and/or chelating agents for soil treatment and precipitation for water cleaning. In industries, the metals, contained in acid-drainage waters, are most of the time precipitated using lime. Such treatments are expensive, and lead to a large quantity formation of metal-hydroxides (Zinck 1997).

Bioremediation processes are divided in two main groups: one group exploits the enzymatic metal reductase activity of the bacteria (direct reduction) and the second group involves the use of hydrogen sulfide, biologically produced to reduce and precipitate metals (indirect reduction).

The metal precipitation, using $H_2S$ produced by sulfate-reducing bacteria, has been proposed in the '80s (Whang et al. 1982). These kind of approaches lead to the selective metal precipitation, such as copper or zinc, sulfate and acidity removal (Hammack et al. 1993; Foucher et al. 2001). The indirect metal reduction by biologically produced $H_2S$ has already been exploited up to industrial scale, but important innovation can be introduced by improving the technical and economical benefit of currently available configurations.

Various technologies for *in situ* clean-up are available. The direct reduction of the metals would be applied to ground water, using bioreactors (pump and treat) and could be applied to soils after excavation, pulping or heaping and inoculation. These techniques are very expensive and are characterized by low metal extraction efficiencies. The concept of *in situ* zones and bio-barriers, using metal reducing bacteria, is an alternative to pump and treat strategies and a novel application of indirect reduction. The installation of sub-surface zones, where the bacterial growth will be induced by injection of substrates, could be a low cost solution. The migrating metals would be intercepted and immobilized by precipitation with biologically produced $H_2S$. The capacity of the soils and sediments together with the biofilms to adsorb, filter out and retain inorganic precipitates would be exploited.

Studies on pure cultures are necessary to understand specific mechanisms of metal reduction. However, application in bioremediation could be done by consortia of different microorganisms, containing mainly sulfate reducing bacteria obtained from contaminated soils (White et al. 1997; Vainstein et al. 2003).

Studies, for *in situ* bioremediation of uranium contaminated sites, have been conducted and shown that the microbial community involved contained *Desulfosphorosinus* spp. and *Clostridium* spp. (Suzuki et al. 2003). U(VI) reduction, in the presence of various sulfate concentration, have been proposed by Spear et al. (2000) in order to reach optimal conditions in a bioremediation process. Moreover, treatment of other metals, using anaerobic bioreactors with SRB community culture, has been described, as for example, the bioremediation of (i) phosphogypsum, waste products from fertilizers industries (Rzeczycka et al. 2001; Karnachuk et al. 2002), since the nitrate concentration is not high (Kowalski et al. 2002), or (ii) lead wastes, from car batteries, to PbS (communly named Galena) (Weijma et al. 2002). In the same manner, the reduction of chromate has been described by an enrichment consortium and an isolate of marine sulfate reducing bacteria (Cheung et al. 2003). Pilot plants developed by

Shell research Ltd. and Budelco BV, using an undefined consortium of SRB, have been used successfully to remove Zn and sulfate (White et al. 1997). Here, the metals were precipitated as sulfides. Acetate, produced by sulfate-reducing bacteria, was removed by methanogenic bacteria present in the consortium. This approach was scaled-up and is able to treat 7,000 $m^3$ per day. Indeed, research on biological approaches of the metal precipitation/immobilisation in contaminated environments are necessary to find out new remediation approaches.

## 6. Conclusion

The importance of microbial metal reduction has been recently highlighted and studies on several microorganisms, which may serve as models, have been conducted. The use of Fe(III)-, sulfur- and sulfate-reducing bacteria provides challenges in the reduction of metals and radionuclides. Recent advances have been made and thanks to the discovery of new bacteria isolated from contamined sites or extremophilic environments, providing new potent tools in bioremediation processes since the chemistry and biology of polluted sites largely influence the bioremediation method to use. Reduction mechanisms of metals and radionuclides using of Fe(III)-, sulfur- and sulfate-reducing microorganisms, are at the discovery stage. Very little information on the enzymatic metal reduction in natural environments is available. Further studies on the biochemistry and microbial ecology of metal reduction would enhance our understanding of the factors controlling the rate and extent of biotechnological processes. The development of new techniques, such as genomic and proteomic approaches, and the availability of environmentally relevant bacteria annotated genome sequence, promises us undoubtedly significant advances in the environmental technology and more specifically in the understanding of the precise mechanims of bacteria-metal interactions *in situ*.

## References

Afkar E, Fukumori Y (1999) Purification and characterization of triheme cytochrome $c_7$ from the metal-reducing bacterium, *Geobacter metallireducens*. FEMS Microbiol Lett 175:205-210

Assfalg M, Bertini I, Bruschi M, Michel C, Turano P (2002) The metal reductase activity of some multiheme cytochromes c: NMR structural characterization of the reduction of chromium(VI) to chromium(III) by cytochrome $c_7$. Proc Natl Acad Sci USA 99:9750-9754

Aubert C, Leroy G, Bianco P, Forest E, Bruschi M, Dolla A (1998a) Characterization of the cytochromes c from *Desulfovibrio desulfuricans* G201. Biochem Biophys Res Commun 242:213-218

Aubert C, Lojou E, Bianco P, Rousset M, Durand MC, Bruschi M, Dolla A (1998b) The *Desulfuromonas acetoxidans* triheme cytochrome $c_7$ produced in *Desulfovibrio desulfuricans* retains its metal reductase activity. Appl Environ Microbiol 64:1308-1312

Barton LL, Plunkett RM, Thomson BM (2003) Reduction of metals and nonessential elements by anaerobes. In: Ljungdahl LG, Adams MW, Barton LL, Ferry JG, Johnson MK (eds) Biochemistry and physiology of anaerobic bacteria, Springer, pp 220-234

Banci L, Bertini I, Bruschi M, Sompornpisut P, Turano P (1996) NMR characterization and solution structure determination of oxidized cytochrome $c_7$ from *Desulfuromonas acetoxidans*. Proc Natl Acad Sci USA 93:14396-14400

Beyerink WM (1895) Ueber *Spirullum desulfuricans* als ursache von sulfatereduction. Zentralbl. Bakteriol. Parasitenkd 1(1-9) 49-59-104-114

Bianco P, Haladjian J, Bruschi M (1986) An electrochemical study of the stability of cytochrome c3 for *Desulfovibrio desulfuricans* Norway. Bioelectrochem Bioenerg 15:57-66

Bianco P (2002) Protein modified and membrane electrode: strategies for the development of biomolecular sensors. J Biotechnol 82:393-409

Blake DA, Jones RM, Blake RC 2[nd], Pavlov AR, Darwish IA, Yu H (2001) Antibody-based sensors for heavy metals ions. Biosens Bioelectron 16:799-809

Brugna M, Giudici-Orticoni MT, Spinelli S, Brown K, Tegoni M, Bruschi M (1998) Kinetics and interaction studies between cytochrome c3 and Fe-only hydrogenase from *Desulfovibrio vulgaris* Hildenborough. Proteins 33:590-600

Bruschi M, Bertrand P, More C, Leroy G, Bonicel J, Haladjian J, Chottard G, Pollock WB, Voordouw G (1992) Biochemical and spectroscopic characterization of the high molecular weight cytochrome c from *Desulfovibrio vulgaris* Hildenborough expressed in *Desulfovibrio desulfuricans* G200. Biochemistry 31:3281-3288

Bruschi M (1994) Cytochrome $c_3$ (Mr26000) isolated from sulfate-reducing bacteria and its relationships to other polyhemic cytochromes from *Desulfovibrio*. Methods Enzymol 243:140-155

Bruschi M, Leroy G, Guerlesquin F, Bonicel J (1994) Amino-acid sequence of the cytochrome $c_3$ (M(r) 26,000) from *Desulfovibrio desulfuricans* Norway and a comparison with those of the other polyhemic cytochromes from *Desulfovibrio*. Biochim Biophys Acta 1205:123-131

Bruschi M, Woudstra M, Guigliarelli B, Asso M, Lojou E, Petillot Y, Abergel C (1997) Biochemical and spectroscopic characterization of two new cytochromes isolated from *Desulfuromonas acetoxidans*. Biochemistry 36:10601-10608

Bruschi M, Michel C, Chardin C (2003) Bioremediation of chromate using sulfate-reducing bacteria: thermodynamic analysis, molecular mechanism and biotechnological applications. Recent Res Devel Applied Microbiol Biotechnol 1:147-160

Caccavo F Jr, Lonergan DJ, Lovley DR, Davis M, Stolz JF, McInerney MJ (1994) *Geobacter sulfurreducens* sp. nov., a hydrogen- and acetate- oxidizing dissimilatory metal-reducing microorrganism. Appl Environ Microbiol 60:3752-3779

Chang IS, Groh JL, Ramsey MM, Ballard JD, Krumholz LR (2004) Differential expression of *Desulfovibrio vulgaris* genes in response to Cu(II) and Hg(II) toxicity. Appl Environ Microbiol 70:1847-1851

Chang YJ, Peacock AD, Long PE, Stephen JR, McKinley JP, Macnaughton SJ, Hussain AK, Saxton AM, White DC (2001) Diversity and characterization of sulfate-reducing bacteria in groundwater at a uranium mill tailings site. Appl Environ Microbiol 67:3149-3160

Chardin B, Dolla A, Chaspoul F, Fardeau ML, Gallice P, Bruschi M (2002) Bioremediation of chromate: thermodynamic analysis of the effects of Cr(VI) on sulfate-reducing bacteria. Appl Microbiol Biotechnol 60:352-360

Chardin B, Giudici-Orticoni MT, De Luca G, Guigliarelli B, Bruschi M (2003) Hydrogenases in sulfate-reducing bacteria function as chromium reductase. Appl Microbiol Biotechnol 63:315-321

Cheung KH, Dong Gu J (2003) Reduction of Chromate ($CrO^{2-}U$) by an enrichment consortium and an isolate of marine sulfate reducing bacteria. Chemosphere S2:1523-1529

Childers SE, Ciufo S, Lovley DR (2002) *Geobacter metallireducens* accesses insoluble Fe(III) oxide by chemotaxis. Nature 416(6882):767-769

Coates JD, Lonergan DJ, Philips EJ, Jenter H, Lovley DR (1995) *Desulfuromonas palmitatis* sp. nov., a marine dissimilatory Fe(III) reducer that can oxidize long-chain fatty acids. Arch Microbiol 164:406-413

Coates JD, Phillips EJ, Lonergan DJ, Jenter H, Lovley DR (1996) Isolation of *Geobacter* species from diverse sedimentary environments. Appl Env Microbiol 62(5):1531-1536

Coates JD, Bhupathiraju VK, Achenbach LA, Mclnerney MJ, Lovley DR (2001) *Geobacter hydrogenophilus*, *Geobacter chapellei* and *Geobacter grbiciae*, three new, strictly anaerobic, dissimilatory Fe(III)-reducers. Int J Syst Evol Microbiol 51:581-588

Coppi MV, Leang C, Sandler J, Lovley DR (2001) Development of a genetic system for *Geobacter sulfurreducens*. Appl Env Microbiol 67:3180-3187

Cummings DE, Snoeyenbos-West OL, Newby DT, Niggemyer AM, Lovley DR, Achenbach LA, Rosenzweig RF (2003) Diversity of Geobacteraceae species inhabiting metal-polluted freshwater lake sediments ascertained by 16S rDNA analyses. Microb Ecol 46:257-269

Czjzek M, Guerlesquin F, Bruschi M, Haser R (1996) Crystal structure of a dimeric octaheme cytochrome $c_3$ (M(r) 26,000) from *Desulfovibrio desulfuricans* Norway. Structure 4(4):395-404

Czjzek M, ElAntak L, Zamboni V, Morelli X, Dolla A, Guerlesquin F, Bruschi M (2002) The crystal structure of the hexadeca-heme cytochrome Hmc and a structural model of its complex with cytochrome $c_3$. Structure (Camb) 10:1677-86

De Luca G, de Philip P, Dermoun Z, Rousset M, Vermeglio A (2001) Reduction of technetium(VII) by *Desulfovibrio fructosovorans* is mediated by the nickel-iron hydrogenase. Appl Environ Microbiol 67:4583-4587

Dolla A, Leroy G, Guerlesquin F, Bruschi M (1991) Identification of the site of interaction between cytochrome $c_3$ and ferredoxin using peptide mapping of the cross-linked complex. Biochim Biophys Acta 1058:171-177

D'Souza SF (2001) Microbial biosensors. Biosensors Bioelectronics 16:337-53

Elias DA, Suflita JM, McInerney MJ, Krumholz LR (2004) Periplasmic cytochrome $c_3$ of *Desulfovibrio vulgaris* is directly involved in $H_2$-mediated metal but not sulfate reduction. Appl Environ Microbiol 70:413-20

Finster K, Coates JD, Liesack W, Pfennig N (1997). *Desulfuromas thiophila* sp. nov., a new obligately sulfur-reducing bacterium from anoxic freshwater sediment. Intl J Syst Bact 47:754-758

Florens L, Bianco P, Haladjian J, Bruschi M, Protasevich I, Makarov A (1995) Thermal stability of the polyheme cytochrome $c_3$ superfamily. FEBS Lett 373:280-284

Foucher S, Battaglia-Brunet F, Ignatiadis I, Morin D (2001) Treatment by sulfate-reducing bacteria of Chessy acid-mine drainage and metal recovery. Chem. Eng. Sc. 56:1639-45

Fournier M, Dermoun Z, Durand MC, Dolla A (2004) A new function of the *Desulfovibrio vulgaris* Hildenborough [Fe] hydrogenase in the protection against oxidative stress. J Biol Chem 279:1787-1793

Forzani ES, Zhang H, Chen W, Tao N (2005) Detection of heavy metals ions in drinking water using high-resolution differential plasmon resonance sensor. Environ Sci Technol 39:1257-62

Gaspard S, Vazquez F, Holliger C (1998) Localization and solubilization of the Iron(III) reductase of *Geobacter sulfurreducens*. Appl Environ Microbiol 64:3188-3194

Giudici-Orticoni MT, Makarov AA, Lobachov VM, Protasevich II, Lexa D, Bruschi M (2003) Conformational properties of multihemic cytochromes c from *Desulfuromonas acetoxidans*. Thermochimica Acta 397:5-12

Goulhen F, Gloter A, Guyot F, Bruschi M (2005) Bioremediation of Chromate: effects of Cr(VI) on the metabolism of sulfate-reducing bacteria. 3rd European Bioremediation Conference Proceedings, Chania, Greece (In press)

Hammack RW, Dvorak DH, Edenborn HM (1993) The use of biogenic hydrogen sulfide to selectively recover copper and zinc from severely contaminated mine drainage. In: Torma AE, Wey JE, Lakshmanan VL (eds) Biohydrometallurgical technologies, The minerals, metals and materials society, Warrendale, PA, pp 631-639

Hobman JL, Wilson JR, Brown NL (2000) Microbial mercury reduction. In: Lovley DL (ed) Environmental microbe-metal interactions, ASM press, Washington DC, pp 177-197

Holmes DE, Finneran KT, O'Neil RA, Lovley DR (2002) Enrichment of members of the family Geobacteraceae associated with stimulation of dissimilatory metal reduction in uranium-contaminated aquifer sediments. Appl EnvironMicrobiol 68:2300-2306

Istok JD, Senko JM, Krumholz LR, Watson D, Bogle MA, Peacock A, Chang YJ, White DC (2004) *In situ* bioreduction of technetium and uranium in a nitrate-contaminated aquifer. Environ Sci Technol 38:468-475

Karnachuk OV, Kurochkina SY, Tuovinen OH (2002) Growth of sulfate-reducing bacteria with solid-phase electron acceptors. Appl Microbiol Biotechnol 58:482-486

Karnachuk OV, Kurochkina SY, Nicomrat D, Frank YA, Ivasenko DA, Phyllipenko EA, Tuovinen OH (2003) Copper resistance in *Desulfovibrio* strain R2. Antonie Van Leeuwenhoek 83:99-106

Kim C, Zhou Q, Deng B, Thornton EC, Xu H (2001) Chromium(VI) reduction by hydrogen sulfide in aqueous media : stoichiometry and kinetics. Environ Sci Technol 35:2219-2225

Kowalski W, Przytocka-Jusiak M, Blaszczyk M, Holub W, Wolicka D, Wesolowska I (2002) Effect of nitrates on biotransformation of phosphogypsum and phenol uptake in cultures of autochthonous sludge microflora from petroleum refining wastewaters. Acta Microbiol Pol 51:47-56

Leang C, Coppi MW, Lovley DR (2003) OmcB, a c-type polyheme cytochrome, involved in Fe(III) reduction in *Geobacter sulfurreducens*. OmcB, a c-type polyheme cytochrome, involved in Fe(III) reduction in *Geobacter sulfurreducens*. J Bacteriol 185(7):2096-2103

Liu C, Gorby YA, Zachara JM, Fredrickson JK, Brown CF (2002) Reduction kinetics of Fe(III), Co(III), U(VI), Cr(VI), and Tc(VII) in cultures of dissimilatory metal-reducing bacteria. Biotechnol Bioeng 80(6):637-649

Lloyd JR, Yong P, Macaskie LE (1998) Enzymatic recovery of elemental palladium by using sulfate-reducing bacteria. Appl Environ Microbiol 64(11):4607-4609

Lloyd JR, Ridley J, Khizniak T, Lyalikova NN, Macaskie LE (1999a) Reduction of technetium by *Desulfovibrio desulfuricans*: biocatalyst characterization and use in a flowthrough bioreactor. Appl Environ Microbiol 65:2691-2696

Lloyd JR, Thomas GH, Finlay JA, Cole JA, Macaskie LE (1999b) Microbial reduction of technetium by *Escherichia coli* and *Desulfovibrio desulfuricans*: enhancement via the use of high-activity strains and effect of process parameters. Biotechnol Bioeng 66:122-130

Lloyd JR, Blunt-Harris EL, Lovley DR (1999c) The periplasmic 9.6-kilodalton c-type cytochrome of *Geobacter sulfurreducens* is not an electron shuttle to Fe(III). J Bacteriol 181:7647-7649

Lloyd JR, Sole VA, Van Praagh CV, Lovley DR (2000) Direct and Fe(II)-mediated reduction of technetium by Fe(III)-reducing bacteria. Appl Environ Microbiol 66:3743-3749

Lloyd JR, Chesnes J, Glausauer S, Bunker DJ, Livens FR, Lovley DR (2002) Reduction of actinides and fission products by Fe(III)-reducing bacteria. Gemicrobiol J 19:103-120

Lloyd JR (2003) Microbial reduction of metals and radionuclides. FEMS Microbiol Rev 27(2-3):411-425

Lloyd JR, Leang C, Hodges Myerson AL, Coppi MV, Cuifo S, Methe B, Sandler SJ, Lovley DR (2003) Biochemical and genetic characterization of PpcA, a periplasmic c-type cytochrome in *Geobacter sulfurreducens*. Biochem J 369:153-161

Lojou E, Bianco P, Bruschi M (1998a) Kinetic studies on the electron transfer between bacterial *c*-type cyrochromes and metal oxides. J Electroanal Chem 452:167-177

Lojou E, Bianco P, Bruschi M (1998b) Kinetic studies on the electron transfer between various c-type cytochromes and iron (III) using a voltametric approach. Electrochim Acta 43:2005-2013

Lojou E, Bianco P (1999) Electrocatalytic reduction of uranium by bacterial cytochromes: biochemical factors influencing the catalytic process. J Electroanal Chem 471:96-104

Londer YY, Pokkuluri PR, Tiede DM, Schiffer M (2002) Production and preliminary characterization of a recombinant triheme cytochrome $c_7$ from *Geobacter sulfurreducens* in *Escherichia coli*. Biochim Biophys Acta 1554:202-211

Lonergan DJ, Jenter HL, Coates JD, Phillips EJ, Schmidt TM, Lovley DR (1996) Phylogenetic analysis of dissimilatory Fe(III)-reducing bacteria. J Bacteriol 178:2402-2408

Lovley DR, Phillips EJ (1992) Reduction of uranium by *Desulfovibrio desulfuricans*. Appl Environ Microbiol 58:850-856

Lovley DR, Giovannoni SJ, White DC, Champine JE, Phillips EJ, Gorby YA, Goodwin S (1993a). *Geobacter metallireducens* gen. *nov*. sp. *nov*., a microorganism capable of coupling the complete oxidation of organic compounds to the reduction of iron and other metals. Arch Microbiol 159:336-344

Lovley DR, Widman PK, Woodward JC, Phillips EJ (1993b) Reduction of uranium by cytochrome $c_3$ of *Desulfovibrio vulgaris*. Appl Environ Microbiol 59:3572-3576

Lovley DR, Phillips EJ (1994) Reduction of chromate by *Desulfovibrio vulgaris* and its $c_3$ cytochrome. Appl Environ Microbiol 60:726-728

Lovley DR (1995) Bioremediation of organic and metal contaminants with dissimilatory metal reduction. J Ind Microbiol 14:85-93

Lovley DR (1997) Microbial Fe(III) reduction in subsurface environments. FEMS Microbiol Rev 20:305-313

Lovley DR, Lloyd JR (2000) Microbes with a mettle for bioremediation. Nature Biotechnol 18:600-601

Magnuson TS, Hodges-Myerson AL, Lovley DR (2000) Characterization of a membrane-bound NADH-dependent Fe(3+) reductase from the dissimilatory $Fe^{3+}$-reducing bacterium *Geobacter sulfurreducens*. FEMS Microbiol Lett 185:205-211

Magnuson TS, Isoyama N, Hodges-Myerson AL, Davidson G, Maroney MJ, Geesey GG, Lovley DR (2001) Isolation, characterization and gene sequence analysis of a membrane-associated 89 kDa Fe(III) reducing cytochrome c from *Geobacter sulfurreducens*. Biochem J 59:147-152

Methe BA, Nelson KE, Eisen JA, Paulsen IT, Nelson W, Heidelberg JF, Wu D, Wu M, Ward N, Beanan MJ, Dodson RJ, Madupu R, Brinkac LM, Daugherty SC, DeBoy RT, Durkin AS, Gwinn M, Kolonay JF, Sullivan SA, Haft DH, Selengut J, Davidsen TM, Zafar N, White O, Tran B, Romero C, Forberger HA, Weidman J, Khouri H, Feldblyum TV, Utterback TR, Van Aken SE, Lovley DR, Fraser CM (2003) Genome of *Geobacter sulfurreducens*: metal reduction in subsurface environments. Science 302:1967-1969

Michel C, Brugna M, Aubert C, Bernadac A, Bruschi M (2001) Enzymatic reduction of chromate: comparative studies using sulfate-reducing bacteria. Key role of polyheme cytochromes c and hydrogenases. Appl Microbiol Biotechnol 55:95-100

Michel C, Giudici-Orticoni MT, Baymann F, Bruschi M (2003a) Bioremediation of chromate by sulfate-reducing bacteria, cytochromes $c_3$ and hydrogenases. Water Air Soil Pollution: Focus 3:161-169

Michel C, Battaglia-Brunet F, Tran Minh C, Bruschi M, Ignatiadis U (2003b) Amperometric cytochrome $c_3$-base biosensor for chromate determination. Biosensors Bioelectronics 19/4:345-352

Nevin KP, Finneran KT, Lovley DR (2003) Microorganisms associated with uranium bioremediation in a high-salinity subsurface sediment. Appl Environ Microbiol 69:3672-3675

Payne RB, Gentry DM, Rapp-Giles BJ, Casalot L, Wall JD (2002) Uranium reduction by *Desulfovibrio desulfuricans* strain G20 and a cytochrome $c_3$ mutant. Appl Environ Microbiol 68:3129-3132

Peacock AD, Chang YJ, Istok JD, Krumholz L, Geyer R, Kinsall B, Watson D, Sublette KL, White DC (2004) Utilization of Microbial Biofilms as Monitors of Bioremediation. Microb Ecol Mar 4 [Epub ahead of print]

Pereira IAC, Pacheco I, Liu MY, Le Gall I, Xavier AV, Texeira M (1997) Multiheme cytochromes from the sulfur-reducing bacterium *Desulfuromonas acetoxidans*. Eur J Biochem 248:323

Pfennig N, Biebl H (1976) *Desulfuromonas acetoxidans* gen. *nov*. and sp. *nov*., a new anaerobic, sulfur-reducing, acetate-oxidizing bacterium. Arch Microbiol 110:3-12

Pignolet L, Auvray F, Fonsny K, Capot F, Moureau Z (1989) Role of various microorganisms on Tc behavior in sediments. Health Phys 57:791-800

Pokkuluri PR, Londer YY, Duke NE, Long WC, Schiffer M (2004) Family of cytochrome c7-type proteins from *Geobacter sulfurreducens*: structure of one cytochrome c7 at 1.45 A resolution. Biochemistry 43:849-859

Postgate JR (1984) The sulphate-reducing bacteria. 2[nd] edition. Cambridge University Press. Cambridge

Radhika V, Milkevitch M, Audige V, Proikas-Cezanne T, Dhanasekaran N (2005) Engineered *Saccharomyces cerevisiae* strain BioS-1, for the detection of water-borne toxic metal contaminants. Biotechnol Bioeng 90:29-35

Roden EC, Lovley DR (1993) Dissimilatory Fe(III) reduction by the marine microorganism *Desulfuromonas acetoxidans.* Appl Environ Microbiol 59:734-742

Rzeczycka M, Mycielski R, Kowalski W, Galazka M (2001) Biotransformation of phosphogypsum in media containing different forms of nitrogen. Acta Microbiol Pol 50(3-4):281-289

Seeliger S, Cord-Ruwisch R, Schink B (1998) A periplasmic and extracellular c-type cytochrome of *Geobacter sulfurreducens* acts as a ferric iron reductase and as an electron carrier to other acceptors or to partner bacteria. J Bacteriol 180:3686-3691

Spear JR, Figueroa LA, Honeyman BD (2000) Modeling reduction of uranium U(VI) under variable sulfate concentrations by sulfate-reducing bacteria. Appl Environ Microbiol 66:3711-3721

Straub KL, Buchholz-Cleven BE (2001) *Geobacter bremensis* sp. *nov*. and *Geobacter* pelophi*l*us sp. *nov*., two dissimilatory ferric-iron-reducing bacteria. Int J Syst Evol Microbiol 51:1805-1808

Suzuki Y, Kelly SD, Kemner KM, Banfield JF (2003) Microbial populations stimulated for hexavalent uranium reduction in uranium mine sediment. Appl Environ Microbiol 69:1337-1346

Tucker MD, Barton LL, Thomson BM (1996) Kinetic coefficients for simultaneous reduction of sulfate and uranium by *Desulfovibrio desulfuricans*. Appl Microbiol Biotechnol 46:74-77

Tucker MD, Barton LL, Thomson BM (1998) Reduction of Cr, Mo, Se and U by *Desulfovibrio desulfuricans* immobilized in polyacrylamide gels. J Ind Microbiol Biotechnol 20:13-19

Utgikar VP, Chen BY, Chaudhary N, Tabak HH, Haines JR, Govind R (2001) Acute toxicity of heavy metals to acetate-utilizing mixed cultures of sulfate-reducing bacteria: EC100 and EC50. Environ Toxicol Chem 20(12):2662-2669

Utgikar VP, Tabak HH, Haines JR, Govind R (2003) Quantification of toxic and inhibitory impact of copper and zinc on mixed cultures of sulfate-reducing bacteria. Biotechnol Bioeng 82:306-312

Vainstein M, Kusch K, Mattusch J, Vatsounina A, Wiessner A (2003) Model experiments on the microbial removal of chromium from contaminated ground water. Water Res 37:1401-1405

Valls M, Atrian S, de Lorenzo V, Fernandez L (2000) Engineering a mouse metallothionein on cell surface of Ralstonia eutropha CH34 for immobilization of hevy metals in soil. Nature Biotechnol 18:661-664

Verma N, Singh M (2005) Biosensors for heavy metals. Biometals 18:121-129

Weijma J, De Hoop K, Bosma W, Dijkman H (2002). Biological conversion of anglesite (PbSO(4)) and lead waste from spent car batteries to galena (PbS). Biotechnol Prog 18:770-775

Whang JS, Young D, Pressman M (1982) Soluble-sulfide precipitation for heavy metals removal from wastewaters. Environ Progress 1:110-113

White C, Sayer JA, Gadd GM (1997) Microbial solubilization and immobilization of toxic metals: key biogeochemical processes for treatment of contamination. FEMS Microbiol Rev 20:503-516

Widdel F, Pfennig N (1991) The genus *Desulfuromonas* and other Gram-negative sulfur-reducing eubacteria. In: Balows A, Trüper HG, Dwarkin M, Harder W, Schleifer KH (eds) The Prokaryotes, pp 3379-3389

Zinck JM (1997) Acid-mine drainage sludge in the canadian mineral industry: physical, chemical, mineralogical and leaching characteristics. In: 4$^{th}$ international conference on acid rock draingage, Vancouver BC, pp 1693-1708

3

# Bioremediation of Soils Polluted with Hexavalent Chromium using Bacteria: A Challenge

Carlo Viti and Luciana Giovannetti

Dipartimento di Biotecnologie Agrarie – Sez. Microbiologia, University of Florence, ITALY, Email: luciana.giovannetti@unifi.it

## 1. Introduction

The contamination of the environment with heavy metals is a serious problem, because industrial activities and sewage sludge applications have largely contributed to a wide spread of these elements in the terrestrial environment. Therefore, over some years, the discarding of solid and/or liquid waste products containing heavy metals due to industrial processes has received a lot of attention, and legislation for the protection of the environment has become more rigid (Benedetti 1998; Chen and Hao 1998).

Chromium, considered as one of the main pollutants in the United States by the Environmental Protection Agency (EPA) (Fig. 1), is widely used in many industrial activities (Fig. 2). Its world production is in the order of 10,000,000 tons per year (Cervantes et al. 2001).

Chromium is able to exist in several oxidation states, ranging from $Cr^{+2}$ to $Cr^{+6}$, but in soils the most stable and common forms are trivalent Cr(III) and hexavalent Cr(VI) species (Fendorf 1995), which display quite different chemical properties and affect organisms in different ways. In fact, in contrast to other metals, the hazard of chromium is dependent on its oxidation state. Hexavalent chromium is water-soluble, highly toxic and mutagenic to most organisms and carcinogenic for humans, while trivalent chromium is essential (in low concentrations) for human and animal nutrition, relatively water-insoluble and less toxic than Cr(VI) (Francisco et al. 2002).

In many countries, chromate, which is the most prevalent form of Cr(VI) present in solid/liquid waste due to human activities, such as electroplating, steel and automobile manufacturing, production of paint pigments and dyes, wood preservation, is a hazardous contaminant (Kamaludeen et al. 2003), because it is a serious threat to human health and it readily spreads beyond the site of initial contamination through aquatic systems and groundwater.

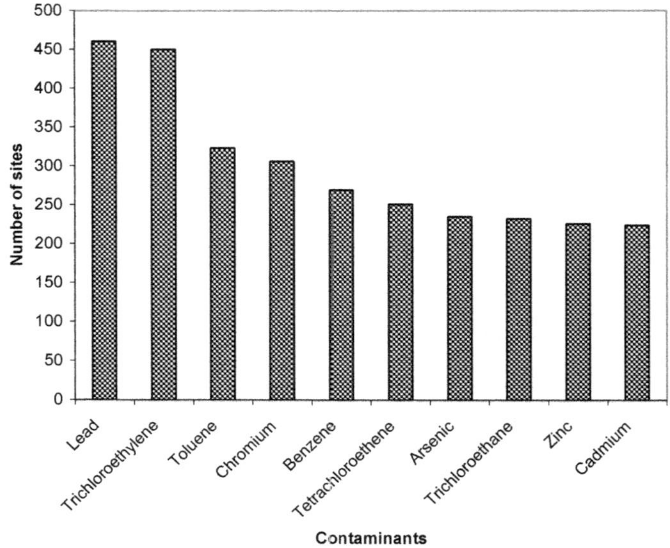


**Fig. 1.** Contaminants most commonly present in all matrices at superfund sites (sources: EPA - Recent developments for *in situ* treatment of metal contaminated soils 1997)

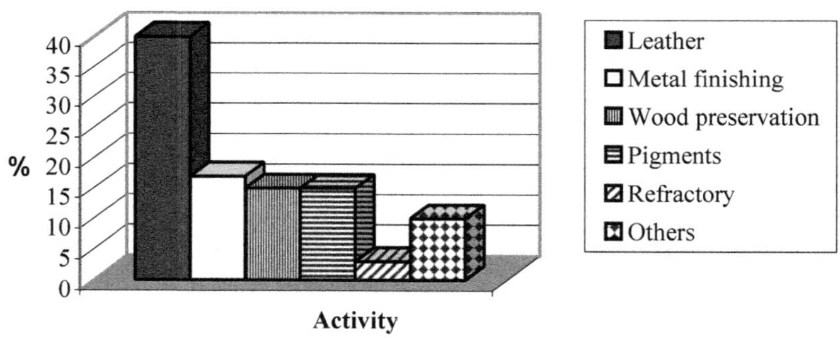


**Fig. 2.** Industrial uses of chromium

The strong impact of hexavalent chromium on the environment and on the human health demands suitable technologies to neutralise the hazard of chromium. The traditional technologies used for the remediation of Cr(VI)-contaminated environments are based on physical and chemical approaches, which require large amounts of chemical substances and energy. Such methodologies have proved prohibitively expensive on a large-scale application at contaminated sites, and also they have generated hazardous by-products (Blowes 2002; Gonzalez et al. 2003). Bioremediation, a strategy that uses living microorganisms, is essentially proposed to clean up the environment from organic pollutants. However, since there is an evidence that several microorganisms possess the capability to reduce Cr(VI) to relatively toxic

Cr(III), bioremediation gives immense opportunities for the development of technologies for the detoxification of Cr(VI)-contaminated soils as an alternative to existing physical-chemical remediation technologies (Roundhill and Koch 2002).

In this review, some of the important efforts, that have been made in the last years in the use of bacteria for potential Cr(VI)-bioremediation of soils, will be summarised.

## 2. Chromium Toxicity

Chromium is an essential micro-nutrient in the diet of animals and humans, as it is indispensable for the normal sugar, lipid and protein metabolism of mammals (Mordenti and Piva 1997). Its deficiency in the diet causes alteration to lipid and glucose metabolism in animals and humans. Chromium is included in the complex named glucose tolerance factor (GFC). On the other hand, no positive effects of chromium are known in plants and microorganisms (Nies 1999; Cervantes et al. 2001). However, elevated levels of chromium are always toxic, although the toxicity level is related to the chromium oxidation state. Cr(VI) not only is highly toxic to all forms of living organisms, mutagenic in bacteria, mutagenic and carcinogenic in humans and animals (Losi et al. 1994b), but also, it is involved in causing birth defects and the decrease of reproductive health (Kanojia et al. 1998). Cr(VI) may cause death in animals and humans, if ingested in large doses (Zayed and Terry 2003). The $LD_{50}$ (dose that causes the death of 50 per cent of a defined animal population) for oral toxicity in rats is from 50 to 100 mg/kg for hexavalent chromium and 1900-3000 mg/kg for Cr(III) (Deflora et al. 1990). Cr(VI) toxicity is related to its easy diffusion across the cell membrane in prokaryotic and eukaryotic organisms and subsequent Cr(VI) reduction in cells, which gives free radicals that may directly cause DNA alterations as well as toxic effects (Arslan et al. 1987; Kadiiska et al. 1994; Liu et al. 1995). Cr(III) has been estimated to be from 10 to 100 times less toxic than C(VI) (Deflora et al. 1990), because cellular membranes appear to be quite impermeable to most Cr(III) complexes. Nevertheless, intracellular Cr(III), which is the terminal product of the Cr(VI)-reduction, forms *in vivo* amino acid nucleotide complexes, whose mutagenic potentiality is not fully known (Roundhill and Koch 2002).

It is well known that prokaryotes are more resistant to Cr(VI) than eukaryotes (Kalamaluden et al. 2003). Toxic effects of chromium on bacteria, algae and plants have been reviewed by Wong and Trevors (1988), Cervantes et al. (2001) and Kamaludeen et al. (2003). On the contrary, scant information is available about the impact of chromium on the structure and diversity of soil microbial communities (Viti and Giovannetti 2001; Viti and Giovannetti 2005). In many studies, it has been difficult to assess the toxicity of chromium to soil microorganisms, because the environments examined were often polluted at the same time with organic pollutants and/or different heavy metals (Turpeinen et al. 2004). In a soil chronically polluted with chromium (about 5000 mg/kg of soil) by leather tannery activity, the oxygenic phototrophic microorganisms and heterotrophic bacterial communities were both affected by chromium (Viti and Giovannetti 2001). Nitrogen-fixing cyanobacteria were not detected in the Cr-contaminated soil using the MPN test, and data obtained from enrichment cultures for nitrogen-fixing cyanobacteria showed that cyanobacteria belonging to the *Nostoc* group were present, but they had a low number of heterocysts (Fig. 3).

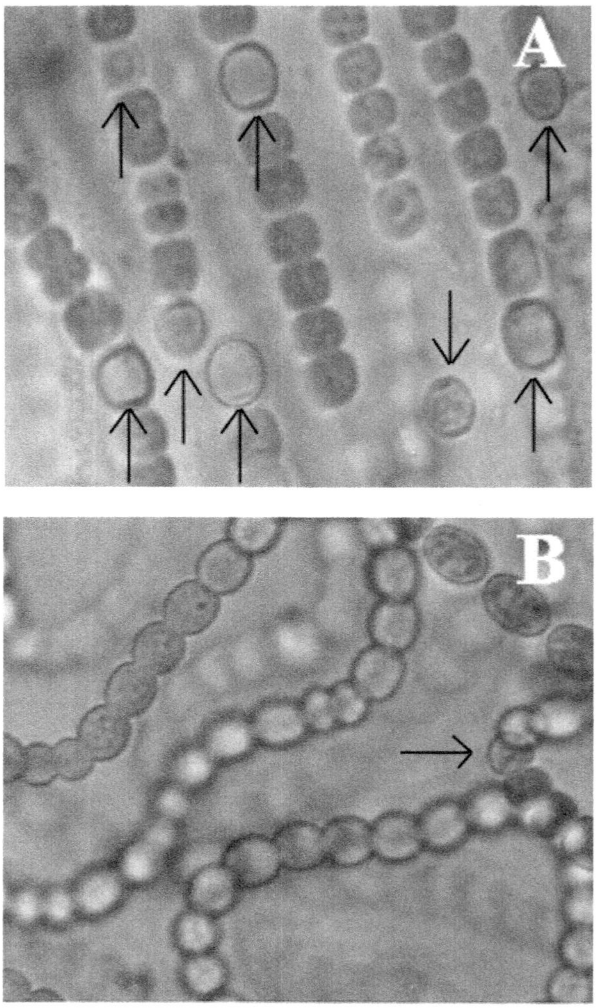


**Fig. 3.** Cyanobacteria belonging to the *Nostoc* group in soil unpolluted (A) and polluted (B) with chromium (Arrows indicate heterocysts)

The size of the cultivable heterotrophic bacterial community was not affected by chromium pollution, but there was a relationship between the percentage of chromate-tolerant bacteria and the level of chromium in the soil. Some authors have found that Gram-positive bacteria were more chromate-tolerant than Gram-negative bacteria (Ross et al. 1981; Viti and Giovannetti 2001; Viti and Giovannetti 2005). Shi et al. (2002), in a study carried out using microcosms in order to establish the long-term effect of chromium upon the activity of the soil microbial community, have shown that chromium negatively affected the microbial activity and led to the accumulation of soil organic carbon. Speir et al. (1995) have found that short-term Cr(VI)-exposure inhibited soil biological properties, such as phosphatase and sulphatase activities, and decreased microbial biomass.

The effects of chromium on soil microbial processes and activities have been reviewed by Kamaludeen et al. (2003).

## 3. Chemical Transformations of Chromium in Soil: Mobility and Bio-availability

In soils, chromium, although it has several oxidation states, occurs mainly with two different oxidation states: Cr(III) and Cr(VI), which have opposite chemical and physical characteristics. The former is the most stable under reduced conditions, it is relatively immobile because it has a strong affinity for negative-charged ions and colloids in soils, and gives sparingly soluble compounds such as $Cr(OH)_3$. Such products dominate at pH values from 4 to 8 (Fendorf 1995). The characteristics of Cr(III) forms limit their bioavailability and mobility in waters and soils. The concentrations of soluble Cr(III) in equilibrium with insoluble compounds are $< 10^{-9}$ M (0.05 parts per billions) in water at pH value 6 to less than $10^{-15}$M at pH value 8 (Richard and Bourg 1991).

Cr(VI) is more soluble, mobile, and bio-available than Cr(III). Cr(VI) is an anion form under most environmental conditions. At higher pH values than 6.4, it is primarily present as chromate ($CrO_4^{-2}$) whereas below pH value 6.4, it is present principally as bichromate ($HCrO_4^-$) (James 2002).

The two different oxidation states of chromium (trivalent and hexavalent forms) can inter-convert and the Cr(VI)-reduction to Cr(III) is favoured. The reduction of Cr(VI) to Cr(III) is of great environmental importance, since Cr(III) is less hazardous. Living organisms, ferrous iron, sulphide and organic matter, have the capacity to reduce hexavalent chromium (James and Bartlett 1983; Fendorf 1995; Kamaludeen et al. 2003). Losi et al. (1994a) have demonstrated that organic matter of the soil plays an important role in the reduction of Cr(VI) to Cr(III) by creating reducing conditions by increasing activities of microbial communities, by acting as an electron donator and by indirectly lowering the oxygen level of the soil (oxygen is depleted through an increase of microbial respiration). Therefore, the presence of an available carbon source to specific bacterial populations is fundamental to alleviate an environment from hazardous forms of chromium.

Although in soils, the reduction of Cr(VI) to Cr(III) is favoured compared to the Cr(III) oxidation, as has been reported above, high concentrations of Cr(VI) may overcome the reducing capability of the environment and thus Cr(VI) may persist in potentially toxic levels (Cervantes et al. 2001). Moreover, a part of Cr(III) can be transformed in Cr(VI) in Bartlett positive soils (Bartlett and James 1979). Bartlett and James (1979) have experimentally demonstrated that the oxidation of Cr(III) to Cr(VI) occurs readily under conditions prevalent in many field soils and the key to Cr(III) oxidation appears to be the presence in soils of oxidised Mn. Oxidised Cr(III) is proportional to the soil content of Mn oxide and to reduced Mn oxide. However, the oxidation of Cr(III) is directly related to its concentration in soil and strongly dependent on Cr(III) forms (Kamaludeen et al. 2003). The influence of pH on oxidation and reduction reactions of chromium in soils is complex, but it is reported that high pH values enhance the oxidative soil power while low pH values enhance reduction reactions. Under laboratory conditions (soils with near-neutral pH values, high levels of Mn oxides and optimal aeration conditions), it has been observed that soluble and freshly precipitated forms of Cr(III), such as $CrCl_3$ and $Cr(OH)_3$, added to soil, may be oxidised up to 15% (James 2002). Therefore, strategies for developing remediation processes of chromium contaminated soils should consider the possibility that certain forms of Cr(III) can be oxidised to Cr(VI). Moreover, caution should be taken regarding the use of hydrogen peroxide for *in situ* remediation of soils contaminated with chemically complex wastes, because mobilisation of Cr(VI) could be a dangerous consequence of using hydrogen peroxide (Rock et al. 2001). Soils contaminated with chromium from chromate ore processing or from electroplating waste released larger amounts of Cr(VI) after treatment with hydrogen peroxide (Rock et al. 2001).

James et al. (1997) developed a Potential Chromium Oxidation Score (PCOS) in order to design remediation by reduction strategies to clean chromium contaminated soils and to predict the levels of mobile Cr(VI) in soils enriched with Cr(III). The PCOS is based on four interacting parameters, solubility and form of Cr(III), reactive soil Mn, soil potential for Cr(VI)-reduction, soil pH as a modifier of the first tree parameters. Such parameters can be quantified and ranked numerically, the sum of their values gives the PCOS. The PCOS ranges from 10 to 40, high scores indicate an elevated probability for Cr(III) oxidation and the persistence of Cr(VI).

## 4. Interaction Between Chromium and Bacteria

Soil contamination by heavy metals is often irreversible and may repress or even kill parts of the microbial community, and it is generally assumed that the exposure to metals leads to the establishment of a tolerant/resistant microbial population (Viti and Giovannetti 2001). The terms resistance and tolerance are often used interchangeably, but their significance is different. Gadd (1992) defined "resistance" as "the ability of a microorganism to survive toxic effects of metal exposure by means of a detoxification mechanism produced in direct response to the metal species concerned" and defined tolerance as "the ability of a microorganism to survive metal toxicity by means of intrinsic properties and or environmental modification of toxicity".

Several bacteria belonging to different taxa have shown resistance/tolerance to Cr(VI) (Table 1). The bacterial chromate resistance is generally linked to plasmids, but it can also be due to chromosomal mutations (Ohta et al. 1971). Chromosomal and plasmid determinants function through different mechanisms. Chromosomal mutation usually affects sulphate transport system (Cervantes and Silver 1992), through which chromate enters the cells of many bacteria (Nies and Silver 1995). Plasmid-determined resistance results in decreased chromate accumulation in cells without involving sulphate transport. The plasmid-coded chromate-resistance has mainly been thought to be based on the chromate efflux (Nies 1999; Cervantes et al. 2001). However, the mechanisms of chromate-resistance have not been yet fully elucidated (Cervantes et al. 2001).

**Table 1.** Tolerance/resistance to Cr(VI) in different bacterial strains (here only main references after 1998 are reported)

| Organisms | Cr(VI) tolerance/ resistance (mg/L) | References |
|---|---|---|
| *Arthrobacter crystallopoites* | 500 | Camargo et al. (2003) |
| *Arthrobacter* sp. | >100 | Megharaj et al. (2003) |
| *Bacillus* sp. | >100 | Megharaj et al. (2003) |
| *Bacillus maroccanus* ChrA21 | 1040 | Viti et al. (2003) |
| *Bacillus* sp. ES29 | 1500 | Camargo et al. (2003) |
| *Bacillus cereus* ES04 | 1500 | Camargo et al. (2003) |
| *Corynebacterium hoagii* ChrB20 | 1144 | Viti et al. (2003) |
| *Bacillus circulans* | 100 | Srinath et al. (2002) |
| *Bacillus megaterium* | 150 | Srinath et al. (2002) |
| *Frankia* strains | 52-91 | Richards et al. (2002) |
| *Ralstonia metallidurans* AE104 | 2-18* | Juhnke et al. (2002) |
| *Pseudomonas* sp. CRB5 | 520 | McLean and Beveridge (2001) |
| *Pseudomonas stutzeri* (two strains) | 52 | Badar et al. (2000) |
| *Escherichia coli* | 10 | Nies (1999) |

*On Tris-buffered mineral medium with different sodium sulphate concentrations

The capability of Cr(VI)-reduction is suggested as an additional chromosome or plasmid resistance mechanism, and represents a potentially useful detoxification process for several bacteria (Komori et al. 1989; Pattanapipitpaisal et al. 2001; Cervantes et al. 2001). Thereby, the bacterial property, that is particularly useful for an effective bioremediation approach, is one which combines high tolerance/resistance with the ability to reduce Cr(VI) to less toxic Cr(III). Microbial Cr(VI)-reduction may occur directly through enzymatic activity (Komori et al. 1989; Losi et al. 1994b; Lovley and Coates 1997) (Fig. 4) or indirectly through producing a compound that can reduce Cr(VI) (Lovley 1993; Fendorf and Li 1996) (Fig. 5).

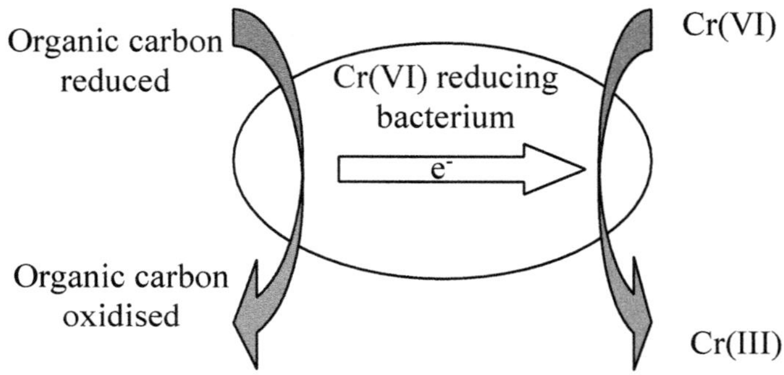


**Fig. 4.** Bacterial direct enzymatic reduction of Cr(VI)

The ability of direct Cr(VI) reduction has been found in many bacterial genera including *Pseudomonas*, *Micrococcus*, *Bacillus*, *Achromobacter*, *Microbacterium Arthrobacter*, *Corynebacterium*. (McLean et al. 2000; Pattanapipitpaisal et al. 2001; McLean and Beveridge 2001; Viti et al. 2003; Megharaj et al. 2003). The capability of Cr(VI)-reduction is not uncommon in Cr(VI)-resistant bacteria of soils. Losi et al. (1994b) found that 9 out of 20 Cr(VI)-resistant bacterial strains, isolated from organic-amended and Cr(VI)-acclimated soils, showed the capability to actively reduce Cr(VI) to Cr(III).

Some bacterial species are capable of both anaerobic and aerobic hexavalent chromium reduction, others in either anaerobic or aerobic conditions (Table 2). The mechanisms through which bacterial strains reduce Cr(VI) to Cr(III) are variable and species dependent (McLean et al. 2000). Anaerobic bacteria may use chromate as a terminal-electron acceptor or reduce chromate in periplasmatic space by hydrogenase or cytocrome $c_3$ (Tebo and Obraztova 1998; Michel et al. 2001; Puzon et al. 2002). In aerobic bacteria, Cr(VI) reduction may be carried out by cellular reducing agents (the primary reductant is glutathione) and NADH-dependent chromate reductase (Suzuki et al. 1992; Shen and Wang 1993; Garbisu et al. 1998). It is yet unknown, although some hypotheses have been formulated, whether enzymatic or non-enzymatic reduction of chromate is dominant in bacterial cells under aerobic conditions, and it also remains unsolved whether the NADH-dependent reductases are specific to chromate. Moreover, it is also unclear whether anaerobic bacterial growth is supported at the expense of chromate as the only electron acceptor. The mechanisms for Cr(VI) reduction might be a secondary utilisation or co-metabolism as suggested for *Shewanella onoidensis* MR-1 (Middleton et al. 2003). Therefore, under anaerobic conditions, Cr(VI)-reduction may be an activity of the reductases that have evolved on other substrates (Kamaludeen et al. 2003).

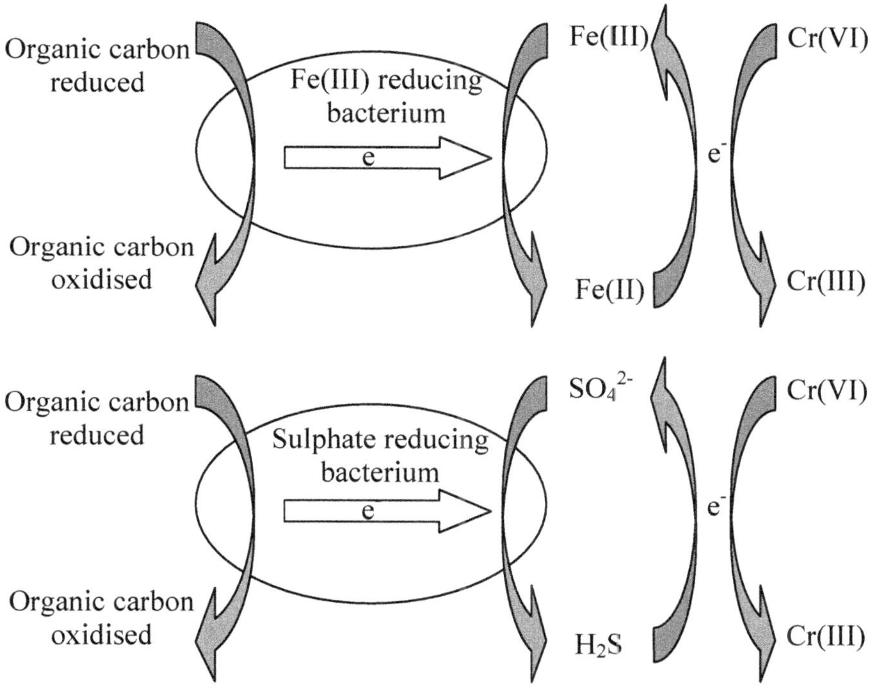


**Fig. 5.** Indirect reduction of Cr(VI) by the activity of iron and sulphate reducing bacteria

The ability of a bacterial strain to reduce hexavalent chromium, although the mechanism of Cr(VI)-reduction may differ from strain to strain, is an attractive feature in order to plan a biological strategy for effective chromate detoxification, but high concentrations of chromate in the environment often can repress the microbial activity and growth (Lovley and Coates 1997; Megharaj et al. 2003). Therefore, before using a selected microorganism or an indigenous microbial mixed-culture for devising bioremediation strategies for Cr(VI)-contaminated soils, there is a need to understand Cr(VI)-resistance mechanisms in these microorganisms. For example *Shewanella* spp., among dissimilatory metal reducing bacteria, are identified as potential attractive candidates for *ex-situ* as well as *in-situ* treatment of chromate contamination (Myers et al. 2000; Daulton et al. 2002). Nevertheless, it has recently been demonstrated that Cr(VI) is toxic even at low concentrations (0.015 mM) to *Shewanella oneidensis* MR-1, a good Cr(VI) reducer (Viamajala et al. 2004). It is important to remember that chromate resistance and reduction are not necessarily interrelated, hexavalent chromium may be reduced by both bacterial Cr(VI)-sensitive and resistant strains and not all Cr(VI)-resistant bacteria reduce hexavalent chromium to trivalent forms.

**Table 2.** Bacterial strains able to reduce Cr(VI) (here only main references after 1998 are reported)

| Organisms | Reduction conditions | References |
|---|---|---|
| *Arthrobacter crystallopoietes* ES 32 | Aerobic | Camargo et al. (2004) |
| *Bacillus maroccanus* ChrA21 | Aerobic | Viti et al. (2003) |
| *Bacillus* sp. ES29 | Aerobic | Camargo et al. (2003) |
| *Corynebacterium hoagii* ChrB20 | Aerobic | Viti et al. (2003) |
| *Cellulomonas* sp. (three strains) | Anaerobic | Sani et al. (2002) |
| *Bacillus* sp. ATCC 700729 | Aerobic | Shakoori et al. (2000) |
| *Pseudomonas* sp. CRB5 | Aerobic and anaerobic | McLean et al. (2000) |
| *Pseudomonas stutzeri* (two strains) | Anaerobic | Badar et al. (2000) |
| *Shewanella oneidensis* MR-1 | Anaerobic | Myers et al. (2000) |
| *Pseudomonas aeruginosa* A2Chr | Aerobic | Ganguli and Tripathi (1999) |

An extensive review about chromium resistance in bacteria, enzymatic mechanisms of microbial Cr(VI)-reduction and factors affecting microbial Cr(VI)-reduction has been published by Chen and Hao (1998).

The reduction of Cr(VI) can also occur indirectly by bacterial activity. For instance, Fe(II) and $HS^-$, metabolic end products of iron and sulphate-reducing bacteria, can catalyse the reduction of Cr(VI) (Fendorf et al. 2000; Wielinga et al. 2001; Nevin and Lovley 2002; Arias and Tebo 2003). The process of indirect reduction of chromate using iron reducing bacteria consists of two reactions (Fig. 5). The Fe(II) produced by reducing bacteria is cycled back to Fe(III) by abiotic chromate reduction. At the ecological level, this process represents a significant role, because it permits the uninterrupted regeneration of the Fe(III), terminal electron acceptor in anaerobic conditions.

In sulphate rich soil environments, when anaerobic conditions are present, such as in flooded compacted soils, the reduction of Cr(VI) by sulphide produced through sulphate reducing bacteria, which couple the oxidation of organic sources to the reduction of sulphate, is an important mechanism to detoxify the environment from hexavalent chromium (Losi et al. 1994b; Pettine et al. 1998).

## 5. Soil Bioremediation Strategies

The remediation of Cr(VI)-contaminated soils, today, is essentially based on physical and chemical approaches, which include excavation or pumping of contaminated material, followed by the addition of reducing chemicals that lead to the precipitation and/or sedimentation of reduced chromium [Cr(III)], less toxic than Cr(VI) and greatly insoluble. Cr(VI) remediation strategies using traditional technologies have been dealt in depth by Higgins et al. (1997) and hence not discussed here in details. However, Table 3 shows the advantages and disadvantages of the main traditional approaches for the remediation of soils contaminated with Cr(VI).

The ability of several microbial groups (bacteria, fungi, microalgae) to reduce Cr(VI) to Cr(III) has been considered of much interest in order to clean up soil/water polluted with chromate. In fact, there is no doubt that the development of an effective biological system to alleviate the environmental problems associated with hexavalent chromium is highly desirable.

Potentially bioremediation is cost-effective and environmentally friendly in comparison with physical-chemical treatments (Lovley and Coates 1997; Chen and Hao 1998; Tseng and Bielefeldt 2002). The Cr(VI) bioremediation of soils can be performed *in situ* or *ex situ* using a bioreactor for treatment of soils or soil wash effluents (Turick and Apel 1997; Lovley and Coates 1997; Kamaludeen et al. 2003). The bioremediation approach offers some advantages compared with traditional techniques (Higgins et al. 1997): i) it can be performed *in situ* without escavation of contaminated soils, ii) it can be applied to sites with high water table, iii) it can allow a continuous Cr(VI) stable reduction process, iv) it does not destroy the site that is to be detoxified.

In contaminated soils with hexavalent chromium, some indigenous bacteria are able to reduce Cr(VI), but the rates of natural attenuation (that is to say without any human interference) of Cr-toxicity are slow and, therefore, unacceptable to devise remediation strategies (Tokunaga et al. 2003). Thus, there is a need to improve *in situ* bioreduction of Cr(VI) to Cr(III). The Cr(VI)-remediation efficiency can be enhanced by introducing in soils selected strains with intrinsic properties, such as high Cr(VI)-resistance and Cr(VI)-reduction capability (bioaugumentation) or stimulating the activity of indigenous Cr(VI)-reducers (biostimulation). In both cases, a strong limitation is that contaminated sites are usually lacking in nutrients and do not permit rapid growth of selected and/or indigenous bacteria and, therefore, their potential bioremediation activities are not fully expressed. A strategy to stimulate the metabolism and proliferation of bacterial Cr(VI)-reducers *in situ* may be the addition of nutrients to the environment (Chen and Hao 1998; Salunkhe et al. 1998; Reddy et al. 2003). Carbon sources, such as organic acids, manure, molasses, have been proposed to improve Cr(VI)-reduction that, otherwise, would be unacceptably slow (Losi et al. 1994a; Higgins et al. 1998; Tokunaga et al. 2003). Reddy et al. (2003) have demonstrated that under laboratory conditions, the nutrient amended

**Table 3.** Advantages and disadvantages of the main traditional approaches for the remediation of soil contaminated with Cr(VI) (from Higgins et al. 1997, modified)

| Approach | Advantages | Disadvantages |
|---|---|---|
| Escavation and off-site disposal | Quick and appropriate for small volumes of soil, completely removes the contaminant | Expensive, may cause health hazard during the excavation, removed soil may need treatment, destroys the site but does not destroy the contaminant |
| Soil washing | Reduces the volume of contaminated material that requires treatment, washing solution may be reused after decontamination treatment | Needs the escavation of soil, may cause health hazard during the excavation, destroys the site but does not destroy the contaminant, generates contaminated water, is not appropriate for all soils |
| Soil flushing | *In situ* technology that does not require escavation, its efficiency may be improved through electrokinetics | Does not destroy the contaminant, generates contaminated water |
| Solidification/ stabilisation *ex situ* | Is relatively inexpensive | May cause health hazard during the excavation, does not destroy the contaminant, increases the volumes of disposal material |
| Solidification/ stabilisation *in situ* | Applicable to sites with high water tables | Does not remove the contaminant, before the treatment Cr(VI) may need to be reduced to Cr(III) in order to minimise potential leaching of pollutants |
| Vitrification (*in situ* or *ex situ*) | Should reduce toxicity, mobility contaminant and volume of polluted soils; the final product can be demonstrated to be non-hazardous, may be performed *in situ* | Demands high energy and technology, requires significant amounts of additive |
| Chemical reduction (*in situ* or *ex situ*) | Reduces toxicity and mobility of the Cr(VI), may be performed *in situ* | Does not remove from the soil the Cr(III) produced, which may be oxidised to Cr(VI); is not appropriated for sites where the level of Cr is to be reduced, requires the control of chemical-physical characteristics of soil, the process of reduction may be slow |

by electrokinetics improved Cr(VI) bioremediation. Moreover, it has been proposed that the organic aromatic pollutants might serve indirectly as carbon sources for microbial Cr(VI) reduction in presence of a mixed culture of appropriate taxa (Shen and Wang 1995). However, the addition of nutrients to Cr(VI)-contaminated soils is a laborious and expensive approach and it may cause problems, because it results in the production of considerable biomass (Tseng and Bielefeldt 2002; Gonzalez et al. 2003). Matin et al. (1995), studying the remediation of trichloroethylene by *Escherichia coli,* in order to reduce the requirement for the external addition of nutrients and biomasses to manageable levels, have applied with success the starvation promoter technology. Such approach permits a decoupling between a high level expression of a gene and the need for rapid growth, giving maximal expression under conditions of slow growth. This innovative technology could also be applied to Cr(VI)-reduction in bioremediation soil processes (Gonzales et al. 2003).

In several studies, the use of bioreactors or biofilms for *ex situ* Cr(VI)-bioremediation of soils or soil wash effluents was assumed (Turick and Apel 1997; Turick et al. 1997; Smith 2001; Ganguli and Tripathi 2002). The possibility to use bioreactor systems for Cr(VI)-bioremediation is cost-effective, but its success has been limited to large-scale decontamination projects (Kamaludeen et al. 2003).

Cr(VI)-bioremediation *ex situ,* such as *in situ,* can be performed using microbial pure cultures or microorganisms consortia. Smith (2001) proposed sulphate-reducing bacteria (mixed-culture) biofilms to treat Cr(VI)-contaminated waterways and soils, suggesting that this system can be used to recover Cr(III) from the reduction and precipitation of Cr(VI). Turick et al. (1997), using a bioreactor where the support was of 6-mm porcelain saddles, developed an anaerobic process for Cr(VI) reduction using an inoculum constituted of enrichment cultures of indigenous microorganisms from Cr(VI)-contaminated soils. Konovalova et al. (2003) have suggested using a membrane bioreactor where *Pseudomonas* cells were immobilised in agar-agar films on the surface of synthetic membranes in order to decrease the toxic action of high chromate concentrations.

Bacterial Cr(VI)-reduction can occur under both aerobic or anaerobic conditions in presence of different electron acceptors, such as oxygen, nitrate, sulphate and ferric iron, but the suitable conditions for Cr(VI) bioremediation are aerobic at higher Cr(VI) concentrations and anaerobic at lower Cr(VI) concentrations (Tseng and Bielefeldt 2002). The former condition is appealing for soil remediation, because it permits to carry out a remediation process without the need of establishing and maintaining anaerobic conditions in soils (Lovley and Coates 1997).

It has been reported that under anaerobic conditions, low concentrations of Cr(VI) can accelerate the growth and activity of sulphate-reducing bacteria, obligate anaerobic heterotrophs, and thereby the indirect Cr(VI)-reduction by evolved sulphide (Kamaludeen et al. 2003). The activity of sulphate-reducing bacteria *in situ* is also enhanced through the addition of sulphate and nutrients, but the sulphide produced promotes not only the reduction of Cr(VI) to Cr(III), but also Mn oxides which can be involved in the reoxidation of Cr(III). Moreover, sulphate-reducing bacteria are particularly sensitive to Cr(VI)-toxicity compared to other bacterial populations (Arias and Tebo 2003). Thereby, in soils where there are high levels of Cr(VI), sulphate-reducing bacteria do not have an important role in Cr(VI) reduction. Marsh et al. (2000) have reported that the production of sulphide by sulphate-reducing bacteria did not occur, when the level of Cr(VI) in sandy sediments was 0.5 mM. Data obtained by Marsh et al. (2000) and Arias and Tebo (2003) should be considered for devising bioremediation strategies for Cr(VI)-contaminated soils. Thus, the use of sulphate reducing bacteria, as has been reported by Losi et al. (1994a), can have some possibilities in *ex situ* detoxification of Cr(VI)-contaminated soil using a bioreactor system instead of *in situ* bioremediation approaches, because all parameters of the processes must be kept under control.

The Cr(VI)-bioremediation approach, being cost-effective and environmentally friendly in comparison to physical-chemical treatments, is very attractive (Lovley and Coates 1997; Chen and Hao 1998; Tseng and Bielefeldt 2002). Nevertheless, to our knowledge, bioremediation strategies for chromate detoxification have yet to be significant on large-scale environmental remediation, mostly because the knowledge of microorganism-chromium interactions is to be deepened. However, there is no doubt that a better understanding of the Cr(VI)-resistance and Cr(VI)-reduction mechanisms, which permit specific bacteria to survive and play their role the in presence of high concentrations of Cr(VI), will result in an adequate biological plan to alleviate the environmental contamination by hexavalent chromium (McLean et al. 2000; McLean and Beveridge 2001; Francisco et al. 2002). Therefore, in order to move from the potential and/or pilot phase to the applied one we need i) to have bacterial strains belonging to different species, selected for Cr(VI)-resistance and capability to Cr(VI)-reduction (a few studies provide quantitative information about Cr(VI) reduction in the presence of high concentrations of hexavalent chromium); ii) to increase knowledge on the mechanisms involved in the processes of resistance and reduction of Cr(VI); iii) to understand how the bacterial kinetics of Cr(VI)-reduction are affected by abiotic factors, such as pH, temperature, electron acceptors and organic substrates.

## 6. Conclusion

Microbial reduction of hexavalent chromium to trivalent chromium, which is relatively insoluble and considerably less toxic, is a potentially valid remediation strategy for chromium-contaminated soils. It could be cost-effective and environmentally friendly in comparison to physical-chemical treatments (DeFilippi and Lupton 1992; Lovley and Coates 1997; Chen and Hao

1998; Tseng and Bielefeldt 2002). Nevertheless, in spite of significant advances that have been made in recent years, some points still need to be studied in depth before applying bioremediation methodologies to large-scale soil reclamation. Many researchers believe there are two strategies for enhancing the applicability of biological systems to clean up environments from hexavalent chromium. One of these two is to deepen the knowledge of the mechanisms involved in the process of strain resistance and how some abiotic factors (initial chromate concentration, pH, temperature, carbon sources, electron acceptors) affect the rate of Cr(VI)-reduction. The capability of indigenous bacteria in reducing Cr(VI) to Cr(III) is to be quantified and the optimal conditions are to be defined in order to improve the ability of specific bacterial strains to play their role under stressful conditions as well as those in polluted-environments. Moreover, the availability of bacterial strains, indigeneous to sites contaminated with chromium, with intrinsic characteristics, will facilitate their utilisation *in situ* bioremediation processes avoiding legal and ethical problems brought up with the introduction of engineered microorganisms into the environment.

The second strategy is to develop engineered protein families and/or strains with improved hexavalent chromium reduction capability in order to utilise them mainly in *ex situ* closed systems. With molecular engineering, it will be possible to enhance Cr(VI)-reduction activities of indigenous bacterial strains that express such activities at high levels under poor nutrient and stressful environmental conditions (Gonzalez et al. 2003). Finally, there is a need to remember that to devise the most suitable bioremediation system in order to detoxify an area successfully, not only the advantages of all available technologies should be taken into consideration, but also the characteristics of the contaminated site. Thereby, a well-netted collaboration among molecular biologists, microbiologists, geochemists and environmental engineers is required in order to bring bioremedition strategies of Cr(VI)-contaminated soils from a promise to their application.

Acknowledgment. This work was partially supported by Ministero dell'Istruzione, dell'Università e della Ricerca (Programmi di ricerca scientifica di rilevante interesse nazionale, 2000).

## References

Arias YM, Tebo BM (2003) Cr(VI) reduction by sulfidogenic and nonsulfidogenic microbial consortia. Appl Environ Microbiol 69:1847-1853

Arslan P, Beltrame M, Tomasi A (1987) Intracellular chromium reduction. Biochim Biophys Acta 931:10-15

Badar U, Ahmed N, Beswick AJ, Pattanapipitpaisal P, Macaskie LE (2000) Reduction of chromate by microorganisms isolated from metal contaminated sites of Karachi, Pakistan. Biotechnol Lett 22:829-836

Bartlett RJ James B (1979) Behavior of chromium in soils: III. oxidation. J Environ Qual 8:31-34

Benedetti A (1998) Defining soil quality: introduction to round table. In: de Bertoldi S, Pinzari F (eds) COST Actions 831, Joint $WC_s$ Meeting. Biotechnology of soil: monitoring conservation and remediation, pp 29-33

Blowes D (2002) Environmental chemistry. Tracking hexavalent Cr in groundwater. Science 295:2024-2025

Camargo FA, Bento FM, Okeke BC, Frankenberger WT (2003) Chromate reduction by chromium-resistant bacteria isolated from soils contaminated with dichromate. J Environ Qual 32:1228-1233

Camargo FA, Bento FM, Okeke BC, Frankenberger WT (2004) Hexavalent chromium reduction by an actinomycete, *Arthrobacter crystallopoietes* ES 32. Biol Trace Elem Res 97:183-194

Cervantes C, Campos-Garcia J, Devars S, Gutierrez-Corona F, Loza-Tavera H, Torres-Guzman JC, Moreno-Sanchez R (2001) Interactions of chromium with microorganisms and plants. FEMS Microbiol Rev 25:335-347

Cervantes C, Silver S (1992) Plasmid chromate resistance and chromate reduction. Plasmid 27:65-71

Chen JH, Hao OJ (1998) Microbial chromium (VI) reduction. Cri Rev Environ Sci Tech 28:219-251

Daulton TL, Little BJ, Lowe K, Jones-Meehan J (2002) Electron energy loss spectroscopy techniques for the study of microbial chromium(VI) reduction. J Microbiol Methods 50:39-54

DeFilippi LJ, Lupton FS (1992) Bioremediation of soluble Cr(VI) using sulphate reducing bacteria. In: Allied Signal Research. National Conference on the Control of Hazardous Materials, San Francisco, CA, pp 138-141

Deflora S, Bagnasco M, Serra D, Zanacchi P (1990) Genotoxicity of chromium compounds - a review. Mutat Res 238:99-172

Fendorf S, Wielinga BW, Hansel CM (2000) Chromium transformations in natural environments: the role of biological and abiological processes in chromium(VI) reduction. Int Geol Rev 42:691-701

Fendorf SE (1995) Surface reactions of chromium in soils and waters. Geoderma 67:55-71

Fendorf SE, Li GC (1996) Kinetics of chromate reduction by ferrous iron. Environ Sci Technol 30:1614-1617

Francisco R, Alpoim MC, Morais PV (2002) Diversity of chromium-resistant and -reducing bacteria in a chromium-contaminated activated sludge. J Appl Microbiol 92:837-843

Gadd GM (1992) Metals and microorganisms: a problem of definition. FEMS Microbiol Lett 100:197-204

Ganguli A, Tripathi AK (1999) Survival and chromate reducing ability of *Pseudomonas aeruginosa* in industrial effluents. Lett Appl Microbiol 28:76-80

Ganguli A., Tripathi AK (2002) Bioremediation of toxic chromium from electroplating effluent by chromate-reducing *Pseudomonas aeruginosa* A2Chr in two bioreactors. Appl Microbiol Biotechnol 58:416-420

Garbisu C, Alkorta I, Llama MJ, Serra JL (1998) Aerobic chromate reduction by *Bacillus subtilis*. Biodegradation 9:133-141

Gonzalez CF, Ackerley DF, Park CH, Matin A (2003) A soluble flavoprotein contributes to chromate reduction and tolerance by *Pseudomonas putida*. Acta Biotechnol 23:233-239

Higgins TE, Halloran AR, Dobbins ME, Pittignano AJ (1998) *In situ* reduction of hexavalent chromium in alkaline soils enriched with chromite ore processing residue. J Air Waste Manage 48:1100-1106
Higgins TE, Halloran AR, Petura JC (1997) Traditional and innovative treatment methods for Cr(VI) in soil. J Soil Contam 6:767-797
James BR (2002) Chemical transformations of chromium in soils: relevance to mobility, bio-availability and remediation. In: The chromium file, International Chromium Development Association, Paris, pp 1-8
James BR, Bartlett RJ (1983) Behaviour of chromium in soils: VI. interactions between oxidation-reduction and organic complexation. J Environ Qual 12:173-176
James BR, Petura JC, Vitale RJ, Mussoline GR (1997) Oxidation-reduction chemistry of chromium: relevance to the regulation and remediation of chromate-contaminated soils. J Soil Contam 6:569-580
Juhnke S, Peitzsch N, Hübener N, Große C, Nies DH (2002) New genes involved in chromate resistance in *Ralstonia metallidurans* strain $CH_3^4$. Arch Microbiol 179:15-25
Kadiiska MB, Xiang QH, Mason RP (1994) In vivo free radical generation by chromium(VI): an electron spin resonance spin-trapping investigation. Chem Res Toxicol 7:800-805
Kamaludeen SP, Megharaj M, Juhasz AL, Sethunathan N, Naidu R (2003) Chromium-microorganism interactions in soils: remediation implications. Rev Environ Contam Toxicol 178:93-164
Kanojia RK, Junaid M, Murthy RC (1998) Embryo and fetotoxicity of hexavalent chromium: a long-term study. Toxicol Lett 95:165-172
Komori K, Wang PC, Toda K, Ohtake H (1989) Factor affecting chromate reduction in *Enterobacter cloacae* strain HO1. Appl Microbiol Biotechnol 31:567-570
Konovalova VV, Dmytrenko GM, Nigmatullin RR, Bryk MT, Gvozdyak PI (2003) Chromium(VI) reduction in a membrane bioreactor with immobilized *Pseudomonas* cells. Enzyme Microb Tech 33:899-907
Liu KJ, Jiang J, Shi X, Gabrys H, Walczak T, Swartz HM (1995) Low-frequency EPR study of chromium (V) formation from chromium (VI) in living plants. Biochem Biophys Res Commun 206:829-834
Losi ME, Amrhein C, Frankenberger WT (1994a) Bioremediation of chromate-contaminated groundwater by reduction and precipitation in surface soils. J Environ Qual 23:1141-1150
Losi ME, Amrhein C, Frankenberger WT Jr (1994b) Environmental biochemistry of chromium. Rev Environ Contam Toxicol 136:91-131
Lovley DR (1993) Dissimilatory metal reduction. Annu Rev Microbiol 47:263-290
Lovley DR, Coates JD (1997) Bioremediation of metal contamination. Curr Opin Biotechnol 8:285-289
Marsh TL, Leon NM, McInerney MJ (2000) Physiochemical factors affecting chromate reduction by aquifer materials. Geomicrobiol J 17:291-303
Matin A, Little CD, Fraley CD, Keyhan M (1995) Use of starvation promoters to limit growth and selectively enrich expression of trichloroethylene- and phenol-transforming activity in recombinant *Escherichia coli*. Appl Environ Microbiol 61:3323-3328
McLean J, Beveridge TJ (2001) Chromate reduction by a pseudomonad isolated from a site contaminated with chromated copper arsenate. Appl Environ Microbiol 67:1076-1084

McLean JS, Beveridge TJ, Phipps D (2000) Isolation and characterization of chromium-reducing bacterium from a chromated copper arsenate-contaminated site. Environ Microbiol 2:611-619

Megharaj M, Avudainayagam S, Naidu R (2003) Toxicity of hexavalent chromium and its reduction by bacteria isolated from soil contaminated with tannery waste. Curr Microbiol 47:51-54

Michel C, Brugna M, Aubert C, Bernadac A, Bruschi M (2001) Enzymatic reduction of chromate: comparative studies using sulfate-reducing bacteria. Key role of polyheme cytochromes c and hydrogenases. Appl Microbiol Biotechnol 55:95-100

Middleton SS, Latmani RB, Mackey MR, Ellisman MH, Tebo BM, Criddle CS (2003) Cometabolism of Cr(VI) by *Shewanella oneidensis* MR-1 produces cell-associated reduced chromium and inhibits growth. Biotechnol Bioeng 83:627-636

Mordenti A, Piva G (1997) Chromium in animal nutrition and possible effects on human health. In: Canali S, Tittarelli F, Sequi P (eds) Chromium environmental issues, Franco Angeli s.r.l., Milan, pp 131-151

Myers CR, Carstens BP, Antholine WE, Myers JM (2000) Chromium(VI) reductase activity is associated with the cytoplasmic membrane of anaerobically grown *Shewanella putrefaciens* MR-1. J Appl Microbiol 88:98-106

Nevin KP, Lovley DR (2002) Mechanisms for Fe(III) oxide reduction in sedimentary environments. Geomicrobiol J 19:141-159

Nies DH (1999) Microbial heavy-metal resistance. Appl Microbiol Biotechnol 51:730-750

Nies DH, Silver S (1995) Ion efflux system involved in bacterial metal resistances. J Ind Microbiol 14:186-199

Ohta N, Galsworthy PR, Pardee AB (1971) Genetics of sulfate transport by *Salmonella typhimurium*. J Bacteriol 105:1053-1062

Pattanapipitpaisal P, Brown NL, Macaskie LE (2001) Chromate reduction and 16S rRNA identification of bacteria isolated from a Cr(VI)-contaminated site. Appl Microbiol Biotechnol 57:257-261

Pettine M, Barra I, Campanella L, Millero FJ (1998) Effect of metals on the reduction of chromium (VI) with hydrogen sulfide. Water Research 32:2807-2813

Puzon GJ, Petersen JN, Roberts AG, Kramer DM, Xun L (2002) A bacterial flavin reductase system reduces chromate to a soluble chromium(III)-NAD(+) complex. Biochem Biophys Res Commun 294:76-81

Reddy KR, Chinthamreddy S, Saichek RE, Cutright TJ (2003) Nutrient amendment for the bioremediation of a chromium- contaminated soil by electrokinetics. Energy Sources 25:931-943

Richard FC, Bourg ACM (1991) Aqueous geochemistry of chromium - a review. Wat Res 25:807-816

Richards JW, Krumholz GD, Chval MS, Tisa LS (2002) Heavy metal resistance patterns of *Frankia* strains. Appl Environ Microbiol 68:923-927

Rock ML, James BR, Helz GR (2001) Hydrogen peroxide effects on chromium oxidation state and solubility in four diverse, chromium-enriched soils. Environ Sci Technol 35:4054-4059

Ross DS, Sjogren RE, Bartlett RJ (1981) Behavior of chromium in soils: IV. toxicity to microorganisms. J Environ Qual 2:145-168

Roundhill DM, Koch HF (2002) Methods and techniques for the selective extraction and recovery of oxoanions. Chem Soc Rev 31:60-67

Salunkhe PB, Dhakephalkar PK, Paknikar KM (1998) Bioremediation of hexavalent chromium in soil microcosms. Biotechnol Lett 20:749-751

Sani RK, Peyton BM, Smith WA, Apel WA, Petersen JN (2002) Dissimilatory reduction of Cr(VI), Fe(III), and U(VI) by *Cellulomonas* isolates. Appl Microbiol Biotechnol 60:192-199

Shakoori AR, Makhdoom M, Haq RU (2000) Hexavalent chromium reduction by a dichromate-resistant gram-positive bacterium isolated from effluents of tanneries. Appl Microbiol Biotechnol 53:348-351

Shen H, Wang YT (1993) Characterization of enzymatic reduction of hexavalent chromium by *Escherichia coli* ATCC 33456. Appl Environ Microbiol 59:3771-3777

Shen H, Wang Y-T (1995) Simultaneous chromium reduction and phenol degradation in a coculture of *Escherichia coli* ATCC 33456 and *Pseudomonas putida* DMP-1. Appl Environ Microbiol 61:2754-2758

Shi W, Bischoff M, Turco R, Konopka A (2002) Long-term effects of chromium and lead upon the activity of soil microbial communities. Appl Soil Ecol 21:169-177

Smith WL (2001) Hexavalent chromium reduction and precipitation by sulphate-reducing bacterial biofilms. Environ Geochem Hlth 23:297-300

Speir TW, Kettles HA, Parshotam A, Searle PL, Vlaar LNC (1995) A simple kinetic approach to derive the ecological dose value, Ed(50), for the assessment of Cr(VI) toxicity to soil biological properties. Soil Biol Biochem 27:801-810

Srinath T, Verma T, Ramteke PW, Garg SK (2002) Chromium (VI) biosorption and bioaccumulation by chromate resistant bacteria. Chemosphere 48:427-435

Suzuki T, Miyata N, Horitsu H, Kawai K, Takamizawa K, Tai Y, Okazaki M (1992) NAD(P)H-dependent chromium(VI) reductase of *Pseudomonas ambigua* G-1: Cr(VI) intermediate is formed during the reduction of Cr(VI) to Cr(III). J Bacteriol 174:5340-5345

Tebo BM, Obraztova AY (1998) Sulfate-reducing bacterium grows with Cr(VI), U(VI), Mn(IV), and Fe(III) as electron acceptors. FEMS Microbiol Lett 162:193-198

Tokunaga TK, Wan J, Firestone MK, Hazen TC, Olson KR, Herman DJ, Sutton SR, Lanzirotti A (2003) *In situ* reduction of chromium(VI) in heavily contaminated soils through organic carbon amendment. J Environ Qual 32:1641-1649

Tseng JK Bielefeldt AR (2002) Low-temperature chromium(VI) biotransformation in soil with varying electron acceptors. J Environ Qual 31:1831-1841

Turick CE, Camp CE, Apel WA (1997) Reduction of Cr(6(−)) to Cr(3(+)) in a packed-bed bioreactor. Appl Biochem Biotech 63:871-877

Turick CE, Apel WA (1997) A bioprocessing strategy that allows for the selection of Cr(VI)-reducing bacteria from soils. J Ind Microbiol Biotechnol 18:247-250

Turpeinen R, Kairesalo T, Häggblom MM (2004) Microbial community structure and activity in arsenic-, chromium- and copper-contaminated soils. FEMS Microbiol Ecol 47:39-50

Viamajala S, Peyton BM, Sani RK, Apel WA, Petersen JN (2004) Toxic effects of chromium(VI) on anaerobic and aerobic growth of *Shewanella oneidensis* MR-1. Biotechnol Prog 20:87-95

Viti C, Giovannetti L (2001) The impact of chromium contamination on soil heterotrophic and photosynthetic microorganisms. Ann Microbiol 51:201-213

Viti C, Giovannetti L (2005) Characterization of cultivable heterotrophic bacterial communities in Cr-polluted and unpolluted soils using biolog and ARDRA approaches. App Soil Ecol (in press).

Viti C, Pace A, Giovannetti L (2003) Characterization of Cr(VI)-resistant bacteria isolated from chromium-contaminated soil by tannery activity. Curr Microbiol 46:1-5

Wielinga B, Mizuba MM, Hansel CM, Fendorf S (2001) Iron promoted reduction of chromate by dissimilatory iron-reducing bacteria. Environ Sci Technol 35:522-527

Wong PTS, Trevors JT (1988) Chromium toxicity to algae and bacteria. In: Nriagu JO, Nieboer E (eds) Chromium in natural and Human Environments, Wiley, New York, pp 305-315

Zayed AM, Terry N (2003) Chromium in the environment: factors affecting biological remediation. Plant Soil 249:139-156

4

# Accumulation and Detoxification of Metals by Plants and Microbes

Rutchadaporn Sriprang[1] and Yoshikatsu Murooka[2]

[1]BIOTEC Center for Genetic Engineering and Biotechnology, National Science and Technology Development Agency, 113 Phaholyothin Rd., Klong 1, Klong Luang, Pathumthani, 12120, THAILAND; [2]Department of Biotechnology, Graduate School of Engineering, Osaka University, 2-1 Yamada-oka, Suita-shi, Osaka 565-0871, JAPAN, Email: murooka@bio.eng.osaka-u.ac.jp

## 1. Introduction

Excessive toxic metal levels in soils pose potential hazards to human and animal health as well as the ecosystem in general. Anthropogenic sources of heavy metal deposition have increased as the result of the Industrial Revolution. Agriculture, mining, smelting, electroplating, and other industrial activities have resulted in the deposition of undesirable concentration of metals, such as As, Cd, Cr, Cu, Ni, Pb and Zn, in the soil.

Although trace metals are important part of the soil ecosystem, the accumulation of these metals may be harmful to people, animals, plants and other organisms contacting the soil or groundwater. Unlike many other pollutants, heavy metals are difficult to be removed from the environment as they cannot be chemically or biologically degraded, and are ultimately indestructible. Now-a-days, various heavy metals constitute a global environmental hazard.

Use of microorganisms and plants for the decontamination of heavy metals has attracted growing attention because of their low cost and high efficiency. Microorganisms could be used to clean up metal contamination by removing metals from contaminated water, sequestering metals from soils and sediments or solubilizing metals to facilitate their extraction.

In this article, we describe how bacteria and plants accumulate and detoxify metal ions, engineering approaches to enhance the metal tolerance, accumulation and detoxification in microorganisms and plants. We also describe bioremediation using symbiosis between plants and microorganisms.

## 2. Phytoremediation

Phytoremediation is the use of plants to remove pollutants from the environment or to render them harmless. Phytoremediation of toxic metals may be of high significance because of the lack of alternative technologies that are affordable and effective. While organic molecules can be degraded in microbial bioremediation, toxic metals can be remediated only by gathering trace amount of dispersed metals in soil or water and removing them from the environment. It may provide an economically viable solution for the remediation of metal-polluted sites. Thus, several sub-sets of metal phytoremediation have been developed and targeted for commercialization.

a) *Phytoextraction*: in which high-biomass, metal-accumulating plants and appropriate soil amendments are used to transport and concentrate metals from the soil into the above-ground shoots, which are harvested with conventional agricultural methods.
b) *Rhizofiltration*: in which plant roots grown in aerated water, precipitate and concentrate toxic metals from the polluted effluents.
c) *Phytostabilization*: in which plants stabilize the pollutants in soil, thus rendering them harmless.
d) *Phytovolatilization*: in which plants extract volatile metals (e.g., mercury and selenium) from soil and volatilize them from foliage.

Here, we focus only on phytoextraction and phytovolatilization strategies. These strategies might become viable alternatives to mechanical and chemical approaches in remediation of metals from the contaminated soils.

### 2.1 Phytoextraction of Metals

Phytoextraction is based on the genetic and physiological capacity of specialized plants to accumulate, translocate, and resist high amounts of metal. The idea of using plants to remove metals from soils came from the discovery of different wild plants that accumulate high concentrations of metals in their foliage. Naturally occurring plants called "metal hyperaccumulators" can accumulate 10-500 times higher levels of metal elements than crops (Chaney et al. 1997). The degree of accumulation of metals such as Ni, Zn, and possibly Cu, observed in hyperaccumulators often reaches 1-5% of their dry weight (Raskin et al. 1997). There is a report that *Brassica* (mustard) species or varieties of *Brassica juncea* (Indian mustard) have an enhanced ability to accumulate metals from hydroponics solution into their above ground (harvestable) parts. These plants concentrate toxic heavy metals (Pb, Cu and Ni) to a level up to several percent of their dried shoot biomass (Kumar et al. 1995).

### *2.1.1 Uptake and Accumulation of Toxic Heavy Metals by Plants*

There are many processes that influence metal accumulation in plants e.g. metal mobilization and uptake from soils, compartmentation and sequestration within the root, efficiency of xylem to load and transport metal, distribution of metal in the aerial parts, sequestration, and storage in leaf cells (Clemens et al. 2002).

**Uptake and bioavailability of heavy metals.** Phytoextraction occurs when heavy metals are ready to be absorbed by roots (bioavailability). Bioavailability depends on metal solubility in soil solution. Some metals, such as Zn and Cd, occur primarily in exchangeable, and readily bioavailable form. Others, such as Pb, occur as soil precipitate, a significantly less bioavailable form. Plants roots increase metal bioavailability by extruding protons to acidify the soil and mobilize the metals. This mechanism has been observed for Fe mobilization in some Fe-deficient dicotyledonous plants (Crowley et al. 1991). Moreover lowering the soil pH affects both metal bioavailability and metal uptake into roots. In *T. caerulescens*, uptake of Mn and Cd was dependent on the soil acidity (Brown et al. 1995).

The formation of metal-chelate complexes prevents precipitation and sorption of the metals thereby maintaining their availability for the plant uptake (Salt and Rauser 1995). Addition of synthetic chelates such as EDTA is very effective in facilitating the plant uptake of Cd, Cu, Ni and Zn (Raskin et al. 1997).

**Transport of heavy metals.** Plants have evolved highly specific mechanisms to take up, translocate, and store macro-nutrients (N, P, K, S, Ca, and Mg) and essential micro-nutrients (Fe, Zn, Mn, Ni, Cu, and Mo). Molecular physiology of the plant transport systems for elemental nutrients and pollutant is still in its infancy. Plant genes encoding metal transporters have been identified and characterized. The IRT1 (iron-regulated transporter) is the first member of the ZIP gene family to be identified. The IRT1 is an Fe(II) transporter that takes up iron from the soil. The *IRT1* was cloned for functional expression in a yeast mutant (*fet3 fet4*) defective for iron uptake (Eide et al. 1996). IRT1 is able to complement the metal uptake defects of the *Saccharomyces cerevisiae zrt1 zrt2* zinc uptake mutants and the *S. cerevisiae smf1* manganese uptake mutant (Korshunova et al. 1999). Although IRT1 was originally identified as the Fe transporter, the studies of complementation and uptake in yeast provided information that IRT1 was able to transport both Mn and Zn in addition to Fe. There are several evidences that point to a role for IRT1 in mediating the accumulation of Cd in iron deficient plants: (1) Cd was shown to compete with Fe uptake in yeast expressing IRT1 (Eide et al. 1996), (2) yeast-expressing IRT1 was more sensitive to Cd (Rogers et al. 2000) than wild type, and (3) plants engineered to over express IRT1 accumulated Cd in greater amounts than wild-type plant (Guerinot 2000). Another member of ZIP protein is zinc transporter (ZIP), which contains *ZIP1*, *ZIP2*, and *ZIP3* genes of *Arabidopsis*. Expression of these genes restored zinc-limited growth of *zrt1 zrt2* yeast mutant

(Grotz et al. 1998). In the plant, *ZIP1* and *ZIP3* are expressed in roots in response to zinc deficiency, thus these genes play a direct role in zinc uptake from the soil. The Zn(II) transport activity of these three proteins is inhibited by Mn(II), Co(II), Cd(II), and Cu(II), indicating that ZIP proteins may transport potentially toxic metals as well as nutrients. From cross-species, microarray transcript profiling reveals high constitutive expression of metal homeostasis gene, such as ZIP6 in shoots and ZIP9 in roots of the zinc hyper accumulator *Arabidopsis halleri* (Weber et al. 2004; Becher et al. 2004). These transporter genes offer a good starting point for the understanding how metals cross membranes.

**Translocation of an element from roots to shoots.** Accumulator plant must have the ability to translocate an element from roots to shoots at high rate. The transport processes are stimulated by metal influx into root and leaf cells, and metal loading into the xylem. Many other factors are also involved in the metalic elements.

**Transporter proteins.** Because of their charge, metal ions cannot move freely across the cellular membrane, which are lipophilic structures. Therefore, membrane proteins must mediate ion transport into cells with transport functions known as transporters. Transporter proteins play an important role in the translocation of an elements, since they contain the binding domains, which bind to specific ions and transfer bound ions from extracellular space through the hydrophobic environment of the membrane into the cell. There is an evidence for higher $Zn^{2+}$ uptake capacity in hyper accumulator, *Thlaspi caerulescens* as compared to the non-hyper accumulating relative *T. arvense* (Lasat et al. 1996). This might be attributable to higher expression levels of $Zn^{2+}$ transporters such as the ZIP member ZNT1 (Pence et al. 2000).

**Chelators.** Cations of heavy metals are often bound to soil particles because of soil cation exchange capacity. The binding affinity of cations reduces cation movement in vascular plants, particularly in the negatively charged cells of the xylem. A solution to this problem is chelation, which means as the process of a cation binding to a compound, resulting in a uncharged complex that can move more freely through a variety of substrates. Several chelators, both natural and synthetic are known to perform this function in soil and plants.

| Natural | Synthetic |
|---|---|
| Phytochelatin (PC) | EDTA (ethylene diamine tetra acetic acid) |
| Metallothionein (MT) | EGTA (ethylene glycol tetra acetic acid) |
| Organic acids | |

The use of specific chemicals, synthetic chelates, has been shown to dramatically stimulate Pb accumulation in plants. These compounds prevent Pb precipitation and keep the metal as soluble chelate-Pb complexes available for uptake into roots and transport within plants. For example, addition of EDTA at a rate of 10 mmol/kg soil, increased Pb accumulation in shoots of maize up to 1.6 wt% of dry biomass (Blaylock et al. 1997). In a subsequent study, Indian mustard exposed to Pb and EDTA in hydroponics solution was able to accumulate more than 1% Pb in dry shoots (Vassil et al. 1998).

Chelation with certain ligands, for example histidine and citrate, appears to route metals primarily to the xylem. Histidine is very important for Ni tolerance and transport in hyper accumulators, since large increases in histidine levels and coordination of Ni with histidine have been reported in the xylem sap of Ni hyper accumulator, *Alyssum lesbiacum* (Kramer et al. 1996). Organic acid, citrate had been also shown to complex with some toxic metals during transport of metals to the shoot of hyper accumulating and non-hyper accumulating plant species (Senden et al. 1992).

Phytochelatins (PCs) are known to play an essential role in the heavy-metal detoxification by chelating heavy metals in the cytosol and sequestering PC-$Cd^{2+}$ complexes in the vacuoles via transport across the tonoplast (Ortiz et al. 1995; Salt and Rauser 1995). In addition, there is an evidence to demonstrate that PCs provide a major mechanism for regulating long distance $Cd^{2+}$ transport in *Arabidopsis*. Transgenic expression of wheat phytochelatin synthase (TaPCS1) cDNA in the *Arabidopsis* PC-deficient mutant, *cad1-3*, revealed the suppression of the heavy metal sensitivity of *cad1-3*. PCs can be transported from roots to shoots and transgenic expression of the *TaPCS1* gene increases long-distance root-to-shoot $Cd^{2+}$ transport and reduces $Cd^{2+}$ accumulation in roots (Gong et al. 2003).

### 2.1.2 Detoxification of Metal Ions by Plants

**Chelation.** Chelation of metals in the cytosol by high affinity ligands is potentially a very important mechanism of heavy-metal detoxification and tolerance. Two major classes of heavy metal chelating peptides are presented in plants, metallothioneins (MTs) and phytochelatins (PCs).

Metallothioneins make up a super family of cysteine-rich metal-chelating proteins. The chelation of divalent or monovalent cations is mediated through the cysteine residues, which are often highly conserved between species. The biological role of MT is focused on the sequestration of toxic heavy metal ions, such as $Cd^{2+}$, in order to prevent them from interacting with other cellular components, and on the homeostatic regulation of essential metal ions, such as $Zn^{2+}$.

MTs are widely distributed among living organisms, and they are fairly well conserved in mammals, plants, and fungi (Butt and Ecker 1987; Huckle et al. 1993). Based on structure, MTs can be subdivided into three classes. Class I includes those polypeptides related to mammalian species (Kagi 1991). These proteins are encoded in structural genes and synthesized by transcription and translation. Mammalian MTs are usually composed of 61 amino acids (molecular mass, 6 to 7 kDa) and lack aromatic amino acids or histidines. Two distinct domains of these proteins coordinate 7 divalent or 12 monovalent metal ions with 20 Cys residues. These metal ions present along the sequence in the form of Cys-X-Cys or Cys-Cys motifs (X is another amino acid residue). Class II MTs originate from yeasts (e.g., *Saccharomyces cerevisiae*, *Candida glabrata*, and *Candida albicans* (Mehra and Winge 1991)), or cyanobacteria [e.g., *Synechococcus sp.* (Olafson et al. 1988)]. A well-known member of class II is the *S. cerevisiae* MT responsible for copper tolerance, called CUP1. This protein contains 12 cysteine residues organized in Cys-X-Cys, Cys-Cys, and Cys-X-X-Cys motifs, which originate eight binding sites for monovalent and four binding sites for divalent metal ions (Weige et al. 1985). In animal and plant species, MTs synthesis is induced by the metal ions, such as Cd, Zn, Hg, Ag and Pb (Kagi 1991). In plant-species, metal-induced expression of MT genes has also been reported in both maize and rice (Chevalier et al. 1995; Hsieh et al. 1995). RNA expression of MT genes in *Arabidopsis* could be induced by copper, and to a lesser degree by Zn and Cd (Zhou and Goldsbrough 1994). Thus, plant MTs may function as metal-binding proteins that can mediate metal tolerance. However, direct evidence that MTs are required for a specific function in metal metabolism, tolerance or another process is currently lacking.

These MTs bind Cd, Zn, Hg, Cu, and Ag. Toyama et al. (2002) demonstrated that $As^{3+}$ bound to MT-2 by ICP-AES and MALDI-TOF-MS. The maximum molar ratio of $As^{3+}$ to human MT-2 is more than 6:1. Hong et al. (2001) developed high yield expression and single step purification of human thionein and metallothionein. hMT was expressed in *E. coli* as an intein (protein splicing element) fusion protein in the absence of added metals and purified by intein-mediated purification with an affinity chitin-binding tag. This procedure constitutes a novel and simple strategy to prepare thionein (T), the metal-free form, or MT when reconstituting T with metals *in vitro*. The yield was 8 mg of T or 6 mg of pure Cd7- or Zn7-MT from 1-liter culture.

Class III metallothioneins are known as phytochelatins (PCs). Phytochelatins are the naturally occurring metal-binding peptides. They are short peptides composed of only three amino acids, namely, Glu, Cys and Gly, with Glu and Cys residues linked through a γ-carboxymide bond. The structure of such peptides can be represented by $(\gamma\text{-Glu-Cys})_n$-Gly, where n ranges from 2 to 11. PCs have been identified in a wide variety of plant species and in some microorganisms (Cobbett 2000). They are structurally related to glutathione [GSH; (γ-Glu-Cys)-Gly] and presumed to be the products of biosynthetic pathway. Numerous physiological, biochemical, and genetic studies have confirmed that GSH is the substrate for PCs biosynthesis. The PCs pathway is involved in the synthesis of γ-Glu-Cys from Glu and Cys by γ-glutamylcysteine synthetase (GCS), then glutathione synthetase (GS) catalyzes the synthesis of GSH. PCs synthesis was presumed to be involved in the transpeptidation of the γ-Glu-Cys moiety of GSH onto initially a second GSH molecule to form $PC_2$ or onto a PC molecule to produce a PC (n+1) oligomer. This γ-Glu-Cys dipeptidyl transpeptidase (EC 2.3.2.15) has been termed PC synthase (PCS). *In vitro*, the activity of the partially purified enzyme was active only in the presence of metal ions. The best activator was Cd followed by Ag, Bi, Pb, Zn, Cu, Hg, and Au cation. The PC biosynthesis continued until the activated metal ions were chelated either by the formed PCs or by the addition of a metal chelator such as EDTA (Loeffler et al. 1989).

**Vacuolar compartmentalization.** Vacuolar compartmentalization appears to be the reason for hypertolerance of natural hyper accumulator plant. The vacuole is generally considered to be the main storage site for metals in yeast and plant cells. The role of Cd detoxification and tolerance is played by the vacuolar compartmentalization, which prevents the free circulation of Cd ions in the cytosol and forces them into a limited area. Cd stimulates synthesis of PCs, which rapidly form a low molecular weight Cd-PC. The Cd-PC complex will be transported into the vacuole by a Cd/H antiport and an ATP-dependent PC-transporter (Salt and Wagner 1993; Salt and Rauser 1995). A gene, which codes for a PC-transporter in yeast was isolated namely *Hmt1* gene. The *Hmt1* gene encodes a member of a family of ATP-binding cassette (ABC) membrane transport proteins that is located in the vacuolar membrane (Ortiz et al. 1992, 1995). The gene product is responsible for transporting Cd-PC complex into the vacuole. Inside the vacuole the Cd-PC complexes acquire acid-labile sulphur ($S^{2-}$) and form a high molecular weight Cd-PC-sulfide complex, that may be essential for Cd resistance in the yeast (Speiser et al. 1992).

Compartmentalization of metals in the vacuole is a part of the tolerance mechanism of some metal hyper accumulators. The Ni hyper accumulator *T. goesingense* enhances its Ni tolerance by compartmentalizing most of the intracellular leaf Ni into the vacuole (Kramer et al. 2000). High-level of metal ion transporter TgMTP1 in *T. goesingense* was proposed to account for the enhanced ability to accumulate metal ions within shoot vacuoles (Persans et al. 2001). Intact vacuoles isolated from tobacco and barley exposed to Zn have been shown to accumulate this metal (Krotz et al. 1989; Burken and Schnoor 1996).

The strategies for uptake, accumulation and detoxification of heavy metals by higher plants are summarized in Figure 1.

### *2.1.3 Ideal Plant for Phytoremediation*

Populations of metal-tolerant hyperaccumulating plants can be found in naturally occurring metal-rich sites (Baker and Brooks 1989). However, these plants are not ideal for phytoremediation since they are usually small and have a low biomass production. In contrast, plants with good growth usually show low metal accumulation capability as well as low tolerance to heavy metals.

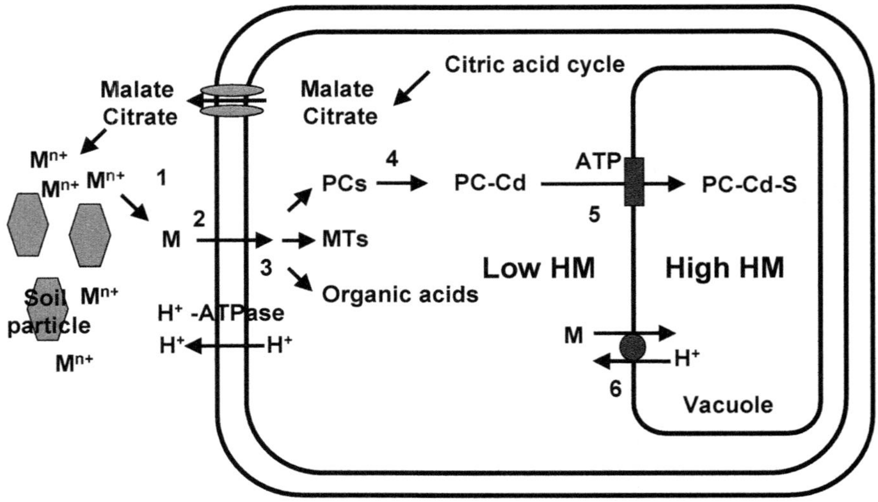


**Fig. 1.** Summary of potential cellular mechanisms for metal uptake, accumulation and detoxification in higher plants. 1. Metal ions are mobilized by secretion of chelators and by acidification of the rhizosphere. 2. Uptake of hydrated metal ions or metal-chelate complexes is mediated by various uptake systems residing in the plasma membrane. 3. Metals are chelated in cytosol by various ligands. 4. PCs form complex with Cd. 5. PC-Cd complex is transported into the vacuole. 6. Metals are transported and accumulated in the vacuole (Modified after Hall 2002; Clemens et al. 2002)

A plant suitable for phytoremediation should be fast growing, develop a large biomass, be tolerant to and accumulate high concentrations of toxic metals in the shoot, and be easily cultivated and harvested (Karenlampi et al. 2000). There are fast-growing hyper accumulators that can produce a large biomass. Examples are the Ni hyperaccumulators *Alyssum bertolonii* and *Berkheya coddii*, which produced 9 and 22 t/ha of shoot dry matter, respectively, in small-scale field experiments (Robinson et al. 1997a, b). The arsenic hyperaccumulator, *Pteris vittata* can also produce a relatively large biomass under favorable climate (Ma et al. 2001). However, fast-growing species that can hyper accumulate Zn, Cd, Cu, Pb and Cr have been not yet reported. Approaches to find metal-tolerant hyperaccumulating plants for phytoremediation involve searching for, and studying natural hyperaccumulators, or developing genetically engineered plants that possess above traits to achieve some of the properties of hyper accumulators. Although most of the cultivated plants are not hyper accumulator for metals, some of them are good candidate of breeding to accumulate toxic metals since their transformation systems have been developed and cultivation conditions in the fields are well known. A winter-growing legume Chinese milk vetch (*Astragalus sinicus*) is widely used as a green manure in rice fields in China and Japan (Murooka et al. 1993). This plant is suitable for use in bioremediation in the rice paddy.

### *2.1.4. Genetic Engineering in the Improvement of Plants for Phytoremediation*

Several criteria must be considered for engineering plant for phytoremediation. First, the plant must be able to solubilize and uptake heavy metals that are tightly bound to soil particles. Second, a mechanism must exist to transfer the heavy metal from the root to the shoot. Third, the heavy metal must be deposited in a compartment where it does not interfere with cellular metabolism.

**Genetic engineered plants for metal uptake and translocation.** In phytoremediation, heavy metal uptake and translocation are essential components of heavy metal hyperaccumulation. Citric and Malic acid are two compounds, which have been shown to complex heavy metals in the plant roots. After loss of one $H^+$, each acid contains a $COO^-$ group which binds to the cation positive charge. Plants secrete acids, which aid in the uptake of non-bioavailable metals. These acids protect cellular function when the acid-Cd complex is brought into the root. Citric acid-metal complexes have been reported to be translocated via the xylem (Senden et al. 1992). If a plant could be genetically altered to produce higher levels of endogenous citric or malic acid, then perhaps phytoextraction could be enhanced.

Free histidine has been found as a metal chelator in xylem exudates in plants that accumulate Ni and the amount of free histidine increases with Ni exposure (Kramer et al. 1996). By modifying histidine metabolism, it might be possible to increase the Ni- accumulating capacity of plants.

The expression of the metal transporter genes, such as the *IRT1* (iron-regulated-transporter) gene, and the wheat $Ca^{2+}$ transporter LCT1 gene mediate the uptake of $Na^+$ and $Cd^{2+}$ in yeast (Schachtman et al. 1997; Clemen et al. 1998). Therefore, the introduction of such genes to plants may enhance the metal ions uptake by the plant roots.

Transporter proteins, isolated from hyperaccumulating species, such as Zn transporter protein (ZNT1) can only uptake Zn, but not the toxic ions (i.e., Cd). Molecular study for alteration of gene for transport of other metals may be useful for phytoextraction. Moreover, several Zn transporters like ZIP1, ZIP3 (Grotz et al. 1998) and IRT1 are expressed in response to metal deficiency. Changing the regulation of the expression of these transporters may modify the uptake of metals to the cells or organelles. By substituting various conserved residues in ZIP family transporters with alanine produces mutant versions of IRT1 that apparently no longer transport $Fe^{2+}$ and $Mn^{2+}$ but retain $Zn^{2+}$ and $Cd^{2+}$ transport activity (Roger et al. 2000). Expression of these genes might enhance metal accumulation in transgenic plants.

**Genetic engineered plants with altered metal tolerance and accumulation.** Increased resistance to metal is another important trait that can improve the efficiency of phytoextraction. As mentioned above that hyper tolerance is essential for the hyper accumulation phenotype to occur in natural hyper accumulators. Hyper tolerance is achieved by internal detoxification and probably involves compartmentation and complexation. With the aim of creating plants that can tolerate and accumulate high levels of toxic metals, various *MT* genes (mouse *MTI*, human *MTI* (alpha domain), human *MTII*, yeast *CUP1*, pea *PsMTA*) were introduced into plants, such as *Nicotiana sp*., *Brassica sp*. or *A. thalina*. Transgenic plants, that express MTs, have been scored for enhanced Cd tolerance, but metal uptake was not markedly altered (Maiti et al. 1988 1989; Misra and Gedamu 1989; Evans et al. 1992; Pan et al. 1994a, b; Hasegawa et al. 1997).

Modification or over-expression of the enzymes that are involved in the synthesis of glutathione and PCs might be a good approach to enhance heavy metal tolerance and accumulation in plants. Over-expression of γ-glutamylcysteine synthetase enhanced $Cd^{2+}$ tolerance and accumulation in Indian mustard (Zhu et al. 1999).

Co-expressed with both *arsC* gene, which encodes arsenate reductase (ArsC) and γ-glutamylcysteine synthetase gene, *Escherichia coli* showed substantially greater tolerance to arsenic than wild type. The transformant accumulated two-to threefold higher concentrations of arsenic in the shoots (Dhankher et al. 2002).

Over-expression of vacuolar transporters and channels involved in metal tolerance from *Saccharomyces cerevisiae* named YCF1 in *A thaliana* significantly increased tolerance towards high concentration of Pb and Cd and led to a more than two fold higher accumulation of these metals in shoots of transgenic plants when compared to control plant (Song et al. 2003). In addition, over expressing of protein that localized to vacuole membrane of poplar named metal-tolerance proteins (MTPs) (cation diffusion facilitator (CDF) family) in *Arabidopsis* confers Zn tolerance (Blaudez et al. 2003). Expression of *Arabidopsis* vacuolar low-affinity $Ca^{2+}/H^{+}$ antiporter, CAX2, in Tobacco (*Nicotiana tabacum*) altered the $Ca^{2+}$, $Cd^{2+}$ and $Mn^{2+}$ content of plants and made transgenic plants more tolerant to $Mn^{2+}$ stress (Hirschi et al. 2000). Thus, introduction of the vacuolar metal transporters into plants may have an important impact on improving phytoremediation.

Introduction of metal binding peptides or proteins involved in intracellular metal sequestration of proteins (MTs, PCs) may increase metal tolerance in plants by prevention of cellular proteins from metal ions. Enhanced accumulation may be achieved by over-expression of plasma membrane transporters under the control of non-metal-responsive promoters. In addition, expression of modified transporters, which altered the metals uptake to the cells or organelles, might enhance metal uptake by plants. Moreover, expression of transporter protein in roots and/ or shoots with an efficient chelator may increase metal ions translocation from roots to shoots.

## 2.2 Phytovolatilization

Phytovolatilization is the transformation of toxic elements into relatively harmless forms. Many elements (e.g. arsenic, mercury, selenium) can exist in a variety of states, including different cationic and oxyanionic species and thio- and organo-metallics. These forms vary widely in their transport and accumulation in plants and in their toxicity to humans and other life forms.

### *2.2.1 Mercury Phytoremediation*

Mercury pollution is a worldwide problem in aquatic environments, resulting from its industrial use in bleaching operations as a catalyst, as a pigment in paints, and in the mining of gold. The Hg(0), becomes problematic, since biological systems can reoxidize it to Hg(II). Microbes present in the sediment capable of converting Hg(II) to methylmercury ($CH3Hg^+$) tend to accumulate in vertebrates and fish. Mercury-resistant bacteria eliminate organomercurals by producing an enzyme, organomercurial lyase (MerB), which catalyzes the protonolysis of the carbon-mercury bond (Begley et al. 1986). The products of this reactions are a less toxic inorganic species, Hg(II), and a reduced carbon compound.

$$R\text{-}CH_2\text{-}Hg^+ + H^+ \xrightarrow{\text{MerB}} R\text{-}CH_3 + Hg(II)$$

These bacteria also synthesize a second enzyme, mercuric ion reductase (MerA), that catalyzes the reduction of the inorganic product, Hg(II), to a volatile and much less reactive elemental form, Hg(0) (Fox and Walsh 1982).

$$RSHg^+ + NADPH \xrightarrow{\text{MerA}} Hg(0) + RSH + NADP^+$$

Hg phytoremediation has been already developed. Yellow poplar expressing a modified *merA*, released ten times more elementary Hg than untransformed plantlets (Rugh et al. 1998).

Transgenic plants expressing MerB were significantly more tolerant to methylmercury and other organomercurials compared with untransformed plants. The MerB plants effectively converted the highly toxic methylmercury to $Hg^{2+,}$ which is about 100 times less toxic in plants (Bizily et al. 1999).

The MerA MerB double-transgenics showed the highest tolerance to organic mercury (up to 10 μM) compared to MerB transgenic (5 μM) and MerA and wild type plants (0.25 μM). The MerA MerB double transgenic plants were 50-fold more tolerant to organic mercury compared to wild type and were shown to volatilize elemental mercury when supplied with organic, whereas the single transgenics and the wild type plant did not. Thus, the MerA MerB double transgenic plants converted organic mercury to elemental mercury, which was released from the plant through volatilization (Bizily et al. 2000).

So far, this system has not been tested in the field conditions. This is, however, the first clear indication that genetic engineering may improve the plant's capacity to phytoremediate metal-polluted soils.

Phytoremediation is recognized as a fast-growing and cost-effective technology to remediate hazardous toxic metals from the contaminated sites. Summary of the processes of phytoaccumulation and phytovolatilization are shown in Figure 2. Accumulation of metal ions is dependent on uptake and bioavailability of heavy metals, transport and translocation of heavy metals from roots to shoot. Detoxification of heavy metals involved chelation, compartmentalization and volatilization. Novel proteins involved transport and translocation of metal ions have been identified and characterized from a variety of organisms. However, a clear role of these proteins yet remains to be elucidated.

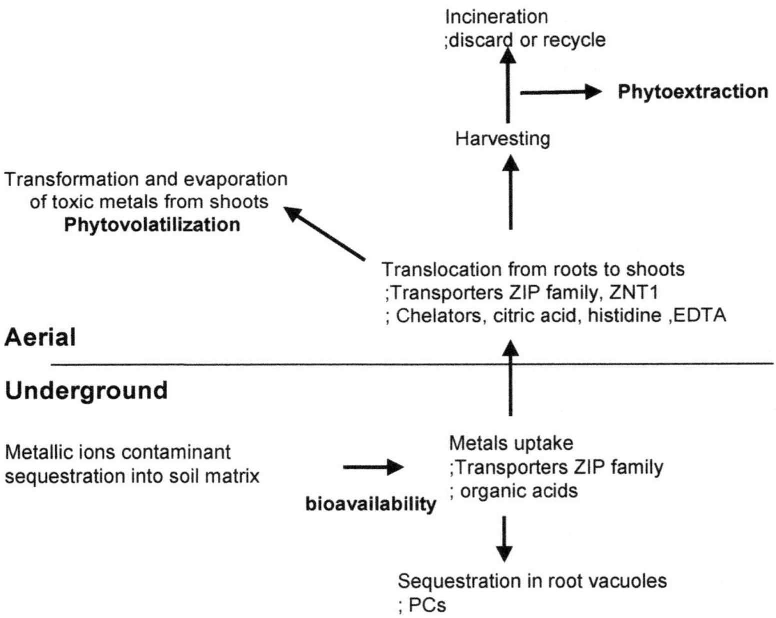


**Fig. 2.** Scheme of metallic ions decontamination in natural phytoremediation processes (modified after Singh et al. 2003)

## 3. Microbial Remediation of Metal-polluted Soils

### 3.1 Expression of Metal-binding Proteins or Peptides in Bacteria

Due to the difficulty in the removal of heavy metals from environment, many researchers attempt to get rid of heavy metals by microbial remediation. A promising way of improving bioremediation processes is to genetically engineer bacterial strains to confer increased abilities to accumulate toxic heavy metals. Attempts to enhance the metal content of bacterial cells have been made by over expressing metal-binding peptides or proteins.

### *3.1.1 Expression of Heterologous Metallothioneins (MTs)*

With the aim of enhancing the tolerance, sequestration or accumulation of heavy metals, bacteria with the high metal-binding capacity of MTs have been widely exploited. MTs from various sources have been expressed intracellularly in *Escherichia coli* monkey (Murooka and Nagaoka 1987), yeast (Berka et al. 1988; Sayers et al. 1993), human (Yamashita et al. 1994; Odawara et al. 1995), and plant (Kille et al. 1990). In many instances, however, problems of the stability and short half-life of the expressed heterologous proteins were encountered. This problem was linked to the high cysteine content of MTs, which might interfere with cellular redox pathways in the cytosol (Raina and Missiakas 1997). The small molecule of MT can be stabilized by fusion to large molecules, such as β–galactosidase. The human MT (hMT) fused to β–galactosidase enhanced uptake of Cd by the recombinant *E. coli.* In the same manner, the increased molecule size of hMT resulted in improved stability and productivity in *E. coli* (Hong et al. 2001). hMT was synthesized with prokaryotic codons and linked by a gly-gly-gly tripeptide linker to form a tetrameric hMT. The tetrameric MTL4 bound 28 gram atom of Cd or Zn (Hong et al. 2000, Murooka et al. 2001). The problem of stability and short half life of intracellular heterologus expression of MTs has been circumvented by the surface display of proteins. The metal-binding proteins have been anchored to the LamB, protein that spans the outer membrane. Yeast and mammalian MTs expressed on the surface of *E. coli* as fusions to LamB, enhanced the metal binding capacity of the cells between 10 - 20-fold (Sousa et al. 1998). Fusion of metallothionein to the autotransporter β-domain of the IgA protease of *Neisseria gonorrhoeae*, which targeted the hybrid protein towards the bacterial outer membrane, was performed on a natural inhabitant of soil bacterium, *Ralstonia eutropha*. The resulting bacterial strain was found to have an enhanced ability for immobilizing $Cd^{2+}$ from external media (Valls et al. 2000).

Expression of both metal transporter proteins and metal binding peptides may enhance strain's ability to accumulate metal ions. There is a report that expression of both $Hg^{2+}$ transport systems (MerT and MerP) and glutathione S-transferase fusion protein of *Saccharomyces cerevisiae* or pea MT in *E. coli* significantly increased the bioaccumulation of $Hg^{2+}$ (Chen and Wilson 1997).

### *3.1.2 Expression of Phytochelatins and Synthetic Phytochelatins*

Phytochelatins are short peptides composed of only three amino acids, namely, Glu, Cys and Gly, with Glu and Cys residues linked through a γ-carboxylamide bond. The structure of such peptides can be represented by $(\gamma\text{-Glu-Cys})_n$-Gly, where n range from 2 to 11. PCs are enzymatically synthesized from glutathione by PC synthase (EC 2.3.2.15) (Cobbett 2000).

Over-expression of PC synthase in bacterial strains appears to be a promising way to improve the heavy metal (such as Cd) or metalloid (such as As) content of organisms for use in bioremediation. There are reports of increasing in Cd accumulation in *Mesorhizobium huakuii* subsp. *rengei* B3 and *E. coli* cells expressing the *Arabidopsis thalina* gene encoding PC synthase (Sriprang et al. 2003; Sauge-Merle et al. 2003). Recently, the synthetic peptide $(\text{Glu-Cys})_n$-Gly, in which Glu and Cys are linked by an α-carboxyamide bond was successfully expressed onto the cell surface using Lpp-OmpA fusion system in *E. coli*, resulting in 15- or 20-fold increases in Cadmium and mercury accumulation (Bae et al. 2000; Bae et al. 2001). However, *E. coli* strains are not suitable for *in situ* soil remediation, since they are not adapted to these environments. A more realistic approach is to engineer soil bacteria that can survive in contaminated environments for an extended period. The surface expression of synthetic PC with 20 cysteines (EC20) using the truncated ice nucleation protein (INPNC) anchor in the robust bacterium, *Moraxella* sp. increases a 10-fold mercury ($Hg^{2+}$) accumulation. The expression of surface protein is more efficient in *Moraxella* sp. than *E. coli* (Bae et al. 2002).

### *3.1.3 Expression of Synthetic Metal-binding Peptides*

Novel metal binding peptides might offer a higher affinity, higher metal-binding capacity and/or specificity and selectivity for a target metal ion than known metal-binding proteins. Peptides with unique binding properties can either be designed *de novo* or selected by screening peptide libraries. Various peptides comprising different sequences of cysteines or histidines have been tested for binding Cd. Recently, metal-binding peptides that contain either histidines $(\text{GHHPHG})_2$ or cysteines (GCGCPCGCG) were engineered to LamB and expressed on the surface of *E. coli*. Surface display of these peptides increased the bioaccumulation of Cd by 4-fold and 2-fold, respectively. Moreover, a $\text{His}_6$ peptide has been expressed on the surface of *E. coli* as a fusion to the OMP LamB. This construct resulted in a 5-fold increase in Cd accumulation, when the peptide was expressed as a single copy and 11-fold increase when expressed in tandem (Sousa et al. 1996; Mejare and Bulow 2001).

## 3.2 Metal and Metalloid Remediation as the Result of Changes in Redox State

Microorganisms can detoxify metals by valence transformation, extracellular chemical precipitation, or volatilization.

Microbial reduction of the highly soluble oxidized form of selenium, $Se^{6+}$, to insoluble elemental selenium, $Se^0$, by microorganisms that conserve energy to support growth from $Se^{6+}$ reduction is a natural mechanism for the removal of selenium from contaminated surface and groundwater. The *Bacillus sp.* SF-1 has been isolated as a selenate-reducing bacterium that can tolerate and efficiently reduce very high concentration of selenate ($Se^{6+}$) (up to about 150 mg-Se/L) into selenite and, subsequently, into elemental Se (Kashiwa et al. 2001).

Enzymatic reduction of Cr(VI) to less mobile and less toxic Cr(III) has been one of the most widely studied forms of metal bioremediation (Lovley 1995; Wang and Shen 1995). The NAD(P)H-dependent chromium reductase, which has ability to reduce Cr(VI), was found in some bacteria such as *Pseudomonas ambigua* (Suzuki et al. 1992), *P. putida* (Ishibashi et al. 1990), *Enterobacter cloacae* (Wang et al. 1989) and *Pseudomonad* (CRB5) (McLean and Beveridge 2001). The Cr (VI) reduction occurs under aerobic and/or anaerobic conditions.

In bioremediation of heavy metals, microorganisms have been mostly used to treat industrial waste streams, with the organisms either immobilized onto different support matrixes or in a free-living state, enclosed in treatment tanks or other kinds of reactor vessels. Subsequently, the metal-loaded biomass can be either disposed appropriately or treated to recover the metals.

## 4. Heavy Metal Bioremediation using "Symbiotic Engineering"

Rhizobia grow slowly for long periods in soil, but if they infect a compatible legume they can grow rapidly; successful infection by a single bacterium can lead to the formation of a nitrogen-fixing nodule on the root of legume, containing over $10^8$ bacterial progeny (Downie 1997). This special character is useful for biotechnological application for the expression of genes, such as metallothionein that sequester heavy metals from contaminated soil. Once symbiosis is established, the heavy metals will be accumulated in nodules. This would be an alternative and less expensive method to remove heavy metals from the soil.

*Mesorhizobium huakuii* subsp. *rengei* strain B3 (Murooka et al. 1993; Nuswantara et al. 1999) is a bacterium that establishes symbiosis with *Astragalus sinicus* (Chinese milk vetch, or renge-soh in Japanese), a legume used as a green manure in rice field in Japan and Southern China, by eliciting the formation of nitrogen-fixing root nodules (Chen et al. 1991). Symbiosis between leguminous plants and rhizobia is initiated when flavonoids and related plant compounds induce the bacteria to produce molecular signals, which stimulate nodule organogenesis (Fisher and Long 1992). Bacteria enter the developing nodule via infection threads and are taken up by plant host cells in an endocytosis-like process. The rhizobia undergo differentiation into a distinct cell type called as bacteroid, which is capable of fixing atmospheric nitrogen into ammonia to be available to the host plants (Mylona et al. 1995).

Likewise, *A. sinicus* is widely used as a natural fertilizer in rice fields during the idle period. It would be more interesting, if one can use this legume plant to increase fertility and at the same time remove heavy metals from the soil. Sriprang et al. (2002) developed a novel plant-bacterial remediation system for heavy metals by the introduction of the chimeric *MTL4* gene to *M. huakuii* subsp. *rengei* B3. This is also the first report that a foreign gene was expressed in bacteroids in the nodules. Murooka proposed this new technology to be called as "Symbiotic Engineering".

### 4.1 Heavy Metal Bioremediation with Oligomeric MTs

Sriplang et al. (2002) developed a plant-bacterial remediation system for heavy metals by the expression of tetrameric hMT (MTL4) in *M. huakuii* subsp. *rengei* B3. The *MTL4* gene (Hong et al. 2000) was fused to the *nifH* and *nolB* promoters, which generated nodule- specific expression of the *MTL4* gene. The expression analysis of the *MTL4* gene was demonstrated in the free-living cells in the presence of $Cd^{2+}$ and $Cu^{2+}$ under the low oxygen condition. The *MTL4* under the *nifH* and *nolB* promoters was expressed and increased the accumulation of $Cd^{2+}$, but not $Cu^{2+}$ in free-living cells. The expression of the integrated *nifH-MTL4* gene in the chromosome of strain B3 was also expressed stably and accumulated $Cd^{2+}$in the bacterial cells. By inoculation of the recombinant B3, *A. sinicus* established symbiosis with the recombinant B3 that was grown in $Cd^{2+}$ and $Cu^{2+}$-polluted soils. The symbionts with recombinant plasmids pNHMT4 and pNBMT4 increased $Cd^{2+}$ accumulation in nodules 2.3 and 6.6- fold, respectively, whereas no significantly increase in $Cu^{2+}$ accumulation was noted. Accumulation of $Cd^{2+}$ in nodules was at the same level in different external Cd concentrations in soils. This might be due to the limitation of the production of the MTL4 protein. The basal level of $Cd^{2+}$ accumulation in nodules by tri-peptide glutathione (GSH) in legume root nodules (Moran et al. 2000) has a crucial role in protecting the plants against xenobiotics, heavy metals and oxidative stress (Noctor and Foyer 1998). By our calculation, one nodule can adsorb as much as 1.4 nmol $Cd^{2+}$. Based on the average nodulation per plant in the rice field (100 nodules), it is estimated that 140 nmol of $Cd^{2+}$ can be removed from soil by one plant containing the recombinant B3.

### 4.2 Heavy Metal Bioremediation with Phytochelatin

The *Arabidopsis* gene for phytochelatin synthase (*AtPCS*) in *M. huakuii* subsp. *rengei* B3 was also expressed (Sriprang et al. 2003). The *AtPCS* gene was expressed under the control of the *nifH* promoter, which regulates the nodule-specific expression of the *nifH* gene. The expression of the *AtPCS* gene was demonstrated in free-living cells under low-oxygen conditions. The

PCS was expressed and catalyzed the synthesis of PCs in strain B3. Cells that expressed the AtPCS gene, whereas no PCs were found in control cells that harbored the empty plasmid, synthesized a range of PCs, with values of n from 2 to 7. The presence of $CdCl_2$ activated PCS and induced the synthesis of substantial amounts of PCs. Cells that contained PCs accumulated 36 nmoles of $Cd^{2+}$/ mg dry weight of cells. The expression of the *AtPCS* gene in *M. huakuii* subsp. *rengei* B3 increased the ability of cells to bind $Cd^{2+}$ by 9- to 19-fold approximately. The PCS protein was detected by immunostaining in bacteroids of mature nodules of *A. sinicus* containing the *AtPCS* gene. When recombinant *M. huakuii* subsp. *rengei* B3 established the symbiotic relationship with *A. sinicus*, the symbionts increased $Cd^{2+}$ accumulation in nodules by 1.5-fold.

### 4.3 Advantages of Bioremediation using Symbiotic Engineering

A limitation of the using microbes for bioremediation is that although the metal was bound microbe, but after decomposition of microbes, the metals are still present in the soils. This consideration suggests that for the majority of metal contaminants, the most effective *in situ* remediation strategies may need to combine microbial methods for binding of metals from soil with methods that can effectively uptake metals from soil and prevent the recycle of metals to soil. Plants uptake such released metals from roots and nodules. Bacteroids in nodules can be easily engineered with metal binding peptides. Expression of both *MTL4* and *AtPCS* genes in B3 strain resulted in the additive accumulation of cadmium in the free-living cells. However, accumulation of cadmium in the nodules, in which the two genes were expressed, was not much increased as compared with each single gene expression. This result suggests that uptake of cadmium into the nodule is very limited. Thus, Murooka et al. (unpublished results) expressed the *Arabidopsis* gene for *AtPCS* and iron-regulated transporter (IRT1) in *M. huakuii* subsp. *rengei* B3. The *AtPCS* gene was integrated in the chromosome under the control of the *nifH* promoter, which regulates the nodule-specific expression of the *nifH* gene. The IRT1 gene was expressed under the control of the *nolB* promoter, which regulates the nodule-specific expression of the *nolB* gene. The presence of single copy of *AtPCS* in the chromosome showed slightly increased in $Cd^{2+}$ accumulation, 2.9 $Cd^{2+}$/ mg dry weight of cells. The presence of multicopy of *AtPCS* in the chromosome showed increased in $Cd^{2+}$ accumulation 20 $Cd^{2+}$/ mg dry weight of cells. The expression of both the *AtPCS* and *IRT1*gene in recombinant *M. huakuii* subsp. *rengei* B3 increased the ability of cells to bind $Cd^{2+}$ 1.7 to 2.5-fold approximately compared to cells expressed only *AtPCS*.

Thus, genetically engineered symbiotic system, "symbiotic engineering" has a great potential for bioremediation of heavy metals-polluted soil. This bioremediation technique can be applicable to use in symbiosis between mycorrhiza and plants.

## 5. Conclusion

Bioremediation is the use of plants and microorganisms to extract sequester or detoxify pollutants. Phytoremediation is the use of plants to clean up chemical-contaminated soils. Bioremediation offers a low-cost method for soil or water remediation and some extracted metals may be recycled for value. This review describes traits of metal- hyper accumulating plants for phytoextraction of metals. The hyper accumulators must have high ability to mobilize and uptake of trace elements/metal ions, into the root, shoot and other viable parts of the plant with the aids of chelators and transporter proteins. Chelation of metal ions by various ligands and vacuole compartmentalization play important role in detoxification in hyper accumulators. Alternatively, phytovolatilization of Hg by plants offer great promise for decontamination of metal ions from soil. Potential transgenic approaches for the development of effective phytoremediation technology have been achieved.

Using of microorganisms to remedy heavy metals has been developed. A promising way of improving bioremediation processes is to genetically engineer bacterial strains to confer increased abilities to accumulate toxic heavy metals. Attempts to enhance the metal content of bacterial cells have been made by over expressing metal-binding peptides or proteins, synthetic metal binding peptides. A novel phytoremediation system using symbiosis between leguminous plants and rhizobia was also developed. This system uses both advantages of plants and microorganisms, particularly engineered genes can be transformed to plants through infection with recombinant microorganisms.

## References

Bae W, Chen W, Mulchandani A, Mehra RK (2000) Enhanced bioaccumulation of heavy metals by bacterial cells displaying synthetic phytochelatins. Biotechnol Bioeng 70:518-524

Bae W, Mehra RK, Mulchandani A, Chen W (2001) Genetic engineering of Escherichia coli for enhanced uptake and bioaccumulation of mercury. Appl Environ Microbiol 67:5335-5338

Bae W, Mulchandani A, Chen W (2002) Cell surface display of synthetic phytochelatins using ice nucleation protein for enhanced heavy metal bioaccumulation. J Inorganic Biochem 88:223-227

Baker AJM, Brooks RR (1989) Terrestrial higher plants which hyperaccumulate metallic elements- a review of their distribution, ecology and phytochemistry. Biorecovery 1:81-126

Becher M, Talke IN, Krall L, Kramer U (2004) Cross species microarray transcript profiling reveals constitutive overexpression of metal homeostasis genes in shoots of the zinc hyperaccumulator *Arabidopsis halleri.* Plant J 37:251-268

Begley TP, Walts AE, Walsh CT (1986) Mechanistic studies of a protonolytic organomercurial cleaving enzyme: bacterial organomercurial lyase. Biochemistry 25:7192-7200

Berka T, Shatzman A, Zimmerman J, Strickler J, Rosenberg M (1988) Efficient expression of the yeast metallothionein gene in *Escherichia coli*. J Bacteriol 170:21-26

Bizily SP, Rugh CL, Summers AO, Meagher RB (1999) Phytoremediation of methylmercury pollution *merB* expression in *Arabidopsis thaliana* confers resistance to organomercurials. Proc Natl Acad Sci USA 96:6808-6813

Bizily SP, Rugh CL, Meagher RB (2000) Phytodetoxification of hazardous organomercurials by genetically engineered plants. Nat Biotechnol 18:213-217

Blaylock MI, Salt DE, Dushenkov S, Zakharova O, Gussman C (1997) Enhanced accumulation of Pb in Indian mustard by soil applied chelating agents. Environ Sci Technol 31:860-865

Blaudez P, Kohler A, Martin F, Sanders D, Chalot M (2003) Poplar metal tolerance protein1 confers Zinc tolerance and is an oligomeric vacuole zinc transporter with an essential leucine zipper motif. The Plant Cell 15:2911-2928

Brown SL, Chaney RL, Angle JS, Baker AJM (1995) Zinc and cadmium uptake by hyperaccumulator *Thlaspi caerulescens* and metal tolerant *Silene vulgaris* grown on sludge-amended soils. Environ Sci Technol 29:1581-1585

Burken JG, Schnoor JL (1996) Phytoremediation:plant uptake of atrazine and role of root exudates. J Environ Eng 122:958-963

Butt TR, Ecker DJ (1987) Yeast metallothionein and applications in biotechnology. Microbiol Rev 51:351-364

Chaney RL, Malik M, Li YM, Brown SL, Brewer EP, Angle JS, Baker AJM (1997) Phytoremediation of soil metals. Curr Opin Biotechnol 8:279-284

Chen S, Wilson DB (1997) Construction and characterization of *Escherichia coli* genetically engineered for bioremediation of $Hg^{2+}$-contaminated environments. Appl Environ Microbiol 63:2442-2445

Chen W, Li GS, Qi YL, Wang ET, Yuan HL, Li L, (1991) *Rhizobium huakuii* sp. nov. isolated from the root nodules of *Astragalus sinicus*. Int J Syst Bacteriol 41:275-280

Chevalier C, Bourgeois E, Pradet A, Raymond P (1995) Molecular cloning and characterization of six cDNAs expressed during glucose starvation in excised maize (*Zea mays* L.) root tips. Plant Mol Biol 28:473-485

Clemens S, Antosiewicz DM, Ward JM, Schachtman DP, Schroeder JI (1998) The plant cDNA *LCT1* mediates the uptake of calcium and cadmium in yeast. Proc Natl Acad Sci USA 95:12043-12048

Clemens S, Palmgren MG, Kramer U (2002) A long way ahead: understanding and engineering plant metal accumulation. Trends Plant Sci 7:309-315

Cobbett CS (2000) Phytochelatins and theirs roles in heavy metal detoxification. Plant Physiol 123:825-832.

Crowley DE, Wang YC, Reid CPP, Szaniszlo PJ (1991) Mechanisms of iron acquisition from siderophores by microorganisms and plants. Plant Soil 130:179-198

Dhankher OP, Li Y, Rosen BP, Shi J, Salt D, Senecoff JF, Sashti NA, Meagher RB (2002) Engineering tolerance and hyperaccumulation of arsenic in plants by combining arsenate reductase and $\gamma$-glutamylcysteine synthetase expression. Nat Biotechnol 20:1140-1145

Downie A (1997) Fixing a symbiotic circle. Nature 387:352-353

Eide D, Broderius M, Fett J, Guerinot ML (1996) A novel iron-regulated metal transporter from plants identified by functional expression in yeast. Proc Natl Acad Sci USA 93:5624-5628

Evans KM, Gatehouse JA, Lindsay WP, Shi J, Tommey AM, Robinson NJ (1992) Expression of the pea metallothionein-like gene *PsMTA* in *Escherichia coli* and *Arabidopsis thaliana* and analysis of trace metal ion accumulation: implications for PsMTA function. Plant Mol Biol 20:1019-1028

Fisher FF, Long SR (1992) Rhizobium-plant signal exchange. Nature 357:655-660

Fox B, Walsh CT (1982) Mercuric reductase. Purification and characterization of a transposon-encoded flavoprotein containing an oxidation reduction active disulfide. J Biol Chem 257:2498-2503

Gong JM, Lee DA, Schroeder JI (2003) Long-distance root-to-shoot transport of phytochelatins and cadmium in Arabidopsis. Proc Natl Acad Sci USA 100:10118-10123

Grotz N, Fox T, Connolly E, Park W, Guerinot ML, Eide D (1998) Identification of a family of zinc transporter genes from Arabidopsis that respond to zinc deficiency. Proc Natl Acad Sci USA 95:7220-7224

Guerinot ML (2000) The ZIP family of metal transporters. Biochim Biophys Acta 1465:190-198.

Hall JL (2002) Cellular mechanisms for heavy metal detoxification and tolerance. J Exp Bot 53:1-11

Hasegawa I, Terada E, Sunairi M, Wakita H, Shinmachi F, Noguchi A, Nakajima M, Yazaki T (1997) Genetic improvement of heavy metal tolerance in plants by transfer of the yeast metallothionein gene (CUP1). Plant Soil 196:277-281

Hirschi KD, Korenkov VD, Wilganowski NL, Wagner GJ (2000) Expression of Arabidopsis CAX2 in tobacco. Altered metal accumulation and increased manganese tolerance. Plant Physiol 124:125-133

Hsieh HM, Liu WK, Huang PC (1995) A novel stress-inducible metallothionein-like gene from rice. Plant Mol Biol 28:381-389

Hong SH, Gohya M, Ono H, Murakami H, Yamashita M, Hirayama N, Murooka Y (2000) Molecular design of novel metal-binding oligomeric human mrtallothioneins. Appl Microbiol Biotechnol 54:84-89

Hong S-H, Toyama M, Maret W, Murooka Y (2001) High yield expression and single step purification of heman thionein/metallothionein. Protein Expres Purifi 21:243-250

Huckle JW, Morby AP, Turner JS, Robinson NJ (1993) Isolation of prokaryotic metallothionein locus and analysis of transcriptional control by trace metal ions. Mol Microbiol 7:177-187

Ishibashi Y, Cervantes C, Silver S (1990) Chromium reduction in Pseudomonas putida. Appl Environ Microbiol 56:2268-2270

Kagi JHR (1991) Overview of metallothionein. Methods Enzymol 205:613-626

Kashiwa M, Ike M, Mihara H, Esaki N, Fujita M (2001) Removal of soluble selenium by a selenate-reducing bacterium *Bacillus sp.* SF-1. J Ferment Bioeng 83:517-522

Karenlampi S, Schat H, Vangronsveld J, Verkleij JAC, van der Lelie D, Mergeay M, Tervahauta AI (2000) Genetic engineering in the improvement of plants for phytoremediation of metal polluted soils. Environ Pollut 107:225-231

Kille P, Winge DR, Harwood JL, Kay J (1990) A plant metallothionein produced in *Escherichia coli*. FEBS Lett 295:171-175

Korshunova Y, Eide D, Clark G, Guerinot M, Pakrasi H (1999) The IRT1 protein from *Arabidopsis thaliana*. Plant Mol Biol 40:37-44

Kramer U, Cotter-Howells JD, Charnock JM, Baker AJM, Smith JAC (1996) Free histidine as a metal chelator in plants that accumulate nickel. Nature 379:635-638

Kramer U, Pickering IJ, Raskin I, Salt DE (2000) Prince Subcellular localization and speciation of nickel in hyperaccumulator and non-accumulator *Thlaspi* species. Plant Physiol 122:1343-1353

Krotz RM, Evangelou BP, Wagner GJ (1989) Relationship between cadmium, zinc, Cd-peptide, and organic acid in tobacco suspension cell. Plant Physiol 91:780-787

Kumar PBAN, Dushenkov V, Motto H, Raskin I (1995) Phytoextraction: the use of plants to remove heavy metals from soils. Environ Sci Technol 29:1232-1238

Lasat MM, Baker A, Kochian LV (1996) Physiological characterization of root $Zn^{2+}$ absorption and translocation to shoots in Zn hyperaccumulator and nonaccumulator species of *Thlaspi*. Plant Physiol 112:1715-1722

Loeffler S, Hochberger A, Grill E, Winnacker EL, Zenk MH (1989) Termination of the phytochelatin synthase reaction through sequestration of heavy metals by the reaction product. FEBS Lett 258:42-46

Lovley DR (1995) Bioremediation of organic and metal contaminants with dissimilatory metal reduction. J Indust Microbiol 14:85-93

Ma LQ, Komar KM, Tu C, Zhang WH, Cai Y, Kennelley ED (2001) A fern that hyperaccumulates arsenic. Nature 409:579

Maiti IB, Hunt AG, Wagner GJ (1988) Seed-transmissible expression of mammalian metallothionein in transgenic tobacco. Biochem Biophys Res Commun 150:640-647

Maiti IB, Wagner GI, Yeargan R, Hunt AG (1989) Inheritance and expression of the mouse metallothionein gene in tobacco. Plant Physiol 91:1020-1024

McLean J, Beveridge TJ (2001) Chromate reduction by a Pseudomonas isolated from a site contaminated with chromate copper arsenate. Appl Environ Microbiol 67:1076-1084

Mehra RK, Winge D (1991) Metal ion resistance in fungi: molecular mechanisms and their regulated expression. J Cell Biochem 45:30-40

Mejare M, Bulow L (2001) Metal-binding proteins and peptides in bioremediation and phytoremediation of heavy metals. Trends Biotecnol 19:67-73

Misra S, Gedamu L (1989) Heavy metal tolerant transgenic *Brassica napus* L. and *Nicotiana tabacum* L. plants. Theo App Genet 78:161-168

Moran JF, Iturbe-Ormaetxe I, Matamoros MA, Rubio MC, Clemente MR, Brewin NJ, Becana M (2000) Glutathione and homoglutathione synthetases of legume; cloning expression and subcellular localization. Plant Physiol 124:1381-1392

Murooka Y, Nagaoka T (1987) Expression of cloned monkey metallothionein in *Escherichia coli*. Appl Environ Microbiol 53:204-207

Murooka Y, Toyama M, Hong S-H, Gohya M, Ono H, Yamashita M, Hirayama N (2001) Genetic design of stable metal-binding biomolecules, Oligomeric metallothioneins. Biocatal Biotransform 19:399-412

Murooka Y, Xu Y, Sanada K, Araki M, Morinaga T, Yokota A (1993) Formation of root nodules by *Rhizobium huakuii* biovar. *Renge* bv. Nov on *Astragalus sincicus* cv. Japan. J Ferment Bioeng 76:38-44

Mylona P, Powlowski K, Bisseling T, (1995) Symbiotic nitrogen fixation. Plant Cell 7:869-885

Noctor G, Foyer CH (1998) Ascorbate and glutathione: keeping active oxygen under control. Annu Rev Plant Physiol Plant Mol Biol 49:249-279

Nuswantara S, Fujie M, Yamada T, Malek W, Inaba M, Kaneko Y, Murooka Y (1999) Phylogenic position of *Mesorhizobium huakuii* subsp. *rengei*, a symbiont of *Astragalus sinicus* cv. Japan. J Biosci Bioeng 87:49-55

Olafson RW, Mccubbin W, Kay C (1988) Primary and secondary structural analysis of a unique prokaryotic metallothionein from *Synechococcus sp.* Cyanobacterium. Biochem J 251:691-699

Ortiz DF, Kreppel L, Speiser DM, Scheel G, McDonald G, Ow DW (1992) Heavy-metal tolerance in the fission yeast requires an ATP-binding cassette-type vacuolar membrane transporter. EMBO J 11:3491-3499

Oritz DF, Ruseitti T, McCue KF, Ow DW (1995) Transport of metal-binding peptides by HMT1, a fission yeast ABC-type B vacuolar membrane protein. J Biol Chem 270:4721-4728

Odawara F, Kurasaki M, Suzuki-Kurasaki M, Oikawa S, Emoto T, Yamasaki F, Arias ARL, Kojima Y (1995) Expression of human metallothionein-2 in *Escherichia coli*: cadmium tolerance of transformed cells. J Biochem 118:1131-1137

Pan A, Tie F, Duau Z, Yang M, Wang Z, Li L, Chen Z, Ru B (1994a) Alpha-domain of human metallothionein IA can bind to metals in transgenic tobacco plants. Mol Gen Genet 242:666-674

Pan A, Yang M, Tie F, Li L, Chen Z, Ru B (1994b) Expression of mouse metallothionein-I gene confers cadmium resistance in transgenic tobacco plants. Plant Mol Biol 24:341-351

Pence NS, Larsen PB, Ebbs SD, Letham DL, Lasat MM, Garvin DF, Eide D, Kochian LV (2000) The molecular physiology of heavy metal transport in the Zn/Cd hyperaccumulation *Thlaspi caerulescens*. Proc Natl Acad Sci USA 97:4956-4960

Persans MW, Nieman K, Salt DE (2001) Functional activity and role of cation-efflux family members in Ni hyperaccumulation in *Thlaspi goesingense*. Proc Natl Acad Sci USA 98:9995-10000

Raina S, Missiakas D (1997) Making and breaking disulfide bonds. Annu Rev Microbiol 51:179-202

Raskin I, Smith RD, Salt DE (1997) Phytoremediation of metals: using plants to remove pollutants from the environment. Curr Opin Biotechnol 8:221-226

Robinson BH, Brooks RR, Howes AW, Kirkman JH, Gregg PEH (1997a) The potential of the high-biomass nickel hyperaccumulator *Berkheya coddii* for phytoremediation and phytomining. J Geochem Explor 60:115-126

Robinson BH, Chiarucci A, Brooka RR, Petit D, Kirkman JH, Gregg PEH, DeDominicis V (1997b) The nickel hyperaccunulator plant *Alyssum bertolonii* as a potential agent for phytoremediation and phytomining of nickel. J Geochem Explor 59:75-86

Rogers EE, Eide DJ, Guerinot ML (2000) Altered selectivity in an Arabidopsis metal transporter. Proc Natl Acad Sci USA 97:12356-12360

Rugh CL, Senecoff JF, Meagher RB, Merkle SA (1998) Development of transgenic yellow poplar for mercury phytoremediation. Nat Biotechnol 16:925-928

Salt DE, Wagner GJ (1993) Cadmium transport across tonoplast of vesicles from oat roots. Evidence for a $Cd^{2+}/H^{+}$ antiport activity. J Biol Chem 268:12297-12302

Salt DE, Rauser WE (1995) MgATP-dependent transport of phytochelatins across the tonoplast of oat roots. Plant Physiol 107:1293-1301

Sauge-Merle S, Cuine S, Carrier P, Lecomte-Pradines C, Luu DT, Peltier G (2003) Enhanced toxic metal accumulation in engineered bacterial cells expressing *Arabidopsis thaliana* phytochelatin synthase. Appl Environ Microbiol 69:490-494

Sayers Z, Brouillon P, Vorgias CE, Nolting HF, Hermes C, Koch MH (1993) Cloning and expression of *Saccharomyces cerevisiae* copper-metallothionein gene in

*Escherichia coli* and characterization of the recombinant protein. Eur J Biochem 212:521-528

Schachtman DP, Kumar R, Schroeder JI, Marsh EL (1997) Molecular and functional characterization of a novel low-affinity cation transporter (LCT1) in higher plants. Proc Natl Acad USA 94:11079-11084

Senden MHMN, Van Paassen FJM, VanDerMeer AJGM, Wolterbeek HTH (1992) Cadmium-citric acid-xylem cell wall interactions in tomato plants. Plant Cell Environ 15:71-79

Singh OV, Labana S, Pandey G, Budhiraja R, Jain RK (2003) Phytoremediation: an overview of metallic ion decontamination from soil. Appl Microbiol Biotechnol 61:405-412

Song WY, Sohn EJ, Martinoia E, Lee YJ, Yang Y, Jasinski M, Forestier C, Hwang I, Lee Y (2003) Engineering tolerance and accumulation of lead and cadmium in transgenic plants. Nat Biotechnol 21:914-919

Sousa C, Cebolla A, de Lorenzo V (1996) Enhanced metalloadsorption of bacterial cells displaying poly-His peptides. Nat Biotechnol 14:1017-1020

Sousa C, Kotrba P, Ruml T, Cebolla A, de Lorenzo V (1998) Metalloadsorption by *Escherichia coli* cells displaying yeast and mammalian metallothioneins anchored to the outer membrane protein LamB. J Bacteriol 180:2280-2284

Speiser DM, Ortiz DF, Kreppel L, Scheel G, McDonald G, Ow DW (1992) Purine biosynthetic genes are required for cadmium tolerance in *Schizosaccharomyces pombe*. Mol Cell Biol 12:5301-5310

Suzuki T, Miyata N, Horitsu H, Kawai K, Takamizawa K, Tai Y, Okazaki M (1992) NAD(P)H-dependent chromium (VI) reductase of *Pseudomonas ambigua* G-1: a Cr(V) intermediate is formed during the reduction of Cr(VI) to Cr(III). J Bacteriol 174:5340-5345

Sriprang R, Hayashi M, Yamashita M, Ono H, Saeki K, Murooka Y (2002) A novel bioremediation system for heavy metals using the symbiosis between leguminous plant and genetically engineered rhizobia. J Biotechnol 99:279-293

Sriprang R, Hayashi M, Ono H, Takagai M, Hirata K, Murooka Y (2003) Enhanced accumulation of $Cd^{2+}$ by a *Mesorhizobium* sp. transformed with a gene from *Arabidopsis thaliana* coding for phytochelatin synthase. Appl Environ Microbiol 69:1791-1796

Toyama M, Yamashita M, Hirayama N, Murooka Y (2002) Interactions of arsenic with human metallothionein-2. J Biochem 132:217-221

Vassil A, Kapulnik Y, Raskin I, Salt DE (1998) The role of EDTA in lead transport and accumulation by Indian mustard. Plant Physiol 117:447-453

Valls M, Atrian S, de Lorenzo V, Fernandez LA (2000) Engineering a mouse metallothionein on the cell surface of *Ralstonia eutropha* CH34 for immobilization of heavy metals in soil. Nat Biotechnol 18:661-665

Wang P, Mori T, Komori K, Sasatsu K, Toda K, Ohtake H (1989) Isolation and characterization of an *Enterobacter cloacae* strain that reduces hexavalent chromium under anaerobic conditions. Appl Environ Microbiol 55:1665-1669

Wang YT, Shen H (1995) Bacterial reduction of hexavalent chromium. J Indust Microbiol 14:159-183

Weber M, Harada E, Vess C, Roepenack-Lahaye EV, Clemens S (2004) Comparative microarray analysis of *Arabidopsis thaliana* and *Arabidopsis halleri* roots identified nicotianamine synthase, a ZIP transporter and other genes as potential metal hyperaccumulation factors. Plant J 37:269-281

Winge DR, Nielson KB, Gray W, Hamer D (1985) Yeast metallothionein: sequence and metal-binding properties. J Biol Chem 260:14464-14470

Yamashita M, Kuwata H, Murakami H, Murooka Y (1994) Genetic design of a gene for human metallothionein II and its expression as an active fusion protein in *Escherichia coli*. J Ferment Bioeng 77:113-118

Zhou J, Goldsbrough PB (1994) Functional homologs of fungal metallothionein genes from Arabidopsis. Plant Cell 6:875-884

Zhu YL, Pilon-Smits EAH, Jouanin L, Terry N (1999) Cadmium tolerance and accumulation in Indian mustard is enhanced by overexpressing gamma-glutamylcysteine synthetase. Plant Physiol 121:1169-1178

# 5

# Role of Phytochelatins in Phytoremediation of Heavy Metals

Erwin Grill[1], Seema Mishra[2], Sudhakar Srivastava[2] and R.D. Tripathi[2]

[1]Technical University of Munich, Centre of Scienctific Research Weihenstephan, Department of Bioscience, Institute of Botany, Am Hochanger-4, 85350 Weihenstephan, GERMANY, [2]Ecotoxicology and Bioremediation Group, National Botanical Research Institute, Rana Pratap Marg, Lucknow 226 001, INDIA, Email: tripathi_rd@rediffmail.com

## 1. Introduction

Heavy metals are defined as metals having density more than 5 g/cm$^3$ (Elmsley 2001). They may include both essential and non-essential metals. For organisms, essential metals, such as Cu, Zn, Co, Mn, Mo etc., play vital role as co-factor in redox reactions, ligand interactions, charge stabilization, charge shielding and water ionization during biocatalysis (Elmsley 2001; Voet and Voet 2004), while non-essential metals, such as As, Cd, Pb and Hg, are not at all required by organisms, instead they interfere with the function of essential metals and enzymes. The supraoptimal level of essential heavy metals and higher levels of non-essential heavy metals, both cause toxicity and their increasing concentration in environment pose threat to living systems. Phytoremediation, an innovative and cost effective technology for environmental cleanup, takes advantage of the natural abilities of the plants to take up, accumulate, store or degrade organic or inorganic substances. It is considered as an environmentally friendly means of reducing the metal load of contaminated substrate and offers cost advantage over the traditional methods, such as landfill, excavation, fixing or leaching. For example, the estimated cost for removal of radionuclide from water using sunflower range from \$2-\$6 per thousand gallons while physical processes cost approximately \$80 for the same (Terry 2003). Phytoremediation also satisfies stringent pollution control board standards at the same time (Chandra Sekhar et al. 2004).

Governments (worldwide) are paying attention for establishing research and demonstration programme for phytoremediation. Several phytoremediation techniques for metals, such as Ni, Co, Cd, Se, Pb, Hg and Zn etc. are commercially available and some other are currently under development. Many demonstration projects in Canada, Europe, and US have given exellent results (US EPA 2000; US Department of Energy 2000). However, phytoremediation is still in its initial stages of research and development. Besides many advantages, phytoremediation also has some limitations. Hyperaccumulating plants often accumulate a specific element only, thus it limits the applicability to site having multiple metal contamination. Many hyperaccumulator plants are relatively rare, with small population that often occurs in remote areas or have restricted distribution. They often have slow growth rate and produce small amount of biomass.

The use of genetic engineering to introduce genes into fast growing cultivars or to increase production of selected plant enzyme may improve this drawback. Thus, a comprehensive knowledge is needed to understand the process of detoxification and storage adopted by plants. Once it is known which pathway is involved, the biotechnology technique can be used to create innovative lines of plants and new gene combinations to increase the efficacy of phytoremediation capabilities of the plants.

This chapter emphasizes on the mechanism of metal tolerance exhibited by the plant with respect to phytochelatin (PC) and deals with all aspects of PC-mediated detoxification like induction of PCs, complexation, sequestration of metals and genetic engineering prospects.

The tolerance of heavy metals in plants includes processes like immobilization, exclusion, chelation and compartmentalization. These mechanisms not only control the uptake and accumulation of essential and non-essential heavy metals but, also detoxify them. Chelation of heavy metal by a ligand followed by subsequent compartmentalization of the ligand-metal complex is thought to be the general heavy metal detoxification mechanism in plants. Several metal chelating plant ligands have been identified including organic acids, amino acids, peptides and proteins, which may complex with metals to detoxify their action. Thus, a role of organic acids has been implicated in the detoxification of Cd (Krotz et al. 1989; Wang et al. 1991; Salt et al. 1995, 1997), Zn (Lasat et al. 1996, 1998; Salt et al. 1999), organic acids and flavonoid type phenolics for Al (Barceló and Paschenrieder 2002; Barceló et al. 2003) and amino acids like histidine for Ni chelation (Krämer et al. 1996, 1997, 2000; Persans et al. 1999) and cysteine for Co (Oven et al. 2002a). Metal ions can be separated into three groups according to their binding preferences i.e. class A (O seeking), class B (N/S seeking) and borderline (intermediate). Class B and borderline metal ions are not separated clearly and include almost all heavy metals. However, affinity of metal ions of class B towards S/N containing ligands increases in the order $Mn^{2+}<Zn^{2+}<Ni^{2+}<Fe^{2+}\cong Co^{2+}<Cd^{2+}<Cu^{2+}<Pb^{2+}$ (Nieboer and Richardson 1980). The affinity of metals towards thiol groups plays an important role in both the homeostasis of essential metal ions and sequestratoin of various non-essential toxic metal ions at the sub-cellular level. Metal ions easily bind to -SH group of the cysteine in cysteine containing ligands, such as glutathione (GSH),

phytochelatins (PCs) and metallothioneins (MTs) (Rauser 1990). Tripeptide GSH plays a role against metal toxicity in several ways. These include direct metal binding, promotion or transfer of heavy metal to other ligand e.g. MT and PC (Freedman et al. 1989), removal of active oxygen species (Inzé and van Montagu 1995), and/or the formation and transport of active heavy metal complexes (Li et al. 1997).

MTs were first discovered in equine renal cortex in 1957 containing large amount of sulfur and cadmium, thus named metallothioneins (Marghoses and Vallee 1957; Kägi and Vallee 1960). Later on, many structurally related proteins were identified in other organisms and were shown to be associated with several metal ions, most commonly Zn, Cu and Cd (Kägi and Kojima 1987). MTs are cysteine rich polypeptides encoded by a family of genes and play an important role in metal detoxification in both animals and many plant species. MTs have been classified on the basis of structural differences into three classes (Rauser 1990) namely MTI, MTII and MTIII (Cobbett and Goldsbrough 2002).

PCs constitute the MTIII group. These are enzymatically synthesized cysteine-rich polypeptides serving similar function as MTs by mediating the high affinity binding and promoting vacuolar sequestration of heavy metals. These are particularly important ligand found in almost all plants and many other organisms (Rauser 1995), playing a lead role in detoxification of heavy metals.

## 2. Phytochelatin

PCs were first idendified and characterized in fission yeast *Schizosaccharomyces pombe* and were termed as cadystins (Murasugi et al. 1981, Kondo et al. 1984). In 1985, it was reported that the major cadmium binding ligands in Cd intoxicated plant cells are composed of (poly γ-glutamylcysteine)-glycine and were termed as phytochelatins (Grill et al. 1985).

Structurally PCs are related to GSH. The general structural formula for PC has been given as $(\gamma\text{-Glu-Cys})_n\text{-Gly}$, where n ranges between 2-11. Thus PCs constitute a number of structural species with increasing repetitions of γ-Glu-Cys units.

PCs contain strongly nucleophilic sulfydryl groups and thus can react with many toxic species within the cell, such as free radicals, active oxygen species, and cytotoxic electrophilic organic xenobiotics and obviously heavy metals (Rabenstein 1989). Their N-terminal and downstream γ-peptidyl bonds probably serve to protect these thiol peptides from general protease action except from specific action of γ-glutamyltranspeptidases. However, the cadmium (or metal) binding peptides formed of both $(\text{Glu-Cys})_n\text{-Gly}$ or $(\gamma\text{-Glu-Cys})_n\text{-Gly}$ have been found indistinguishible (Bae and Mehra 1997; Satofuka et al. 2001).

PCs are found in many higher plants, several fungi, including *Schizosaccharomyces pombe, Candida glabrata* and *Mucor racemosus* (Grill et al. 1986b; Mehra et al. 1988; Miersch et al. 2001), algae, bryophytes, pteridophytes and gymnosperms (Gekeler et al. 1988, 1989). A thorough survey of plant kingdom for ability to bind metal through PC or iso-PC formation revealed that over 200 plant species investigated ranging from algae to orchids produce these metal complexing peptides. Various examples showing synthesis of PC in response to metals have been summarized in Table 1.

Recently it has been reported that *Salix viminalis* does not synthesise PCs upon exposure to heavy metals. In various clones of *Salix,* having different metal tolerence, exposed to heavy metals (Cd, Cu, Ni, Pb & Zn), for both short and long-term durations, no detectable level of PCs was synthesised (Landberg and Greger 2004). This plant is, thus, supposed to be first exception in plant kingdom that would fail to complex heavy metals by PCs. Earlier azuki bean (*Vigna angularis*) was also reported not to synthesize PC upon exposure to Cd besides having GSH (Inouhe et al. 2000). Later, it was found that only hGSH can be detected in this plant and homophytochelatins (hPCs) are formed when azuki beans are challenged with heavy metal such as Cd (Oven et al. 2001).

In some plants and microorganisms, a few structural variants of PCs have been identified. PCs fall into five main classes containing γ-Glu-Cys repeats but different C-terminal residues. These are canonical PC, $[\gamma\text{-Glu-Cys}]_n$-Gly with C-terminal glycine, homo-PC, [iso-(PC)-β-alanine] with C-terminal β-alanine, hydroxymethyl-PC, [iso-(PC)-Serine] with C-terminal serine, Iso-PC, [iso-(PC)-Glu] with C-terminal glutamic acid, and des-Gly PC, [des-(Gly-PC)] (Zenk 1996). The distribution and abundance of these classes differ between species (Rauser 1999) with exception of canonical and des-Gly PC, which are ubiquitous in PC containing organisms.

Homo-PCs are present in many legumes, such as *Vicia faba, Pisum sativum, Phaseolus vulgaris, Glycine max* etc. *Glycine max* have been shown to contain homo-GSH and produce high amount of homo-PCs and but not PC upon exposure to $Cd^{2+}$ (Grill et al. 1986a). Oven et al. (2002b) concluded that the presence of the substrate (GSH and its isoforms) and not the specificity of the enzyme determine the nature of PCs synthesized i.e. PC or hPC in any particular species.

Besides detoxification of heavy metals, Chen et al. (1997) also suggested some other essential functions of PCs in tomato cells and plants. PCs have been suggested as possible sulfur carriers during sulfate reduction and sulfur metabolism (Steffens 1990). Zn-induced PC synthesis not only enhances algal tolerance towards heavy metals such as Cd, Hg, Cu, Pb and arsenite, but also towards oxidative stress caused by hydrogen peroxide or paraquat serving as a strong scavanger of hydrogen peroxide and superoxide radicals (Tsuji et al. 2002). PCs seem to be also involved in transport of metals from root to shoot (Gong et al. 2003).

**Table 1.** Induction of Phytochelatins in various plants from algae to higher plants in response to different metals

| Plant Species | Metal | Species of PCs | Remark | Reference |
|---|---|---|---|---|
| **Algae** | | | | |
| *Dunaliella tertiolecta* | Cd and Zn | $PC_2$ - $PC_6$ | Algal cells were exposed to 200 μM Zn or 400 μM Cd for 24 h. Amount of phytochelatins synthesized was greater in response to Zn though PCS was more strongly activated by Cd. PCs showed to play a very strong role in mitigation of oxidative stress. | Tsuji et al. (2002); Tsuji et al. (2003) |
| *Dunaliella tertiolecta* | Cd and Zn | $PC_2$ - $PC_5$ | Alga was exposed to Cd (100-600 μM) and Zn (100 -600 μM) for 24 hr. Interestingly amount of PC synthesized was significantly higher in Zn treatment compared to Cd. | Hirata et al. (2001) |
| *Euglena gracilis* (Euglenophyceae), *Fragillaria crotonensis* (Chrysophyceae), *Navicula pelliculosa* (Bacillariophyceae), *Bumilleriopsis filiformi* (Xanthophyceae), *Chlamydomonas reinhardii, Chlorella fusca, Monoraphidium minutum* (, *Scenedesmus acutiformis, Stichococcus bacillaris* (Chlorophyceae), *Sargassum muticum* (Phaeophyceae), *Porphyridium cruentum* (Rhodophyceae) | Cd | In varying amounts and content from $PC_2$ - $PC_6$ | The algae were exposed to 20 μM Cd during their growth period for 2 to 10 days | Gekeler et al. (1988) |
| *Pheodactylum tricornutum* | Cd and Pb | $PC_2$ - $PC_6$ | 10 μM Cd or Pb treatment for 6 h. Metal-$PC_n$ complexes were identified. | Scarano and Morelli (2002) |

| | | | | |
|---|---|---|---|---|
| *Stichococcus bacillaris* | As | $PC_2$ - $PC_3$ | After 24 h in response to 100 μM As(V) or As(III) | Pawlik-Skowrońska et al. (2004) |
| *Stichococcus bacillaris* | Pb | $PC_2$ - $PC_4$ and their des-Gly derivatives | Cells were exposed to 10 μM Pb for 24 h. Level of thiol peptides was measured at different pH (5-8.5), and various concentrations of hardness cations (Ca, Mg), orthophosphate, chloride, citrate and humic acid. | Pawlik-Skowrońska (2002) |
| *Stigeoclonium tenue* | Zn | $PC_2$ - $PC_4$, and Novel Phytochelatin -related peptiedes (P1-P3) | Zn-tolerant alga produced, after long exposure period of 6 weeks to 30 μM Zn, phytochelatins (approximately 6 μmol SH per g DW) and novel thiol peptides (approximately 31 μmol SH per g DW). Synthesis of the novel-thiol peptides was 22-fold higher in tolerant strains than sensitive strains. These novel peptides contained one cysteine residue more than phytochelatins and differed from each other by one γ-Glu-Cys unit. | Pawlik-Skowrońska (2003) |
| *Thalassiosira weissflogii* | Cd, Pb, Cu and Ni | $PC_2$ - $PC_4$ | At Cd concentrations of < 1pM and 1nM or less concentrations for other metals | Ahner et al. (1994) |
| **Fungi** | | | | |
| *Neurospora crassa* | Cd | $PC_2$ | Grown in 100 μM Cd for 7 days | Kneer et al. (1992) |
| *Saccharomyces cerevisiae* | Cd | $PC_2$ | In response to 250 μM Cd | Kneer et al. (1992) |
| *Schizosaccharomyces pombe* | Cd | $PC_2$ - $PC_8$ | Yeast cells were treated with 1 mM Cd for 24 h | Grill et al. (1986a) |
| *Mucor racemosus, Articulospora tetracladia* | Cd, Cu, Zn | $PC_2$, $PC_3$ | 100 μM of Cd, Cu and Zn | Miersch et al. (2001) |

| | | | | |
|---|---|---|---|---|
| **Bryophyta** | | | | |
| *Marchantia polymorpha* | Zn | $PC_2$, $PC_3$ | Zn, 100 μM for 4 days | Gekeler et al. (1989) |
| **Pteridophyta** | | | | |
| *Selaginella viticulosa* (Selaginellales), *Lycopodium viticulosa* (Lycopodiales), *Equisetum giganteum* (Equisetales), *Azolla filiculoides* (Hydropteridales) | Cd, or Zn | In varying amounts and contents ranging from $PC_2$ - $PC_6$ | In natural condition substrate irrigated with 1 mM Al, 1 mM Zn, and 100 μM Cd after 4 days, In *Equisetum* suspension culture tissue was exposed to 100 μM Cd after 2 days | Gekeler et al. (1989) |
| **Lichens** | | | | |
| *Xanthoria parietina, Physconia grisea, Physia adscendens* | Cd, Pb, and Zn | $PC_2$ - $PC_4$, des-Gly-$PC_2$ | Exposed to 18, 36 and 54 μM concentrations of three metals separately for 24 hr. Only the photobiont partner (green alga *Trebouxia)* was able to synthesize PCs, not the mycobiont partner. | Pawlik-Skowrońska et al. (2002) |
| **Gymnosperms** | | | | |
| *Cycas revoluta* (Cycadatae), *Gingko biloba* (Gingkoatae), *Abies grandis, Abies alba, Picea abies, Pinus pinea, Pinus sylvestris* (Pinales) | Cd | In varying amounts and contents ranging from $PC_2$ - $PC_5$ | Differentiated plants/ suspension culture cells were exposed to 100 μM Cd for 3 days | Gekeler et al. (1989) |
| **Angiosperms** | | | | |
| *Anthemis arvensis* (Asterales), *Sinapsis alba* (Capparales), *Linum usitatissimum* (Geraniales), *Laurus nobilis* (Laurales), *Cannabis sativa* (Urticales), *Minuartia verna* (Caryophyllales), *Eucommia ulmoides* | Cd | In varying amounts and contents ranging from $PC_2$ - $PC_6$ | - | Gekeler et al. (1989) |

| | | | | |
|---|---|---|---|---|
| (Eucommiales), *Capsicum annuum* (Solanales), *Viola calaminaria* (Violales), *Rauvolfia serpentine* | | | | |
| *Brassica juncea* (Brassicaceae) | Cd | - | 6 week old plants were exposed to 25 μM Cd for 5 days | Heiss et al. (2003) |
| *Brassica oleracea, Lycopersicon esculentum, Zea mays, Silene cucubalus, Eichhornia crassipes, Agrostis tenuis* | Cd | $PC_2$ - $PC_6$ in *B. oleracea, L. esculentum,* $PC_2$ - $PC_5$ in *S. cucubalus,* and $PC_2$ - $PC_4$ in others | *B. oleracea* and *E. crassipes* were treated for 3 weeks with 90 μM Cd, and others were exposed for 3 days to 20 μM Cd. | Grill et al. (1987) |
| *Cicer arietinum* | Cd, As | $PC_2$ - $PC_5$ and $hPC_2$, $hPC_3$ | In response to 5 μM Cd exposure for 3 days in 5 day old seedlings | Gupta et al. (2002) and (2004) |
| *Cicer arietinum, Astragalus spp., Coronilla varia, Galega officinalis, Lens culinaris, Lotus ornithopodioides, Melilotus alba, Ononis natrix, Trifolium spp., Trigonella spp.* | Cd | Various hPCs and PCs synthesized | Plants exposed to 20 μM Cd for 4 days | Grill et al. (1986a) |
| *Cuscuta reflexa* | Cd | $PC_3$ - $PC_4$ | Upto 500 μM Cd in Callus, seedlings after 4 day | Srivastava et al. (2004) |
| *Cytisus striatus* | As (V) | $PC_2$ - $PC_5$ | Non-metallicolous and mine plants were exposed for 7 days in presence of different concentration of Phosphorus to a maximum of 64 μM As (V). | Bleeker et al. (2003) |
| *Holcus lanatus* | As | $PC_2$ - $PC_4$ | Different clones were exposed for 7 days to their | Hartley-Whitaker et |

| | | | | |
|---|---|---|---|---|
| | | | own arsenate $EC_{50}$ concentration, that was more than 1000 μM for the most tolerant clone, and only 3 μM for the most sensitive clone. | al. (2001) |
| *Hydrilla verticillata* | Cd | $PC_2$ - $PC_3$ | The plants were exposed to 2.5 and 10 μM Cd for 72 h. | Tripathi et al. (1996) |
| *Hydrilla verticillata* | Pb | $PC_2$ - $PC_3$ | The plants were exposed to 2.5 and 10 μM Pb for 24 and 96 h. | Gupta et al. (1995) |
| *Hydrilla verticillata* and *Vallisneria spiralis* | Hg | $PC_2$ - $PC_3$ | The plants were exposed to 0.25 and 1.0 μM Hg for 24 and 96 h. Synthesis of phytochelatins was observed in both roots and leaves. PC-Hg complexes were also reported in both plants. | Gupta et al. (1998) |
| *Lemna gibba* (Arales), *Phoenix dactylifera* (Arecales), *Asparagus officinales* (Asparagales), *Ananas comosus* (Bromeliales), *Commelina graminifolia* (Commelinales), *Cyperus esculentus* (Cyperales), *Triticum aestivum* (Poales), *Musa ensete* (Zingiberales) | Cd | In varying amounts and contents ranging from $PC_2$ - $PC_5$ | Exposed to 20 μM Cd for 3 days | Gekeler et al. (1989) |
| *Lycopersicon esculentum* | Cd | $PC_2$ - $PC_5$ | Suspension-Cells were exposed to 400 μM Cd | Inouhe et al. (1991) |
| *Nicotiana rustica* | Cd | Cadmium binding peptides (CdBPs) | Exposed to 20 μM Cd for 1 week | Vogeli-Lange and Wagner (1990) |
| *Oryza sativa* | Cu, Cd | $PC_2$ - $PC_3$ | *In vitro* study characterizing PCS of rice seedlings | Yan et al. (2000) |
| *Phaseolus vulgaris* | Pb | $hPC_2$ - $PC_4$ | After 96 h in response to 1 mM Pb | Piechalak et al. (2002) |

| *Phaseolus vulgaris, P. coccineus, P. aureus, P. lunatus, P. multifloris, Canavalia ensiformis, Cajanus cajan, Dolichos lablab, Glycine max, Glycine clandestina, Erythrina crista-galli, E. melanacantha, E. coralloides* | Cd | Various hPCs synthesized | Plants exposed to 20 μM Cd for 4 days | Grill et al. (1986a) |
|---|---|---|---|---|
| *Pistia strattiotes* | Cd | $PC_2$ - $PC_3$ | The plants were exposed to 2.5 and 10 μM Cd for 96 h. Synthesis of phytochelatins was observed in both roots and shoots. | Rai et al. (1995) |
| *Pisum sativum* | Pb | $PC_2$ - $PC_4$, $hPC_4$ | After 96 h in response to 1 mM Pb | Piechalak et al. (2002) |
| *Pteris vittata* | As | $PC_2$ | Fern plants were provided with 500 ml of 13.3 mM Sodium Arsenate solution at interval of 2 weeks for five times. The presence of a novel As complex was also reported. | Zhang et al. (2004) |
| *Pteris vittata* | As | Unidentified thiol | Plants were treated with 0-600 mg/Kg Sodium Arsenate for a period of 1-21 days. Arsenic accumulation and increase in the acid soluble thiol content was correlative. | Cai et al. (2004) |
| *Rauvolfia serpentina* | Cd | $PC_2$ – $PC_7$ | Suspension cells were exposed to 200 μM Cd | Kneer and Zenk (1997) |
| *Rauvolfia serpentina* | As (V), As (III) | $PC_2$ - $PC_4$ | Cell suspension culture were exposed to 100 μM Arsenite or Arsenate for 4 d. | Schmöger et al. (2000) |
| *Rosa canina* | Cd | $PC_2$ - $PC_8$ | Suspension cells were exposed to 200 μM Cd for 4 days | Grill et al. (1986a) |

| | | | | |
|---|---|---|---|---|
| *Rubia tinctorum* | Ag, $As^{3+}$, $As^{5+}$, Cd, Cu, Ga, Hg, In, Ni, Pb, Pd, Se, and Zn | Various PC species induced in different concentrations by different metals | Cell cultures were exposed for 3 days with various metals in concentration ranging from 10 - 1000 μM | Maitani et al. (1996) |
| *Silene cucubalus* | Cd | $PC_2$ - $PC_3$ | *In vitro* study, crude extracts of cells contained 1 mM glutahione and 0.1 mM Cd ions. | Löeffler et al. (1989) |
| *Silene vulgaris* | Zn | $PC_2$ - $PC_4$ | Zn-sensitive and Zn-tolerant plants were exposed their respective $EC_{50}$ Zn concentration for 3 d. Surprisingly sensitive plants synthesized more PCs than tolerant plants. | Harmens et al. (1993) |
| *Silene vulgaris* | Cd | $PC_2$ - $PC_4$ | The root tips of Cd-tolerant plants exhibit a lower rate of PC production accompanied by a lower rate of longer chain PC synthesis than those of Cd-sensitive plants. At the same Cd exposure level, stable Cd-PC complexes are more rapidly formed in the roots of Cd-sensitive plants than in those of tolerant plants. Thus tolerance is not correlated to the PC production. | de Knecht et al. (1994) |
| *Tamarindus indica, Bauhinia purpurea, Caesalpinia sappan, Cassia angustifolia* (Caesalpiniaceae), *Acacia karroo, Neptunia oleracea, Albizzia lophanta, Mimosa pudica* (Mimoseae), *Onobrychis viciifolia, Lonchocarpus violaceus, Pterocarpus officinalis,* | Cd | In varying amounts and contents $PC_2$ (and higher homologues), and $hPC_2$ (and higher | - | Gekeler et al. (1989) |

| | | | | |
|---|---|---|---|---|
| *Astragalus gummifer, Crotolaria crassipes, Lotus ornithopodioides, Ononis natrix, Canavalia ensiformis, Clitoria ternatea, Erythrina coralloides, Baptisia australis, Sophora japonica, Melilotus alba Lathyrus ochrus* (Fabaceae) | | homologues) | | |
| *Thlaspi caerulescens* | Cd | $PC_2$ - $PC_4$ | The plants were exposed to 500 μM Pb for 14 d. Synthesis of phytochelatins was observed in both roots and leaves. Upon BSO treatment PC synthesis reduced drastically. | Wójcik et al. (2005) |
| *Vallisneria spiralis* | Pb | $PC_2$ - $PC_3$ | The plants were exposed to 2.5 and 10 μM Pb for 24 and 96 h. Synthesis of phytochelatins was observed in both roots and leaves. | Gupta et al. (1999) |
| *Vicia faba* | Pb | $PC_2$ - $PC_4$ | After 96 h in response to 1 mM Pb | Piechalak et al. (2002) |
| *Vigna angularis* | Cd | des-Gly- $PC_2$, $hPC_2$ - $hPC_3$ | Plants were grown for 7 days in presence of 10 μM Cd. | Oven et al. (2001) |
| *Zea mays* | Cd | $PC_2$ - $PC_4$, $hPC_2$ - $hPC_3$, des-Gly$PC_2$ - des-Gly $PC_4$ | Maize roots were exposed to 3 μM Cd for 2 and 7 day. Duration caused no change in profile. | Rauser and Meuwly 1995 |

## 3. Biosynthesis of Phytochelatins

PCs are synthesized enzymatically by using GSH as a substrate. The enzyme catalysing the reaction is specifically called as γ-glutamylcysteine dipeptidyl transpeptidase (EC 2.3.2.15), given the trivial name Phytochelatin synthase (PCS) (Grill et al. 1989). Use of GSH as a substrate for PC formation is consistent with the finding of PC-deficient mutants of *S. pombe* and *A. thaliana*, both of which are deficient in GSH (Mutoh and Hayashi 1988, Cobbett et al. 1998).

Synthesis of PC is induced by the entry of metal ion into the cell. The induction of PC biosynthesis is reported by a variety of metals. PC inducing metals are $Ag^{+}$, $As^{5+}$, $Au^{+}$, $Bi^{3+}$, $Cd^{2+}$, $Cu^{2+}$, $Hg^{2+}$, $Ni^{2+}$, $Pb^{2+}$, $Sb^{3+}$, $Se^{4+}$, $Sn^{2+}$, $Te^{4+}$, $W^{6+}$, $Zn^{2+}$, $Fe^{2+}$, $Ga^{3+}$, $In^{3+}$, $Pd^{2+}$ and $La^{3+}$ (Grill et al. 1989; Maitani et al. 1996; He et al. 2004). Cd was found to be best activator metal tested followed by Ag, Bi, Pb, Zn, Cu, Hg and Au cations (Grill et al. 1989). However, in some cases, other metals like Zn (Tsuji et al. 2003) dominated Cd in inducing the synthesis PCs. They include majority of soft metals, which have high affinity for -SH groups, whereas hard metals like $Mg^{2}$, can not induce PC, because they do not show preference for a soft ligand like -SH. $Fe^{2+}$ is also reported to induce PC synthesis in tomato (Chen et al. 1997) but complexation of $Fe^{2+}$ to PC has not yet been reported.

The reaction catalyzed by PCS is given as:

$$(\gamma\text{-Glu-Cys})_n\text{-Gly} + (\gamma\text{-Glu-Cys})_n\text{-Gly} \xrightarrow{\text{PCS}} (\gamma\text{-Glu-Cys})_{n+1}\text{-Gly} + (\gamma\text{-Glu-Cys})_{n-1}\text{-Gly}\ (n \geq 1)$$

Specific metals or metal species may produce distinct response of phytochelatin formation and total thiol content. There has been found a distinct change in the amount of total thiol content in response to $As^{3+}$ and $As^{5+}$ and organic monomethylarsonic acid (MMA). In *Pteris* plants, MMA was the strongest inducer of thiols, followed by $As^{5+}$ and $As^{3-}$ (Tu et al. 2004). Hg abundantly induced $PC_2$ (Maitani et al. 1996, Grill et al. 1987), which may be attributed to linear configuration of Hg in coordination compounds. Cu favoured the synthesis of $PC_2$ whereas Cd synthesized $PC_3$ and $PC_4$ predominantly in *B. juncea* (Heiss et al. 2003). This difference was also seen in a study on *R. serpentina* cell cultures (Grill et al. 1987). Ca does not induce PCS activity, however in presence of Cd, Ca treatment has been shown to increase the PCS gene (LsPCS1) expression and to enhance plant tolerance towards Cd and its accumulation (He et al. 2004).

There are some fragmentary reports of some novel thiol peptides that are supposed to be related to PC and play a role in metal detoxification in these organisms. Zn-tolerant alga, *Stigeoclonium tenue,* produced PC (approximately 6 μmol SH per g DW) and three novel thiol peptides (approximately 31 μmol

SH per g DW), designated as P1, P2 and P3, after long exposure period of 6 weeks to 30 μM Zn. Synthesis of the novel-thiol peptides was 22-fold higher in tolerant strains than sensitive strains. These novel peptides contained one cysteine residue more than PC and differed from each other by one γ-Glu-Cys unit (Pawlik-Skowrońska 2003). Arsenic hyperaccumulator plant *Pteris vittata* produced an unidentified thiol and a novel complex was also isolated upon exposure to As (Zhang et al. 2004; Cai et al. 2004). The concentration of this unidentified thiol showed a very strong and positive correlation with As concentration in leaflet and rachis. The synthesis of unidentified thiol is specific to As toxicity and is not synthesized upon exposure to other metals like Cd, Cu, Pb, Hg, and Se. Transgenic *Arabidopsis,* overexpressing PCS, and wild type plants exposed to As resulted in expression of many unknown thiol products among which three were produced in 6-16 fold higher amounts than wild type (Li et al. 2004).

### 3.1 Characteristics of Phytochelatin Synthase Enzyme

Phytochelatin synthases (PCSs) have been characterized in a few plants to date. They share some of the common characteristics, but differ with each other substantially in some properties. Grill et al. (1989) firstly characterized PCS from *Silene cucubalus.* They stated it to be a protein complex of molecular mass approximaetly 100 kDa, having pH optima of 7.9, temperature optima of 35°C and isoelectric pH of 4.8. The Km for GSH was 6.7 mM in presence of 0.1 mM Cd. The enzyme was supposed to be constitutive and to be active also as a 50 kDA protomer. The purified enzyme showed a specific activity of 463 pkat/mg protein.

PCSs have been characterized in plants like *Arabidopsis, Triticum, Lycopersicon, Brassica, Oryza, Lactuca, Glycine max,* and in yeast *Schizosaccharomyces pombe* (Chen et al. 1997; Ha et al. 1999; Vatamaniuk et al. 1999; Clemens et al. 1999; Yan et al. 2000; Oven et al. 2002b; Heiss et al. 2003; He et al. 2004).

Tomato PCS shows pH optima of 8.0 and temperature optima of 35°C. Km value for GSH was 7.7 mM in presence of 0.5 mM Cd. PCS is present constitutively in tomato plants, however the enzyme shows regulated activity in cell culture in absence of Cd. In some plants PCS was present in roots and stem, but not in leaves and fruits. These data suggested some other role of PCS other than metal binding (Chen et al. 1997). PCS from fungus *Schizosaccharomyces pombe* (SpPCS) has been characterized by Ha et al. (1999). The enzyme is constitutively expressed, having a total size of 414 amino acids. The C-terminal region in AtPCS1 and SpPCS has 10 and 7 Cys residues, respectively.

Wheat PCS (TaPCS1) gene encodes protein of approximately 55 kDa of 500 amino acids showing transcriptional regulation upon exposure to Cd (Clemens et al. 1999). Rice PCS enzyme has a molecular mass of 100 kDa with an isoelectric point of 4.0 and pH optima of 7.5. However, the temperature optima of this enzyme is 55°C, which is very high. The enzyme is thermotolerant and is unstable under refrigeration (4 or -20°C) (Yan et al. 2000). Homophytochelatin synthase from *Glycine max* (GmhPCS1) has a pH optimum of 8.2±0.2, similar to AtPCS1. The temperature optimum is 35°C in both cases. The Km value for GSH was determined 15 mM for GmhPCS1 and 11 mM for AtPCS1, *Arabidopsis* PCS (Oven et al. 2002b). PCS of *B. juncea* has molecular mass of 54 kDa containing a total of 485 amino acids. The enzyme is constitutively synthesized, however longer duration treatments could cause an increase in protein levels which was due to post-transcriptional regulation (Heiss et al. 2003).

PCS of *Arabidopsis thaliana* (AtPCS1) has a molecular mass of 55 kDa containing 485 amino acids (Vatamaniuk et al. 1999). The enzyme is localized in leaves (at a very high frequency in leaf trichomes), roots, cotyledons, and stems, but not in root tips and root hairs. The absence of PCS in root hairs and root tips is supposed to be due to the low vacuolation, whereas presence of highly active biosynthesis of GSH and also 90-95% volume occupation of total cell volume by vacuoles is responsible for a very high amount of PCS in leaf trichomes. A second homologue of PCS gene in *A. thaliana* (AtPCS2) has also been characterized which has 84% homology with AtPCS1. Catalysing the production of Cd-PC compelxes might not be the physiological function of AtPCS2. The expression of AtPCS2 is weak in both shoots and roots of *Arabidopsis* as compared to that of AtPCS1 due to both low promoter activity and a low efficiency of translation of AtPCS2 mRNA (Lee and Kang 2005). Further, localization in a cellular compartment with significantly less available cadmium than the cytosol could explain why no PCs were formed upon cadmium exposure in *cad1-3* plants (Cazalé and Clemens 2001).

Recent studies also indicated the presence of PCS like protein in some cyanobacteria and nematode, *Caenorhabditis elegans* on the basis of database searches. The *Caenorhabditis* gene designated as CePCS1 encodes a hypothetical polypeptide of 371 amino acids (Clemens et al. 1999; Ha et al. 1999; Vatamaniuk et al. 1999).

The predicted PCS product of *Nostoc* alr0975 contains the conserved N-terminal domain but not the variable C-terminal domain found in eukaryotic PCSs. The recombinant alr0975 protein expressed in *E. coli* strongly catalysed the first step of PC synthesis where GSH is converted to γ-Glu-Cys by cleavage of Gly. The protein, however, only weakly catalysed the second step of the PC synthesis namely the transfer of γ-Glu-Cys moiety to an acceptor GSH molecule to form PCs (Tsuji et al. 2004). The alr0975 protein has only one conserved cysteine residue out of five in the N-terminal domain of PCS found in eukaryotes and this may explain why the protein showed very weak PCS activity.

However, further studies conducted on alr0975 protein by Harada et al. (2004) showed no PCS activity of the protein. Instead, this protein catalyzed only the conversion of GSH to γ-Glu-Cys. Unlike PC synthesis, the conversion of GSH to γ-Glu-Cys is not dependent on activation by metal cations. No evidence was found for the accumulation of PCs in *S. pombe* or *E. coli* expressing alr0975, or in cyanobacteria even after prolonged exposure to $Cd^{2+}$.

The database searches reveal other cyanobacterial sequences similar to alr0975 of *Nostoc* sp. PCC7120. It suggests that the proteins encoded by cyanobacterial genes may be progenitor or more primitive forms of PCS and may represent an early stage in the evolution of enzyme in photoautotrophic organisms (Tsuji et al. 2004). Moreover C-terminal domain of PCS varies much more widely among plant species. Thus the eukaryotic PCS may have evolved from the cyanobacterial protein by acquiring more Cys residues and C-terminal fusion with the another domain.

The homology between different PCSs could provide information about the phylogenetic evolution of enzyme. At the amino acid level, PCS from *B. juncea* (BjPCS1) displays 90% sequence identity with PCS from *A. thaliana* (AtPCS1) and *Thlaspi caerulescens* (TcPCS1). The PCS proteins from *A. thaliana* (AtPCS2), *Glycine max* (GmhPCS1), *Triticum aestivum* (TaPCS1), *Typha latifolia* (TlPCS1), and *Athyrium yokoscense* (AyPCS1) shared 77%, 65%, 59%, 55% and 50% sequence identity, respectively with BjPCS1, whereas only 30% similarity could be seen with PCS from *Caenorhabditis elegans* (CePCS1) and *Schizosaccharomyces pombe* (SpPCS) (Heiss et al. 2003). The putative protein product of gene alr0975 from *Nostoc* sp. PCC7120 has 36% identity to AtPCS1 (Tsuji et al. 2004). The two PCS from *Arabidopsis,* AtPCS1 and AtPCS2 are 84% identical (Cazalé and Clemens 2001). Heiss et al. (2003) confirmed, by comparing complete protein sequences of many PCSs, the earlier observation that sequence conservation in the putative catalytic N-terminal domain is much higher than in the variable C-terminal domain such as all analysed sequences displayed an N-terminal conserved motif with the consensus sequence [Q-T-G-x-G-H-F-S-P-x(11)-L-I-[LM]-D-V-A-R-E-K-Y-P-[PC]-[HY]-W-x(2)-L].

PCSs seem to perform some other cellular function. Characterization of enzymatic properties of PCS argues for two cellular functions of the enzyme: the formation of heavy metal binding peptides as a part of heavy metal detoxification system and secondly the degradation of glutathione-S-conjugates in the detoxification pathway of xenobiotics. Purification of glutathione-S-conjugates catabolising activity from the cell suspension culture of *S. cucubalus* indicated that PCS catalysed the first step of the pathway i.e. removal of carboxy terminal residue of the tripeptide GSH to give rise to S-Glu-Cys derivative in the plants (Grill et al. 1989). Heterologously expressed *Arabidopsis thaliana* PCS efficiently converted S-bimaneglutathione to Sbimane-Glu-Cys (Beck et al. 2003). No further products, such as S-derivative phytochelatin, were observed. Mechanistically the formation of phytochelatin is the result of γ-Glu-Cys transpeptidation onto GSH or derivatives thereof while the catabolic function reflects transpeptidation of S-Glu-Cys adducts onto the acceptor molecule water. Thus the dipeptidyl transferase seems to fulfill besides the established function in heavy metal detoxification, a role in GSH metabolism of green plants and possibly in other organisms expressing a functional PCS.

### 3.2 Mechanism of Activation of Enzyme

Susceptibility of PCSs to activation by heavy metals is physiologically very crucial, as it is this activation specificity that organisms produce PCs upon exposure to heavy metals, which are able to poison other enzymes. To elucidate the mechanism of PCS activation, various experiments have been conducted by different groups (Grill et al. 1989; Löeffler et al. 1989; Klaphek et al. 1995; Ha et al. 1999). Initially it has been proposed that PCS is activated by heavy metals and kinetic analysis of PCS catalysed reaction indicated that synthesis of PCs consists of two distinct steps; i- formation of γ-Glu-Cys concomitant with the cleavage of glycine from GSH, ii- transfer of γ-Glu-Cys unit from the enzyme to acceptor molecule i.e. GSH or oligomeric PC peptides ($PC_n$). The two-step reactions may be given as:

Step (1): γ-Glu-Cys-Gly + PCS ⟶ γ-Glu-Cys-PCS + Gly
Step (2): γ-Glu-Cys -PCS + (γ-Glu-Cys $)_n$-Gly ⟶ (γ-Glu-Cys$)_{n+1}$-Gly + PCS ($n \geq 1$)

Identification of PCS gene in various organisms provided an additional information regarding the mechanism of PCS activation. *Arabidopsis cad1-5* mutant which lacks the C-terminal domain of AtPCS1 could generate 33% of PC synthesis compared to wild type *in vivo* (Howden et al. 1995). These results indicate that the N-terminal domain of PCS is the catalytic domain and is essential for the generation of PCs and that the C-terminal domain is not absolutely required for catalysis. Cobbett (1999) proposed a model for the mechanism of PCS activation in which PCS invokes direct metal binding at several sites in the enzyme. It is proposed that the strongly conserved N-terminal half of the enzyme is responsible for core catalysis and that activation arises from the binding of metal ions to Cys residues, in this domain (Cobbett 2000). The presence of five conserved Cys residues, two of which are vicinal, and consequently optimally disposed for the co-ordination of ions, such as $Cd^{2+}$, $Cu^{2+}$, and/or $Hg^{2+}$ in the N-terminal halves of eukaryotic PCSs, is consistent with this notion, as is the observation that the three most extreme *Arabidopsis cad1* (Howden et al. 1995) alleles have amino acid substitutions in this region

(Ha et al. 1999). An extension of this model, proposed to ascribe a role to the more sequence-divergent C-terminal half of the molecule and to account for the properties of the least extreme *cad1* allele, *cad1-5* - a nonsense mutation causing premature termination and deletion of the C-terminal segment, is the concept of a C-terminal metal-sensing domain whose multiple Cys residues bind heavy metals and bring them into contact with the putative activation site within the N-terminal, catalytic half of the molecule.

A substantially different mechanism has been proposed based on a study conducted with recombinant *Arabidopsis* AtPCS1, in which metal binding to the enzyme is not primarily responsible for catalytic activation, but rather a Cd-$GS_2$ complex is the substrate used (Vatamaniuk et al. 2000). More specifically, Cd-$GS_2$ thiolate complex (or Cd-$PC_n$ complex) and free GSH can act as γ-Glu-Cys acceptor and donor, respectively, in the AtPCS1 catalysed dipeptidyl transfer. The complexes formed between heavy metals and thiol compounds are among the most stable known complex (Rabenstein 1989). Under the conditions in which PCS catalyses high rates of PC synthesis from GSH, the concentration of free $Cd^{2+}$ is very low and more than 98% of the total $Cd^{2+}$ added to the reaction medium is associated with GSH as the bidentate thiolate, bisglutathionato cadmium (Cd.$GS_2$). When assayed in media devoid of metal salts, AtPCS1 catalyses the net synthesis of S-alkyl-PCs from S-alkyl glutathione derivatives (Vatamaniuk et al. 2000). This suggested that blocked thiols are also substrates in which both free GSH and its metal thiolate are required as donor and acceptor, respectively. However, this reaction was metal dependent in analyses with homogenous enzyme preparations (Oven et al. 2002b). This analysis and a subsequent AtPCS1 characterisation (Beck et al. 2003) clearly support the requirement of heavy metal ions for PCS activity.

The observed activation of PCS by $Mg^{2+}$ (Vatamaniuk et al. 2000) is not easily explainable as PCS is activated essentially by metal-thiolate interaction. According to Pearson's rule, $Mg^{2+}$ (a hard metal) does not show preference for -SH group (soft ligand), thereby it would not be able to form a thiolate complex.

According to the model, the PC synthesis is terminated when metal-PC complexes are removed from cytosolic pool into the vacuole (Vatamaniuk et al. 1999), which could not explain the observed termination of PC biosynthesis reaction *in vitro* (Löeffler et al. 1989). It was suggested earlier that PC biosynthesis terminates when GSH or apo-PCs compete with thiolates for high affinity sites of the enzyme or when maximum substrate inactive metal-PC complexes are formed (Grill et al. 1989). The synthesis of PCs was terminated by addition of EDTA or apo-PCs in the reaction media *in vitro* and the enzyme showed immediate inactivation. It is concluded that activity of PCS is regulated by the reaction product, PCs (Löeffler et al. 1989).

Oven et al. (2002b) showed that the presence of thiols, in the metal containing PCS reaction mixture, was decisive for AtPCS1 activation. In the absence of thiols, free metal ion cannot activate PCS even if blocked thiols (other than metal blocked e.g. S-methyl-GSH) are present. However, other heavy metal thiolate complexes, for instance, those of cadmium with 2-mercaptoethanol or cysteine, that are not substrate for the enzyme, contribute to the activation of *Glycine max* hPCS1 (GmhPCS1) strongly arguing for the participation of metal ions via interfering with a metal activation domain of the enzyme.

Recently, Vatamaniuk et al. (2004) confirmed that PCS is a dipeptidyltransferase by using radioactive isotope labeled substrates and showed that the first step of PC synthesis involves γ-Glu-Cys acylation at two different sites within the enzyme. At first step PCS is acylated by γ-Glu-Cys independent of Cd, with simultaneous cleavage of Gly from GSH. On the other hand, Cd dependent γ-Glu-Cys acylation of the enzyme takes place at the second step and γ-Glu-Cys acylation at both sites is essential for net synthesis of PC.

The recent identification of PCS-like proteins in several prokaryotes having high homology to the N-terminal domain of eukaryotes PCS and absence of four out of five of the conserved cysteine residues in the eukaryotic PCS sequence provided an additional tool for understanding the mechanism of enzyme activation (Tsuji et al. 2004). The comparative study and functional analysis of various mutants of NsPCS1 and AtPCS1, led Tsuji et al. (2005) to propose that:

a. Presence of heavy metal ion is essential for the first step of reaction catalysed by AtPCS1, but not for the NsPCS1.
b. The amineterminal region 1-221 contains the catalytic domain of the PCS.
c. Out of five-conserved cysteine residues in N-terminal domain, Cys 56 (in eukaryote) or Cys 70 (in prokaryote) is associated with the first step of PC synthesis.
d. C-terminal region of AtPCS1 stabilizes the N-terminal region and maintains its active state.
e. The divergence in AtPCS1 and NsPCS1 in respect to the activation by heavy metal may be due to differences in their three-dimensional structure. NsPCS1 may be able to maintain an active conformation in absence of heavy metal, while AtPCS1 requires direct binding to Cd or Cd-$GS_2$ complex for the folding into an active confirmation.

Thus, it is proposed that in absence of heavy metal, PCS adopts an inactive conformation and binding of metal-thiolate complex induces its folding into a three dimensional active conformation in which thiol reductants may contribute by reducing the intramolecular disulfide bonds. In the active PCS, a donor molecule, such as free GSH or PCs binds to Cys 56 and γ-Glu-Cys unit is cleaved which is immediately transferred to an acceptor molecule.

Further, since NsPCS1 catalyzes the deglycination of GSH to form γ-Glu-Cys as major product and weakly synthesizes PCs despite having only 22-30%

sequence identity with N-terminal domains of eukaryotic enzyme, it thus contains catalytic domain of eukaryotic PCS. In this backdrop, the stereo structure of NsPCS1 in its native and γ-Glu-Cys acylated forms have been recently presented (Vivares et al. 2005). Crystal structure revealed that PCS belong to papains family of cysteine proteases and is a dimer. The catalytic action involves a triad of Cys-70, His-183 and Asp-201 in prokaryotic PCS, which is equivalent to Cys-56, His-162 and Asp-180 in eukaryotic PCS. An oxyanion hole, comprising of Cys-70 and Gln-64 in prokaryote, is involved in deglycination of the GSH, the donor molecule for the first step of the PC synthesis. Subsequently, an ideally placed water molecule can attack the thioester bond and liberates γ-Glu-Cys. For transpeptidation, an acceptor GSH should bind in a putative site close enough to the first GSH binding site to allow PC synthesis (Vivares et al. 2005, Rea 2006). Further, structural studies on eukaryotic PCS might reveal the structural reasons why eukaryotic PCSs are more efficient in PC synthesis than the prokaryotic enzyme.

### 3.3 Domain Organisation of Phytochelatin Synthase Enzyme

No structural information on eukaryotic PCS enzyme is yet available. It is known that the active site region is located in a more conserved N-terminal region of PCS whereas various, but supposedly less critical roles, have been proposed for the C-terminal region (Cobbett 2000). To gain insight in metal binding domain of PCS enzyme, a thorough study has been done through peptide scan technique on two diverse PCSs, SpPCS and TaPCS1. These were synthesized and incubated with $^{109}$Cd and based on Cd binding pattern, the distinct binding sites and binding motifs have been localized. A strong correlation was found between binding activity and degree of conservation among known PCSs. The functional role of several cysteine suggested the presence of five functionally essential cysteine residues in the N-terminal catalytic part of PCS and additional binding sites at the C-terminal domain though not essential for activity. The detection of Cd even in presence of millimolar concentration of GSH or a vast excess of the non-activating divalent cation, such as Co, suggests that the affinity of binding site in PCS proteins localized by peptide scanning could be sufficiently high to be of relevance *in vivo*. This is in agreement with the notion that Cd binding occurs in both the essential catalytic N-terminal half of PCS enzyme as well as the C-terminal "sensor" half (Maier et al. 2003).

A limited proteolysis analysis of the PCS enzyme from *Arabidopsis* (AtPCS1) has given insight into the structural/functional organization of PCS (Ruotolo et al. 2004). Two N-terminal fragments ending at positions 372 (PCS_Nt1) and 283 (PCS_Nt2) were produced sequentially upon $V_8$ protease digestion, without any detectable accumulation of corresponding C-terminal fragments. The two N-terminal fragments were functionally characterized and the results of *in vivo* and *in vitro* functional assays reveal that the core PCS_Nt2 fragment is biosynthetically active in the presence of Cd ions and supports phytochelatin formation at the rate that is albeit five fold lower than that of full length AtPCS1. The loss of C-terminal region, however, substantially decreases the thermal stability of the enzyme and impairs PCs formation in the presence of certain heavy metals e.g. Hg and Zn, but not Cd and Cu. The differential catalysis phenomenon was shared by PCS_Nt2 and by its precursor fragments PCS_Nt1, which on the other hand was almost as stable and biosynthetically active (in presence of Cd) as the full-length enzyme. AtPCS1 thus appears to be composed of a protease resistant (and hence presumably highly structured) N-terminal domain, flanked by an intrinsically unstable C-terminal region. The most upstream part of such a region (positions 284-372) is important for enzyme stabilisation, whereas its most terminal part (373-485) appears to be required to determine enzyme responsiveness to a broader range of heavy metals.

## 4. Mechanism of Action of Phytochelatins

### 4.1 Formation of Metal-Phytochelatin Complexes

PC-metal complexes have been revealed by gel filtration chromatography in various plants maninly with Cd e.g. *Rauvolfia serpentina* (Grill et al. 1985), *Chlorella fusca* (Gekeler et al. 1988), tobacco (Vogeli-Lange and Wagner 1990), *Neurospora crassa* (Kneer et al. 1992), *Brassica juncea* (Speiser et al. 1992), *Silene vulgaris* (de Knecht et al. 1994), Maize (Rauser and Meuwly 1995), *Silene cucubalus* (Kneer and Zenk 1997), *Phaeodactylum tricornutum* (Scarano and Morelli 2002) and also with Pb e.g. *Hydrilla verticillata* (Gupta et al. 1995), *Vallisneria spiralis* (Gupta et al. 1999), *Phaeodactylum tricornutum* (Scarano and Morelli 2002), As e.g. *Rauvolfia serpentina* (Schmöger et al. 2000), and Hg e.g. *Hydrilla verticillata* and *Vallisneria spiralis* (Gupta et al. 1998). In *Rubia tinctorum* cultures, Maitani et al. (1996) reported the induction of PCs and formation of metal-PC complexes of Ag, Cd and Cu. Cu was also bound to PCs induced by other metals like As, Ag, and Cd.

Regarding stoichiometries and crystallographic structures of metal-PC complex, many important studies have been done performing *in vitro* studies with metal (e.g. Cd, Pb, Ag, Hg, and Cu) and PCs.

UV/visible and circular dichroism (CD) spectroscopy studies of binding of Pb(II) to $PC_2$, $PC_3$ and $PC_4$ revealed that $PC_2$ and $PC_3$ bound one metal ion per peptide molecule, whereas $PC_4$ formed two distinct species with stoichiometries of one and two Pb(II) ions per peptide molecule, respectively. The optical spectra of $Pb(II)_1$-($\gamma$-Glu-Cys)$_4$-Gly were similar to those of $Pb(II)_1$-($\gamma$-Glu-Cys)$_3$-Gly, whereas the spectra of $Pb(II)_2$-($\gamma$ -Glu-Cys)$_4$-Gly were similar to those of $Pb(II)_1$-($\gamma$-Glu-Cys)$_2$-Gly. Pb(II) may thus exhibit multiple coordination in longer chain PCs (Mehra et al. 1995).

The *in vivo* and *in vitro* studies on As complexation by PCs demonstrated a stoichiometry of metal to Cys residues provided by PCs of approximately 1 to 3. The formation of reconstituted As-$PC_2$ complex and corresponding mass signal identified by ESI-MS analyses is in perfect accordance with the structural model of three thiol groups provided by two $PC_2$ molecules that coordinate As (Schmöger et al. 2000). Earlier NMR structural analyses of As-GSH complexes generated by the incubation of the tripeptide with arsenite revealed a coordination of $As^{3+}$ by three peptide molecules (Scott et al. 1993). Arsenate also coordinated in the same way due to reduction of $As^{5+}$ to $As^{3+}$ by GSH (Jocelyn 1972; Schmöger et al. 2000). In bacteria and yeasts mechanism of $As^{5+}$ detoxification involves its reduction to $As^{3+}$ by arsenate reductases and then its subsequent transport to vacuole or exclusion (Ghosh et al. 1999). Though arsenate reductases have not yet been discovered in plants, the same mechanism is supposed to take place.

Analysis of the biochemical fate of As in *Brassica juncea* revealed (Pickering et al. 2000) storage of As as an $As^{3+}$-tris-thiolate complex in the shoot. In root also majority of As occurred as $As^{3+}$-tris-thiolate complex, which is indistinguishable from that found in shoot and from $As^{3+}$-tris-GSH. The thiolate donors are thus supposed to be GSH or PC. These studies implied that the As:PC ratio as 1:3 in bound form.

The stoichiometry of Ag(I) and PC is strongly pH dependent , at neutral pH, $PC_2$, $PC_3$ and $PC_4$ bind 1.0, 1.5 and 4 equivalents of Ag(I), respectively, however, at lower pH (pH 5.0 or lower) binding capacity increases and approaches to 1:1 ratio of Ag/SH. Similar binding of Ag(I) with GSH was also found. The increased Ag(I) binding to PCs at lower pH is of more physiological significance, as these peptides accumulate in vacuole in acidic pH (Mehra et al. 1996a). $PC_2$ and Hg(II) binding stoichiometry is also reported to be 1:1 by optical spectroscopic studies. However $PC_3$ binds to Hg(II) as two distinct species having stoichiometries of around 1.25 and 2.0 Hg(II) per peptide molecule. Similarly $PC_4$ also shows two distinct binding species with stoichiometries around 1.25 and 2.5 as observed by UV/visible spectroscopy and CD Spectroscopy. The Hg(II) binding stoichiometry was found pH independent. The RP-HPLC studies showed a GSH mediated transfer of Hg(II) to PCs and that of lower PCs to higher PCs (Mehra et al. 1996b).

Sulfide ions play an important role in efficacy of Cd detoxification by PC in some plants and yeasts (*S. pombe* and *Candida glabrata).* The incorporation of sulfide into high molecular weight complexes increases the amount of Cd per PC molecule and also the stability of complex. Some complexes with high ratio of sulfide and Cd consist of aggregates of 20Å diameter particles which themselves consist of CdS crystallite core coated with PCs (Dameron et al. 1989). Characterization of Cd sulfide nanocrystallites (CdSNCs) isolated from *S. pombe* and *Candida glabrata* showed the particles to consist of Cd, PCs and sulfide with diameter of approximately 20±3 Å and 18 Å, respectively. Ratios of sulfide to Cd were 0.7 and 0.6 for CdS from *C. glabrata* and *S. pombe* respectively. *S. pombe* CdSNCs did not easily coalesce and CdSNCs capped with (Glu-Cys)$_3$Gly or (Glu-Cys)$_4$Gly were more resistant to accretion than those capped with (Glu-Cys)$_2$Gly. *C. glabrata* CdSNCs were less stable than those of *S. pombe* at extreme pH. PCs were very effective in controlling the size of CdSNCs or preventing accretion. Further, PC capped CdSNCs protected NCs from oxygen radical-mediated dissolution. The CdS-PCs formed *in vitro* appears to be indistinguishable from those formed *in vivo* (Mehra and Tripathi 2000).

Metal binding capacity of PCs is typically increased upon sulfide incorporation. PC$_2$, the smallest of the PCs, typically incorporated approximately 0.8 sulfide ions per Cd(II). It has been suggested that the amount of sulfide incorporated may depend on the affinity of Cd(II) for SH groups of the PC involved. Thus, PC$_2$ with lower affinity for Cd(II) incorporates significantly more sulfide than PC$_4$ which presumably have higher affinity for Cd(II) (Mehra and Tripathi 2000).

PC formed CdS crystallites are of uniform size as indicated by the similar optical properties, whereas in contrast incorporation of sulfide to Cd-GSH led to formation of variety of GSH capped CdS (GSH-CdS) complexes that differed in sulfide/ Cd(II) ratios, optical spectroscopic properties and Cd(II)-binding capacity of GSH and these GSH-CdS complex behaved like semiconductor nanocrystallites (Bae and Mehra 1998). CdS-PC complexes also reduced methylviologen, which confirms its nanocrystalline nature (Dameron and Winge 1990). Additionally, electron microscopic and XRD studies showed that the size of these crystallites were typically in the 2 nm range. PCs themselves are not able to form crystallites larger than 2 nm but replace GSH from larger particles without changing the size of the particles (Bae and Mehra 1997).

### 4.2 Sequestration to the Vacuole/Transport of Metal-Phytochelatin Complex

In both plants and yeasts, PC-Metal complexes are finally sequestered into the vacuole. Ortiz et al. (1992) isolated a gene designated as *hmt1* (heavy metal tolerance) from *Schizosaccharomyces pombe.* The *hmt1* gene encodes a vacuolar protein having sequence identity with the family of ABC (ATP-binding cassette) type transport proteins. HMT1 is an ATP dependent transporter of both apoPC and phytochelatin-Cd$^{2+}$ complexes. This is essential for Cd tolerance but has not been found to transport Cd to the vacuole (Ortiz et al. 1995).

A Yeast Cadmium Factor (YCF1) gene conferring cadmium resistance has been isolated. It encodes a ABC type protein which was shown to be a MgATP energised vacuolar glutathione-S-conjugate transporter responsible for the vacuolar sequestration of organic compounds after their S-conjugation with GSH (Li et al. 1996). Further studies revealed that YCF1 selectively mediates

MgATP energised vacuolar transportation and accumulation of bis-(glutathionato)-Cd (Cd-$GS_2$) complexes (Li et al. 1997).

In tobacco plants exposed to Cd, almost all of Cd and PC accumulated were confined to vacuole (Vogeli-Lange and Wagner 1990). An Mg-ATP dependent and proton gradient independent activity similar to that of HMT1 capable of transporting both PC and PC-Cd complexes has been identified in oat root (Salt and Rauser 1995). Plant genes encoding this function have not yet been identified.

## 5. Characterization and Regulation of Phytochelatin Synthase Gene

### 5.1 Characterization of Gene

For the first time, PCS genes have been characterized in *Arabidopsis*, S. *pombe*, and wheat (Vatamaniuk et al. 1999; Ha et al. 1999; Clemens et al. 1999). After that, the gene has been characterized from other plants, and even animals and prokaryotes as given in Table 2.

**Table 2.** Characterization of phytochelatin synthase gene plants

| PCS gene Characterized | Plant | References |
|---|---|---|
| AtPCS1 | *Arabidopsis thaliana* | Vatamaniuk et al. (1999) |
| TaPCS1 | *Triticum aestivum* | Clemens et al. (1999) |
| SpPCS | *Schizosaccharomyces pombe* | Ha et al. (1999) |
| CePCS1 | *Caenorhabditis elegans* | Vatamaniuk et al. (2001) |
| AtPCS2 | *Arabidopsis thaliana* | Cazalé and Clemens (2001) |
| GmhPCS1 | *Glycine max* | Oven et al. (2002b) |
| BjPCS1 | *Brassica juncea* | Heiss et al. (2003) |
| Alr0975 (PCS like protein) | *Nostoc sp.* PCC 7120 | Tsuji et al. (2004) |

Database searches identified a PCS like gene in a nematode *Caenorhabditis elegans*. Functional analysis established it as a PCS gene. This was the first report of PCS in animal (Vatamaniuk et al. 2001). Heterologous expression of CePCS1 in Cd hypersensitive *S. cerevisiae,* confers increased Cd tolerance and intracellular PC biosynthesis (Vatamaniuk et al. 2001) and expression of the same clone in *S. pombe* PCS deficient mutant suppress Cd hypersensitivity and restores Cd induced PC accumulation. A targeted suppression of CePCS1 in *C. elegans* leads to severe toxicity and even death of organism at higher concentration of Cd. This suggests contribution of PCs in metal detoxification at the level of whole organism.

Recently a gene encoding a PCS like protein identified from cyanobacteria, *Nostoc sp.* PCC7120, has been termed as alr0975. It is reported for the first time from prokaryotes (Tsuji et al. 2004), however, PC synthesis could not be shown conclusively (Harada et al. 2004).

EST sequencing programme demonstrates that PCS genes are present in number of species that have not yet been reported to synthesize PCs, such as *Dictyostelium discoideum* (slime molds), *Phytophthora sojae* (oomycetes) (Accession Nos. BE584918 and BE584958) and *Ciona intestinalis* (chordate) (Accession No. BW266987, BW255339). Survey done by Gekeler et al. (1989) and such database sequences further strengthen the presence of PCSs from fungi, algae to higher plants and also in animal kingdom from nematode model (Vatamaniuk et al. 1999) to model chordate (Tsuji et al. 2004).

## 5.2 Regulation of Gene

PCS is thought to be constitutively expressed in plants (Grill et al. 1989; Howden et al. 1995; Chen et al. 1997) and there is self-regulation of its activity by heavy metals in *Arabidopsis* (Zenk 1996; Cobbett 2000). However, as the PCS gene from various plants has been cloned and characterised, there came somewhat conflicting reports on the transcriptional regulation of PCS gene.

Clemens et al. (1999) reported that TaPCS1 was regulated at the transcriptional level after observing his results on 4 day old wheat seedlings treated for 6 h with 100 μM Cd. Analysis of TaPCS1 expression in roots indicated increased level of mRNA on exposure to Cd.

On the other hand, Ha et al. (1999) and Vatamaniuk et al. (2000) indicated that AtPCS1 did not exhibit transcriptional regulation by Cd, however, they used 10-day and 21-day old seedlings for their study. Recently, Lee and Korban (2002) conducted a study to analyze transcriptional regulation of AtPCS1 at various stages of plant development using transgenic *Arabidopsis* and wild type plants. They showed an increase in AtPCS1 promotor as evident by GUS activity, which decreased as the plants grew up to 15 days. The steady state level of AtPCS1 mRNA showed a 2-fold increase in the wild type treated plants, demonstrating a transcriptional regulation of AtPCS1 by Cd. They also showed an increase in the amount of AtPCS1 protein in transgenic lines during Cd exposure. Though they found it difficult to explain why transcriptional regulation of AtPCS1 appears during early stages of plant growth and then disappears, however, they could correlate such a response with higher sensitivity of *cad2* mutants in early phases of development to Cd which decreases as *cad2* mutants grow and 15 day old plant show same sensitivity to Cd as the wild type plants.

Lee et al. (2002) presented an evidence for an intron mediated increase of AtPCS1 mRNA after cadmium exposure. AtPCS1 promoter fusion with genomic AtPCS1 sequence but not with AtPCS1 cDNA sequence showed an increase in AtPCS1 mRNA accumulation after cadmium exposure. In a study by Heiss et al. (2003) in *B. juncea*, an increase of PC protein observed in leaf after prolonged Cd treatment could not be related to BjPCS1 mRNA. It was assumed that the increase in BjPCS protein is due to posttranscriptional regulation. A heavy metal induced increase of endogenous PCS protein in plants is thought to be reported for the first time in *B. juncea*. The results suggest a high expression of PCS in vascular tissues in *B. juncea*.

This suggests that PCS expression and activity may be moderately regulated at diverse levels.

A schematic representation of metal detoxification pathways has been presented in Figure 1.

## 6. Evolutionary Aspects of Phytochelatin Synthase

Presence and conservation of functional PCS throughout the plant kingdom is difficult to explain because heavy metals, although being ubiquitous, are mostly present at negligible concentrations in the environment (Schat et al. 2002) despite the contribution of man-made activities. Why did PCS evolve for the detoxification of heavy metals? The question arises whether its role has primarily been in essential ion homeostasis and possibly in degradation of xenobiotic. There are the further evidence for possible additional functions of PCS-related proteins in GSH metabolism (Beck et al. 2003) and provide a lead as to the evolutionary history of PCS (Tsuji et al. 2004).

A number of essential metals, such as Zn, Ni, Cu, Fe, Mo, Mn, are known to induce PC synthesis (Grill et al. 1987; Chen et al. 1997), however their detoxification by PCs is not established (Schat et al. 2002; Brune et al. 1995). Hence, PCs are supposed to play a role in homeostasis of metals like Zn and Cu (Zenk 1996). A well-established role of PCs is the detoxification of non-essential metals and metalloids with relatively high affinities to sulfur, such as Cd, Hg (Howden and Cobbett 1992; Gupta et al. 1998) and As (Schmöger et al. 2000; Hartley-Whitaker et al. 2001). However, in metal hyperaccumulator plants, metal detoxification involves some other mechanisms independent of PC synthesis (de Knecht et al. 1992, 1994, 1995; Ebbs et al. 2002; Schat et al.

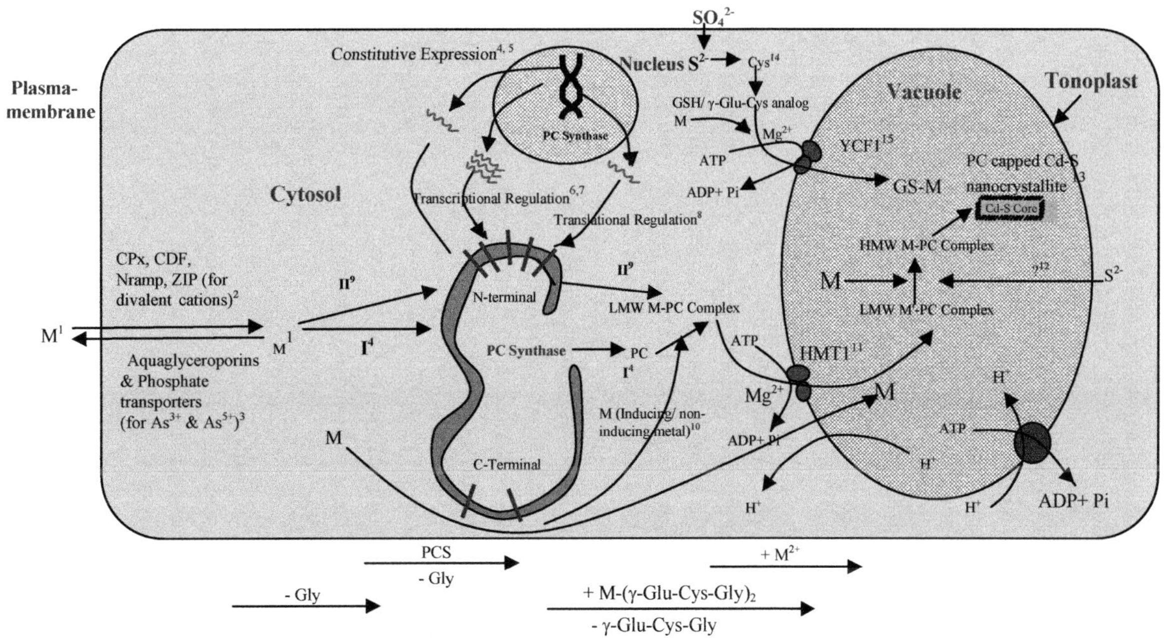


**Fig. 1:** Schematic representation of PC mediated metal detoxification. Metal or metalloid enters into the cytosol by different type of transorters. Within the cytosol metal activates PCS either by direct interaction (I) or by interaction of metal-glutathione complex with the enzyme (II). PCS enzyme may be present constitutively or may be regulated at the level of transcription or translation. PC-metal complex (LMW) thus formed is transported into vacuole by ATP energized ABC type transporters. Inside vacuole HMW complexes are formed from LMW complexes by incorporation of sulfide ions. HMW complexes may be stabilized in the form of nanocrystallites having metal-sulfide core capped by PCs. Superscript denote references which are given as follows. [1]- Zenk et al., 1996, [2]- Hall and Williams, 2002, [3]- Abedin et al., 2002, [4]- Grill et al., 1989, [5]- Vatamaniuk et al., 1999, [6]- Clemens et al., 1999, [7]- Lee and Korban, 2001, [8]- Heiss et al., 2003, [9]- Vatamaniuk et al., 2000, [10]- Maitani et al., 1996, [11]- Ortiz et al., 1992, 1995, [12]-Mehra and Tripathi, 2000, [13]- Dameron et al., 1990, [14]- Saito, 2004, [15]- Li et al., 1997

2002; Cai et al. 2004). Cu, Cd, As and Zn-hyperaccumulating plants accumulated low amounts of PCs not correlated with metal abundance. In Zn/Cd hyperaccumulator, *Thlaspi caerulescens,* the level of PCs has been found 2-3 fold lower inspite of having >10 fold higher concntration of leaf Cd compared to *T. arvense,* a non-accumulator plant (Ebbs et al. 2002). In a study on As hyperaccumulator *Pteris vittata,* formation of LMW thiol was not found sufficient to bind all As accumulated inspite of having a positive correlation between PCs and metal (Cai et al. 2004). In several studies, Cu did not apparently induce PCs untill the threshold exposure level for acute toxicity exceeded (De Vos et al. 1992; Rijstenbil et al. 1998; Rijstenbil and Gerringa 2002). These studies may suggest that significant PC levels are induced when the capacity of non-PC based homeostasis/detoxification system is exhausted (de Knecht et al. 1995; Schat et al. 2002). Schat et al. (2002) observed the role of PCs in Cu, Cd, Zn, As, Ni, and Co tolerance in non-metallicolous and metallicolous, hypertolerant populations of *Silene vulgaris, Thlaspi caerulescens, Holcus lanatus,* and *Agrostis castelana.* Based on plant-internal PC-thiol to metal molar ratios, the metals'- tendency to induce PC accumulation, decreased in the order As/Cd/Cu>Zn>Ni/Co, and was consistently higher in non-metallicolous plants than in hypertolerant ones, except for the case of As. The sensitivities to Cu, Zn, Ni and Co were consistently unaffected by BSO treatment, both in non-metallicolous and hypertolerant plants, suggesting that PC-based sequestration is not essential for constitutive tolerance or hypertolerance to these metals. However, BSO dramatically increased As sensitivity, both in non-adapted and As-hypertolerant plants indicating that GSH- and PC-based sequestration is essential for both normal constitutive tolerance and adaptive hypertolerance to this metalloid. Naturally selected As hypertolerance in *Holcus lanatus* was found to be associated with enhanced rates of PC accumulation and increased PC-thiol to As molar ratios in roots, suggesting that PC synthesis might be essential for hypertolerance to As, at least (Hartley-Whitaker et al. 2001).

Cai et al. (2004) suggested that mechanism of As detoxification in Chinese Brake fern might be more complex than simple chelation of As anions by the thiols. Other mechanisms of detoxification/tolerance of metals include volatilization (Meagher 2000), cell wall binding (Salt et al. 1997) chelation with organic acids (Wang et al. 1991; Salt et al. 1995, 1997; Krotz et al. 1989), direct transport to vacuoles by antiporter systems (Salt and Wagner 1993) and reduced uptake of metal (Hartley-Whitaker et al. 2001).

Steffens (1990) suggested that the energetic cost associated with sulphate reduction and PC synthesis would make this mechanism of Cd tolerance evolutionary prohibitive. Thus formation of a huge amount of PCs to chelate all the metals does not look like a simple solution to the problem. Considering all these studies, the exact role of PCs is still very elusive and limited in proposed roles of homeostasis and/or detoxification.

A role for PCs in metal transport from root to shoot (Gong et al. 2003) and stabilization of metal complexes inside vacuoles has been postulated. Ebbs et al. (2002) further suggested that incorporation of Cd as CdS into HMW complexes would allow a greater number of Cd atoms to be detoxified per molecule of PC than LMW complexes formed with PC-Cd. LMW thiols may only play a transport role by facilitating the transport of As into the vacuole where As may form a more stable aggregation with sulfide and organic acids (Cobbett 2000; Cai et al. 2004). If PC could act as a chelator involved in transport of metal from root to shoot and then for sequestration inside the vacuole where the metal complex dissociates partly into PCs and metal/metalloid ions and may be degraded into precursor molecules, which are shuttled back to cytoplasm (Hartley-Whitaker et al. 2001; Li et al. 2004), then a few molecules of PC would detoxify exceeding amounts of metals. This hypothesis of detoxification and transporter role for PCs also looks more attractive from an energetic perspective.

The question of evolution of PCS becomes another twist when we think of presumed additional function of PCS, namely in degradation of glutathione-S-conjugates. These two mechanisms have common initial step i.e. cleavage of glycine from GSH or glutathione-S-conjugate, both catalysed by PCS resulting in the formation of γ-Glu-Cys or γ-Glu-Cys-S-conjugate (Beck et al. 2003). In second and final step PC synthesis involves transpeptidation of γ-Glu-Cys into GSH or derivatives thereof, whereas transfer of γ-Glu-Cys-S-conjugate into smaller molecules like water occurs during detoxification of xenobiotics.

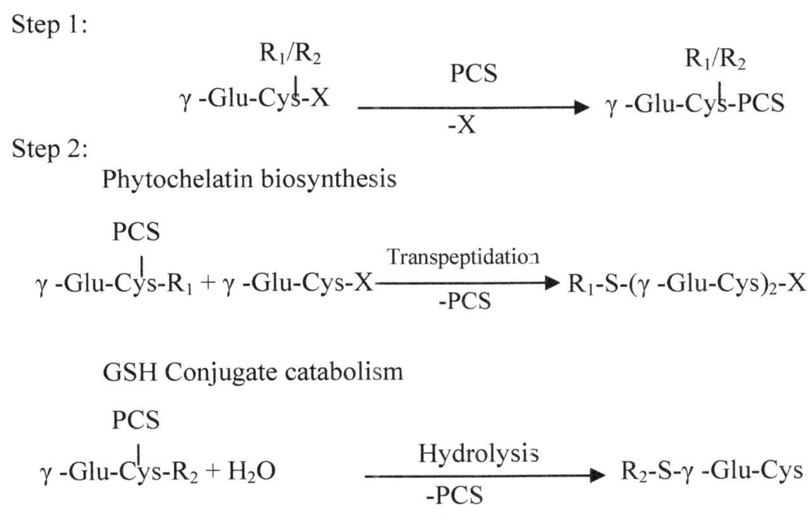


R1= metal, linear hydrocarbons such as methyl and hexyl-residues
R2= bulky groups like cyclic hydrocarbons

Steric hindrance may be the main regulatory element in the second step of the reaction. Thus it is uncertain, which is the more ancient cellular function of PCS and which one has been aquired later in the process of evolution. However, the prokaryotic PCS-like enzyme has the GSH hydrolyzing activity not the PC forming indicating that is adaptive. Its evolution might have occurred either for PC synthesis or for detoxification of xenobiotics or other undiscovered funcions and during evolution, it possibly evolved multiple functions.

## 7. Genetic Engineering for Enhancing Phytoremediation Potential

The important features of an effective phytoremediator plant are that the plant should have high biomass production, efficient mechanism for metal accumulation and detoxification, fast growth and a short life cycle. Hyperaccumulators accumulate metal to an extremely high concentration without suffering any toxic effect, thus they may appear as good candidates for phytoremediation. But their slow growth and low biomass is a limitation for this purpose. However, hyperaccumulators may provide a source of genes involved in metal uptake, translocation and sequestration for enhancing phytoremediation. Transfer of these genes into a suitable candidate plant is a strategy for engineering of plants with improved phytoremediation traits. Transfer or overexpression of such genes may lead to enhanced metal uptake, translocation, sequestration, or intracellular targeting (Eapen and D'souza 2005). To date, a few attempts have been made using enzymes of sulfur/ PC metabolism in this regard, which have been summarized in Table 3.

Overexpression of two enzymes γ-glutamycysteine synthetase (γ-ECS) or GSH synthetase (GS) in transgenic Indian mustard resulted in accumulation of higher levels of GSH and PC. They showed enhanced Cd tolerance and accumulation and also extracted more Cd, Cr, Cu, Pb and Zn than wild plants (Zhu et al. 1999a,b). Transgenic *Nicotiana* plants overexpressing cytosolic cysteine synthase gene of rice showed greater growth and produced more PC in shoots upon Cd exposure than wild-type plants though Cd accumulation was 20% lower in transgenics (Harada et al. 2001).

Bennett et al. (2003) conducted a green house experiment using transgenic Indian mustard plants overexpressing adenosine triphosphate sulfurylase (APS) or γ-glutamylcysteine synthetase (γ-ECS) or GSH synthetase (GS). The ECS and GS transgenic plants accumulated 1.5-fold more Cd and 1.5- to 2-fold more Zn compared to control while APS plants did not. γ-ECS transgenics also accumulated 2.4- to 3-fold more Cu, Cr and Pb compared to wild plants. Transgenic Indian mustard plants overproducing PC accumulated significantly high level of Zn and Cd in contaminated soil from Leadville, Colorado.

**Table 3.** Recombinant genes of PC/ GSH metabolism in relation to bioremediation of heavy metal ions

| Transgenic made | Gene transformed/ overexpressed | Bioremediator metal | Remark | Reference |
|---|---|---|---|---|
| Poplar hybrid *Populus tremula* x *Populus alba* | Bacterial gene *gsh1* for γ-glutamylcysteine synthetase | - | Transgenics showed foliar contents of γ-EC and GSH, 10- and 3-fold high respectively. A supply of exogenous supply of cysteine caused a further increase. | Noctor et al. (1996) |
| *Brassica juncea* | *E. coli gsh1* gene encoding γ-glutamyl cysteine synthetase (γ-ECS) | Cd | Transgenic seedlings showed increased tolerance to Cd and higher concentrations of PCs, γ-GluCys, glutathione and total non-protein thiols | Zhu et al. (1999a) |
| *Brassica juncea* | *E. coli gsh2* gene encoding glutathione synthetase | Cd | Transgenic plants accumulated significantly high Cd than wild type with shoot Cd concentration up to 25% higher and total Cd accumulation per shoot was up to 3-fold higher. The plants also had higher concentration of GSH, PCs, thiols, sulfur and calcium. | Zhu et al. (1999b) |
| *Nicotiana tabaccum* | Rice Cytosolic Cysteine synthase (RCS1) | Cd | Transgenics plant exhibited 3-fold higher activity of cysteine synthase than WT plants. Upon Cd exposure they showed greater growth and produced more phytochelatins in shoots than WT plants though Cd accumulation was 20% lower in transgenics. | Harada et al. (2001) |
| *Escherichia coli* | Two types of transformations done, in one coexpression of $Hg^{2+}$ transporter system (MBP-both MerP and MerT transporters of Hg) with (Glu- | Hg | Both approaches were effective and both types of transformed bacteria showed a maximum accumulation of 230 nmol Hg/mg dw. Results suggest that bioaccumulation by bacterial biosorbents with surface-expressed metal-binding | Bae et al. (2001) |

| | | | | |
|---|---|---|---|---|
| | Cys)$_{20}$Gly (EC20) was tested, and in other expression of EC20 on cell surface was tested | | peptides may be useful as a universal strategy for the cleanup of heavy metal contamination. | |
| *Arabidopsis thaliana* | *E. coli arsC* gene encoding Arsenate reductase (*SRS1p/ArsC*) and *γ-ECS* encoding γ-glutamyl- cysteine synthetase (*ACT2p/γ-ECS*) | As | Plants expressing only arsenate reductase were found hypersensitive to arsenate, whereas expressing only γ-ECS were tolerant to arsenic compared with wild type. But the plants expressing both the gene were found substatially tolerant to arsenic, accumlating 4- to 17-fold greater fresh shoot weight and accumulating 2- to 3-fold more arsenic per gram of tissue than wild type or plants expressing single gene. | Dhanker et al. (2002) |
| *Arabidopsis thaliana* | *A. thaliana* PCS (AtPCS1) | Cd | Transgenic lines showed 12-25 fold higher accumulation of AtPCS1 mRNA, 1.3 to 2.1 fold higher PC accumulation under 85 μM Cd stress for 3 days. However transgenics were hypersensitive to Cd and Zn, which was due to reduced availability of GSH. | Lee et al. (2003) |
| *Arabidopsis thaliana* | Wheat PCS (TaPCS1) | Cd | Transgenic plants complemented the Cd, Hg and As sensitivities of *cad1-3* mutant. PCs were detected in roots and rosette leaves and stems however long distance TaPCS1 mRNA transport was not observed. Transgenic plants showed less Cd accumulation in roots and enhanced long distance Cd transport into rosette leaves and stems. | Gong et al. 2003 |
| *Brassica juncea* | Adenosine triphosphate | Cd, Zn, Cu, | The ECS and GS transgenic plants accumulated | Bennett et al. (2003) |

| | | | | |
|---|---|---|---|---|
| | sulfurylase (APS) or γ-glutamylcysteine synthetase (γ-ECS) or glutathione synthetase (GS) | Pb | 1.5-fold more Cd and 1.5- to 2-fold more Zn compared to control while APS plants did not. γ-ECS transgenics also accumulated 2.4- to 3-fold more Cu, Cr and Pb compared to wild plants. | |
| *Nicotiana glauca* | Wheat PCS (TaPCS1) | Pb and Cd | Transgenic plants were more tolerant to Pb and Cd developing seedling roots 160% longer than wild type plants. In addition seedlings of transformed plants grown in mining soils containing high level of Pb (1572 ppm) accumulated double concentration of Pb than wild type plants | Gisbert et al. (2003) |
| *Escherichia coli* | *Arabidopsis thaliana* PCS (AtPCS) | Cd and As | A marked accumulation of PCs was observed *in vivo* together with a decrease in the cellular glutathione content. When bacterial cells expressing AtPCS were placed in the presence of heavy metals like Cd and As, cellular metal content were increased 20- and 50-fold respectively. | Sauge-Merle et al. (2003) |
| *Arabidopsis thaliana* | Overexpression of AtPCS1 | Cd and As | Transgenic plants were highly resistant to As accumulating 20-100 times more biomass on 250 and 300 μM As. These plants significantly synthesized $PC_2$-$PC_4$ and other unidentified thiols specially three thiols designated as a,b and c. However these plants were hypersensitive to Cd treatment. | Li et al. (2004) |

*Nicotiana glauca* is widely distributed, fast-growing, high biomass producing and a herbivore-repulsive plant. Gisbert et al. (2003) used this plant to overexpress wheat gene encoding PCS (TaPCS1). The transgenics showed greatly increased tolerance to metals, such as Pb and Cd, developing seedling roots 160% longer than wild type plants. In addition, seedlings of transformed plants grown in mining soils containing high levels of Pb (1572 ppm), accumulated double concentration of this heavy metal than wild type. Transgenic *Arabidopsis* plants overexpressing AtPCS1 showed 12- to 25-fold higher accumulation of AtPCS1 mRNA, and also higher PC accumulation under Cd stress, however transgenics were hypersensitive to Cd and Zn (Lee et al. 2003). In another study, transgenic *Arabidopsis* plants overexpressing AtPCS1 were found to be highly resistant to As, accumulating 20-100 times more biomass exposed to 250 and 300 μM As. These plants significantly synthesized $PC_2$-$PC_4$ and other unidentified thiols. However these plants were hypersensitive to Cd treatment (Li et al. 2004). Transgenic expression of TaPCS1 showed suppression of the heavy metal sensitivity of the *cad1-3* mutant, increase in long distance root to shoot transport of Cd, and reduction of Cd accumulation in root. The protection mechanism was attributed to maintaining a low Cd content in root by transporting extra Cd to shoot (Gong et al. 2003). These studies point to the question why did only expression of TaPCS1 significantly enhance root to shoot Cd transport besides the fact that WT plant also synthesized PCs in roots. It may be that transgenic expression of TaPCS1 may result in increased PC accumulation in unique cells, such as vascular parenchyma, leading to more accumulation of PCs in these cells which would further augment Cd or PC-Cd loading into vascular transport pathway. In addition, a recombinant PCS protein alone has sufficient enzymatic activity required for PC synthesis. As the transgenic TaPCS1 protein differs in amino acid sequence (55% homology) with amino acids from the native AtPCS1 protein, it is likely that recombinant protein acts constitutively and more independently from a possible regulatory network in *Arabidopsis.*

Recently a novel bioremediation system called symbiotic engineering using symbiosis between leguminous plants and rhizobia was developed. The metallothionein gene (*MTL4*) (Murooka et al. 2001) and AtPCS were fused to *nifH* promotor, generating nodule specific expression of these genes in *Mesorhizobium haukii* strain B3 infecting *Astragalus sinicus* (Sriprang et al. 2004). AtPCS expression in *M. haukii* subsp. *regeni* strain B3 resulted in 9- to 19-fold increased ability of cells to bind cadmium. When the recombinant strain B3 established symbiotic relationship with *Astragalus sinicus,* the symbionts increased the Cd accumulation by 1.5-fold. The expression of both AtPCS and *MTL4* resulted in enhanced Cd uptake by legumes. Further the expression of AtPCS and an iron regulated transporter, *IRT1* in the recombinant strain B3 increased the ability of cells to bind Cd up to 2.5-fold compared to cells only expressing AtPCS. In the rice paddy soil addition of recombinant strain enhanced the accumulation of Cd in roots and nodules of *A. sinicus* (Murooka et al. 2005).

Somatic cell hybrid produced between *B. juncea,* a high biomass Pb accumulator plant, and *T. caerulescens,* a known Zn and Ni hyperaccumulator, showed increased resistance to Pb, Ni and Zn and total amount of Pb phytoextracted was much greater because of the high biomass produced (Gleba et al. 1999; Dushenkov et al. 2002).

Vacuolar sequestration is the compartmentational detoxification mechanism afforded by PC and other ligands. Hence engineering vacuolar transporter genes, such as *hmt1* or *YCF1*, is a second-generation approach for phytoremediation (Tong et al. 2004). Tissue specific overproduction of a functional transporter in transgenic plant might be a mean to alter the tissue localization of the heavy metal to sequester them away from consumable part of the crop plant leading in order to increase food safety.

Dhanker et al. (2002) made transgenic *Arabidopsis* plant by co-expressing *E. coli* Ars C gene (SRSIp/ArsC), encoding Arsenate reductase, and *E. coli* γ-ECS gene (ACT 2p/γ-ECS), encoding γ-glutamylcysteine synthetase. These plants accumulated 4- to 17-fold greater fresh shoot weight under metal exposure and showed higher arsenic accumulation, 2- to 3-fold more arsenic per gram of tissue than wild plants or transgenic plants expressing γ-ECS or ArsC alone. Yeast YCF1 protein when overexpressed in *Arabidopsis thaliana,* enhanced Pb and Cd tolerance (Song et al. 2003).

## 8. Phytochelatin as a Biosensor

PCs have been used in electrochemical biosensors and they provide rapid, simple and low-cost on-field determination of heavy metals. Synthetic PCs, $(\text{Glu-Cys})_{20}\text{Gly}$ (EC20), fused to maltose binding domain were expressed in *E. coli* and purified for construction of the novel capacitance biosensor. The biosensor was able to detect Hg, Cd, Pb, Cu, and Zn ions in concentration range of 100 fM-10 mM, and the order of sensitivity was $S_{Zn} > S_{Cu} > S_{Hg} >> S_{Cd} \cong S_{Pb}$. The biological sensing element of the sensor could be regenerated using EDTA and the storage stability of the biosensor was 15 days (Bontidean et al. 2003). A new heavy metal biosensor based on interaction of heavy metal ions (Cd and Zn) with PCs showed a detection limit of Cd and Zn of about 1.0 and 13.3 pmole in 5 μl, respectively (Adam et al. 2005).

## 9. Conclusion

Metal induced PC synthesis is known throughout the plant kingdom, in some fungi as well as in animals and PC-based metal detoxification is an important mechanism in several plants. However, in some metal hyperaccumulator plants, metal tolerance and detoxification are not PC dependent and involve other processes. PC dependent accumulation and detoxification can be used for metal phytoremediation from contaminated sites. Overexpression of enzymes related to PC synthesis, such as γ-ECS, GS and PCS, or enzymes related to sulfur metabolism like, sulfur transporters, APS sulfurylase and cysteine synthase or overexpression of vacuolar tranporters of PC-metal complexes bear the promise to result in the development of efficient phytoremediator plant. Studies with overexpression of some of these genes have generated promising results in that respect, both in the laboratory and under field conditions. However, in some studies, overexpression led to hypersensitivity towards the metal probably due to insufficient sequestration or enhanced uptake. On the other hand, overexpression of PCS and a vacuolar transporter are prime examples of a second-generation approach. The novel bioremediation system called symbiotic engineering involving advantage of both rhizobia and leguminous plants using many useful genes like AtPCS, *MTL4* (metallothionein gene) and *IRT1* (iron regulated transporter) may provide another valuable bioremediation tool. Identification of PCS gene in prokaryotes shed light on its evolutionary history and provided a tool for understanding the mechanism of PCS catalysed reaction. Understanding the mechanistics in detail may contribute to the development of a good phytoremediator transgenic with the features of fast growth, high biomass and improved removal of metals. From the present knowledge, it looks like that PCS perform an additional function in plants involving detoxification pathway of GS-conjugates of organic xenobiotics. Besides, PCs may also act as biosensors of heavy metal pollution.

## References


Abedin MJ, Feldmann J, Meharg AA (2002) Uptake kinetics of arsenic species in rice plants. Plant Physiol 128:1120-1128

Adam V, Zehnalek J, Petrlova J, Potesil D, Sures B, Trnkova L, Jelen F, Vitecek J, Kizek R (2005) Phytochelatin modified electrode surface as a sensitive heavy metal ion biosensor. Sensors 5:70-84

Ahner BA, Price NM, Morel FMM (1994) Phytochelatin production by marine phytoplankton at low free metal ion concentrations: Laboratory studies and field data from Massachusetts Bay. Proc Natl Acad Sci USA 91:8433-8436

Bae W, Mehra RK (1997) Metal-binding characteristics of phytochelatin analog (Glu-Cys)$_2$Gly. J Inorg Biochem 68:201-210

Bae W, Mehra RK (1998) Properties of glutathione- and phytochelatin-capped CdS bionanocrystallites. J Inorg Biochem 69:33-43

Bae W, Mehra RK, Mulchandani A, Chen W (2001) Genetic engineering of *Escherichia coli* for enhanced uptake and accumulation of mercury. Appl Environ Microbiol 67:5335-5338

Barceló J, Poshenrieder C (2002) Fast root growth responses, root exudates, and internal detoxification as clues to the mechanisms of aluminium toxicity and resistance: a review. Environ Exp Bot 48:75-92

Barceló J, Poschenrieder C, Tolrá RP (2003) Importance of phenolics in rhizosphere and roots for plant-metal relationships. In: Gobran G (ed) Extended Abstracts 7th ICOBE Upsala 15-19 June, pp 162-163

Beck A, Lendzian K, Oven M, Christmann A, Grill E (2003) Phytochelatin synthase catalyses key step in turnover of glutathione conjugates. Phytochemistry 62:423-431

Bennett LE, Burkhead JL, Hale KL, Terry N, Pilon M, Pilon-Smits EAH (2003) Analysis of transgenic Indian Mustard plants for phytoremediation of metal contaminated mine tailings. J Environ Qual 32:432-440

Bleeker PM, Schat H, Vooijs R, Verkleij JAC, Ernst WHO (2003) Mechanisms of arsenate tolerance in *Cytisus striatus*. New Phytol 157:33-38

Bontidean I, Ahlqvist J, Mulchandani A, Chen W, Bae W, Mehra RK, Mortari A, Csöregi E (2003) Novel synthetic phytochelatin-based capacitive biosensor for heavy metal ion detection. Biosensors and Bioelectronics 18:547-553

Brune A, Urbach W, Dietz KJ (1995) Differential toxicity of heavy metals is partly related to a loss of preferential extraplasmic compartmentation: a comparison of Cd-stress, Mo-stress, Ni-stress, and Zn-stress. New Phytol 129:403-409

Cai Y, Su J, Ma LQ (2004) Low molecular weight thiols in arsenic hyperaccumulator *Pteris vittata* upon exposure to arsenic and other trace elements. Environ Pollut 129:69-78

Cazalé A-C, Clemens S (2001) *Arabidopsis thaliana* expresses a second functional phytochelatin synthase. FEBS Lett 507:215-219

Chandra Sekhar K, Chary NS, Kamala CT, Anjaneyulu Y (2004) Utilization of plant metal interactions for environmental management: From a general disbelief to universal acceptance. Proc Indian Sci Acad B70:13-30

Chen J, Zhou J, Goldsbrough PB (1997). Characterization of phytochelatin synthase from tomato. Physiol Plant 101:165-172

Clemens S, Kim EJ, Newmann D, Schroeder J (1999) Tolerance to toxic metals by a gene family of phytochelatin synthases from plants and yeast. EMBO J 18:3325-3333

Cobbett CS (1999) A family of phytochelatin synthase genes from plant, fungal and animal species. Trends Plant Sci 4:335-343

Cobbett CS (2000). Phytochelatins and metallothioneins: roles in heavy metal detoxification and homeostasis. Plant Physiol 123:825-832

Cobbett C, Goldsbrough P (2002) Phytochelatins and Metallothioneins: Roles in heavy metal detoxification and homeostasis. Annu Rev Plant Biol 53:159-182

Cobbett CS, May MJ, Howden R, Rolls B (1998) The glutathione-defecient, cadmium-sensitive mutant, *Cad2-1*, of *Arabidopsis thaliana* is deficient in γ-glutamylcysteine synthetase. Plant J 16:73-78

Dameron CT, Reese RN, Mehra RK, Kortan AR, Carroll PJ, Steigerwald ML, Brus LE, Winge DR (1989) Biosynthesis of cadmium sulfide quantum semiconductor crystallites. Nature 338:596-598

Dameron CT, Winge DR (1990). Characterization of peptide-coated cadmium-sulfide crystallites. Inorg Chem 29:1343-1348

de Knecht JA, Koevoets PLM, Verkleij JAC, Ernst WHO (1992) Evidence against a role for phhytochelatins in naturally selected increased cadmium tolerance in *Silene vulgaris* (Moench) Garcke. New Phytol 122:681-688 de Knecht JA, van Dillen M, Koevoets PLM, Schat H, Verkleij JAC, Ernst WHO (1994) Phytochelatins in cadmium-sensitive and cadmium-tolerant *Silene vulgaris*. Chain length distribution and sulfide incorporation. Plant Physiol 104:255-261

de Knecht JA, Van Baren N, ten Bookum WM, Wong Fong, Sang HW, Koevoets PLM, Schat H, Verkleij JAC (1995) Synthesis and degradation of phytochelatins in cadmium-sensitive and cadmium-tolerant *Silene vulgaris*. Plant Sci 106:9-18

De Vos CHR, Vonk MJ, Vooijs R, Schat H (1992) Glutathione depletion due to copper-induced phytochelatin synthesis causes oxidative stress in *Silene cucubalus*. Plant Physiol 98:853-858

Dhanker OP, Li Y, Rosen BP, Shi J, Salt D, Senecoff JF, Sashti NA, Meagher RB (2002) Engineering tolerance and hyperaccumulation of arsenic in plants by combining arsenate reductase and γ-glutamylcysteine synthetase expression. Nature Biotechnol 20:1140-1145

Dushenkov S, Skarzhinskaya M, Glimelius K, Gleba D, Raskin I (2002) Bioengineering of a phytoremediation plant by means of somatic hybridization. Int J Phytoremediat 4:117-126

Eapen S, D'Souza SF (2005) Prospects of genetic engineering of plants for phytoremediaton of toxic metals. Biotechnol Adv 23:97-114

Ebbs S, Lau J, Ahner B, Kochian L (2002) Phytochelatin synthesis is not responsible for Cd tolerance in the Zn/Cd hyperaccumulator *Thlaspi caerulescens* (J&C Prestl.). Planta 214:635-640.

Elmsley J (2001) Nature's building blocks. An A-Z guide to the elements. Oxford University Press, Oxford, UK

Freedman JH, Ciriolo MR, Peisach J (1989) The role of glutathione in copper metabolism and toxicity. J Biol Chem 264:5598-5605

Gekeler W, Grill E, Winnacker E-L, Zenk MH (1988) Algae sequester heavy metals via synthesis of phytochelatin complexes. Arch Microbiol 150:197-202

Gekeler W, Grill E, Winnacker E-L, Zenk MH (1989) Survey of the Plant Kingdom for the ability to bind heavy metals through phytochelatins. Z Naturforsch 44c:361-369

Ghosh M, Shen J, Rosen BP (1999) Pathway of As(III) detoxification in *Sachharomyces cerevisiae*. Proc Natl Acad Sci USA 96:5001-5006

Gisbert C, Ros R, De Haro A, Walker DJ, Bernal P, Serrano R, Navarro-Aviñó J (2003) A plant genetically modified that accumulates Pb is especially promising for phytoremediation. Biochem Biophys Res Comm 303:440-445

Gleba D, Borisjuk NV, Borisjuk LG, Kneer R, Poulev A, Skarzhinskaya M et al. (1999) Use of plant roots for phytoremeidation and molecular farming. Proc Natl Acad Sci USA 96:5973-5977

Gong J-M, Lee DA, Schroeder JI (2003) Long-distance root-to-shoot transport of phytochelatins and cadmium in *Arabidopsis*. Proc Natl Acad Sci USA 100:10118-10123

Grill E, Winnacker E-L, Zenk MH (1985) Phytochelatins: the principal heavy-metal complexing peptides of higher plants. Science 230:674-676

Grill E, Gekeler W, Winnacker E-L, Zenk MH (1986a) Homo-phytochelatins are heavy metal-binding peptides of homo-glutathione containing Fabales. FEBS Lett 205(1):47-50

Grill E, Winnacker E-L, Zenk MH (1986b) Synthesis of seven different homologous phytochelatins in metal-exposed *Schizosaccharomyces pombe* cells. FEBS Lett 197:115-120

Grill E, Winnacker E-L, Zenk MH (1987) Phytochelatins, a class of heavy-metal-binding peptides from plants, are functionally analogous to metallothioneins. Proc Natl Acad Sci USA 84:439-443

Grill EL, Löeffler S, Winnacker E-L, Zenk MH (1989) Phytochelatins, the heavy-metal-binding peptides of plants, are synthesized from glutathione by a specific γ-glutamylcysteine dipeptidyl transpeptidase (phytochelatin synthase). Proc Natl Acad Sci USA 86:6838-6842

Gupta M, Rai UN, Tripathi RD, Chandra P (1995). Lead induced changes in glutathione and phytochelatin in *Hydrilla verticillata*. Chemosphere 30(10):2011-2020

Gupta M, Tripathi RD, Rai UN, Chandra P (1998). Role of glutathione and phytochelatin in *Hydrilla verticillata* (l.f.) Royle and *Vallisneria spiralis* L. under mercury stress. Chemosphere 37:785-800

Gupta M, Tripathi RD, Rai UN, Haq W (1999) Lead induced synthesis of metal binding peptides (Phytochelatins) in submerged macrophyte *Vallisneria spiralis* L. Physiol Mol Biol Plants 5:173-180

Gupta DK, Tohoyama H, Joho M, Inouhe M (2002) Possible roles of phytochelatins and glutathione metabolism in cadmium tolerance in chickpea roots. J Plant Res 115:429-437

Gupta DK, Tohoyama H, Joho M, Inouhe M (2004) Changes in levels of phytochelatins and related metal-binding peptides in chickpea seedlings exposed to arsenic and different heavy metal ions. J Plant Res 117:253-256

Ha S-B, Smith AP, Howden R, Dietrich WM, Bugg S, Connell MJO, Goldsbrough PB, Cobbett CS (1999) Phytochelatin synthase genes from Arabidopsis and the yeast *Schizosaccharomyces pombe*. Plant Cell 11:1153-1163

Hall JL, Williams LE (2002) Transition metal transporters in plants. J Exp Bot 54:2601-2613

Harada E, Choi YE, Tsuchisaka A, Obata H, Sano H (2001) Transgenic tobacco plants expressing a rice cysteine synthase gene are tolerant to toxic levels of cadmium. J Plant Physiol 158:655-661

Harada E, von Roepenack-Lahaye E, Clemens S (2004) A cyanobacterial protein with similarity to phytochelatin synthases catalyses the conversion of glutathione to γ-glutamylcysteine and lacks phytochelatin synthase activity. Phytochemistry 65:3179-3185

Harmens H, Hartog PRD, Ten Bookum WM, Verkleij JAC (1993) Increased zinc tolerance in *Silene vulgaris* (Moench) Garcke is not due to increased production of phytochelatins. Plant Physiol 103:1305-1309

Hartley-Whitaker J, Ainsworth G, Vooijs R, Bookum WT, Schat H, Meharg AA (2001) Phytochelatins are involved in differential arsenate tolerance in *Holcus lanatus*. Plant Physiol 126:299-306

He Z, Li J, Zhang H, Ma M (2004) Different effects of calcium and lanthanum on the expression of phytochelatin synthase gene and cadmium absorption in *Lactuca sativa*. Plant Science 168:309-318

Heiss S, Wachter A, Bogs J, Cobbett C, Rausch T (2003) Phytochelatin synthase (PCS) protein is induced in *Brassica juncea* leaves after prolonged Cd exposure. J Exp Bot 54:1833-1839

Hirata K, Tsujimoto Y, Namba T, Ohta T, Hirayanagi N, Miyasaka H, Zenk MH, Miyamoto K (2001) Strong induction of phytochelatin synthesis by zinc in marine green alga *Dunaliella tertiolecta*. J Biosci Bioeng 92:24-29

Howden R, Cobbett CS (1992) Cadmium-sensitive mutants of *Arabidopsis thaliana*. Plant Physiol 99:100-107

Howden R, Goldsbrough PB, Andersen CR, Cobbett CS (1995) Cadmium-sensitive, *cad1* mutants of *Arabidopsis thaliana* are phytochelatin deficient. Plant Physiol 107:1059-1066

Inouhe M, Mitsumune M, Tohoyama H, Joho M, Murayama T (1991) Contributions of cell wall and metal-binding peptides in suspension-cultured cells of tomato. Bot Mag Tokyo 104:217-229

Inouhe M, Ito R, Ito S, Sasada N, Tohoyama H, Joho N (2000) Azuki bean cells A hypersensitive to Cadmium and do not synthesise phytochelatins. Plant Physiol 123:1029-1036

Inzé D, van Montagu M (1995) Oxidative stress in plants. Curr Opin Biotechnol 6:153-158

Jocelyn PC (1972) Biochemistry of the SH group. The occuuence, chemical properties, metabolism and biological function of thiols and disulfides. Academic press, London

Kägi JHR, Vallee BL (1960) Metallothionein: a Cadmium- and Zinc-containing Protein from Equine Renal Cortex. J Biol Chem 235:3460-3465

Kägi JHR, Kojima Y (1987) Chemistry and biochemistry of metallothionein. Experientia Suppl 52:25-61

Klapheck S, Schlunz S, Bergmann L (1995) Synthesis of phytochelatins and homo-phytochelatins in *Pisum sativum*L. Plant Physiol 107:515-521

Kneer R, Zenk MH (1997) The formation of Cd-phytochelatin complexes in plant cell cultures. Phytochemistry 44:69-74

Kneer R, Kutchan TM, Hochberger A, Zenk MH (1992) *Saccharomyces cerevisiae* and *Neurospora crassa* contain heavy metal sequestering phytochelatin. Arch Microbiol 157:305-310

Kondo N, Imai K, Isobe M, Goto T, Murasugi A, Wada-Nakagawa C, Hayashi Y (1984) Cadystin A and B, major unit peptides comprising cadmium binding peptides induced in a fission yeast-separation, revision of structures and synthesis. Tetrahedron Lett 25:3869-3872

Krämer U, Cotter-Howells JD, Charnock JM, Baker AJM, Smith JAC (1996) Free histidine as a metal chelator in plants that accumulate nickel. Nature 379:635-638

Krämer U, Smith RD, Wenzel WW, Raskin I, Salt DE (1997) The role of metal transport and tolerance in nickel hyperaccumuloation by *Thlaspi caerulescens* Hálácsy. Plant Physiol 115:1641-1650

Krämer U, Pickering IJ, Prince RC, Raskin I, Salt DE (2000) Subcellular localization and speciation of nickel in hyperaccumulator and non-accumulator *Thlaspi* species. Plant Physiol 122:1343-1353

Krotz RM, Evangelou BP, Wagner GJ (1989) Relationships between Cd, Zn, Cd-peptide and organic acid in tobacco suspension cells. Plant Physiol 91:780-787

Landberg T, Greger M (2004) No phytochelatin ($PC_2$ and $PC_3$) detected in *Salix viminalis*. Physiol Plant 121:481-487

Lasat MM, Baker AJM, Kochian LV (1996) Physiological characterization of root $Zn^{2+}$ absorption and translocation to shoots in Zn hyperaccumulator and nonaccumulator species of *Thlaspi*. Plant Physiol 112:1715-1722

Lasat MM, Baker AJM, Kochian LV (1998) Altered zinc compartmentation in the root and stimulated Zn absorption both play a role in Zn hyperaccumulation in *Thlaspi caerulescens*. Plant Physiol 118:875-883

Lee S, Korban SS (2002) Transcriptional regualation of *Arabidopsis thaliana* phytochelatin synthase (*AtPCS1)* by cadmium during early stages of plant development. Planta 215:689-693

Lee S, Kang BS (2005) Expression of *Arabidopsis* phytochelatin synthase 2 is too low to complement an *AtPCS1*-defective *Cad1-3* mutant. Mol Cells 19:81-87

Lee S, Moon JS, Domier LL, Korban SS (2002) Molecular characterization of phytochelatin synthase expression in transgenic *Arabidopsis*. Plant Physiol Biochem 40:727-733

Lee S, Moon JS, Ko T-S, Petros D, Goldsbrough PB, Korban SS (2003) Overexpression of Arabidopsis phytochelatin synthase paradoxically leads to hypersensitivity to cadmium stress. Plant Physiol 131:656-663

Li Z-S, Szczypka M, Lu Y-P, Thiele DJ, Rea PA (1996) The yeast cadmium factor protein (YCF1) is a vacuolar glutathione-S-conjugate pump. J Biol Chem 271:6509-6517

Li Z-S, Lu Y-P, Thiele DJ, Rea PA (1997) A new pathway for vacuolar cadmium sequestration in *Saccharomyces cerevisiae*: YCF1-mediated transport of *bis* (glutathionato) cadmium. Proc Natl Acad Sci USA 94:42-47

Li Y, Dhanker OP, Carreira L, Lee D, Chen A, Schroeder JI, Balish RS, Meagher RB (2004) Overexpression of phytochelatin synthase in *Arabidopsis* leads to enhanced arsenic tolerance and cadmium hypersensitivity. Plant Cell Physiol 45:1787-1797

Löeffler S, Hochberger A, Grill E, Winnacker E-L, Zenk MH (1989) Termination of the phytochelatin synthase reaction through sequestration of heavy metals by the reaction product. FEBS Lett 258:42-46

Maier T, Yu C, Küllertz G, Clemens S (2003) Localization and functional characterization of metal-binding sites in phychelatin synthases. Planta 218:300-308

Maitani T, Kubota H, Sato K, Yamada T (1996) The composition of metals bound to class III metallothionein (phytochelatin and its desglycyl peptide) induced by various metals in root cultures of *Rubia tinctorum*. Plant Physiol 110:1145-1150

Marghoses M, Vallee BL (1957) A cadmium protein from equine kidney cortex. J Am Chem Soc 79:4813-4814

Meagher RB (2000) Phytoremediation of toxic elemental and organic pollutants. Curr Opin Plant Biol 3:153-162

Mehra RK, Tarbet EB, Gray WR, Winge DR (1988) Metal-specific synthesis of 2 metallothioneins and γ-glutamyl-transferase peptides in *Candida glabrata*. Proc Natl Acad Sci USA 85:8815-8819

Mehra RK, Kodati R, Abdullah R (1995) Chain length-dependent Pb(II)-coordination in phytochelatins. Biochem Piophys Res Comm 215:730-738

Mehra RK, Tran K, Scott GW, Mulchandani P, Saini SS (1996a) Ag(I)-binding to phytochelatins. J Inorg Biochem 61:15-142

Mehra RK, Mielat J, Kodati VR, Abdullah R, Hunter TC, Mulchandani P (1996b) Optical spectroscopic and reverse-phase HPLC analyses of Hg(II)-binding to phytochelatins. Biochem J 314:73

Mehra RK, Tripathi RD (2000) Phytochelatins and metal tolerance. In: Agarwal SB, Agarwal M (eds) Environmental Pollution and Plant Responses, Lewis Publishers, Boca Raton, FL, USA

Miersch J, Tschimedbalshir Barlocher F, Grams Y, Pierau B, Schierhorn A, Kraus GJ (2001) Heavy metals and thiol compounds in *Mucor racemosus* and *Articulospora tetracladia.* Mycol Res 105:883-889

Murasugi A, Wada C, Hayashi Y (1981) Purification and unique properties in UV and CD spectra of Cd-binding peptides 1 from *Schizosachharomyyces pombe.* Biochem Biophys Res Commun 103:1021-1028

Murooka Y, Toyama M, Hong S-H, Gohya M, Ono H, Yamashita M, Hirayama N (2001) Genetic Design of Stable Metal-Binding Biomolecules, Oligomeric Metallothioneins. Biocatalysis Biotransformation 19:399-412

Murooka Y, Ike A, Sriprang R, Yamashita M (2005) Bioremediation for heavy metals through symbiosis between leguminous plants and rhizobia. Abstract in Third International Conference on Plants and Environmental Pollution (ICPEP-3), pp 3

Mutoh N, Hayashi Y (1988) Isolation of mutants of *Schizosachharomyces pombe* unable to synthesize cadystin, small Cd-binding peptides. Biochem Biophys Res Commun 151:32-39

Nieboer E, Richardson DHS (1980) The replacement of the nondescript term 'heavy metals' by a biologically and chemically significant classification of metal ions. Environ Pollut Ser B 1:3-26

Noctor G, Strohm M, Jouanin L, Kunert K-J, Foyer CH, Rennenberg H (1996) Synthesis of glutathione in leaves of transgenic poplar overexpressing γ-glutamylcysteine synthetase. Plant Physiol 112:1071-1078

Ortiz DF, Kreppel L, Speiser DM, Scheel G, McDonald G, Ow DW (1992) Heavy metal tolerance in fission yeast requires an ATP-binding cassette-type vacuolar membrane transporter. EMBO J 11:3491-3499

Ortiz DF, Ruscitli T, McCue KF, Ow DW (1995) Transport of metal binding peptides by HMT1: a fission yeast ABC-type vacuolar membrane protein. J Biol Chem 270:4721-4728

Oven M, Raith K, Neubert RHH, Kutchan TM, Zenk MH (2001) Homo-phytochelatins are synthesized in response to cadmium in Azuki beans. Plant Physiol 126:1275-1280

Oven M, Grill E, Golan-Goldhirsh A, Kutchan TM, Zenk MH (2002a) Increase of free cysteine and citric acid in plant cells exposed to cobalt ions. Phytochemistry 60:467-474

Oven M, Page JE, Zenk MH, Kutchan TM (2002b) Molecular characterization of the homo-phytochelatin synthase of soyabean *Glycine max*. J Biol Chem 277:4747-4754

Pawlik-Skowrońska B (2002) Correlations between toxic Pb effects and production of Pb-induced thiol peptides in the microalga *Stichococcus bacillaris*. Environ Pollut 119:119-127

Pawlik-Skowrońska B (2003) When adapted to high zinc concentrations the periphytic green alga *Stigeoclonium tenue* produces high amounts of novel phytochelatin-related peptides. Aquatic Toxicol 62:155-163

Pawlik-Skowrońska B, Sanità di Toppi L, Favali MA, Fossati F, Pirszel J, Skowroński T (2002) Lichens respond to heavy metals by phytochelatin synthesis. New Phytol 156:95-102

Pawlik-Skowrońska B, Pirszel J, Kalinowska R, Skowroński T (2004) Arsenic availability, toxicity and direct role of GSH and phytochelatins in As detoxification in the green alga *Stichococcus bacillaris* Aquatic Toxicol 70:201-212

Persans M, Yan X, Patnoe JMML, Krämer U, Salt DE (1999) Molecular dissection of histidine's role in nickel hyperaccumulation in *Thlaspi goesingense* (Hálácsy). Plant Physiol 121:1117-1126

Pickering IJ, Prince RC, George MJ, Smith RD, George GN, Salt DE (2000) Reduction and coordination of arsenic in Indian Mustard. Plant Physiol 122:1171-1177

Piechalak A, Tomaszewska B, Baralkiewicz D, Malecka A (2002) Accumulation and detoxification of lead ions in legumes. Phytochemistry 60:153-162

Rabenstein DL (1989) Metal complexes of glutathione and their biological significance. In: Dolphin D, Poulson R, Avramovic O (eds) Glutathione: Chemical, Biochemical and Medical aspects. John Wiley & Sons, New York, pp 147-186

Rai UN, Tripathi RD, Gupta M, Chandra P (1995) Induction of phytochelatins under cadmium stress in water lettuce (*Pistia stratiotes*). J Environ Sci Hlth 30(9):2007-2026

Rauser WE (1990) Phytochelatins. Annu Rev Biochem 59:61-86

Rauser WE (1995) Phytochelatins and related peptides. Structure, biosynthesis and function. Plant Physiol 109:1141-1149

Rauser WE (1999) Structure and function of metal chelators produced by plants: the case for organic acids, amino acids, phytochelain and metallothioneins. Cell Biochem Biophys 31:19-48

Rauser WE, Meuwly P (1995) Retention of cadmium in roots of maize seedlings. Plant Physiol 109:195-202

Rea PA, Vatamaniuk OK, Rigden DJ (2004) Weed, worms, and more. Papain's long lost cousin, phytochelatin synthase. Plant Physiol 136:2463-2474

Rea PA (2006) Phytochelatin synthase, papain's cousin, in stereo. PNAS 103:507-508

Rijstanbil JW, Haritonidis S, Malea P, Seferlis M, Wijnholds JA (1998) Thiol pools and glutathione redox ratios as possible indicators of copper toxicity in the green macroalgae *Enteromorpha* spp. from the Scheldt Estuary (SW Netherlands, Belgium) and Thermakos Gulf (Greece, N Aegean Sea). Hydrobiologia 385:171-181

Rijstenbil JW, Gerringa LJA (2002) Interactions of algal ligands, metal complexation and availability, and cell responses of the diatom *Ditylum brightwelli* with a gradual increase in copper. Aquatic Toxicol 56:115-131

Ruotolo R, Peracchi A, Bolchi A, Infusini G, Amoresano A, Ottonello S (2004) Domain organization of phytochelatin synthase. Functional properties of truncated enzyme species identified by limited proteolysis. J Biol Chem 279:14686-14693

Saito K (2004) Sulfur assimilatory metabolism. The long and smelling road. Plant Physiol 136:2443-2450

Salt DE, Wagner GJ (1993) Transport of Cd in tonoplast vesicles from oat roots. Evidence for a Cd/H antiport activity. J Biol Chem 268:1297-12302

Salt DE, Rauser WE (1995) MgATP-dependent transport of phytochelatins across the tonoplast of oat roots. Plant Physiol 107:1293-1301

Salt DE, Prince RC, Pickering IJ, Raskin I (1995) Mechanisms of cadmium mobility and accumulation in Indian mustard. Plant Physiol 109:1427-1433

Salt DE, Pickeing IJ, Prince RC, Gleba D, Smith RD, Raskin I (1997) Metal accumulation by aquacultured seedlings of Indian mustard. Environ Sci Technol 31:1636-1644

Salt DE, Prince RC, Baker AJM, Raskin I, Pickering IJ (1999) Zinc ligand in the metal hyperaccumulator *Thlaspi caerulescens* as determined using X-ray absorption spectroscopy. Environ Sci Technol 33:713-717

Satofuka H, Fukui T, Takagi M, Atomi H, Imanaka T (2001) Metal-binding properties of phytochelatin-related peptides. J Inorg Biochem 86:595-602

Sauge-Merle S, Cuiné S, Carrier P, Lecomte-Pradines C, Luu D-T, Peltier G (2003) Enhanced toxic metal accumulation in engineered bacterial cells expressing *Arabidopsis thaliana* phytochelatin synthase. Appl Environ Microbiol 69:490-494

Scarano, G, Morelli, E (2002) Characterization of cadmium- and lead- phytochelatin complexes formed in a marine microalga in response to metal exposure. BioMetals 15:145-151

Schat, H, Llugany, M, Vooijs, R, Hartley-Whitaker, J, Bleeker, PM (2002) The role of phytochelatins in constitutive and adaptive heavy metal tolerances in hyperaccumulator and non-hyperaccumulator metallophytes. J Exp Bot 53:2381-2392

Schmöger MEV, Oven M, Grill E (2000) Detoxification of arsenic by phytochelatins in plants. Plant Physiol 122:793-801

Scott N, Hatlelid KM, MacKenzie NE, Carter DE (1993) Reactions of arsenic (III) and arsenic (V) species with glutathione. Chem Res Toxicol 6:102-106

Song WY, Sohn EJ, Martinoia E, Lee YJ, Yang YY, Jasinski M et al. (2003) Engineering tolerance and accumulation of lead and cadmium in transgenic plants. Nat Biotechnol 21:914-919

Speiser DM, Abrahamson SL, Banuelos G, Ow DW (1992) *Brassica juncea* produces a phytochelatin-cadmium-sulfide complex. Plant Physiol 99:817-821

Sriprang R, Hayashi M, Ono H, Takagi M, Hirata K, Murooka Y (2003) Enhanced accumulation of $Cd^{2+}$ by *Mesorhizobium* sp. transformed with a gene from *Arabidopsis thaliana* coding for phytochelatin synthase. Appl Environ Microbiol 69:1791-1796

Srivastava S, Tripathi RD, Dwivedi UN (2004) Synthesis of phytochelatins and modulation of antioxidants in response to cadmium stress in *Cuscuta reflexa* - an angiospermic parasite. J Plant Physiol 161:665-674

Steffens JC (1990) The heavy metal-binding peptides of plants. Annu Rev Plant Physiol Mol Biol 41:53-575

Terry M. (2003) Phytoremediation of heavy metales from siols. Adances in Biochemical Engineering/Biotechnology 78:97-123

Tong Y-P, Kneer R, Zhu Y-G (2004) Varuolar compartmentalization: a second-generation approach to engineering plants for phytoremediation. Trends Plant Sci 9:7-9

Tripathi RD, Rai UN, Gupta M, Chandra P (1996) Induction of phytochelatins in *Hydrilla verticillata* (l.f.) Royle under cadmium stress. Bull Environ Contam Toxicol 56:505-512

Tsuji N, Hirayanagi N, Okada M, Miyasaka H, Hirata K, Zenk MH, Miyamoto K (2002) Enhancement of tolerance to heavy metals and oxidative stress in *Dunaliella tertiolecta* by Zn-induced phytochelatin synthesis. Biochem Biophys Res Commun 292:653-659

Tsuji N, Hirayanagi N, Iwabe O, Namba T, Tagawa M, Miyamoto S, Miyasaka H, Takagi M, Hirata K, Miyamoto K (2003) Regulation of phytochelatin synthesis by zinc and cadmium in marine green alga, *Dunaliella tertiolecta.* Phytochemistry 62:453-459

Tsuji N, Nishikori S, Iwabe O, Shiraki K, Miyasaka H, Takagi M, Hirata K, Miyamoto K (2004) Characterization of phytochelatin synthase-like protein encoded by alr0975 from a prokaryote, *Nostoc* sp. PCC 7120. Biochem Biophys Res Commun 315:751-755

Tsuji N, Nishikori S, Iwabe O, Matsumoto S, Shiraki K, Miyasaka H, Takagi M, Miyamoto K, Hirata K (2005) Comparative analysis of the two-step reaction catalysed by prokaryotic and eukaryotic phytochelatin synthase by an ion-pair liquid chromatography assay. Planta 222(1):181-191

Tu S, Ma LQ, MacDonald GE, Bondada B (2004) Effects of arsenic species and phosphorus on arsenic absorption, arsenate reduction and thiol formation in excised parts of *Pteris vittata* L. Environ Exp Bot 51:121-131

US EPA (2000) Introduction to phytoremediation. EPA Document #EPA/600/R-99/107 U.S.Environmental protection agency office of research and development, Woshington DC

US Department of energy (2000) Proceedings from the workshop on phytoremediation of inorganic contaminants, Idaho National Engineering and Environmental Laboratory Document # INEEL/EXT-2000-00207, IDHO Falls, ID

Vatamaniuk OK, Mari S, Lu YP, Rea PA (1999) AtPCS1, a phytochelatin synthase from Arabidopsis: isolation and *in vitro* reconstitution. Proc Natl Acad Sci USA 96:7110-7115

Vatamaniuk OK, Mari S, Lu YP, Rea PA (2000) Mechanism of heavy metal ion activation of phytochelatin (PC) synthase: blocked thiols are sufficient for PC synthase-catalysed transpeptidation of glutathione and related thiol peptides. J Biol Chem 275:31451-31459

Vatamaniuk OK, Bucher EA, Ward JT, Rea PA (2001) A new pathway for heavy metal detoxification in animals. Phytochelatin synthase is required for cadmium tolerance in *Caenorhabditis elegans.* J Biol Chem 276:20817-20820

Vatamaniuk OK, Mari S, Lang A, Chalasani S, Demkiv LO, Rea PA (2004) Phytochelatin synthase, a dipeptidyl transferase that undergoes multisite acylation with c-glutamylcysteine during catalysis. J Biol Chem 279:22449-22460

Vivares D, Arnoux P, Pignol D (2005) A papain-like enzyme at work: native and acyl-enzyme intermediate structures in phytochelatin synthesis. PNAS 102:18848-18853

Voet D, Voet JG (2004) Biochemistry, Ed. 3. John Wiley & Sons, New York

Vogeli-Lange R, Wagner GJ (1990) Subcellular localization of cadmium and cadmium binding peptides in tobacco leaves: Implication of a transport function for cadmium binding peptides. Plant Physiol 92:1086-1093

Wang J, Evangelou BP, Nielsen MT, Wagner GJ (1991) Computer simulated evolution of possible mechanisms for quenching heavy metal ion activity in plant vacuoles. Plant Physiol 97:1154-1160

Wójcik M, Vangronsveld J, Tukendorf A (2005) Cadmium tolerance in *Thlaspi caerulescens* I. Growth parameters, metal accumulation and phytochelatin synthesis in response to cadmium. Environ Exp Bot 53:151-161

Yan S-L, Tsay C-C, Chen Y-R (2000) Isolation and characterization of phytochelatin synthase in rice seedlings. Proc Natl Sci Counc ROC (B) 24:202-207

Zenk MH (1996) Heavy metal detoxification in higher plants: a review. Gene 179:21-30

Zhang W, Cai Y, Downum KR, Ma LQ (2004) Arsenic complexes in the Arsenic hyperaccumlator *Pteris vittata* (Chinese Brake fern). J Chromato A 1043:249-254

Zhu YL, Pilon-Smits EAH, Tarun AS, Weber SU, Jouanin L, Terry N (1999a) Cadmium tolerance and accumulation in Indian Mustard is enhanced by overexpressing γ-glutamylcysteine synthetase. Plant Physiol 121:1169-1177

Zhu YL, Pilon-Smits EAH, Jouanin L, Terry N (1999b) Overexpression of glutathione synthetase in Indian mustard enhances cadmium accumulation and tolerance. Plant Physiol 119:73-79

6

# Metal Resistance in Plants with Particular Reference to Aluminium

B.P. Shaw, V.K. Jha and B.B. Sahu

Institute of Life Sciences, Nalco Square, Bhubaneswar 751 023, Orissa, INDIA,
Email: b_p_shaw@yahoo.com

## 1. Introduction

### 1.1 Metals

The term "metal" designates an element which is a good conductor of electricity and whose electric resistance is directly proportional to absolute temperature. In addition to this distinctive characteristic, metals share several other typical physical properties, such as high thermal conductivity, density, malleability and ductility (Forstner and Wittmann 1979). Several nonmetallic elements exhibit one or more of these properties. And hence, the only feature that defines a metal unambiguously is the electric conductivity, which decreases with increase in temperature. There are, of course, elements in the periodic table, like boron, silicon, germanium, arsenic and tellurium, which show electric conductivity, but their electric conductivity is low, and it increases with the rise in temperature. These are termed metalloids (or half-metals) situated between metals and non-metals in the periodic table (Forstner and Wittmann 1979).

Metals constitute more than 50 % of the elements present in the earth's crust; out of 110 elements known today 69 are metals, excluding the element of the trans-uranium series (Shaw et al. 2004). Their relative abundance, however, differ greatly at a region over the globe, and the region at which a metal is found in high concentration serves as the source of the metal. The variation observed is not only natural but also man-made; metals present in the earth's crust are mined and extracted by the human beings to meet the requirement of their day to day life leading to their accumulation at some regions. Metals remaining present in high concentration in the earth's crust do not pose any threat to the environment until the landmass of the region is used for agro-industry. This is because they remain tightly bound to their Lewis components as sulfides, oxides, or carbonates, as the case may be (see below), and the ore particles also remain tightly packed along with the particle of the soil, which makes them highly immobilized. It is only the mining of the ore, and subsequent uses of the extracted metals that lead to far and wide contamination of the environment. From the figures of the crustal abundance of important metals and their production per annum (Table 1), the magnitude of contamination or pollution by metals as a result of anthropogenic activities may be imagined.

**Table 1**. World wide metal production and uses

| Metal | Crustal abundance (mg/kg) | Yearly production (x 1000 tonnes) | Major uses | Principal ores |
|---|---|---|---|---|
| Al | 83000 | 16200 | In making cable and wire for high voltage electric transmission and various parts of autos, aircraft, electrical equipment. | Bauxite, $Al_2O_3$ |
| As | 1.80[a] | 50 | In making alloys for bullets and shot, storage batteries, herbicides, insecticides and wood preservatives | Arsenide |
| Bi | 0.20 | 4 | Used in phamaceuiticals, electronics, cosmetics and pigments, and as catalyst | Principally in flue dust as $Bi_2S_3$, during smelting of Pb, Zn or Cu |
| Cr | 110 | 10800 | Used in metal plating, making stainless steel, wear-resistant and cutting-tool alloys, and used as an anticorrosive | Chromite, $FeOCr_2O_3$ |
| Cd | 0.2 | 19 | Used in electroplating, making Ni/Cd batteries, alloys, control rods in nuclear reactor and pigments, and as stabilizer of polyvinyl chloride (PVC) plastic | Greenockite, CdS |
| Cu | 63 | 8700 | Mainly used in making alloys and electrical products, the only wire used in windings in generator, motors and transformer | As metal sulfides and oxides |
| Au | 0.0035 | 1.61 | Used in jewelry, and is the basis of currency | Calavarite ($AuTe_2$), Petzite [$(Ag,Au)_2Te$] |
| Fe | 58000 | 508000 | Most widely produced metal, usually as steel, also used in many alloys for special purposes | Hematite, $Fe_2O_3$, goethite, $Fe_2O_4.H_2O$, magnetite, $Fe_3O_4$ |

| | | | | |
|---|---|---|---|---|
| Pb | 12 | 3400 | Making storage batteries, petrol additive, pigments, ammunition, cable sheathing | Galena, PbS |
| Mn | 1300 | 22000 | Used as oxygen and sulfur scavenger in steel, manufacture of alloys, dry cells, chemicals | Found mainly as oxides |
| Hg | 0.089 | 6 | Used as cathode in chlor-alkali cells, and also used in making paints, electrical apparatus, fungicides | Cinnabar, HgS |
| Mo | 1.30 | 89 | In making alloys, pigments chemicals, lubricants, and as catalyst | Molybdenite, $MoS_2$, wulfenite, $PbMoO_4$ |
| Ni | 89 | 800 | Used in making coins, storage battery, alloys, and as catalyst | Pentlandite $[(Fe,Ni)_9S_8]$, Nicolite (NiAs) |
| Se | 0.075 | 1.6 | In electronics, glass, pigments, photocopying | Mainly as clausthalite, PbSe, crrokesite $(Cu,Tl,Ag)_2Se$ |
| Ag | 0.075 | 14 | Finds uses mainly in making photographic materials and jewelry | Found with sulfide minerals |
| Sn | 1.70 | 190 | Used in coatings, solders, in making bearing alloys, bronze | Cassiterite, Stannite |
| Ti | 6400 | 4200 | Mainly used in making aircraft parts, and their engine, also in making valve, pumps, paint pigments | As oxide, $TiO_2$ |
| V | 140 | 32 | Used in making strong steel alloy | Primarily occurs as V(III) in igneous rocks |
| Zn | 94 | 7200 | Widely used in making brass (alloy), paint pigments, in galvanization | Found as sulfides, oxides and silicates |

(Source: Manahan 1990; Ochiai 1977; Fergusson 1990; Evans 1995; Chaterjee 1993; Wedepohl 2000)

### 1.2 Classification of Metals: the HSAB Principle

A metal in a chemical reaction reacts as an electron pair acceptor (Lewis acid) with an electron pair donor (Lewis base) to form various chemical groups, such as an ion pair, a metal complex, a co-ordination compound, or a donor-acceptor complex. The reaction may be generalized as follows:

$$M + L \rightarrow M{:}L$$

M represents the metal ion, L the ligand, and M:L the product (complex). The stability of the complex will depend on the magnitude of the equilibrium constant, $K_{ML}$, also called the stability constant.

$$K_{ML} = [ML]/[M]\,[L]$$

The larger the magnitude of the $K_{ML}$ the more stable will be the product (ML) in the solution.

Pearson (1968a,b) has classified the metal acceptors and the ligand donors into “hard” and “soft” categories to explain the stability of the product complex (also see http://chemistry.uttyler.edu/~coe/lectures/ num16.ppt). The chief criteria for such classification are electron mobility or polarizability (the degree to which the electron cloud is distorted by interaction with a charge or electric field), electron negativity (a measure of the power of an atom to attract electron to itself in a covalent bonding), and ionic charge density. A hard acceptor is characterized by low polarizability, low electronegativity and large positive charge density (high oxidation state and small radius), and the opposite is true for a soft acceptor. A hard donor on the other hand is characterized by low electron mobility or polarizability, but high electronegativity and a high negative charge density, and the reverse constitutes the characteristics of a soft donor. In between the two groups lie the intermediate donors and acceptors (Table 2).

**Table 2.** Different metal/ligand acceptors and donors

| | Hard | Intermediate | Soft |
|---|---|---|---|
| Acceptors | $H^+$, $Na^+$, $K^+$, $Be^{2+}$, $Mg^{2+}$, $Ca^{2+}$, $Mn^{2+}$, $Al^{3+}$, $Cr^{3+}$, $Co^{3+}$, $Fe^{3+}$, $As^{3+}$ | $Fe^{2+}$, $Co^{2+}$, $Ni^{2+}$, $Cu^{2+}$, $Zn^{2+}$, $Pb^{2+}$ | $Cu^+$, $Ag^+$, $Au^+$, $Tl^+$, $Hg_2^{2+}$, $Pd^{2+}$, $Cd^{2+}$, $Pt^{2+}$, $Hg^{2+}$, $CH_3Hg^+$ |
| Donors | $H_2O$, $OH^-$, $F^-$, $Cl^-$, $PO_4^{3-}$, $SO_4^{2-}$, $CO_3^{2-}$, $O_2^{2-}$ | $Br^-$, $NO^{2-}$, $SO_3^{2-}$ | $SH^-$, $S^{2-}$, $RS^-$, $CN^-$, $SCN^-$, CO, $R_2S$, RSH, $RS^-$ |

(after Pearson 1968a; R= alkyl or aryl group)

Experimental evidences suggest that hard acceptors prefer to bind hard donors and soft acceptors prefer to bind soft donors to form stable compound (Pearson 1968a,b; Ahrland 1968). This is called HSAB (hard soft acids and bases) principle. The HSAB principle is very much in work in nature: some metals occur in the earth’s crust as ores of oxide and carbonate, whereas other metals occur as sulfides. This is because hard acids, like $Mg^{2+}$, $Ca^{2+}$ and $Al^{3+}$, form strong bond with the hard bases, like $O_2^{2-}$ or $CO_3^{2-}$, and conversely softer acids, like $Hg_2^{2+}$ or $Hg^{2+}$ and $Pb^{2+}$, prefer soft bases like $S^{2-}$.

### 1.3 Metal Pollution: Some Facts

It is important to realize that metal "pollution" represents a subtly different form of pollution than do many other forms of contamination. The primary source of heavy metals in the environment is from naturally occurring geo-chemical materials; all metals occur to varying extent within all components of the environment. Although this occurrence may be enhanced by a human activity, this activity is not itself the source of a metal, rather it is the cause of an elevated occurrence. Hence, heavy metal 'pollution' of environment does not represent a unique occurrence of a metal within ecosystem, rather represents an increase in concentration of the metal relative to the natural occurrence of the element.

Literatures on the contamination of environment by metals are enormous (see Shaw et al. 2004, and the references therein). But majority of the studies have been associated with various industrial and agricultural activities. However, generally speaking, agricultural or industrial activities result in more diffuse contamination of the environment than does the natural occurrence. Nevertheless, in many cases of naturally high occurrence of heavy metals there is often close link with human-derived contamination (e.g. mining, smelting).

### 1.4 Metal Contamination of Soil: The Associated Agricultural Problems

Although the figures of yearly production of important metals (Table 1) are of much environmental concern, these are of little importance so far as contamination of soil is concerned. This is because the use of metals as industrial produce by mankind remain only confined to the cities and suburban areas, which may constitute only less than 10 to 15 % of the total inhabitable land mass. More importantly the metals used by the mankind as industrial produce mostly find their way into aquatic environment through the drainage system and run-off water during the rainy season from where their return to the atmosphere and landmass through bio-geochemical cycling is very slow (Fergusson 1990). Furthermore, it may also be noted that the use of metals, like of Hg and As, as components of pesticides in agriculture has been nearly discontinued, and the contamination of the land mass by these through agricultural practices is now only a history. Also, the use of fertilizers although may result in contamination of the environment by various metals present in them (Misra and Mani 1991; Dean et al. 1972), this is unlikely to be of much significance as these (metals) are continuously removed from the soil along with each harvest.

Mining of the earth for ore is the first step towards increasing contamination of the landmass by various metals depending upon the type of the ore. The mining operation let the ore particles loose, otherwise bound tightly among each other, remaining virtually immobile. And they become prone to be blown away by wind contaminating a vast area around the mine, particularly in the windward direction. Besides, the mining operation leaves stretches of mined lands devoid of vegetation, because of their high metal contents. This problem of contamination of agricultural uncontaminated agricultural lands is going to increase further with increase in the area of mining and the mining operation; it is generally in practice to use only the ores rich in metal for its cost-effective extraction, but when the currently available stock of the metal rich ores comes to an end the ores less rich in metal content may eventually be processed to meet the requirement of the man-kind leading to spatial increase in metal contaminated/polluted agricultural and other lands.

Processing of the ores for the extraction of metals is the second major step during which metals find their way into land mass; the metals escaping out of the chimneys of smelters are ultimately deposited in agricultural fields or other land, which may be far away from the smelting unit. Atmospheric metal enrichment, leading subsequently to pollution of soil, is also associated with other higher temperature anthropogenic activities, like burning of fossil fuels, production of cements, etc. Despite modern technological advances smelting operation and fossil fuel burning in industries continue to be important source of metals to the terrestrial environment (Shaw et al. 2004).

## 2. Phytotoxicity of Al and Agricultural Losses

The environmental and agricultural problems associated with Al needs special mention. The two sources of metals to the terrestrial environment described above hold true for this metal also. But, Al as such occurs in high levels in soil, which may be appreciated from its high crustal abundance (Table 3); it is the most abundant metal and third most common element in the earth's crust. Al is mostly found as oxide or silicate precipitates that are not toxic to plants. However, in acidic soil (pH < 5.0) Al speciates to soluble octahedral hexahydrate form, $Al(H_2O)_6^{3+}$, commonly called $Al^{3+}$ (Kochian 1995), which is phytotoxic. Thus, wherever the soil pH is acidic the Al present may cause serious agricultural losses. It has been estimated that approximately 40% of the world's cultivated lands, and up to 70% of the potentially arable lands are acidic (Haug 1984), which speaks of the gravity of environmental problems and economical losses associated with Al contamination of soil.

The typical visible toxicity symptoms of Al ($Al^{3+}$) in plants are thickening of root tips and inhibition of root growth (Delhaize and Ryan 1995; Kochian 1995). Besides, stunting, dark green leaves, purpling of stems, leaves and leaf vein, yellowing and death of leaf tips, curling of young leaves and collapse of growing points of petioles, etc. have also been observed in plants exposed to Al. At cellular level Al ions interact with lipid components of the plasma membrane leading to increase in its rigidity, disruption of its integrity, failure of $Ca^{2+}$ homeostasis and inhibition of signal transduction (Akeson et al. 1989; Tamas et al. 2004; Kochian 1995; Matsumoto 2000). Al toxicity has also been reported to be mediated via the formation of reactive oxygen species (ROS, see below), which causes peroxidative damage of cellular membranes (Cakmak and Horst 1991; Horst et al. 1992). The oxidative stress theory of Al toxicity is further strengthen by the observation of Boscolo et al. (2003) that Al stress induces dose- and time-dependent formation of ROS (reactive oxygen species) and subsequent protein oxidation in Al-sensitive maize inbred line, but not in Al-tolerant line. It has further been reported that the induction of oxidative stress by Al in plants may be a result of stimulation of the pro-oxidant nature of the endogenous phenolic compounds by the element.

**Table 3.** Common elements in the earth' crust

| Elements | Relative abundance (weight per cent) |
|---|---|
| Oxygen | 46.60 |
| Silicon | 27.72 |
| Aluminum | 8.13 |
| Iron | 5.00 |
| Calcium | 3.63 |
| Sodium | 2.83 |
| Potassium | 2.59 |
| Magnesium | 2.09 |
| Titanium | 0.44 |
| Hydrogen | 0.14 |
| Phosphorus | 0.12 |
| Manganese | 0.1 |
| All other elements | 0.61 |
| Total | 100.00 |

(after Mason 1958)

## 3. Aluminum Tolerant Crop Plants

Plants by virtue of their stationary status, unlike animals, cannot migrate to avoid unfavourable fluctuation or changes in their environment, and hence they must change their metabolic activities suitably, which would allow them to cope with the changing environment, otherwise perish. The resulting changes in their metabolism is called as "stress response", which may enable the plant to survive under the condition of stress, either for a short time only, known as acclimation, or the changes induced may be good enough to support continuous growth of the plant, known as adaptation. It is the latter quality, which is being or may be exploited to finding the solution to increasing metal contamination of the land masses, and form the basis of "phytoremediation". Phytoremediation may be achieved by growing plants over a number of years the aim is to either remove the pollutants from the contaminated matrix or to alter the chemical and physical nature of the contaminants within the soil so that they no longer present a risk to human health and the environment (Cunningham and Ow 1996). Thus the plants resistant to heavy metals can be used under the concept of phytoremediation in one or more of the following ways: i) to remove the metals from the soil, ii) to chelate the metals in the soil and bind the soil particles tight so that their erosion by wind, and so also further contamination of the land in the windward direction is prevented, and iii) to make possible the use of the metal contaminated land for agriculture.

It is explicit that the plants to be used under the first category, i.e. for the removal of metals from soil, should be hyperaccumulator of the metals contaminating the land, and that to be used under the second category may or may not be a hyperaccumulator, but should be resistant to the metals present in the soil and should be able to grow well with good rooting system. And for the plants to be used under the third category it is necessary that they besides being resistance to the metals contaminating the soil do not take-up and accumulate them in their tissues, otherwise the agricultural products would be highly contaminated with the metals.

Researches on understanding the mechanism of metal tolerance dates back to as early as 1950s when only ecological and physiological differences between plants from metal enriched and non-contaminated habitats were being studied (Bradshaw 1952; Jowett 1958). But the investigation gained momentum only in the late 1960s when time- and cost-effective technique for the analysis of metals, atomic absorption spectrophotometry, was developed (Ernst et al. 1992, and the references therein). However, during the period the research was mainly concentrated upon the uptake of metals, and their cellular compartmentation (Peterson 1969; Reilly 1967). It is from 1970s that the physiological and genetical aspects of metal tolerance were started being studied using the rewarding approach of comparison of metal tolerant and non-tolerant cultivars of a species, or even isogenic line of a species, which differed as far as possible only in resistance to one or more metals (Strange and Macnair 1991; Schat and Ten Bookum 1992). So far as Al is concerned, currently our understanding on tolerance of plants to the metal narrows down basically to two categories: 1) resistance by exclusion of the metals, and 2) resistance by uptake, but subsequent sequestration of the metals to inactive form inside the cells. In addition, however, there is another emerging concept in this field; Al tolerance involving the antioxidative machinery, which would be worth discussing.

### 3.1 Resistance as a Result of Exclusion of Metals

With regard to Al it has been found that the root apex (root cap, meristem, and elongation zone) accumulates more Al and attracts greater physical damage than the mature root tissues (Delhaize and Ryan 1995, Miyasaka and Hawes 2001). In fact, only the apical 2 to 3 mm of maize roots (root cap and meristem) need to be exposed to Al for the growth to be inhibited (Ryan et al. 1993). Moreover, when Al is selectively applied to the elongation zone or to the whole root except the apex, growth is unaffected (Ryan et al. 1993). And the reason of the toxic manifestation has been related to the movement of Al into symplasm in the root apex (Trice et al. 1992; Lazof et al. 1994), although the polyvalent ions like $Al^{3+}$, which is the major ionic species at acidic pH, is virtually insoluble in lipid bilayer. The conclusion is based on the fact that the root tips of the Al-resistant cultivar of wheat always accumulates less Al in both apoplasmic and symplasmic pools when compared to the Al-sensitive genotype grown in the same condition, as demonstrated using Al-fluorescent dye morin (Trice et al. 1992). This was also demonstrated by Delhaize et al. (1993a) using near-isogenic wheat (*Triticum aestivum* L.) line differing in Al tolerance at a single locus *Alt1* (aluminium tolerance), and Larsen et al. (1998) using Al-resistant (*alr*) mutants of *Arabidopsis thaliana*. Secondly, the differential Al sensitivity in wheat correlates with the concentration of Al in the root meristem (Rincon and Gonzales 1992). Lazof et al. (1994) using secondary-ion mass spectroscopy (MS) detected Al in the symplasm of soybean (*Glycine max*) root after only 30 min of exposure to Al while root growth inhibition requires about 60 min, suggesting that entry of Al occurs into cells before root growth is inhibited, and that entry into the symplasm is probably a must for Al to produce its toxic effect.

### *3.1.1 Exclusion Due to Increase in Rhizosphere pH*

Al has a complex chemistry. It hydrolyzes in solution such that the trivalent Al species, $Al(H_2O)_6^{3+}$, dominates in acidic condition (pH<5), which deprotonates to form $Al(OH)^{2+}$ and $Al(OH)^{+}$ species as the pH increase (Martin,`1988; Mortell and Motekaitis 1989). At near-neutral pH the solid phase $Al(OH)_3$, or gibbsite, occurs, whereas $Al(OH)_4^-$, or aluminate, dominates in alkaline condition. Out of the three soluble forms in acidic pH, it is the octahedral hexahydrate form, $Al^{3+}$, which is believed to be the primary phytotoxic species (Kochian 1995). Hence, explicit is that the pH of the growth medium would determine greatly the toxicity of Al. This led Foy et al. (1965) to propose an Al-exclusion mechanism that involves increase in rhizosphere pH; increase in the pH of rhizosphere would reduce the concentration of $Al^{3+}$ in favour of the less-toxic Al species. Since then there have been many studies to establish the relationship between Al resistance and transient increase in pH of the growth solution for several species including wheat, barley, pea, rye and triticale (Mugwira et al. 1978, 1976; Foy et al. 1967; Klimashevsky and Bernadskaya 1973; Mugiwira and Patel 1977). However, the first attempt to demonstrate any role of increase in rhizosphere pH in Al tolerance was made by Miyasaka et al. (1989) using a self-developed micro-electrode for measuring the pH along the root surface of Al tolerant (Atlas 66) and Al-sensitive (Scout) cultivars of wheat (*Triticum aestivum* L.). They observed increase in pH of the root apical rhizosphere by 0.15 unit relative to bulk solution in "Atlas 66" grown in complete nutrient solution with or without Al, and in 'Scout' grown without Al. 'Scout' grown with Al showed a slight decrease in pH. However, they concluded that the difference observed in the apical rhizosphere pH between the two cultivars in presence of Al should not account for difference in Al tolerance, and that the difference could be the consequence of $Al^{3+}$ tolerance rather than the cause of $Al^{3+}$ tolerance.

Degenhardt et al. (1998) adapted a molecular-genetic approach to check the relationship, and also used a vibrating microelectrode to measure the rhizosphere pH. They used an Al resistant mutant of *Arabidopsis thaliana*, *alr* 104, which did not exhibit any organic acid secretion (described later). The pH measurement at the root surface of wild type and *alr*-104 grown over $Al^{3+}$ revealed a difference of 0.1 to 0.15 unit along the root apex (between 0 and 500 μm from the root tip, the region of maximum $H^+$ influx), which was not observed when grown without $Al^{3+}$ (Fig. 1). They later on performed a root growth assay to assess the Al resistance of *alr*-104 and wild type in a strongly pH buffered nutrient solution. It was observed that increasing the solution pH from 4.4 to 4.5 significantly increased Al-resistant in wild type, which confirmed the idea that increase in $H^+$ influx accounted for a greater Al-resistance in *alr*-104. Furthermore, they also found that the difference in Al resistance between wild type and *alr*-104 disappeared when the roots were grown in pH-buffered medium, suggesting that the Al resistance in *alr*-104 is mediated by pH change in rhizosphere. The experiment provided first evidence of possible rhizosphere pH dependent Al tolerance in plants.

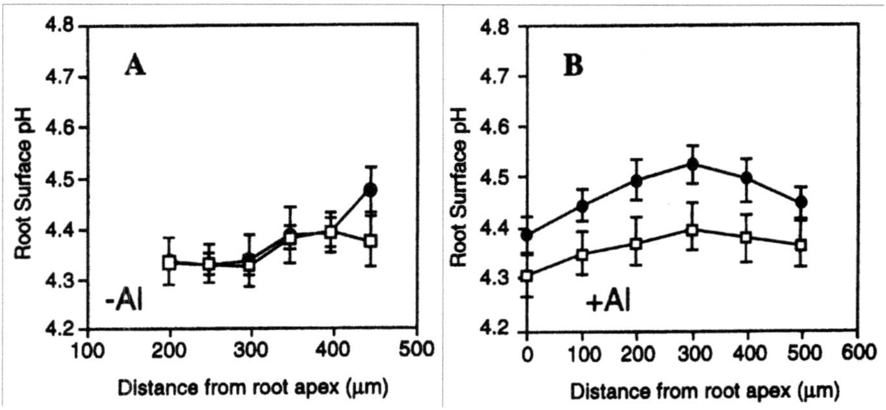


**Fig. 1.** Influence of $Al^{3+}$ exposure on rhizosphere pH along the surface of *A. thaliana* root tips in wild type (□) and *alr*-104 mutant resistant to Al (●). The mutant did not show enhanced root organic acid release in the absence (A) or presence (B) of 300 μM $AlCl_3$ (Source: Degenhardt et al. 1998)

### 3.1.2 Exclusion by Efflux of Organic Acids

Hue et al. (1986) demonstrated that addition of citric, oxalic or tartaric acid to the hydroponic solution alleviated the inhibitory effect of $Al^{3+}$ on root elongation in cotton. The antagonistic effect of chelating agents on $Al^{3+}$ toxicity has also been demonstrated for corn (Berlett and Riego 1972), ryegrass (Muchovej et al. 1988) and sorghum (Shuman et al. 1991). It was known earlier that several plant species excreted organic acids (citric acid and other) from their root in response to P deficiency (Gardner et al. 1983; Lipton et al. 1987; Dinkelaker et al. 1989). Later on the cell cultures of carrot (*Daccus carota* L.) and tobacco (*Nicotiana tabacum* L.) selected for $Al^{3+}$ tolerance were also shown to posses enhanced ability to excrete citric acid in response to Al treatment (Ojima et al. 1984, 1989; Ojima and Ohira 1988; Koyama et al. 1990). These led to the development of hypothesis that the organic acid secretion might be involved in Al exclusion mechanism.

**Citric acid.** Miyasaka et al. (1991) using differentially Al-resistant cultivars of snapbean (*Phaseolus vulgaris*) demonstrated that Al-resistant cultivar excreted a higher level of citric acid into the rhizosphere, 70 times more, than the Al-sensitive cultivar in response to $Al^{3+}$ stress. Besides, the tolerant cultivar secreted 10 times more citric acid than the sensitive cultivar even in the absence of Al. This led them to suggest that the resistance to $Al^{3+}$ in the Al-tolerant cultivar could be due to decrease in the active form ($Al^{3+}$) of Al around the rhizosphere as a result of its complex formation with the acid. But they also noticed the formation of Al-phosphate precipitates, which could have caused P deficiency, and the latter is known to trigger organic acid secretion (Ojima et al. 1989). Thus, the relationship between citric acid secretion and Al exclusion remained unclear. Nevertheless, enhanced secretion of citric acid has also been observed in several Al-resistant maize lines (Pellet et al. 1995; Kollmeier et al. 2001) and in an Al-resistant species, *Cassia tora* L. (Ma et al. 1997a) in response to Al, which further supports the existence of relationship between citric acid secretion and Al exclusion. A possible role of citric acid in Al resistance further stems from the observation that tobacco and papaya plants genetically engineered for over production of citric acid by introducing a citrate synthase gene from *Pseudomonas aeruginosa* shows increased Al resistance (Fuente et al. 1997). Also, the Al-resistant mutants, *alr*-108, *alr*-128 and *alr*-131, of *A. thaliana* show enhanced cellular exudation of citrate, malate and pyruvate than the wild type upon exposure to $Al^{3+}$ although the enhanced exudation of citric acid is not sustained for a long period (Larsen et al. 1998).

**Malic acid.** Delhaize et al. (1993a,b) used genetic approach to prove the relationship between Al tolerance and organic acid secretion. They used near-isogenic wheat (*Triticum aestivum* L.) lines, which showed 5 to 10-fold difference in Al tolerance, and differed in Al tolerance at single locus (*Alt1*). The test species, however, excreted malic acid and succinic acid instead of citric acid, and the malic acid excretion was 5- to 10-fold greater in the Al-tolerant (ET3) seedlings than in the Al-sensitive (ES3) seedlings despite the cellular content of the acid remaining nearly unchanged and similar. Significant correlation between Al-triggered malate release, Al resistance, and Al exclusion from the root apex was observed (Delhaize et al. 1993a). It was proposed that the release of malic acid from roots exposed to Al could be the Al tolerance mechanism encoded by *Alt1* locus (Delhaize et al. 1993b). This is because: a) there occurred a consistent correlation of the *Alt1* locus with malic acid excretion in the population of seedlings segregating for Al tolerance; b) Al stimulated malic acid excretion within 15 min, consistent with observation that Al tolerance is apparent after short exposure to $Al^{3+}$; c) malic acid excretion was localized at root apices, the primary site of $Al^{3+}$ toxicity; and d) malic acid added to nutrient solution was found to ameliorate $Al^{3+}$ toxicity. They also demonstrated that the low external inorganic phosphorous (Pi) conditions did not stimulate malic acid excretion over 24 h, and high external Pi concentration did not prevent $Al^{3+}$ from stimulating malic acid secretion. Basu et al. (1994) later on observed similar difference in malate efflux from roots of several cultivars differing in Al tolerance, and Ryan et al. (1995b) after screening 36 different wheat cultivars for Al resistance proposed that Al-stimulated malate efflux might be a general mechanism for Al tolerance in wheat. The view is further substantiated from the observation that the inhibition of malate exudation results in enhanced accumulation of Al in the Al-resistant wheat (cv Atlas) upon exposure to the metal (Osawa and Matsumoto 2001).

Concomitant with malate excretion, Basu et al. (1994) also observed enhanced *de novo* synthesis of the organic acid, which is consistent with the data that the matate content of Al-tolerant root apices is replenished over five-times during the initial 2 h of Al exposure (Delhaize et al. 1993b). Furthermore, it was observed that although the root apices of Al-tolerant seedlings synthesized more malate in response to Al than the root apices from the Al-sensitive seedlings, the root apices of both the genotype showed similar activities of phosphoenolpyruvate carboxylase and malate dehydrogenase, the two enzymes important in malate synthesis (Ryan et al. 1995a). Since the root apices of Al-sensitive and Al-tolerant genotype showed nearly similar malic acid contents, whether exposed to Al or not (Delhaize et al. 1993b), and they had same capacity to synthesize the acid (Ryan et al. 1995a), it was hypothesized that the difference in efflux probably lied in their relative ability to transport malate across the plasma membrane in response to $Al^{3+}$ (Delhaize et al. 1993b; Ryan et al. 1995a), the cytoplasm pool being replenished by fresh synthesis (Basu et al. 1994).

Taking into consideration all the observations, Delhaize and Ryan (1995) proposed a working model for the transport of malic acid across the membrane (Fig. 2). Malate exists primarily as divalent anion (malate$^{2-}$) in the cytoplasm, and if transported out of the cell in this form electroneutrality must be maintained either by an equivalent uptake of anions or by an equivalent efflux of cations.

Ryan et al. (1995a) and Kollmeier et al. (2001) showed that excretion of malate in fact is accompanied by efflux of $K^+$. Zhang et al. (2001) further observed that the efflux of $K^+$ in the Al-tolerant line of wheat (ET8) is maintained not because of insensitivity of the $K^+$ outward rectifying channel to $Al^{3+}$ suggested by Kollmeier et al. (2001). Rather, $Al^{3+}$ inhibits the $K^+$ outward rectifying channel in ET8 strongly. Later on, however, the inhibited channel, or additional $K^+$ outward rectifying channel is activated in which cAMP is involved (Zhang et al. 2001). The movement of malate$^{2-}$ could be mediated by anion channels in the plasma membrane. The evidence to this was provided by Ryan et al. (1995a); the rapid release of malate in response to $Al^{3+}$ was inhibited by anion channel antagonists, anthracene-6-carboxylic acid (A-9-C) and niflumic acid (NIF). The existence of malate permeable channel and its activation by Al in wheat has also been confirmed by Zhang et al. (2001) using anion channel antagonists. Furthermore, it has been observed that in Al-tolerant maize cultivar (cv ATP-Y) the malate channel is permeable to citrate as well (Kollmeier et al. 2001).

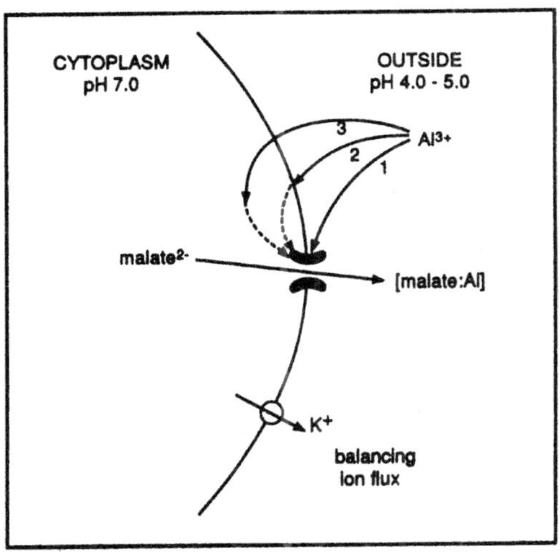


**Fig. 2.** A hypothetical scheme showing how $Al^{3+}$ interacts with a malate-permeable channel (hatched structure) in plasma membranes to stimulate malate efflux. The three mechanisms suggested (numbered arrows) are explained in the text. Electroneutrality is maintained by efflux of $K^+$ (Source: Delhaize and Ryan 1995)

Delhaize and Ryan (1995) proposed three ways in which Al, probably as $Al^{3+}$, could trigger the opening of the putative malate$^{2-}$ permeable channel: 1) Al may interact directly with the channel protein causing a change in conformation, increasing its mean open time or conductance; 2) It may interact with a specific receptor on the membrane surface or with the membrane itself, which through a series of secondary messages in the cytoplasm could change the channel activity; and 3) It may enter the cytoplasm and alter the channel activity either directly by binding with the channel or indirectly through a signal transduction pathway. They further suggested that the *Alt1* locus could code for a malate$^{2-}$ permeable channel that is responsive to $Al^{3+}$ or for a component of the pathway that regulates the activity of the putative channel leading to enhanced excretion of malate$^{2-}$ in the Al-tolerant cultivar, but not in the Al-sensitive one. Recently it has been seen that the exudation of malate in the root of Al-tolerant cultivar of wheat (cv Atlas) is inhibited by K-252a, a broad range inhibitor of protein kinases, suggesting that the opening of the channel is preceded by protein phosphorylation (Osawa and Matsumoto 2001). Treatment of the root apices by K-252a prior to exposure to $Al^{3+}$ also leads to enhanced accumulation the metal. The interaction of Al with the malate channel is thus likely through the 2$^{nd}$ or 3$^{rd}$ pathway proposed by Delhaize and Ryan (1995).

**Oxalic acid.** Ma et al. (1997b) and Zheng et al. (1998a) observed Al tolerance in buckwheat to be much greater than that in the Al-tolerant cultivar of wheat (cv Atlas 66), and found this to be a result of secretion of oxalic acid, the simplest dicarboxalic acid, from the root apex, the Al sensitive region. The secretion was specific to $Al^{3+}$ stress, as neither exposure to $La^{3+}$ nor phosphorous (P) deficiency resulted in any enhanced secretion of the acid. They also observed that the secretion of oxalic acid in response to $Al^{3+}$ was inhibited in the presence of anion channel inhibitor, phenylglyxol (PG), with subsequent inhibition of root elongation by as much as 40%, suggesting that the secretion of oxalic acid might be contributing to high Al resistance in buckwheat. The secretion was, however, not inhibited by NIF or A-9-C, which inhibited the secretion of malic acid in wheat; the secretion of oxalic acid probably occurs through the anion channel, which differs in characteristics from the malate$^{2-}$ anion channel in wheat, and hence the tolerance mechanism in wheat and buckwheat could be mediated through different gene function.

Although the secretion of organic acid as mechanism of Al resistance is well established in many plants, it is still not clear why there occurs difference in the requirement of the type of organic acid to be secreted by the plants to achieve the resistance. Based on the Al-detoxifying capacity in a plant species, organic acids can be grouped into strong (citric, oxalic and tartaric), moderate (malic, malonic and salicyclic) and weak (succinic, lactic, formic, acetic and phthalic) (Hue et al. 1986). Using 1:1 ratio of organic acid to Al experimentally it has been proved that for a species (corn) the detoxifying capacity is in the order citric>oxalic>malic (Zheng et al. 1998b). The difference in capacity of organic acids in ameliorating Al toxicity is attributable to their different stability constant with Al (stability constant: Al-citrate>Al-oxalate>Al-malate), which probably results in different activity of free $Al^{3+}$.

### *3.1.3 Necessity of Continuous Secretion of Organic Acids*

Irrespective of the type of organic acid secretion by a species for detoxification of Al, it is, however, necessary that continuous secretion of the acid at a high level is maintained for Al-resistance (Zheng et al. 1998b). According to the total amount of organic acids secreted, three patterns are observed with different cultivars differing in Al-resistance/-sensitivity (Zheng et al. 1998b): 1) the amount secreted is very low during the treatment (wheat cv Scout 66, oat)- sensitive; 2) the amount of secretion is high at the initial phase of exposure, but gradually decreases with duration of treatment (wheat cv Atlas 66, rape oilseeds)- moderately tolerant to tolerant; and 3) the amount of secretion is maintained at a high level during the whole period of Al-treatment (buckwheat and radish)- highly tolerant. The categorization, however, may not be strict, particularly for the sensitive category, as the tolerance mechanism other than organic acid secretion may provide tolerance to the species against the metal (Taylor 1991; Pellet et al. 1996). Furthermore, as is known, it is not necessary that all the Al molecules in solution need to be detoxified, rather it is the concentration of Al around the root apex, possibly just at the cell plasma membrane are to be reduced. In this context, the mucilage exuded by root cap may be of much importance as it will increase the unstirred layer around the root apex helping the root to maintain the organic acid concentration sufficient to protect the root cap (Henderson and Ownby 1991). And hence, Al tolerance of plant may also be determined by its ability to exude mucilage around the root cap. The view is substantiated further from the observation that the root border cells (the living cells surrounding the root apices) of Al-tolerant cultivar (cv Dade) of *Phaseolus vulgaris* produces a thicker mucilage layer than the Al-sensitive cultivar (cv Romano) in response to Al treatment (Miyasaka and Hawes 2001).

### 3.2 Resistance Mediated by Intracellular Sequestration

Al is also known to be sequestered inside the cell by complex formation with organic acids converting the metal to almost inactive and non-toxic forms. At least two organic acids are known to function as chelators. One is citric acid (Ma et al. 1997c): nearly two-third Al in hydrangea leaves remain present in the cell sap in soluble form as Al-citrate complex at a 1:1 molar ratio of Al to citrate, a non-toxic form of Al. Another acid, which has been reported to form intracellular complex with Al is oxalic acid (Ma et al. 1998). About 90% Al in buckwheat remain present as soluble oxalate-Al complex in the symplasm, and the intracellular concentration of Al detected is as high as 2 mM. The complex occurs in molar ratio of 1:3, Al:oxalate. Oxalic acid can form three species of complexes with Al at an Al to oxalic acid molar ratio of 1:1, 1:2 and 1:3, but 1:3 Al-oxalate complex is the most stable, with a stability constant of 12.4 (Nordstrom and May 1996). This stability constant is much higher than that of Al-citrate (8.1) or Al:ATP (10.9), meaning that formation of 1:3 Al-oxalate complex can prevent binding of Al to cellular components, thereby detoxifying Al very effectively. The report is in contrast to the order of stability constant for Al-organic acid complexes: Al-citrate>Al-oxalate>Al-malate (Zheng et al. 1998b). It is, however, not known whether the Al complexes of citrate or oxalate remain located in cytoplasm or in the vacuole.

## 3.3 Antioxidative System in Al Tolerance

### 3.3.1 Oxygen and Reactive Oxygen Species

Oxygen, which appeared in the earth's atmosphere mainly as a product of photosynthesis, is a two-edge sword for aerobic organisms: it enables efficient energy production by enzymatic combustion of organic compounds, but at the same time leads to damage of aerobic cells due to the formation of reactive oxygen species (Bartosz 1997). The reactive oxygen species (ROS) generally encountered are superoxide radical ($O_2^{\bullet-}$), hydrogen peroxide ($H_2O_2$), hydroxyl radical ($HO^{\bullet}$) and singlet oxygen ($^1O_2$). These are called ROS because they are more prone to participate in chemical reactions than the molecular $O_2$. The greater reactivity of two of them, $O_2^{\bullet-}$ and $HO^{\bullet}$, is because of their "free radical" nature: a free radical is any species capable of independent existence that contains one or more unpaired electrons (Fig. 3).

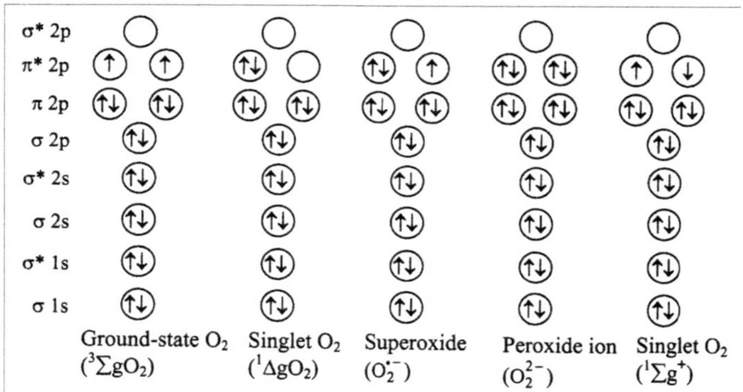


**Fig. 3.** Electronic configuration of oxygen molecule and its derivatives. In covalent compounds the atomic orbitals interact to form molecular orbitals and the electrons occupy these molecular orbitals. The number of the molecular orbitals are twice that of the number of atomic orbitals; for example interaction of 3 2p orbitals ($2p_x$, $2p_y$ and $2p_z$) of two oxygen atoms will result in the formation of 6 2p orbitals, 3 bonding, designated as $\sigma 2p_x$, $\pi 2p_y$ and $\pi 2p_z$, and 3 antibonding, designated as $\sigma{*}2p_x$, $\pi{*}2p_y$ and $\pi{*}2p_z$ (please note one of the 2p orbitals forms σ bond and the other two π bonds). The energy of the antibonding orbitals is higher than the respective bonding orbitals, and that of the σ* is greater than of π*. The energy of the molecular 1s, 2s and 2p bonding and antibonding orbitals is in the order $\sigma 1s < \sigma{*}1s < \sigma 2s < \sigma{*}2s < \sigma 2p < \pi 2p < \pi{*}2p < \sigma{*}2p$. The orbital with lowest energy level is filled first (Aufbau principle), and all orbitals with equal energy levels receive one electron before any receives two (Hund's rule). Presence of electrons in the antibonding orbitals energetically cancels the bonding of the respective bonding orbital(s). For example the presence of two electrons, one each in the two antibonding orbitals, cancels out one of the π2p bonding orbitals, and hence two oxygen items are effectively joined by a double bond. In fluorine three bonding and two antibonding 2p orbitals are occupied, and hence two fluorine atoms in the fluorine molecule are effectively bond by only a single bond

Going by the definition of free radical, $O_2$ in fact itself qualifies as a free radical (Fig. 3); the ground state $O_2$ has two unpaired electrons, one each in the $2P\pi^*$ (antibonding) molecular orbitals (Halliwell and Guttridge 1985). But its reactivity is restricted because both the unpaired electrons are in the same spin, and thus it must receive only one electron at a time, making the molecule to react only sluggishly with many non-radicals.

### *3.3.2 Oxidative Stress in Plants*

Significant quantities of ROS are in fact commonly produced in various compartments or organelles even under normal condition. To countermine the toxicity of ROS, living organisms posses highly efficient defense system, called antioxidative or antioxidant system, comprising of both non-enzymatic and enzymatic constituents (Fig. 4). The non-enzymatic antioxidants are generally small molecules that include the tripetide glutathione, cysteine, hydroxyquinone, ascorbate (vitamin C), the lipophilic antioxidant $\alpha$-tocopherol, carotenoid pigments, alkaloids, and a variety of other compounds (Larson 1988). The enzymatic antioxidant components include the enzymes capable of removing, neutralizing or scavenging ROS, such as catalase (Cat), peroxidase (Px), ascorbate peroxidase (APx), superoxide dismutase (SOD), glutathione reductase (GR), monodehydroascorbate reductase (MDHAR) and dehydroascorbate reductase (DHAR). The non-enzymatic and enzymatic components work in close co-ordination for an effective removal of ROS. Thus, a sort of balance is maintained between their formation and destruction. A shift in the balance between the prooxidative and antioxidative reactions in favour of the former, or inhibition of the functioning of the antioxidative system will lead to accumulation of the toxic ROS, otherwise called oxidative stress.

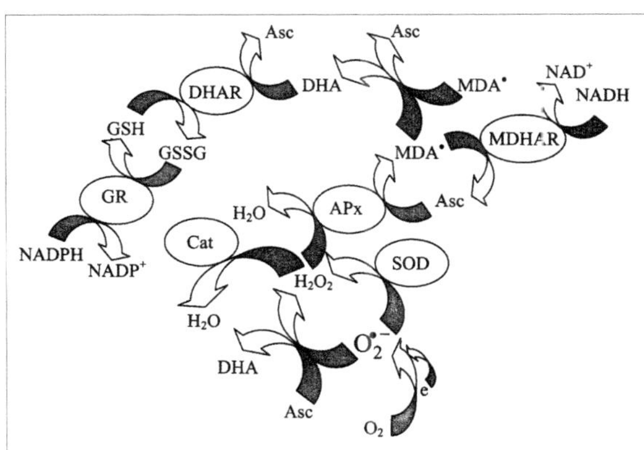


**Fig. 4.** Coordinated functioning of various antioxidative components in plants. Apx, ascorbate peroxidase; Cat, catalase; SOD, superoxide dismutase; MDHAR, monodehydroascorbate reductase; DHAR, dehydroascorbate reductase; GR, glutathione reductase; Asc, ascorbic acid; MDA$^{\bullet}$, monodehydroascorbate radical; DHA, dehydroascorbate; GSH, glutathione; GSSG, oxidized glutathione

### 3.3.3 Al Toxicity Due to Oxidative Damage

Oxidative toxicity of metals in plant is relatively a recent concept, although it is well documented in animal system; studies have shown that the metals such as iron, copper, cadmium, chromium, lead, mercury, nickel and vanadium exhibit ability to produce ROS, resulting in lipid peroxidation, DNA damage, depletion of sulfhydryls, and altered calcium homeostasis (Stoh and Bagchi 1995). The growing bodies of evidences, nevertheless, do suggest that in plants also the toxicities associated with metals may be due at least in part to oxidative damage (Stoh and Bagchi 1995; Shaw 1995a,b; Shaw and Rout 1998; Maksymiec 1997; Lidon and Henriques 1993). In fact, plants, the most of which are green, face greater danger of oxidative damage upon exposure to metals than animals because of the photosynthetic process they carry on, as stated earlier.

The threat of oxidative damage of living tissues by Al, or metals in general, may be due to two reasons: 1) as a result of enhancement in production of ROS, and 2) as a result of slowing down or inhibition of the removal/scavenging of ROS. Both would lead to enhanced accumulation of ROS. And so far as plant is concerned, metals may enhance generation of ROS by interfering with the respiratory processes, similar to that in animal system, and in addition by impairing the photosynthetic processes, specific to plant.

Although all ROS are more or less highly reactive and are toxic to living organisms, the ultimate damaging effect is, however, mainly by $^1O_2$ and $HO^\bullet$. While $^1O_2$ is a result of input of energy, $HO^\bullet$ is produced as a result of Haber-Weiss reaction, i.e. reaction of $O_2^{\bullet-}$ with $H_2O_2$ in the presence of Fe (Shaw et al. 2004). Both the species are extremely reactive, reacting instantly at the site of their generation; while the extreme reactivity of $^1O_2$ is due removal of its spin restriction (Fig. 3), the reactivity of $HO^\bullet$ is a result of having an unpaired electron in its outer orbital (Shaw et al. 2004). Their rapid and non-specific reaction leads to damage of all classes of bio-molecules including lipid, protein, enzyme and DNA (Breen and Murphy 1995; Fridovich 1978; Stadtman 1992; Asada 1992, 1994). Reaction of $HO^\bullet$ and $^1O_2$ with unsaturated fatty acids causes peroxidative degradation of essential lipids in the plasma membrane or the intracellular organelles leading to rapid desiccation and cell death (Halliwell and Gutteridge 1985). Intracellular membrane damage in turn can affect respiratory activity in mitochondria, cause pigment breakdown, and loss of carbon-fixing ability in chloroplasts. Damage to proteins and DNA can often lead to irreparable metabolic dysfunction and cell death (Bartosz 1997; Halliwell and Gutteridge 1985).

Haber-Weiss reaction signifies that the greater the generation of $O_2^{\bullet-}$, the higher will be the chances of formation $HO^\bullet$, and in turn greater would be the chances of peroxidative damage of the membrane lipids. Considering this relationship of $O_2^{\bullet-}$ generation and lipid peroxidation, although not a direct one, the elevated level of MDA (malnodialdehyde) observed in the root tips of soybean exposed to Al for 24 h (Cakmak and Horst 1991) could be a result of enhanced generation of $O_2^{\bullet-}$ due to impairment of mitochondrial electron transport chain by the metals. That oxidative damage through Haber-Weiss reaction could be the prime route of Al toxicity in plants is further supported from the fact that presence of 100 μM Fe(II) along with even only 100 μM Al reduced the viability of cultured tobacco cells by as much as 90% with concomitant highly significant increase in lipid peroxidation, while Al alone was not toxic to the cells even at 300μM concentration (Ono et al. 1995; Yamamoto et al. 1997). Enhanced accumulation of MDA, the end product of lipid peroxidation has also been reported in the leaves of plants exposed to Al (Guo et al. 2004; Kuo and Kao 2003). However, the source of ROS, mitochondria or chloroplast, was not investigated.

The oxidative damage by metals by inhibition of removal of ROS is mediated through the inhibition of the functioning of one or more of the enzymes and/or depletion of one or more of the antioxidant molecules of the antioxidative system by the metals. Several reports are available to substantiate this view. For example Cakmak and Horst (1991) observed that the increase in the MDA content of the root tips of soybean (Glycine max) exposed to Al for 48 h was concomitant with significant decrease in catalase activity. They also observed a significant increase in SOD activity. They concluded that while the increase of SOD activity resulted in enhanced formation of $H_2O_2$, the latter accumulated as a result of decrease of the catalase activity, and the accumulated $H_2O_2$ was mostly consumed in oxidative processes leading to enhanced accumulation of MDA. It is well established that $H_2O_2$ gives rise to highly reactive $HO^{\bullet}$ radical through Fenton reaction (Shaw et al. 2004). Working on rice, Kuo and Kao (2003) observed similar relationship between MDA accumulation and the activity of the antioxidative enzymes in response to Al.

The concept of inhibition of removal of ROS (because of inhibition of the antioxidative enzymes) as the cause of oxidative damage by Al, or metals in general, is, however, not widely accepted, and the oxidative damage due the inhibition of removal of ROS is considered to be of much less significance than that due to additional generation of ROS. This is because highly significant increase in the level of MDA has been reported despite no significant change in the activity of the $H_2O_2$ scavenging enzyme, catalase in the root tips of soybean exposed to Al for 24 h (Cakmak and Horst 1991), or even upon increase in the $H_2O_2$ scavenging enzyme, peroxidase in the leaves of Al-sensitive genotype of barley upon 40 days of exposure to the metal (Guo et al. 2004). In fact, the activity of the antioxidative enzymes has been reported to increase significantly in response to environmental stress in general, and the increase in their activity is considered as a circumstantial evidence of induction of oxidative stress by an environmental stress (Foyer et al. 1994; Polle et al. 2000; Kangasjarvi et al. 1994; Rout and Shaw 2001).

### *3.3.4 Possible Role of the Antioxidative System in Al Tolerance*

It is generally being considered that virtually all the biochemical effects of metals may ultimately lead to oxidative damage of cells and tissues (Shaw et al. 2004). And hence, arguments are also being placed that heavy metal tolerance could be to some extent also be linked with reactive oxygen scavenging capability of a plant species (Stroinski 1999). But there is little direct evidence to prove this hypothesis. Indirect evidences, however, do suggest such relationship. For example Cakmak and Horst (1991) observed significant increase the activity of peroxidase, one of the $H_2O_2$ scavenging enzymes, in soybean root in response to Al treatment with concomitant increase in MDA content, indicating that plants respond to the oxidative stress by increasing the activity of one or more of their antioxidative enzymes. Subsequently Ezaki et al. (1996) reported Al-stress induced appearance of two cationic peroxidases and two moderately anionic peroxidases in tobacco cells. They also produced evidence that at least one of the isoenzyme was produced by enhanced expression of *pAL201* gene, and opined the possibility of the isoenzyme to have some function in Al resistance. It has also been observed that Al-resistant plant genotype accumulates less MDA and shows greater increase in the activity of the SOD and peroxidase than the –sensitive one (Guo et al. 2004). Furthermore, recently Darko et al. (2004) opined that among the antioxidative enzymes catalse and glutathione-*S*-transferase (GST) might be important for the detoxification of reactive oxygen species in the Al-tolerant wheat lines as the activities of these enzymes were significantly higher in the Al-tolerant plants than in their Al-sensitive genotype.

The observations of various workers presented above although do suggest an active involvement of the antioxidative components in Al tolerance, besides the possible involvement of the other processes, it must be kept in mind that contradictory observations have also been reported (Shaw et al. 2004). Furthermore, the database in support of the involvement of antioxidative machinery in Al tolerance, or in metal tolerance in general, is very limited, particularly the observation from the studies involving metal-tolerant and -sensitive varieties of a species. Hence, at this stage it will be premature to draw a definite conclusion in favour of the involvement of antioxidative system in Al/metal tolerance in plants, and it will be wise if at present the idea is treated only as a supposition. Nevertheless, it would be worth mentioning that *Arabidopsis* transgenic line, AtPox(4-1) showing enhanced expression of peroxidase shows significantly less lipid peroxidation upon exposure to Al and greater tolerance to the metal than the non-transgenic plant (Ezaki et al. 2001).

## 4. Conclusion

Thus, we see that metals, including aluminium, are nature's gift to mankind, and the modern civilization would not have developed without bringing them into use. But at the same time they are very toxic to the living organism, and hence suitable measures must be taken to prevent excessive exposure of mankind to them, and to immobilize them in the areas of their "hot-spots". It is increasingly been realized that this can be achieved by the use of plants, through phytoremediation. So far as Al is concerned, its presence at elevated levels in soil, which is likely due to its high crustal abundance, particularly in acidic soil, which constitutes nearly 40 % of the arable lands the world over, is also associated with agricultural losses. And hence, keeping in view the agronomic importance of Al toxicity, it is necessary to improve Al tolerance of crop plants. However, since many genes could be involved in the tolerance process, it is prerequisite to have a clear understanding of the metabolic pathways leading to tolerance before attempting to improve tolerance of a crop for the metal using biotechnological approach. And in this regard it is encouraging to note that resistance to Al is mostly due to its exclusion, which is a highly required character for a crop plant so that trophic-level transfer of the metal is avoided.

The exclusion mediated Al resistance is mostly due to secretion of organic acids (by the root apex), which form complexes with the metal making it unavailable to the plants, and/or due to influx of $H^+$, which increases the rhizosphere pH causing $Al^{3+}$ species in proximity with the root to get converted to less toxic and less available forms. Nevertheless, resistance to Al due to intracellular complex formation with oxalic acid has also been reported. Resistance of plants to Al by its exclusion is in contrast to the reports available for the metal in general where the resistance is achieved by their intracellular sequestration inside the vacuoles, which is believed to be mostly mediated through the formation of complex with phytochelatins, the non-translationally synthesized low molecular wt polypeptides. Further, the involvement of antioxidative machinery in Al tolerance is increasingly being advocated, as for the heavy metals, but is not sufficiently substantiated. The exclusion based tolerance of a few crop plants to Al is although encouraging from the point of few of utilizing the information for engineering tolerance to the metal in the crop of interest, the information is, however, not adequate in this regard also. Study is totally lacking on how the plant (root apex) perceive the presence of the metal (Al) in the soil. It is only after acquiring the knowledge on this signal transduction mechanism that it may be possible to achieve the goal of over-coming Al-associated agronomic losses.

Acknowledgments. The authors express their gratitude to the Director, Institute of Life Sciences and to CSIR and UGC, New Delhi for financial support in preparation of the manuscript.

## References

Ahrland S (1968) Thermodynamics of complex formation between hard and soft acceptors and donors. Nature and scope of the classification of acceptors and donors as hard and soft. Struct Bonding 5:118-123

Akeson MA, Munns DN, Burau RG (1989) Adsorption of $Al^{3+}$ to phosphatidylcholine vesicles. Biochim Biophys Acta 986:33-40

Asada K (1994) Production and action of active oxygen species in photosynthetic tissues. In: Foyer CH, Mullineaux PM (eds) Causes of Photooxidative Stress and Amelioration of Defense Systems in Plants, CRC Press, Boca-Raton Ann-Arbor London Tokyo, pp 77-104

Asada K (1992) Production and scavenging of active oxygen in chloroplasts. In: Scandalios JG (ed) Photoinhibition, Cold Spring Harbor Laboratory Press, Cold Spring Harbor, pp173-192

Bartosz G (1997) Oxidative stress in plants. Acta Physiol Plant 19:47-64

Basu U, Godbold D, Taylor GJ (1994) Aluminum resistance in *Triticum aestivum* associated with enhanced exudation of malate. J Plant Physiol 144:747-753

Berlett RJ, Riego DC (1972) Effect of chelation on the toxicity of aluminum. Plant Soil 37:419-423

Boscolo PRS, Menossi M, Jorgea RA (2003) Aluminium-induced oxidative stress in maize. Phytochemistry 62:181-189

Bradshaw AD (1952) Populations of *Agrostis tenuis* resistant to lead and zinc poisoning. Nature 169:1098.

Breen AP, Murphy JA (1995) Reactions of oxyl radicals with DNA. Free Rad Biol Med 18:1033-1077

Cakmak I, Horst WJ (1991) Effect of aluminium on lipid peroxidation, superoxide dismutase, catalase, and peroxidase activities in root tips of soybean (*Glycine max*). Physiol Plant 83:463-468

Chaterjee KK (1993) An Introduction to Mineral Economics, Wiley Eastern Limited, Bombay, New Delhi, Calcutta, Hyderabad, Banglore, Guwahati, Lucknow, Madras, Pune

Cunningham SD, Ow DW (1996) Promises and prospects of phytoremediation. Plant Physiol 110:715-719

Darko E, Ambrus H, Stefanovits-Banyai E, Fodor J, Bakos F, Barnabas B (2004) Aluminium toxicity, Al tolerance and oxidative stress in an Al-sensitive wheat genotype and in Al-tolerant lines developed by in vitro microspore selection. Plant Sci 166:583-591

Dean JG, Bosqui FL, Lanouette VH (1972) Removing heavy metals from waste water. Environ Sci Technol 6:518-525

Degenhardt J, Larsen PB, Howell SH, Kochian LV (1998) Aluminum resistance in the Arabidopsis mutant *alr*-104 is caused by an aluminum-induced increase in rhizosphere pH. Plant Physiol 117:19-27

Delhaize E, Ryan PR (1995) Aluminum toxicity and tolerance in plants. Plant Physiol 107:315-321

Delhaize E, Craig S, Beaton CD, Bennet RJ, Jagadish VC, Randall PJ (1993a) Aluminum tolerance in wheat (*Triticum aestivum* L.) I. Uptake and distribution of aluminum in root apices. Plant Physiol 103:685-693

Delhaize E, Ryan PR, Randall PJ (1993b) Aluminum tolerance in wheat (*Triticum aestivum* L.). II. Aluminum-stimulated excretion of malic acid from root apices. Plant Physiol 103:695-702

Dinkelaker B, Romheld V, Marschner H (1989) Citric acid excretion and precipitation of calcium citrate in the rhizosphere of white lupin (*Lupinus albus* L.). Plant Cell Environ 12:285-292

Ernst WHO, Verkleij JAC, Schat H (1992) Metal tolerance in plants. Acta Bot Neerl 41:229-248

Evans AM (1995) Ore, mineral economics and mineral exploration. In: Evans AM (ed) Introduction to Mineral Exploration, Chap. 1, Blackwell Science, Oxford

Ezaki B, Katsuhara M, Kawamura M, Matsumoto H (2001) Different mechanisms of four aluminum (Al)-resistant transgenes for Al toxicity in *Arabidopsis*. Plant Physiol 127:918-927

Fergusson JE (1990) The Heavy Elements: Chemistry, Environmental Impact and Health Effects. Pergamon Press, Oxford New-York Beijing Fankfurt Sao-Paulo Sudney Tokyo Toronto

Forstner U, Wittmann GTW (1979) Metal Pollution in the Aquatic Environment. Springer-Verlag, New York

Foy CD, Burns GR, Brown JC, Fleming AL (1965) Differential aluminum tolerance of two wheat varieties associated with plant-induced pH changes around their roots. Soil Sci Soc Am Proc 29:64-67

Foy CD, Fleming AL, Burns GR, Arninger WH (1967) Characterization of differential aluminum tolerance among varieties of wheat and barley. Soil Sci Soc Am Proc 31:513-52

Fridovich I (1978) The biology of oxygen radicals. Science 201:875-880

Fuente JM, Ramirez-Rodriguez V, Cabrera-Ponce JL, Herrera-Estrella L (1997) Aluminum tolerance in transgenic plants by alteration of citrate synthesis. Science 276:1566-1568

Gardner WK, Barber DA, Parbery DG (1983) The acquisition of phosphorus by *Lupinus albus* L. III. The probable mechanism by which phosphorus movement in the soil/root interface is enhanced. Plant Soil 70:107-124

Guo T, Zhang G, Zhou M, Wu F, Chen J (2004) Effects of aluminium and cadmium toxicity on growth and antioxidant enzyme activities of two barley genotypes with different al resistance. Plant Soil 258:241-248

Halliwell B, Gutteridge JMC (1985) Free Radical in Biology and Medicine. Clarendon Press, Oxford

Haug A (1984) Molecular aspects of aluminum toxicity. Crit Rev Plant Sci 1:345-373

Henderson M, Ownby JD (1991) The role of root cap mucilage secretion in aluminum tolerance in wheat. Curr Top Plant Biochem Physiol 10:134-141

Horst WJ, Asher CJ, Cakmak I, Szulkiewica P, Wissemeier AH (1992) Short-term responses of soybean roots to aluminium. J Plant Physiol 140:174-178

Hue NV, Craddock GR, Adams F (1986) Effect of organic acids on aluminum toxicity in subsoils. Soil Sci Soc Am J 50:28-34

Klimashevsky EL, Bernadskaya ML (1973) The activity of ATPase and acid phosphatase in the root growth zones of two pea varieties with different tolerance to toxic Al ions. Sov Plant Physiol 20:257-263

Kochian LV (1995) Cellular mechanisms of aluminum toxicity and resistance in plants. Annu Rev Plant Physiol Plant Mol Biol 46:237-260

Kollmeier M, Dietrich P, Bauer CS, Horst WJ, Hedrich R (2001) Aluminum activates a citrate-permeable anion channel in the aluminum-sensitive zone of the maize root apex. A comparison between an aluminum-sensitive and an aluminum-resistant cultivar. Plant Physiol 126:397-410

Koyama H, Ojima K, Yamaya T (1990) Utilization of anhydrous aluminum phosphate as a sole source of phosphorus by a selected carrot cell line. Plant Cell Physiol 31:173-177

Kuo MC, Kao CH (2003) Alumiunium effects on lipid peroxidation and antioxidative enzyme activities in rice leaves. Biol Plant 46:149-152

Larsen PB, Degenhardt J, Tai C-Y, Stenzler LM, Howell SH, Kochian LV (1998) Aluminum-resistant Arabidopsis mutants that exhibit altered patterns of aluminum accumulation and organic acid release from roots. Plant Physiol 117:9-18

Larson RA (1988) The antioxidants of higher plants. Phytochemistry 27:969-978

Lazof DB, Goldsmith JG, Rufty TW, Linton RW (1994) Rapid uptake of aluminum into cells of intact soybean root tips. A microanalytical study using secondary ion mass spectroscopy. Plant Physiol 106:1107-1114

Lidon FC, Henriques FS (1993) Copper-mediated toxicity in rice chloroplasts. Photosynthetica 29:385-400

Lipton DS, Blanchar RW, Blevins DG (1987) Citrate, malate, and succinate concentration in exudates from P-sufficient and P-stressed Medicago sativa L. seedlings. Plant Physiol 85:315-317

Ma JF, Hiradate S, Matsumoto H (1998) High aluminum resistance in buckwheat. II. Oxalic acid detoxifies aluminum internally. Plant Physiol 117:753-759

Ma JF, Kiradate S, Nomoto K, Iwashita T, Matsumoto H (1997c) Internal detoxification mechanism of Al in hydrangea. Identification of Al form in the leaves. Plant Physiol 113:1033-1039

Ma JF, Zheng SJ, Matsumoto H (1997a) Specific secretion of citric acid induced by Al stress in *Cassia tora* L. Plant Cell Physiol 38:1019-1025

Ma JF, Zheng SJ, Hiradate S, Matsumoto H (1997b) Detoxifying aluminum with buckwheat. Nature 390:569-570

Maksymiec W (1997) Effect of copper on cellular processes in higher plants. Photosynthetica 34:321-342

Manahan SE (1990) Environmental Chemistry. Lewis Publishers, Boston

Martell AE, Motekaitis RJ (1989) Coordination chemistry and speciation of Al (III) in aqueous solution. In: Lewis I, Timothy E (eds) Environmental Chemistry and Toxicology of Aluminium, Lewis Publishers, Chelsea, MI, pp 3-17

Martin RB (1988) Bioinorganic chemistry of aluminum. In: Siegel H, Siegel A (eds) Metal Ions in Biological Systems: Aluminum and its Role in Biology, vol 24, Marcel Dekker, New York, pp 2-57

Mason B (1958) Principle of Geochemistry. John Wiley & Sons, Inc., New York

Matsumoto H (2000) Cell biology of aluminium toxicity and tolerance in higher plants. Int Rev Cytol 200:1-46

Misra SG, Mani D (1991) Soil pollution. Ashish Publishing House, New Delhi

Miyasaka SC, Hawes MC (2001) Possible role of root border cells in detection and avoidance of aluminum toxicity. Plant Physiol 125:1978-1987

Miyasaka SC, Buta JG, Howell RK, Foy CD (1991) Mechanism of aluminum tolerance in snapbeans. Root exudation of citric acid. Plant Physiol 96:737-743

Miyasaka SC, Kochian LV, Shaff JE, Foy CD (1989) Mechanism of aluminum tolerance in wheat. An investigation of genotypic differences in rhizosphere pH, $K^+$, and $H^+$ transport, and root-cell membrane potentials. Plant Physiol 91:1188-1196

Muchovej RCM, Allen VG, Martens DC, Muchovej JJ (1988) Effects of aluminum chelates in nutrient solution on the growth and composition of ryegrass. J Plant Nutr 11:117-129

Mugwira LM, Patel SU (1977) Root zone pH changes and ion uptake imbalances by triticale, wheat, and rye. Argon J 69:719-722

Mugwira LM, Elgawahry SM, Patel SU (1976) Differential tolerances of triticale, wheat, rye and barley to aluminum in nutrient solution. Agron J 68:782-786

Mugwira LM, Elgawahry SM, Patel SU (1978) Aluminum tolerance in triticale, wheat and rye as measured by root growth characteristics and aluminum concentration. Plant Soil 50:681-690

Nordstrom DK, May HM (1996) Aqueous equilibrium data for mononuclear aluminum species. In: Sposito G (ed) Environment Chemistry of Aluminum, CRC Press, Boca Raton FL, pp 39-80

Ochiai E-I (1977) Bioinorganic Chemistry: An Introduction. Allyn and Bacon, Boston London Sydney Toronto

Ojima K, Ohira K (1988) Aluminum-tolerance and citric acid release from a stress-selected cell line of carrot. Commun Soil Sci Plant Anal 19:1229-1238

Ojima K, Abe H, Ohira K (1984) Release of citric acid into the medium by aluminum-tolerant carrot cells. Plant Cell Physiol 25:855-858

Ojima K, Koyama H, Suzuki R, Yamaya T (1989) Characterization of two tobacco cell lines selected to grow in the presence of either ionic Al or insoluble Al-phosphate. Soil Sci Plant Nutr 35:545-551

Ono K, Yamamoto Y, Hachiya A, Matsumoto H (1995) Synergistic inhibition of growth by Al and iron of tobacco (*Nicotiana tabacum* L.) cells in suspension culture. Plant Cell Physiol 36:115-125

Osawa H, Matsumoto H (2001) Possible involvement of protein phosphorylation in aluminum-responsive malate efflux from wheat root apex. Plant Physiol 126:411-420

Pearson R (1968a) Hard and soft acids and bases, HSAB, Part I, Fundamental principles. J Chem Educ 45:581-587

Pearson R (1968b) Hard and soft acids and bases, HSAB, Part II, Underlying theories. J Chem Educ 45:643-648

Pellet DM, Grunes DL, Kochian LV (1995) Organic acid exudation as an aluminum-tolerance mechanism in maize (*Zea mays* L.). Planta 196:788-795

Pellet DM, Papernik LA, Kochian LV (1996) Multiple aluminum-resistance mechanisms in wheat: Roles of root apical phosphate and malate exudation. Plant Physiol 112:591-597

Peterson PJ (1969) The distribution of zinc-65 in *Agrostis tenuis* Sibth. and *A. stolonifera* L. tissues. J Exp Bot 20:863-875

Reilly C (1967) Accumulation of copper by some Zambian plants. Nature 215:667-668

Rincon M, Gonzales RA (1992) Aluminum partitioning in intact roots of aluminum-tolerant and aluminum-sensitive wheat (*Triticum aestivum* L.) cultivars. Plant Physiol 99:1021-1028

Ryan PR, Delhaize E, Randall PJ (1995a) Characterization of Al-stimulated efflux of malate from the apices of Al-tolerant wheat roots. Planta 196:103-110

Ryan PR, Delhaize E, Randall PJ (1995b) Malate efflux from root apices: evidence for a general mechanism of Al-tolerance in wheat. Aust J Plant Physiol 22:531-536

Ryan PR, DiTomaso JM, Kochian LV (1993) Aluminium toxicity in roots: an investigation of spatial sensitivity and the role of the root cap. J Exp Bot 44:437-446

Schat H, Ten Bookum WM (1992) Genetic control of copper tolerance in *Silene vulgaris*. Heredity 68:219-229

Shaw BP (1995a) Changes in the levels of photosynthetic pigments in Phaseolus aureus Roxb. exposed to Hg and Cd at two stages of development: a comparative study. Bull Environ Contam Toxicol 55:574-580

Shaw BP (1995b) Effects of mercury and cadmium on the activities of antioxidative enzymes in the seedlings of Phaseolus aureus Roxb. Biol Plant 37:587-596

Shaw BP, Rout NP (1998) Age-dependent responses of *Phaseolus aureus* Roxb. to inorganic salts of mercury and cadmium. Acta Physiol Plant 20:85-90

Shaw BP, Sahu SK, Mishra RK (2004) Heavy metal induced oxidative damage in terrestrial plants. In: Prasad MNV (ed) Heavy Metal Stress in Plants – From Biomolecules to Ecosystems, Springer-Verlag, Heidelberg, pp 84-126

Shuman LM, Wilson DO, Ramseur EL (1991) Amelioration of aluminum toxicity to sorghum seedlings by chelating agents. J Plant Nutr 14:119-128

Stadtman ER (1992) Protein oxidation and aging. Science 257:1220-1224

Stohs SJ, Bagchi D (1995) Oxidative mechanisms in the toxicity of metal ions. Free Rad Biol Med 18:321-336

Strange J, Macnair MR (1991) Evidence for a role of cell membrane in copper tolerance of *Mimulus guttatus* Fischer ex DC. New Phytol 119:383-388

Tamas L, Simonovicova M, Hurrova J, Mistrik I (2004) Aluminium stimulated hydrogen peroxide production of germinating barley seeds. Environ Exp Bot 51:281-288

Taylor GL (1991) Current views of the aluminum stress response: the physiological basis of tolerance. Curr Top Plant Biochem Physiol 10:57-93

Trice KR, Parker DR, DeMason DA (1992) Operationally defined apoplastic and symplastic aluminum fractions in root tips of aluminum-intoxicated wheat. Plant Physiol 100:309-318

Wedepohl KH (2000) The composition of the upper earth's atmosphere and the natural cycles of selected metals. Metals in natural raw materials. Natural resources. In: Merian E (ed) Metals and Their Compounds in the Environment: Occurrence, Analysis and Biological Relevance, Chap. 1, John Wiley & Sons, Inc., New York

Yamamoto Y, Hachiya A, Matsumoto H (1997) Oxidative damage to membranes by a combination of aluminum and iron in suspension-cultured tobacco cells. Plant Cell Physiol 38:1333-1339

Zhang W-H, Ryan PR, Tyerman SD (2001) Malate-permeable channels and cation channels activated by aluminum in the apical cells of wheat roots. Plant Physiol 125:1459-1472

Zheng SJ, Ma JF, Matsumoto H (1998a) High aluminum resistance in buckwheat. I. Al-induced specific secretion of oxalic acid from root tips. Plant Physiol 117:745-751

Zheng SJ, Ma JF, Matsumoto H (1998b) Continuous secretion of organic acids is related to aluminum resistance during relatively long-term exposure to aluminum stress. Physiol Plant 103:209-214

7

# Bioremediation of Metals: Microbial Processes and Techniques

K. Ramasamy, Kamaludeen and Sara Parwin Banu

Centre for Plant Molecular Biology, Tamil Nadu Agricultural University, Coimbatore 641 003, Tamil Nadu, INDIA, Email: ramasamytnau@yahoo.com

## 1. Introduction

Bioremediation is a technology that uses metabolic processes to degrade or transform contaminants, so that they remain no longer in harmful form. In some cases, the contaminant is the primary part of the metabolic process, acting as a main source of carbon and energy for the microbial cell. In others, it is transformed into a second substance, serves as a primary energy or carbon source. This co-metabolism process may be purely fortuitous, and the microorganism gains nothing from the process. In case of metals, it is only the biotransformation process that was exploited widely as a bioremediation strategy. After the use of super bug in cleaning up oil spills, there has been numerous successful stories of bioremediation technique in clean-up of vast areas of contaminated environments (USGS 1997).

This chapter focuses on the role of metal-microbial relationships, microbial processes governing bioremediation and various techniques available for metal-contaminated sites. This chapter also throws light on bioremediation techniques used exclusively for chromium-contaminated soils and possible future developments in the field of bioremediation.

## 2. Metals and Microbes

Metals play an integral role in the life processes of microbes. Some metals, such as Cr, Ca, Mg, Mn, Cu, Na, Ni and Zn are essential as micronutrients for various metabolic activities and for redox processes. Toxicity of metals occur through the displacement of essential metals from their active binding sites or through ligand interactions (Bruins et al. 2000). Most of the metal ions enter the microbial cell to have a physiological toxic effect. Many divalent metal cations like $Mn^{2+}$, $Fe^{2+}$ and $Zn^{2+}$ are very similar in structure. Also, the structure of oxyanions, such as chromate, resembles that of sulphate. Thus, to be able to differentiate between very similar metal ions, the microbial uptake systems have to be tightly regulated. Usually the microbes have solved this problem by using two types of uptake systems of metal ions. One is fast, non-specific and driven by chemiosmotic gradient across the cytoplasmic membrane of the bacteria (Nies 1999). The second type of uptake system has high substrate specificity, is slower and often uses ATP hydrolysis as the energy source and is only produced by the cell in times of need (Nies and Silver 1995).

Though, there are specific uptake systems, high concentrations of non-essential metals may be transported into the cell by a constitutively expressed non-specific system. This open gate is the one reason why metal ions are toxic to microbes. As a consequence, microbes have been forced to develop metal ion homeostasis factors and metal resistance determinants (Nies and Silver 1995; Nies 1999). As metal ions cannot be degraded or modified like toxic organic compounds, there are six possible mechanisms for a metal resistance system; i) exclusion by permeability barrier, intra- and extra-cellular sequestration, active efflux pumps, enzymatic reduction and reduction in the sensitivity of cellular targets to metal ions (Ji and Silver 1995; Nies and Silver 1995, Bruins et al. 2000). One or more of these resistance mechanisms allows microorganisms to function in metal contaminated environments.

## 2.1 Metals Microbe Interactions and Periodic Table

In recent days, efforts are made to depict metabolism in the context of the full constellation of chemical elements. Most metabolism databases also deal only with limited number of chemical elements, principally C, H, O, N, P and S. With both biological functions and chemical properties in mind, one permutation was arranged by Wackett et al. (2004) as depicted in Figure 1A&B.

A key feature of the spiral element depicted in Fig. 1B is the centrality of hydrogen as more than 60% of the microbiological biomass is $H_2O$, most microbial enzymes effect $H^+$ transfer, $H^+$ gradients are widely used in ATP generation and H-bonding is crucial for the stability of major biomacromolecules. Also, most of the prokaryotes are known to contain hydrogenases. Next elements in the series are C, O, N and S, which are often bonded, together in structural and metabolic compounds. Elemental cations (Na, K, Ca, Mg, Na) also play a major role in microbial metabolism therein affecting the nature of metal species prevalent. Though rubidium and barium are not of concern as radioactive pollutants, their absence resulted in some abnormal growth functions (Bruce and Duff 1968). Chloride, the major element anion is also present in soil, water and microbial cells in the form of elemental chlorine. Chloride is required by some halophiles for their metabolism. Though chlorine oxyanions are mainly used as disinfectants some bacteria can use perchlorate as terminal electron acceptors (Coates et al. 1999). The transition elements like Zn function normally as enzyme catalysts. In brief, the diversity of prokaryotes is so enormous that it can access most of the metals in the periodic table in either oxidized or reduced form based on their need. A better understanding of most chemistry and microbial metabolism has to be unravelled in detail for a thorough understanding of nature of metals in natural environments, which is main task to be resolved in contaminated environments.

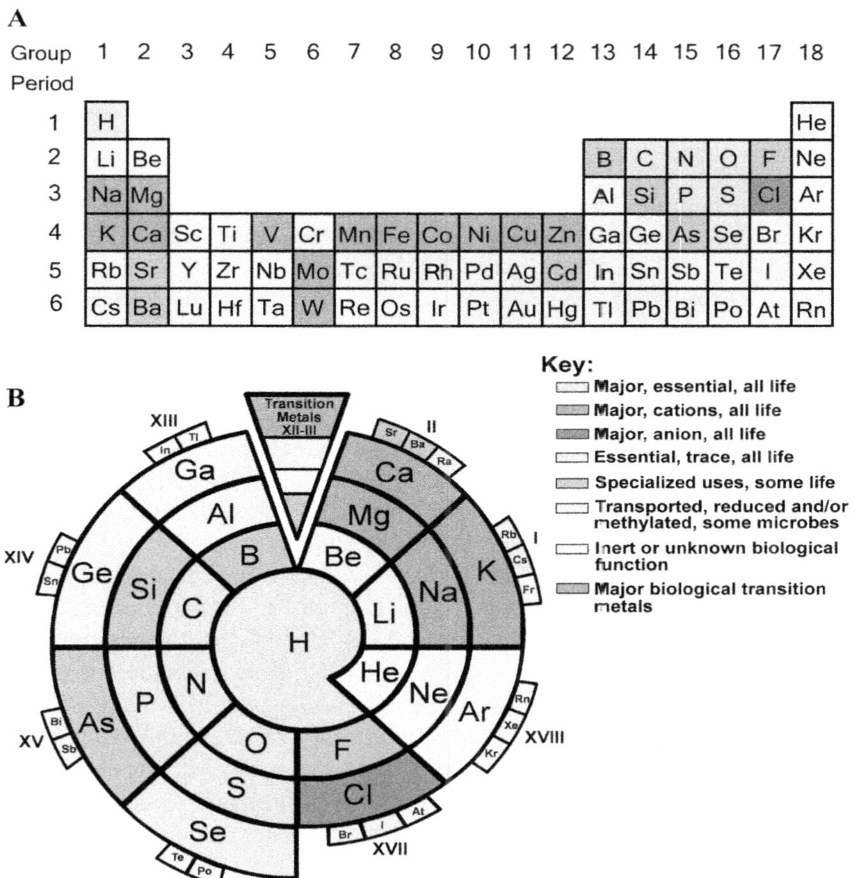


**Fig. 1.** Periodic representation of elements. A. Conventional periodic table B. Spiral representation of elements clustering prominent elements in biological systems

### 2.2 Metal Contaminated Environments

Mineral rock weathering and anthropogenic sources provide two of the main types of metal inputs to soils. According to Ross (1994), the anthropogenic sources of metal contamination can be divided into five major groups; metalliferous mining and smelting, industrial source, atmospheric deposition, agriculture and waste disposal practices. In India, the contamination is mainly due to industrial activities and indiscriminate waste disposal practices.

## 2.3 Microbial Transformations of Metals

Microbial transformation of metals serve various functions. Generally, microbial transformations of metals occur either by redox conversions of inorganic forms or conversions from inorganic to organic forms and *vice versa* (Tebo et al. 1997). On the other hand, reduction of metals can occur through dissimilatory metal reduction, where microbes utilize metals as terminal electron acceptors for anaerobic respiration (Lovley and Coates 1997). In addition, microbes may possess reduction mechanisms that are not coupled to respiration, but instead are thought to impart metal resistance. For example, aerobic and anaerobic reduction of Cr(VI) to Cr(III) (Cifuentes et al. 1996; Fude et al. 1994; Ramasamy 2000), reduction of Se(VI) to elemental Se (Lloyd et al. 2001), reduction of U(VI) to U(IV) (Chang et al. 2001) and reduction of Hg(II) to Hg(0) (Brim et al. 2000) are widespread detoxification mechanisms among microbes. Microbial methylation plays an important role, because methylated compounds are often volatile. Mercury, Hg(II) can be biomethylated by a number of different bacterial species (*Pseudomonas* sp., *Escherichia* sp., *Bacillus* sp., *Clostridium* sp.) to gaseous methyl mercury (Pongratz and Heumann 1999). This is the most toxic and most accumulated form of Hg (Nikunen et al. 1990). Also biomethylation of arsenic to gaseous arsines (Gao and Burau 1997), selenium to volatile dimethyl selenide (Dungan and Frankenberger 2000) and lead to dimethyl lead has been observed in various contaminated environments. In addition to redox-conversions and methylation reactions, acidophilic iron and sulfur oxidizing bacteria are able to leach high concentrations of As, Cd, Cu Co and Zn from contaminated soils. On the other hand metals can be precipitated as insoluble sulfides indirectly by the metabolic activity of sulphate reducing bacteria (White et al. 1997). Sulphate reducing bacteria are anaerobic heterotrophs utilising a range of organic substrates with $SO_4^{2-}$ as the terminal electron acceptor. The half reaction reduction potentials are given in Table 1 which is of great significance in natural environments.

**Table 1.** Microbially significant half reaction reduction potentials

| Redox pairs | | $E_0$ (V) |
|---|---|---|
| $O_2 + 4H^+ + 4e^-$ | $\rightarrow 2\,H_2O$ | +1.229 |
| $MnO_2 + 4H^+ + 2e^-$ | $\rightarrow Mn^{2+} + 2H_2O$ | +1.208 |
| $NO_3^- + 2H^+ + 2e^-$ | $\rightarrow NO_2^- + H_2O$ | +0.94 |
| $Fe^{3+} + e^-$ | $\rightarrow Fe^{2+}$ | +0.77 |
| $SO_4^{2-} + 4H^+ + 2e^-$ | $\rightarrow H_2SO_3 + H_2O$ | +0.20 |
| $2H^+ + 2e^-$ | $\rightarrow H_2$ | 0.0 |

(Tinoko et al. 1985)

In summary, microbial processes can either solubilise metals, thereby increasing their bioavailability and potential toxicity or immobilize them and thereby reduce the bioavailability of metals. These biotransformations are important components of biogeochemical cycles of metals and may be exploited in the bioremediation of metal contaminated soils (Lovley and Coates 1997, Lloyd and Lovley 2001).

## 3. Microbial Processes Affecting Bioremediation of Metals

Bioremediation of metals is achieved through biotransformation. There are atleast three major microbial processes that influence the bioremediation of metals (Fig. 2).

- Biosorption and bioaccumulation
- Biologically catalysed immobilization and
- Biologically catalysed solubilisation

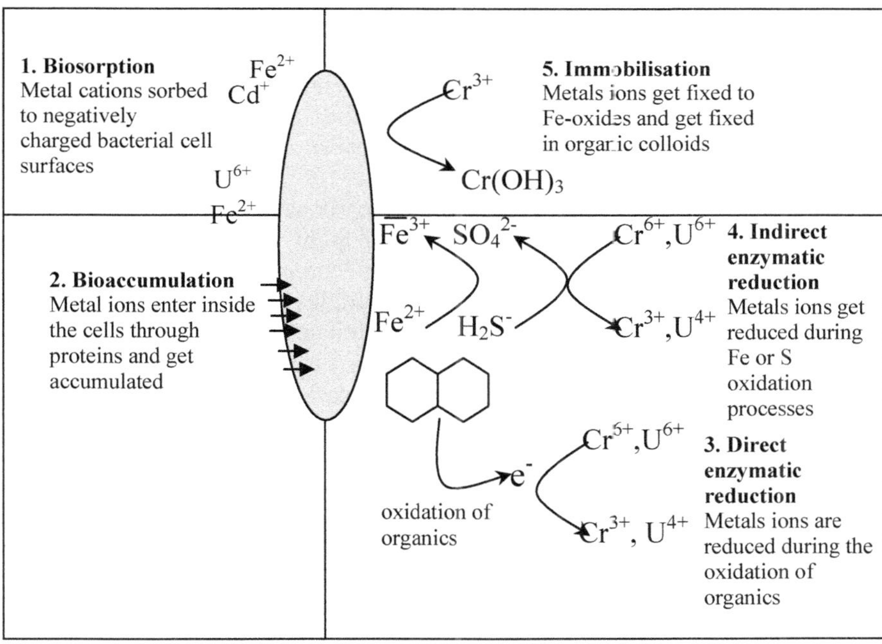


**Fig. 2.** Microbial processes used in bioremediation technologies

### 3.1 Biosorption and Bioaccumulation

Biosorption is the sequestration of the positively charged metal ions to the negatively charged cell membranes and polysaccharides secreted in most of the bacteria on the outer surfaces through slime and capsule formation. Bioaccumulation is the retention and concentration of a substance by an organism. The metals are transported from the outside of the microbial cell through the cell membrane and into the cell cytoplasm. The metal is sequestered and becomes immobile inside the cell (Losi et al. 1994).

### 3.2 Biologically Catalysed Immobilization

Metal reducing microorganisms reduce a wide variety of metals. Direct enzymatic reduction involves the use of oxidized form of metals as electron acceptors {Cr(VI), U(VI)}. The oxidized forms of these metals are highly soluble and hence pose the danger of groundwater contamination. However, the reduced forms are highly insoluble and precipitated. Studies have also found that bioreduction of hexavalent chromium can occur in aerobic and anaerobic conditions. A number of Cr(VI)-reducing microbial strains have recently been isolated from chromate-contaminated waters, soils, and sediments, including *Oscillatoria* sp., *Arthrobacter* sp., *Agrobacter* sp., *Pseudomonas aeruginosa* S128, *Chlamydomonas* sp. (algae), *Chlorella vulgaris* (algae), *Zoogloea ramigera*, and anaerobic sulfate-reducing bacteria (Kamaludeen et al. 2003). A wide range of bacteria, including *Enterobacter cloacea* and all known metal reducing bacteria, reduce the highly soluble chromate ion to Cr(III), which under appropriate conditions precipitates as $Cr(OH)_3$ (Komori et al. 1989 1990).

Metal-reducing organisms reduce uranyl carbonate, which is exceedingly soluble in carbonate-bearing groundwater, to highly insoluble U(IV), which precipitates from solution as the uranium oxide mineral uraninite. Recently, scientists have had success in microbial binding of U(VI), which is then converted by the living cells to U(IV) and precipitated intracellularly (Lovley et al. 1993; Anderson et al. 2003, GNN 2003).

Although some microorganisms can enzymatically reduce heavy metals directly, indirect reduction of soluble contaminants may be more feasible in natural sedimentary and subsurface environments. This indirect immobilization could be accomplished by metal-reducing and sulfate-reducing bacteria. This can be achieved by coupling the oxidation of organic compounds or hydrogen to the reduction of ferric iron [Fe(III)], Mn(IV), or sulfate ($SO_4^{2-}$) (Lovley and Phillips 1988; Lovley et al. 1989). In this way, iron(III) is reduced to iron(II), manganese(IV) to manganese (II), and $SO_4^{2-}$ to hydrogen sulfide ($H_2S$). The reduced form then chemically interacts with the contaminants and forms separate or multicomponent insoluble species. The most reactive of these reduced forms are Fe(II) and $H_2S$. Ferrous iron [Fe(II)], which is generated by the enzymatic activity of iron reducing and some fermentative bacteria, can reduce multivalent metals such as uranium and chromium. The use of Fe(II) as an electron donor for reduction and precipitation of chromium contaminated soils have been widely studied using chemically iron barriers and also by using

Fe-reducing bacteria (Buerge and Hug 1998; Wielenga et al. 2001). Sulfate-reducing bacteria also may be stimulated to produce a chemically reactive redox barrier. Hydrogen sulfide generated by sulfate-reducing bacteria could chemically reduce the contaminant directly or indirectly in the case of sulfide minerals such as pyrite that would be chemically stable for extended periods of time.

### 3.3 Biologically Catalysed Solubilisation

Solubilization of biosorbed and co-precipitated metals also can occur by direct or indirect microbial processes. However, the solubilization of toxic heavy metals from co-precipitates requires at least partial solubilization of the oxide mineral itself. Bacteria can catalyze the dissolution of iron oxide minerals by direct and indirect mechanisms. As previously described, metal-reducing bacteria enzymatically reduce and, under proper environmental conditions, solubilize oxide minerals. Such dissolution reactions have been shown to release cadmium, nickel, and zinc into solution during reduction of goethite (a form of iron oxide) by an anaerobic *Clostridium* species. Direct reduction of iron oxide precipitates by metal-reducing bacteria has been shown to release soluble radium from uranium mine tailings. Metal-reducing bacteria also can promote the mobilization of insoluble forms of some heavy metals.

## 4. Bioremediation Options for Metal Contaminated Sites

In the recent years, there is tremendous increase in utilization of bioremediation invariably for all types of pollutants starting from rare metals to radionuclides. Native microorganisms in any contaminated site are acclimatized and were capable of transforming the toxic metals to their oxides or hydroxides. Some of the promising and successful bioremediation techniques are given as below:

### 4.1 Intrinsic Bioremediation

This technique has gained popularity, as the contaminant in the place itself and cuts down the excavation cost. Intrinsic bioremediation is done *in-situ* and relies on naturally occurring biological processes carried out by indigenous microorganisms. Intrinsic bioremediation is a component of natural attenuation, which includes physical and chemical processes (Hinchee and Wilson 1995). This technique is very successful in organically polluted soil especially with PAHs. However, promising results have been obtained with intrinsic bioremediation of selenium polluted agricultural drainage water in marsh lands (NABIR).

### 4.2 Biostimulation

Biostimulation is the addition of nutrients (usually sources of carbon, nitrogen, phosphorus), oxygen or other electron donors or acceptors. These amendments serve to increase the number or activity of naturally occurring microorganisms available for bioremediation. Amendments can be added in either liquid or gaseous form, *via* injection. Liquids can be injected into shallow or deep aquifers to stimulate the growth of microorganisms involved in the bioremediation.

### 4.3 Bioaugmentation

Bioaugmentation is the addition of microorganisms that can biotransform or biodegrade a particular contaminant. This process can be enhanced by the continuous addition of microorganisms to a bioreactor for the above-ground treatment of groundwaters. Commercial inoculants of enriched cultures consisting of one or more microbial species have been successfully used to colonize contaminated environments where the intrinsic microbial communities act on metals.

Bioremediation depends on the presence of the appropriate microorganisms in the correct amounts and in combinations and in the appropriate environmental conditions. Microorganisms already living in contaminated environments are often well adapted to survival in the presence of existing contaminants and to the temperature, pH and Eh of the site. These indigenous microbes tend to utilize the nutrients and electron acceptors that are available in-situ, provided moisture is present. Presence of moisture acts as a vehicle to transport both microbes and dissolved substances, including contaminants and their breakdown products .

Bioremediation works either by transforming or degrading contaminants to less hazardous chemicals or innocuous substances. In case of metal contaminated sites, the microbes interact with metals and transform them from one chemical form to another by changing their oxidation state through addition or removal of electrons. In some bioremediation strategies, the solubility of the transformed metal increases, thus increasing the mobility of the contaminant and allowing it to be more easily flushed out of the environment. In other strategies, the opposite will occur, and the transformed metal may precipitate out of solution, leading to immobilization. Both kinds of transformation present opportunities for bioremediation – either to immobilize them in place or accelerate their removal. Microorganisms can also influence the contaminant behaviour by changing the acidity of the system in the vicinity thereby altering the extent of metal mobility.

***Ex-situ* bioremediation**. *Ex situ* bioremediation usually refers to the above ground treatment in which soils have been excavated and washed or sediments have been extracted from subsurface and then decontaminated. *Ex-situ* bioremediation methods also try using genetically engineered microorganisms recently.

Another key application of bioremediation is at the forefront of a contaminant plume where a permeable biobarrier can be established. Contaminated groundwater is pumped to the surface and mixed with nutrients, then injected upgradient of the contaminant plume to biostimulate degradation of the contaminant *in-situ* by the indigenous organisms.

### 4.4 Composting

Composting is another process used to soil biopiles that utilizes the heat generated during composting (USACE 1998). Bulking agents like wood chips and straw are added to enhance air movements through biopiles. This is widely used technology for recycling solid waste in industries in India. Composting in windrows, prepared beds holds a number of possibilities for bioremediation of metals by degrading organic chelating agents, altering pH, redox potential and production of surfactants.

### 4.5 Slurry Bioreactor and Sediment Washing

Slurry bioreactors are stirred tank within which biotransformation takes place in an aerated environment (Agathos and Reineke 2002). Sediment washing relies on reducing the volume of contaminated sediment by solubilising readily desorbed contaminants. Through rinsing, excavated sediments are screened to remove large debris and screened sediments are treated in bioreactor.

## 5. Bioremediation of Chromium Contaminated Soils

Remediation of soils, water and sediments, contaminated with metal and organic pollutants, has been studied extensively in the last two to three decades and several treatment techniques are available for remediation of soils contaminated with chrome wastes. In Tamil Nadu, the problem due to tanneries is very acute in the northern region. This is mainly due to crowding of hundreds of tanneries located in nearby places. Studies reveal that the groundwater Cr(VI) concentrations were > 20 mg/L in groundwater samples. A special case study was done to remediate soils around this site. This section deals with the various techniques available for Cr bioremediation and associated problems.

Traditional and innovative methods to manage Cr(VI) contaminated soils have been reviewed (Higgins et al. 1997). The techniques chosen are mainly based on the feasibility and cost at that particular location and the concentration of Cr(VI) present in the polluted soils. Though the total Cr concentration is important, in remediation technologies, utmost consideration is given to Cr(VI) levels, because of its carcinogenic nature. The risk-based soil clean-up level guideline (USEPA 1996) is 390 mg Cr/kg based on the ingestion pathway and the soil screening level is 270 mg Cr(VI) / kg for human exposure by inhalation. But, there is no comparable soil screening level for Cr(III) as such. Also the permissible limit for Cr(VI) in potable water is 0.05 mg/L as per USEPA (1996).

The selection of the remediation depends on: 1) the size, location and history of the site, 2) soil characteristics like structure, texture, pH etc., 3) the type, physical and chemical state of the contaminants, 4) the degree of contamination, 5) the desired final land use and 6) the technical and financial means available.

Advances in understanding the chemistry and toxicity of Cr compounds have led to efforts to remediate the Cr-contaminated soil. Some of the important techniques used are excavation and disposal, soil washing, soil flushing, solidification (*ex situ* and *in situ*), vitrification, chemical and biological reduction and phytoremediation.

The advantages of using bioremediation over other methods are compared in Table 2. All the methods listed have their own advantages and disadvantages. The selection of the most appropriate technology is based on the concentration of Cr(VI) present in the polluted soils, nature of contamination, feasibility and cost at that particular location. Compared to all the methods, bioremediation have been widely used, because they are economical and also do not generate further waste into the environment. The main aim of current remediation techniques is irreversible reduction of Cr(VI) to Cr(III) and its hydroxides. Reduction of Cr(VI) can be achieved by incorporation of organic matter, Fe-containing salts and organic acids (James et al. 1997). The Cr(VI) reduction reactions are as under:

1. Reduction with Fe and Fe compounds

$$Fe + CrO_4^{2-} + 0.5\ H_2O \longrightarrow Fe(OH)_3 + 0.5\ Cr_2O_3$$

$$6\ Fe^{2+} + 2\ CrO_4^{2-} + 13\ H_2O \longrightarrow 6\ Fe(OH)_3 + Cr_2O_3 + 8H^+$$

2. Reduction by organic compounds (e.g., hydroquinone)

$$1.6\ C_6H_6O_2 + CrO_4^{2-} + 2H^+ \longrightarrow 0.5\ Cr_2O_3 + 1.5\ C_6H_4O_2 + 2.5H_2O$$

A wide range of microorganisms have been demonstrated to have Cr reducing ability (cited in Kamaludeen et al. 2003). These properties are harnessed in bioremediation, wherein the microbial strains are multiplied to desired population and pumped into soil/sediments to promote Cr reduction. The efficiency can be enhanced, if the organic matter content and nutrient availability of the soil are sufficient to promote the growth of the introduced microflora. In *in situ* techniques, nutrients will be pumped along with aeration to promote the Cr reduction. Some of the Cr-reducing bacteria and algae have been efficiently used in the treatment of Cr-rich waste water (Fude et al. 1994; Losi et al. 1994a; Cifuentes 1996). However, success was limited in complex soils.

**Table 2.** Comparison of various methods available for remediation of chrome contaminated soils

| Method | Advantages | Disadvantages | Cost (US $/ tonnes) |
|---|---|---|---|
| Excavation and offsite disposal | Appropriate for small volumes of soil and quick | Makes Cr(VI) airborne and hence related health hazard, can be expensive especially for deep materials | 100-200 |
| Soil washing | Used where there is a high concentration of Cr | Makes Cr(VI) airborne, generates contaminated water | 50-200 |
| Soil flushing | *In situ* technique used for spills | Generates contaminated water | 75-200 |
| Solidification | Relatively inexpensive | Cr (VI) should be reduced first, may require soils dewatering | 40-100 |
| Vitrification | Reduces and immobilises Cr (VI) | Very expensive, high energy requirement | 350-400 |
| Chemical reduction | Mainly *ex situ* processes | Requires high quantity of reducing agents, sometimes generate lots of chemical waste | 75-100 |
| Bioremediation | *In situ*, applicable for sites where there is Cr (VI) leaching | Does not remove Cr, required controlled conditions and process is slow | 20-100 |
| Phytoremediation | *In situ* remediation | Does not remove the Cr | - |

(Higgins et al. 1997)

Anaerobic sulphate reducing and methanogenic bacteria possess inherent abilities to sorb more than 90% of chromium to its cell biomass. Small scale bioreactors studies indicate the potential use of *Methanosarcina* and *Methanobacterium* in reducing the Cr toxicity (Ramasamy 2000).

Recently, for treatment of Chromite Ore Processing Residue (COPR), a technique involving the use of organic-rich acidic manure along with chrome reducing microbes to effectively reduce the Cr(VI) in the waste has been developed (Fig. 3). This layer acts as a sandwich and the Cr(VI), leaching out of the waste, is effectively reduced in the organic layer, thereby preventing further contamination of groundwater (James 1997; Higgins 1997).

As described by Losi et al. (1994), the bioremediation of the soil is achieved by a direct or indirect biological reduction of Cr(VI). Most of the direct microbial reduction would be expected on surface soils where aeration favours the enzymatic reduction. In the sub-surface layers, indirect biological reduction of Cr(VI) involving $H_2S$ is predominant and very effective. The $H_2S$, diffused into inaccessible soil pores, promotes the reduction of Cr(VI) and also Mn oxides, involved in reoxidation. *In situ* stimulation of sulphate reducing bacteria may be achieved by addition of sulphate and nutrients. This method has shown some promise for remediation of Cr(VI) contaminated soils when applied to an anaerobic bioreactor system (Losi et al. 1994). Turick et al. (1996) have confirmed the usefulness of anaerobic chromate reducing strains in the reduction and sedimentation of tannery wastes. There is an evidence to suggest that organic contaminants, such as aromatic compounds, are suitable electron donors for Cr(VI) reduction (Shen et al. 1996). Chromium-reducing microbes may then be able to simultaneously remediate organic contaminants as well.

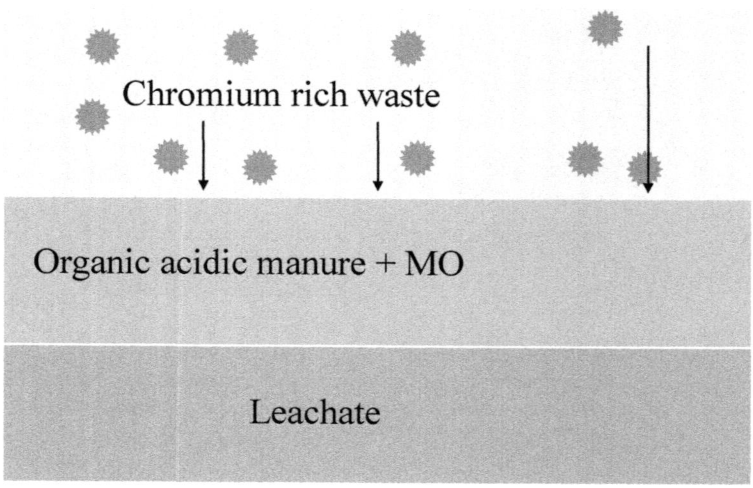


**Fig. 3.** Bioremediation of COPR contaminated soil using organics and microorganisms

## 6. Future Thrust – Do We Really Need to Do More?

Harnessing the tremendous potential of microbes is a great task. Though, the success of bioremediation has been assessed under laboratory conditions, there is no conclusive evidence that bioremediation also works effectively under natural environments. The success stories were very few under Indian conditions. Most research examining metal microbes interactions is conducted using laboratory strains and yet more fundamental questions remain unanswered regarding natural populations of bacteria in contaminated sites.

- How do the bacteria behave under natural conditions with different metals in the soil matrix ?
- How do the mixed microbial population sequester or release metals

- Do the biotransformed metals remain immobile throughout ?

By improving the knowledge and understanding of the structure of natural communities , scientists will be able to answer these issues and setbacks. Also, the anaerobic scavengers were not tapped efficiently in the bioremediation field. Since these archaebacteria can naturally thrive under extreme conditions, they will have special mechanisms to clean-up even hazardous pollutants of globe. The best example is *Geobacter* sp.

Future research is to be carried out to integrate the experimental approach for data collection and mathematical modeling to achieve better prediction. Experimental data generated by the scientists of different disciplines are needed, for incorporation in different approaches to test their efficacy in bioremediation.

## 7. Conclusion

Bioremediation has developed from the laboratory to a fully commercialised technology over the last 30 years in many industrialised countries. However, the rate and the extent of development has varied from country to country. A successful bioremediation scheme relies on the management of soil microbial populations capable of catabolising the contaminants. The role of soil microbiota in the biochemical conversion of organic and inorganic contaminants has been realised, priority research needs have been identified and effort has been made to understand the ecological, biochemical and genetic basis of microbial contaminant degradation, with a view to enhancing microbial capabilities and thus designing more effective bioremediation processes.

## References

Agathos SN, Reineke W (2002) Biotechnology for the Environment: Soil Remediation, Focus on Biotechnology series volume 3B, Kluwer Academic Publishers, Dordrecht, pp 140

Anderson RT, Vrionis HA, Ortiz-Bernad I, Resch CT, Long PE, Dayvault R, Karp K, Marutzky S, Metzler DR, Peacock A, White DC, Lowe M, Lovley DR (2003) Stimulating the in situ activity of *Geobacter* species to remove uranium from the groundwater of a uranium-contaminated aquifer. Appl Environ Microbiol 69:5884-5891

Brim H, McFarlan SC, Ferdickson JK, Minton KW, Zhai M, Wackett LP, Daly MJ (2000) Engineering *Deinococcus radiodurans* for metal remediation in radioactive mixed waste environments. Natl Biotech 8:85-90

Bruce DL, Duff DC (1968) Requirement of potassium or rubidium for biosynthesis of pigment by *Serratia marcescens*. J Bacteriol 96:278-279

Bruins MR, Kapil S, Oehme FW (2000) Microbial resistance to metals in the environment. Ecotoxicol Environ Saf 45:198-207

Buerge IJ, Hug SJ (1998) Influence of organic ligands on chromium(VI) reduction by iron(II). Environ Sci Technol 32:2092-2099

Chang YJ, Peacock AD, Lang PE, Stephen JR, McKinley JP, MacNaughton SJ, Hussain AKMA, Saton AM, White D (2001) Diversity and characterization of sulphate reducing bacteria in groundwater uranium mill tailings site. Appl Environ Microbiol 67:3149-3160

Cifuentes FR, Lindemann WC, Barton LL (1996) Chromium sorption and reduction in soil with implications to bioremediation. Soil Sci 161:233-241

Coates JD, Michaelidon U, Bruce RA, O'Connor SM, Crespi JN, Achenbach LA (1999) Ubiquity and diversity of dissimilatory (per) chlorate reducing bacteria. Appl Environ Microbiol 65:5234-5241

Dungan RS, Frankenberger WT Jr (2000) Factors affecting the volatalisation of dimethyl selenide by *Enterobacter cloacae* SLD 1a-1. Soil Biol Biochem 32:1353-1358

Fude L, Harris B, Urrutia MM, Beveridge TJ (1994) Reduction of Cr(VI) by a consortium of sulfate-reducing bacteria (SRB III). Appl Environ Microbiol 60:1525-1531

Gao S, Burau RG (1997) Environmental factors affecting rates of arsine evolution from and mineralization of arsenicals in soil. J Environ Qual 26:753-763

GNN (Genome News Network) (2003) Super microbe cleans up Uranium. *http://www.genomenewsnetwork.org/articles/12_03/geobacter.shtml*

Higgins TE, Halloran AR, Petura JC (1997) Traditional and innovative treatment methods for Cr (VI) in soil. J Soil Contam 6:767-797

Hinchee RE, Wilson JT (eds) (1995) Intrinsic Bioremediation. Bioremediation 3:1

James BR, Petura JC, Vitale RJ, Mussoline GR (1997) Oxidation-reduction chemistry of chromium: Relevance to the regulation and remediation of chromate contaminated soils. J Soil Contam 6:569-580

Ji G, Silver S (1995) Bacterial resistance mechanisms for heavy metals of environmental concern. J Indus Microbiol 14:61-75

Kamaludeen SPB, Arunkumar R, Avudainayagam S, Ramasamy K (2003) Bioremediation of chromium contaminated environments. Ind J Expt Biol 41:972-985

Komori K, Wang P, Toda K, Ohtake H (1989) Factors affecting chromate reduction in *Enterobacter cloacae* strain HO1. Appl Microbiol Biotechnol 31:567-570

Komori K, Rivas R, Toda K, Ohtake H (1990) Biological removal of toxic chromium using an *Enterobacter cloacae* strain that reduces chromate under anaerobic conditions. Biotechnol Bioeng 35:951-954

Lloyd JR, Lovley DR (2001) Microbial detoxification of metals and radionuclides. Curr Opin Biotechnol 12:248-253

Lloyd JR, Mabbett AN, Williams DR, Macaskie LE (2001) Metal reduction by sulphate reducing bacteria: physiological diversity and metal specificity. Hydrometallurgy 59:327-337

Losi ME, Amrhein C, Frankenberger WT (1994) Environmental biochemistry of chromium. Rev Environ Contam Toxicol 136:92-121

Losi ME, Amrhein C, Frankenberger WT (1994a) Bioremediation of chromate contaminated groundwater by reduction and precipitation in surface soils. J Environ Qual 23:1141-1150

Lovley DR, Phillips EJP (1988) Novel mode of microbial energy metabolism: organic carbon oxidation coupled to dissimilatory reduction of iron or manganese. Appl Environ Microbiol 54:1472-1480

Lovley DR, Baedecker MJ, Lonergan DJ, Cozzarelli IM, Philips EJP (1989) Oxidation of aromatic contaminants coupled to microbial iron reduction. Nature 339:297-300

Lovley DR, Widman PK, Woodward JC, Philips EJ (1993) Reduction of uranium by cytochrome c3 of *Desulfovibrio vulgaris*. Appl Environ Microbiol 59:3572-3576

Lovley DR, Coates JD (1997) Bioremediation of metal contamination. Curr Opin Biotechnol 8:285-289

NABIR (Natural and Accelerated Bioremediation Research Programme) Primer, Office of the Biological Environmental Research, Office of Science, US Department of Energy

Nies DH (1999) Microbial metal resistance Appl Microbiol Biotechnol 51:730-750

Nies DH, Silver S (1995) Ion efflux system involved in bacterial metal resistance. J Indus Microbiol 14:186-199

Nikunen E, Leinonen R, Kultamaa A (1990) Environmental properties of chemicals. Ministry of the Environment, Environmental Protection Department, Research Report 91:885-889

Pongratz R, Heumann KG (1999) Production of mehylated mercury, lead and cadmium by marine bacteria as a significant natural source for atmospheric heavy metals in polar regions. Chemosphere 39:89-102

Ramasamy K (2000) Towards the better management of tannery waste contaminated soils, Conference Proceedings, ACAIR. Canberra

Ross S (1994) Toxic metals in soil-plant systems. John Wiley & Sons, UK

Shen H, Pritchard PH, Sewell GW (1996) Microbial reduction of Cr(VI) during anaerobic degradation of benzoate. Environ Sci Technol 30:1667-1674

Tebo BM, Ghiorse WC, van Waasbergen LG, Siering PL, Caspi R (1997) Bacterially mediated mineral formations: insights into manganese (II) oxidation from molecular genetic and biochemical studies. Rev Minerology 35:255-266

Tinoko I Jr, Sauer K, Wang JC (1985) Physical Chemistry: Principles and Applications in Biological Sciences, Second edition, Prentice Hall, Inc, New Jersey

Turick CE, Apel WA, Carmiol NS (1996) Isolation of hexavalent chromium-reducing anaerobes from hexavalent-chromium-contaminated and noncontaminated environments. Appl Microbiol Biotechnol 44:683-688

USACE (1998) Bioremediation of soils using windrow composting, Guide specifications for military constructions, CEGS - 02191

USEPA (1996) Test methods for evaluating solid wastes, physical/chemical methods (Method 7199) SW-846. 3rd edition. In: Office of the Solid Waste and Emergency Response, Washington DC

USGS (1997) Bioremediation: Nature's Way to a Cleaner Environment, *http://water.usgs.gov/wid/html/bioremed.html*

Wackett LP, Dodge AG, Ellis LBM (2004) Microbial genomics and the periodical table. Appl Environ Microbiol 70:647-655

Wielinga B, Mizuba MM, Hansel CM, Fendorf S (2001) Iron promoted reduction of chromate by dissimilatory iron-reducing bacteria. Environ Sci Technol 35:522-527

White C, Sayer JA, Gadd GM (1997) Microbial solubilisation and immobilization of toxic metals: key biochemical processes for treatment of contamination. FEMS Microbiol Rev 20:503-516

# 8

# Phytoremediation of Metals and Radionuclides

Susan Eapen, Shraddha Singh and S.F. D'Souza

Nuclear Agriculture & Biotechnology Division, Bhabha Atomic Research Centre, Mumbai 400 085, INDIA, Email: eapenhome@yahoo.com

## 1. Introduction

The air, water and soil have been contaminated as a result of industrial revolution and increased urbanization of the landscape. Excavation and deposition of contaminated soil in depositories are of common occurrence and physico-chemical methods are normally used for the remediation of contaminants. Recently, bioremediation - the use of biological agents for remediation of soils and solutions has received a lot of attention (Suresh and Ravishankar 2004). In our laboratory, a variety of biological systems of microbes and plant organs are being investigated for the treatment of heavy metal and radionuclide waste (Bhainsa and D'Souza 1999; Sar and D'Souza 2001 2002; Melo and D'Souza 2003; Eapen et al. 2003). Phytoremediation - the use of plants for environmental clean-up, offers an attractive, environmental friendly and cost-effective approach to remediate metal and radionuclide polluted solutions and soil (Entry et al. 1997, Zhu and Shaw 2000) (Table 1). Plants have constitutive (present in most phenotypes) and adaptive (present only in tolerant phenotypes) mechanisms for accumulation or tolerance of high contaminant concentration in their rhizosphere. A phytoremediation system capitalizes on the synergistic relationship among plants, micro-organisms, water and soil that have evolved naturally in wetlands and upland sites over millions of years. This approach makes use of the plant's ability to extract, concentrate and metabolize materials from air, water and soil (Salt et al. 1995). Plants can be described as solar-driven pumping stations (Cunningham et al. 1995) and possess homeostatic mechanisms to maintain the correct concentrations of essential metal ions in different cellular compartments and to minimize the damage from exposure to non-essential metal ions.

Phytoremediation is an umbrella term which covers several plant-based approaches for cleaning up contaminated environments and includes phytoextraction, the accumulation of high concentrations of metals in plant biomass; rhizofiltration, removal of contaminants from aqueous wastestreams by adsorption into plant roots; phytovolatalization, which includes volatilization into the air through plants, phytodetoxification, which involves the ability of plants to change the chemical species to a less toxic form and phytostabilization, where plants immobilize contaminants chemically and physically at the site, thereby preventing their movement to the surrounding areas.

**Table 1.** Advantages and disadvantages of Phytoremediation

| |
|---|
| Cost |
| - Low capital and operational costs |
| - Metal recycling in case of phytoextraction |
| Performance |
| - Not capable of 100% reduction |
| - Low concentration of waste- it is very effective |
| - May not be applicable to all types of waste |
| - Only applicable to surface soil |
| Others |
| - Aesthetically pleasing |
| - Environmentally non-destructive |
| - Public acceptance |

## 2. Metals in Soils

Enhanced anthropogenic activities and increased industrialization like mining, smelting, electroplating and agriculture have contributed to an increase in the deposition of undesirable concentrations of metals such as Cd, Cr, Cu, Ni, Pb and Zn in the soil and water (Singh et al. 2004). Metal concentrations in soil range from < 1mg/kg to as high as 100,000 mg/kg, depending on the material and deposition event. The risk and the regulatory limits for each metal varies (Table 2). Solubility of metal is dependent on soil characterstics and is strongly influenced by pH of the soil and degree of complexation with soluble ligands (Norvell 1984). Different metals in soil can exist as discrete particles or be associated with different soil components like exchangeable ions sorbed onto inorganic select phase surfaces, non-exchangeable ions sorbed onto inorganic solid phase surfaces, insoluble inorganic metal compounds (oxides, hydroxides, phosphates, or carbonates), metal complexed with soluble or insoluble inorganic material and metals bound in silicate materials.

Metal uptake is an essential component of the plant nutrition. Metals, which are taken up by plants are those which exist as soluble components in the soil solution or are easily desorbed or solubilized by root exudates. Only a small portion of the total metal content in the soil is normally taken up by plants. For effective phytoextraction, it is essential to have abundant source of soluble metal and conditions of soil can be altered to increase metal solubility and availability. By decreasing the pH below 5.5, metal availability for plant roots can be enhanced. However, growth of plants at low pH may be inhibited because of increased Al solubility and subsequent toxicity. Lead in soil is normally unavailable for plant uptake and solubilization through addition of chelating agents like EDTA complexes the free metal ion in the solution, allowing further dissolution of the sorbed or precipitated phases until an equilibrium between complexed metal, free metal and insoluble phases occurs (Norwell 1991).

**Table 2.** Regulatory guidelines for metals and radionuclides

| Element | Concentration range (µg/kg) | Regulatory limit (mg/kg) |
|---|---|---|
| **Metals** | | |
| Lead | 1000-6,900,000 | 600 |
| Cadmium | 100-345,000 | 100 |
| Arsenic | 100-102,000 | 20 |
| Chromium | 5.1-3,950,000 | 100 |
| Mercury | 0.1-1,800,000 | 270 |
| Copper | 30-550,000 | 600 |
| Zinc | 150-5,000,000 | 1,500 |
| **Radionuclides** | | |
| Uranium | 0.2-16,000 (µg/g) | |
| Cesium | 0.2-46,900 (µg/g) | |
| Plutonium | 0.00011-3,500,000 pci/kg | |
| Strontium | 0.03-540,000 pci/kg | |

Plant species differ in their ability to accumulate metals from contaminated soils and some plant species have an inherent ability to accumulate high levels of toxic metals (Sinha et al. 2002). Plants are called as hyperaccumulators when they can accumulate more than 0.1% Pb, Co, Cr or more than 1% Mn, Ni or Zn in plant shoots when grown in their natural habitats (Brooks et al. 1979, 1980, Baker and Brooks 1989). More than 400 plant species are so far known to be hyperaccumulators of metals, belonging to Euphorbiaceae, Brassicaceae, Asteraceae and Rubiaceae (Table 3).

Different species of *Alyssum,* such as *A. bertolonii*, *A. murale* and *Thlaspi goesingense* and *Hybanthus floribundus* are known to take up high levels of Ni (Minguzzi and Vergnano 1948, Doksopulo 1961, Severne and Brooks 1972), while *Viola sp., Thlaspi caerulescens* and *T. rotundifolium* are recognized as accumulators of zinc (Rascio 1977, Barry and Clark 1978). *Thlaspi caerulenscens* has been also found to accumulate high concentrations of Cd. Similarly, *Crotolaria cobalticola* accumulated high concentrations of Co from cobalt rich soils of Zaire (Brooks et al. 1980). High concentration of Cr was detected in the leaves of *Diccoma nicolifera* and *Sutera fodina* growing near a chrome mine in Zimbabwe (Wild 1974). *Astragalus* species were found to accumulate high concentrations of selenium (Christopher et al. 2003) and chinese brake fern *Pteris vittata* is known to take up high concentrations of arsenic (Ma et al. 2001). However, many of these hyperaccumulator plants show slow growth rate and low biomass and hence cannot be used for commercial phytoextraction.

**Table 3.** Selected examples of hyperaccumulators of different metals

| | Concentration (mg/kg) |
|---|---|
| **A. Nickel** | |
| *Berkheya codii* (Asteraceae) | 11,600 |
| *Pentacalia spp.* (Asteraceae) | 16,600 |
| *Senecia spp.* (Asteraceae) | 11,000 |
| *Alyssium spp.* (Brassicaceae) | 1280-29,400 |
| *Bornmuellera spp.* (Brassicaceae) | 11,400-31,200 |
| *Thlaspi spp.* (Brassicaceae) | 2000-31,000 |
| *Psychotria coronata* (Rubiaceae) | 25,540 |
| **B. Zinc** | |
| *Thlaspicaerulescence* (Brassicaceae) | 43,710 |
| *Thlaspi rotundifolium* (Brassicaceae) | 18,500 |
| *Dichopetalum gelonioides* (Brassicaceae) | 30,000 |
| **C. Cadmium** | |
| *Thlaspi caerulescens* (Brassicaceae) | 2,130 |
| **D. Lead** | |
| *Minuartia verna* (Caryophyllaceae) | 20,000 |
| *Agrostis tenuis* (Poaceae) | 13,490 |
| *Festuca ovina* (Poaceae) | 1,750 |
| **E. Cobalt** | |
| *Haumaniastum robertii* (Lamiaceae) | 10,232 |
| *Aeollanthus subacaulis* (Lamiaceae) | 4,300 |
| *Crotolaria cobalticola* (Fabaceae) | 30,100 |
| **F. Copper** | |
| *Ipomoea alpina* (Convolvulaceae) | 12,300 |
| *Aeollanthus subacaulis* | 13,700 |
| **G. Manganese** | |
| *Maystenus bureaviana* (Celastraceae) | 19,230 |
| *Maystenus sebertiana* (Celastraceae) | 22,500 |
| *Macadania Neurophylla* (Proteaceae) | 55,200 |
| **H. Selenium** | |
| *Astragalus racemosus* (Leguminosae) | 1,49,200 |
| *Lecithis ollaria* (Lecithidiaceae) | 18,200 |

## 3. Radionuclides

Radioactive contamination of the environment can be due to emissions and accidental spills from operations typical of nuclear fuel cycle like mining

($^{220}$Rn), milling ($^{238}$U, $^{230}$Th, $^{226}$Ra, $^{310}$Pb) and fall out from nuclear testing ($^{131}$I, $^{90}$Sr, $^{137}$Cs, Pu) and accidents like Chernobyl disaster in Ukraine in 1986. Naturally occurring radionuclides, such as U, Rn, Ra and Th, may be brought to the surface of the Earth by extraction processes such as oil drilling. Problems associated with remediation of soil, ground water and wastewater with radionuclides are similar to those with metals. However, one of the important factors is the radioactive decay component in the selection of appropriate technology. Selection of suitable technology for the remediation of soil and aqueous streams contaminated with radionuclides is based on the environmental chemistry of each element, type of deposition and the rate of radioactive decay. A variety of physico-chemical methods for treatment of radionuclide contamination include removal of top soil, soil washing, leaching with chelating agents, flocculation and reverse osmosis-ultrafiltration. Recently, there has been a spark of interest in the biological methods for radionuclide removal. Phytoremediation, a novel plant-based technology, is being tested for a variety of radioactive contaminated sites, especially for treatment of low level radionuclides in large areas.

Phytoremediation is not commercially used for decontamination of radioactive sites. However, it has been successfully tested for remediation of uranium from wastewater in Ashtabula site and Fernald site, both at Ohio, USA. Remediation of $^{137}$Cs from soil at Brookhaven National lab, NY and $^{90}$Sr and $^{137}$Cs from a pond near Chernobyl, Ukraine, through plants has also been studied. While the technology can be used for removal of groundwater and surface water contamination, radionuclides from soils are more difficult to be decontaminated. Specific amendments and treatment of the soil may increase the rate of transfer of radionuclide in to the plant available forms.

$^{137}$Cesium (half life 32 years) is one of the most important constituents of fallouts and is also a consequence of spills and accidents. Cesium binds tightly to soils and in the soil after Chernobyl accident, 60-90% of $^{137}$Cs was found to be unavailable for plant uptake. Beet (*Beta vulgaris*), quinoa (*Chenopodium quinoa*), red pigweed (*Amaranthus retroflexus*) and russian thistle (*Salsola kali*) are known to remove $^{137}$Cs (Arthur 1982; Broadley and Willey 1997). Water hyacinth (*Eichornia crassipes*) was found to take up $^{137}$Cs and a 60-fold increase in medium activity resulted in a 17-fold increase in accumulation levels (Jayaraman and Prabhakar 1982). Monterey pine and Pondorosa pine seedlings grown on spiked medium were shown to take up 6-8% of $^{137}$Cs in 4 weeks (Entry et al. 1993). Dushenkov et al. (1999) found a drastic reduction in $^{137}$Cs in solutions in which sunflower plants were grown hydroponically. $^{137}$Cs could also be taken up by the leaf surface and transported to roots and subsequently to the soil (Zehnder 1995). Studies in the ponds near the vicinity of Chenobyl, Ukraine, showed that sunflower plants grown hydroponically in the pond could take up 90% of $^{137}$Cs (from 80Bq/L $^{137}$Cs) in 12 days. It was estimated that 55 kg of dry sunflower biomass could remove the entire radioactivity in the pond in the Chernobyl having $9.2\times10^{6}$ Bq $^{137}$Cs and

$1.4x10^8$ Bq $^{90}Sr$ (Dushenkov et al. 1999). *Amaranthus retroflexus* was shown to accumulate high concentrations of $^{137}Cs$ from soil in experiment conducted at Brookhaven National Laboratories (BNL), NY. Cornish et al. (1997) conducted field trials at BNL soil and found that Indian mustard and corn could remove high amounts of $^{137}Cs$. Studies at Argonne National Lab (ANL), West site in Idaho showed that $^{137}Cs$ removal using phytoremediation may take upto 4-7 years for complete removal. Idaho National and environmental laboratory used *Kochia scoparia* plants for soil contaminated with $^{137}Cs$ and the harvested plant matter was treated and disposed off at disposal facilities (http://www.incl.gov/facilities/ant-w-status.shtml). Field and bench studies on phytoremediation of Cs are shown in the Table 4.

**Table 4.** Studies on phytoremediation of cesium

| Radionuclides | Sites | Type of study | Reference |
|---|---|---|---|
| $^{137}Cs$ | Brookhaven, National lab N.Y.-soil | Bench, greenhouse, field | Cornish et al. 1997; Lasat et al. 1997 |
| $^{137}Cs$ | Argonne National lab, soil | Bench, greenhouse, field | Idaho Dept. of Health and Welfare 1998 |
| $^{137}Cs$ and $^{90}Sr$ | Chernobyl Ukraine, surface water | Greenhouse, field | Dushenkov et al. 1999 |

Uranium is a naturally occurring radionuclide and consists of $^{234}U$, $^{235}U$ and $^{238}U$ and is a key element of the nuclear fuel cycle. Nuclear reactor operations, weapons research, nuclear fuel productions and waste reprocessing have resulted in uranium concentration in surface soils and groundwater. Under acidic conditions, uranyl ($UO_2^{2+}$) is the prominent U species, while hydroxide complexes such as $UO_2OH^+$, $UO_2\ (OH)_2^{2+}$ and phosphate complexes form under natural conditions (Langmuir 1978). Uranyl ($UO_2^{2+}$) cation is taken up more readily by plants compared to carbonate and U complexes (Ebbs et al. 1998). Cornish et al. (1995) conducted experiments to phytoremediate U from soil at the Fernald site in Ohio and at a uranium waste dumps in Montana, USA. Chelating agents like citric acid, and other organic acids that are present in the root exudates of plants have been shown to help in the uptake of uranium. Huang et al. (1998) found that addition of 20 m mol/ kg citric acid increased the uptake of U and its accumulation in shoots in *Brassica* species and *Amaranth.* Ebbs et al. (1998) observed that tepary bean and beet showed the greatest accumulation of uranium and addition of citric acid increased U accumulation by a factor of 14. A commercial scale pilot rhizofiltration system set up at Ashtabula site (Dushenkov et al. 1997) containing wastewater (20-870 μg/L), considerably reduced the U concentration in wastewaters with 95% being removed in 24 h. The bench and field studies on rhizofiltration of uranium is given in Table 5.

**Table 5.** Studies on remediation of Uranium

| Site | Type of study | Reference |
|---|---|---|
| Ashtabula OH, wastewater | Pilot Rhizofiltration | Dushenkov et al. 1997 |
| Ashtabula OH, soil | Bench Phytoextraction | Huang et al. 1998 |
| Ferland, OH, soil | Green house | Cornish et al. 1995 |

Strontium 90 ($^{90}$Sr) – a fission product with a half life of 28 years is very mobile and is available to plant uptake. Water hyacinth could take up $^{90}$Sr depending on the pH (highest at 9 and lowest at 4) with 80-90% activity confined to the roots (Jayaraman and Prabhakar 1982). Dushenkov et al. (1999) found that hydroponically grown sunflower reduced Sr concentrations from 200 to 35 μg/l within 48 h and it was further reduced to 1μg/l. Plants such as *Salsola kali* (Blanchfield and Hoffman 1984) and Atriplex (Wallace and Romney 1972), are known to accumulate $^{90}$Sr substantially. Monterey pine and Pondorosa pine seedlings also accumulated high concentrations of $^{90}$Sr (Entry et al. 1993), when grown on artificially contaminated medium. Studies by Phytotech Inc and International Institute of Cell Biology, Kiev, showed that sunflower plants could effectively remove strontium from ponds at Chernobyl with bioaccumulation concentration of 600 for both shoots and roots. However, very little information is available on the removal of Sr from soil of the site.

Plutonium isotopes are present in the environment as a consequence of nuclear weapons testing, fuel reprocessing facilities and accidental releases and include $^{239\text{-}240}$Pu, $^{241}$Pu and $^{238}$Pu. North Atlantic Sargassum was shown to have a high affinity for plutonium with a concentration factor of 21,000 over the marine water (Noshkin 1972). Plutonium uptake by plants appears to vary with plant species, tissue, age and soil characterstics (Garland et al. 1987).

Tritium (half life 12.3 years) occurs naturally when cosmic radiation reacts with gases in the upper atmosphere. Natural tritium combines with oxygen to form water and reaches earth's surface as rain. Tritium also results as a component of nuclear weapons, reactors and nuclear test explosions and contaminates groundwater. Tritium, since it is directly incorporated into water, is taken up by plants which later on release trace amounts of tritium through foliage. Tritium incorporated in water is used by plants for transpiration (IAEA 1981). The tritium phytoremediation project using trees has effectively reduced tritium concentration in waste discharges at Argonne National Laboratory site in Illinois, U.S. However, modeling studies are needed to assess the hazard posed by tritium.

## 4. Phytoextraction

Phytoextraction refers to the use of metal accumulating plants that translocate and concentrate chemical elements from the soil to roots and finally in the above ground shoots and leaves. Phytoextraction exploits vascular plant's natural ability to take up a variety of chemical elements through the root system, deliver these elements to the vascular tissues and to transport and compartmentalize in different organs. Above-ground biomass loaded with metals/radionuclides is harvested, processed for volume reduction and further element concentrations and safely recycled to reclaim metals of economic importance or disposed off as waste in the case of radionuclides. Phytoextraction offers cost advantages over alternative schemes of soil excavation and treatment or disposal. Major limiting factor for phytoextraction are lower metal availability in soil and poor metal translocation from root to shoots. Application of soil amendments could eliminate the limiting steps in metal phytoextraction. Addition of soil amendments increased the metal availability in solutions more than 10-fold for $^{137}Cs$ and 100-fold for Pb and U (Huang et al. 1997 1998). In order to use this practically, it is essential to have vigorously growing plant (>3 tons dry matter/ha-yr) which cause easily harvested and that accumulates large concentrations of metal in the harvestable portions (> 1000mg/kg metal). This technique has been effectively used by Phytotech Inc. (USA) for removal of Pb and Cd from contaminated soil. Excessive selenium in agricultural soils is also successfully remediated by plants using this technology (Banuelos 1993).

Successful phytoextraction of radionuclides depends on the bioavailability of radionuclides in soil, the rate of uptake by the plant roots and efficiency of radionuclide transport through the vascular system. However, not every site is conducive to phytoremediation as a result of excessively high contaminant concentration, which may be unsuitable for the plant growth. Only phytoextraction of $^{137}Cs$, $^{90}Sr$ and $^{235,238}U$ is approaching field application (Dushenkov et al. 1999, Huang et al. 1998), being an element specific and site specific technology. It is possible to formulate a general approach to develop a phytoextraction process for radionuclides, even though numerous challenges have to be overcome to ensure a substantial flux of radionuclide from soil to the aboveground biomass. The radionuclide uptake by plant roots need not necessarily result in translocation to shoots. The majority of $^{137}Cs$ taken up by plants tends to be localized in the roots (Clint and Dighton 1992). Ebbs et al. (1998) demonstrated in hydroponic U uptake studies at pH 5, that the uranyl ($UO_2^{2+}$) cations were more readily taken up and translocated by plants than hydroxyl (pH 6) and carbonate (pH 8) U complexes. Formation of stable U-phosphate complexes in roots may prevent U translocation to aboveground plant parts. In contrast to Cs and U, almost 80% of $^{90}Sr$ taken up the plant, is usually localized in the shoots.

Radionuclides such as $^{90}Sr$, $^{95}Nb$, $^{99}Tc$, $^{106}Ru$, $^{144}Ce$, $^{226,228}Ra$, $^{239\text{-}240}Pu$, $^{241}Am$, $^{228,230,232}Th$, $^{244}Cm$ and $^{237}Np$, were tested for phytoremediation (Dushenkov 2003). A pilot scale phytoextraction project was conducted in the Chernobyl Exclusion Zone (Dushenkov et al. 1999). Three sequential mustard crops were used to obtain noticeable decrease in $^{137}Cs$ activity that was reduced from an average of 2558 Bq/kg to an average of 2239 Bq/kg. In one growing season, areas having $^{137}Cs$ levels>3000 Bq/kg decreased from 29.4% of the total plot area before treatment to 7.7% after treatment. After the final harvest of the phytoremediation crop, areas having $^{137}Cs$ levels<2000 Bq/kg increased to 33.3% compared to 27.4% before treatment. Some of the plants, which can be used for phytoextraction are listed in Table 6.

**Table 6.** Plants with potential for the phytoextraction of various metals and radionuclides

| Metal | Plant species | Reference |
|---|---|---|
| Cd | *Brassica juncea* | Kumar et al. 1995; Huang et al. 1997; Ebbs et al. 1997; Salt et al. 1995 |
| Cr | *B. juncea* | Kumar et al. 1995; Huang et al. 1997 |
| $^{137}Cs$ | *Amaranthus retroflexus* L.; *B. juncea, B. oleracea* L.; *Phalaris arundinacea* L.; *Phaseolus acutifolius* A.Gray. | Lasat et al. 1997, 1998; Negri and Hinchman 2000 |
| Cu | *B. juncea* | Ebbs and Kochian 1997 |
| Ni | *B. juncea* | Ebbs and Kochian 1997 |
| Pb | *B. campestris* L.; *B. carinata* A. Br.; *B. juncea*; *B. napus* L.; *B. nigra* (L.) Koch.; *Helianthus annuus* L.; *Pisum sativum* L.; *Zea mays* L. | Begonia et al. 1998; Blaylock et al. 1997; Ebbs and Kochian 1998 |
| Se | *B. napus* L.; *Festuca arundianacea* Schreb; *Hibiscus cannabinus* L. | Bañuelos et al. 1997 |
| U | *B. chinensis* L; *B. juncea*; *B. narinosa* L., *Amaranthus* spp. | Huang et al. 1998 |
| Zn | *Avena sativa*; *B. juncea*; *B. napus* L. *Hordeum vulgare*, *B. rapa* | Ebbs et al. 1997; Ebbs and Kochian 1998 |

## 5. Rhizofiltration

Rhizofiltration is the use of plant roots to sorb, concentrate or precipitate metal contaminants from solutions. The ideal plant for rhizofiltration should have the capacity to remove maximum amount of toxic metal from contaminated streams coupled with easy handling. An ideal plant used for rhizofiltration should produce significant amount of root biomass with large surface area when grown hydroponically, should be able to take up high concentration of toxic metal and tolerate high amount of toxic metal in roots. Nutrients can be supplied to the plant through artificial soil mixture kept on the top of the hydroponic system (feeder layer). Indian mustard plants were capable of removing Pb from aqueous solutions in the range of 4 to 500 mg/l (Dushenkov et al. 1995). The roots of Indian mustard could effectively remove Cd, Cr, Cu, Ni and Zn. Sunflower plants, tested in the batch experiments in a growth chamber significantly, reduced the concentrations of Cd, Cr, Cu, Mn, Ni and Pb within an hour of treatment. Most cationic species of toxic metals were removed from solutions at least initially and more rapidly in comparison with anionic ones.

Rhizofiltration has been successfully employed by Phytotech Inc. using sunflower at a US Dept of energy (DOE) pilot project with uranium wastes at Ashtabula, Ohio and water from a pond near Chernobyl nuclear plant in Ukraine. In batch experiments with hydroponically grown sunflower plants (Dushenkov et al. 1997), it was shown that concentrations of Cs, U and Sr in contaminated water were significantly reduced within a few hours. Uranium concentration was reduced 10 fold in 1 h while Cs concentration showed a decrease after 6 h and within 24 h, almost all the Cs was removed. Strontium concentration was reduced to 35μg/l within 48 h and at the end of 4 days, it was further reduced to 1μg/l. Sunflower roots concentrated uranium from solution by upto 10,000 fold. Rhizofiltration is proved to be a feasible approach for removing radionuclides from aqueous streams. However, it requires optimization and economic evaluation against conventional technologies.

## 6. Phytostabilization

Phytostabilization is stabilizing process for contaminated soils and sediments in place using vegetation, thus preventing the migration of toxic metals. This is applicable for metal contaminants of waste sites where the best option is to immobilize them *in situ*. Low level of radionuclides also can be maintained this way. Metal cations are most tightly bound and form strong complexes with -H groups on the surface of minerals and hydrous oxides in waste materials. Metals can also bind to the organic material. Addition of manure, digested sewage sludge, straw etc. to inorganic waste sites may help in binding of metals. Supplementation of lime (CaO) and limestone ($CaCO_3$) may help in neutralizing acid soils so as to help in binding of cationic metals with inorganic wastes. Anions such as arsenate and chromate can form surface complexes on hydrous oxides.

Unlike plants chosen for phytoextraction, candidate plants for phytostabilization should be poor translocators of metal contaminants to above ground tissues of plants. The plants should be capable of tolerating high level of metal contaminants and should have efficient growth with dense root system and canopies. Plants which are most suitable for soil conservation are suitable for phytostabilization. Mine tailing at Superfund site in South Dakota with upto 1000 mg/kg of arsenic and also lower concentrations of cadmium and smelter in Kansas with 200,000 mg/kg of zinc could be phytostabilized by decreasing vertical migration of leachate to groundwater using hybrid poplar trees (Hse 1996).

Phytostabilization is particularly suitable for radionuclide-contaminated sites, where one of the alternatives is to hold contaminants in place to prevent secondary contamination and exposure. Capturing radionuclides *in situ* is often the best alternative at sites with low contamination levels or vast contaminated areas where a large scale removal action or other *in situ* remediation is not feasible. This can result in a considerable risk reduction, especially if radionuclides with relatively short half–lives are involved. Plant roots also help to minimize water percolation through soil, thus reducing radionuclide leaching. Phytostabilization may be useful in controlling tailings in uranium mining areas. However, phytostabilization does not remove the radioactivity from the site which has the potential risk of radiation exposure to wild life and humans.

## 7. Phytovolatilization

Phytovolatilization exploits a plant's ability to transpire large amounts of water and is currently used for $^{3}H$ remediation. Phytoremediation of $^{3}H$ through irrigation of forest area has been investigated at Savannah River Site (SRS) for consideration as part of a system to reduce the discharge of $^{3}H$ from the Burial Ground Complex southwest plume. This system is a combination of hydraulic control and enhanced evapotranspiration. Tritium contaminated water is collected, moved to a location upgradient of the discharge point and used to irrigate plants.

## 8. Design of Phytoremediation System

Design of a phytoremediation system will depend on the various parameters, such as the type of contaminant, concentration, clean up required, condition of the site and selection of plant. Phytoextraction has a different design requirement compared to phytostabilization. Most important parameters will include selection of suitable plants, planting density and pattern, contaminant uptake, clean up time required, ground water capture zone and transpiration rate.

Plants generally used for phytoextraction include sunflower and Indian mustard for lead and sunflower and aquatic plants for radionuclides . Recovery of metals from vegetation will depend on recovery from the ash or use of wet extraction techniques. If the metal is for disposal, they will have to be concentrated into a much smaller volume for ultimate disposal/ storage. Aquatic plants include emergent, submerged and floating species. It is easier to harvest emergent populations, while submerged species have more biomass in contact with the solution. Some of the plants generally used for phytostabilization, phytoextraction and rhizofiltration are given in Table 7 and the critical success factors are included in Table 8.

**Table 7.** Phytoremediation applications for metals and radionuclides

| Application | Media | Contaminants | Plants/Character |
|---|---|---|---|
| Phytostabilization | Sediments, Soil | Pb, Cd, Zn, As, Cu, Cr, Se, U | -Trees which transpire large amounts of water for hydraulic control<br>-Grasses with fibrous roots to stabilize soil erosion<br>-Dense root systems needed to sorb/bind |
| Phytoextraction | Sediments, Soil | Pb, Cd, Zn, Ni, Cu, EDTA addition for Pb, Citric acid addition for U | -Sunflower<br>-Indian mustard<br>-Rapeseed<br>-Amaranthus<br>-Chenopodium |
| Rhizofiltration | Groundwater, Wastewater, Created wetland | Pb, Cd, Zn, Ni, Cu, $^{137}Cs$, $^{90}Sr$, U | -Sunflower<br>-Indian mustard<br>-Aquatic plants- Emergent- water hyacianth, Duckweed Submerged plants- Hydrilla, |

**Table 8.** Critical success factors for Phytoremediation

| Phytoremediation process | Critical factors | Conditions for success | Basis for success | Data required | Type of plants |
|---|---|---|---|---|---|
| Phytostabi-lization | Immobili-zation<br>Hydraulic control<br>Soil stabilization | Good roots & biomass<br>Immobile chemicals | Roots hold soil<br>Immobilize metals | Fate and toxicity | Trees, Grasses, Legumes |
| Phytoextra-ction | High biomass<br>Accumulati on in harvestable portion of plants | > 3 tons dry matter/acre/year<br>> 1000 mg/kg of metal | Vigorous growth | Fate and toxicity | Terrestrial plants<br>Aquatic plants |
| Rhizofiltration | Sorption/filt ration by roots | Plant densities 200-1000 gm/m$^2$ | Roots sorb and immobilize contaminants | Fate and toxicity | Aquatic plants<br>-Submerged<br>-Emergent |

### 8.1 Laboratory to Pilot Scale Studies

The sequence of information needed typically range from hydroponic studies to small pot studies with soil from the site in a green house to plot studies (15x15m). Different concentrations of contaminants can be used for toxicity studies. In the last 5 years, about 20 projects, which include field applications of phytoremediation of radionuclides were initiated in USA, Belarus, Ukraine, UK, Yugoslavia, Czech Republic and China.

### 8.2 Plant Density and Pattern

Hybrid poplar-1000 to 2000 per acre are planted normally. Willow and cottonwood belonging to Salix family can also be used for this purpose. The average life time of hybrid poplar is about 30 years and every 4-6 years, the above ground biomass can be cut and removed and new shoots will grow from the cut stem.

### 8.3 Irrigation and Maintenance

Irrigation of the plants ensures a vigorous growth of the plant. Hydrologic modeling may be required to estimate the rate of percolation to groundwater under irrigated conditions. After initial irrigation, irrigation can be discontinued provided the area receives sufficient rains. Agronomic inputs such as addition of NPK, addition of soil conditioners like straw, manure etc should be taken into account. Costs of fertilizer, monitoring of vegetation mowing, pruning, harvesting and replanting should also be included. For phytostabilization, phosphate fertilizers or rock phosphate are effective in binding lead and zinc. In case of phytoextraction, chelates such as EDTA (0.5-10ug EDTA/kg soil) have been added in soils to ensure effective plant uptake (Raskin 1996).

### 8.4 Cost

Phytoremediation is very cost-effective in comparison with other technologies. It is aesthetically pleasing and public acceptance is high (Table 1). Although phytoremediation offers cost advantages, the time period required for clean up is important. Mathematical modeling and monitoring are necessary to demonstrate the effectiveness of the technology to regulatory agencies.

## 9. Challenges for Phytoremediation

As the technology of phytoremediation emerges, so do its challenges. The technology of phytoremediation is still in research and development phase and there are some technical barriers, which need to be addressed. Most heavy metal accumulating plants have a small biomass and are slow growing. To make phytoremediation a viable technology, there is a need to either find fast growing (as yet undiscovered) hyperaccumulators or engineer common plants with hyperaccumulator genes for higher metal accumulation. Conventional breeding and biotechnology have been used to correct these shortcomings by transferring desired traits from metal hyperaccumulator plants to selected high biomass producing non accumulator species. For phytoremediation to be possible, heavy metals must be within the plant's root zone, biologically adsorbed and bioavailable. Attempts are being made to maximize heavy metal concentrations in the plant tissues that grow fast and to isolate genes for metal uptake, which can be potentially transferred to other high yielding biomass plants.

## 9.1 Genetic Engineering of Plants for Metal Tolerance and Accumulation

Several genes are involved in metal uptake, translocation, sequestration and transfer of these genes into candidate plants will result in developing transgenic plants with enhanced ability for metal uptake/accumulation.

Transfer of metallothionin genes have been achieved in several plants. Transfer of human MT-2 gene to tobacco and oil seed rape resulted in plants with enhanced Cd tolerance (Pan et al. 1994). Enhanced Cu accumulation was obtained in *Arabidopsis thaliana* with a pea MT gene (Evans et al. 1992). Transfer of yeast CUP1 gene resulted in 16-fold higher accumulation of cadmium in cauliflower plants (Hasegawa et al. 1997). Similarly, ransfer of two genes for production of γ-glutamylcysteine synthase or glutathione synthase showed enhanced tolerance/accumulation of Cd (Zhu et al. 1999a,b). De la Fuenta et al. (1997) obtained plants with enhanced Al tolerance by overexpression of citrate synthase which resulted in enhanced production of metal chelator-citric acid. Introduction of metal transporter genes also enhances accumulation of metals in plants as in case of *A. thaliana* having Zn-transporter-ZAT gene from *T. goesingense* resulting in 2-fold accumulation of Zn in roots. Likewise, increased Fe tolerance was obtained by overexpression of At Nramp/gene (Curie et al. 2000).

Introduction of merA and merB genes resulted in transgenic *A. thaliana* plants which could phytovolatalize mercury (Bizily et al. 2002). Dhankher et al. (2002) also developed transgenic *Arabidopsis* plants which could take up arsenate by introducing arsenic reductase and γ-glutamyl cysteine synthetase genes. Transport of oxyanion arsenate to above ground, reduction to arsenite and sequestration to thiol peptide complexes by transfer of E. coli ars c and γ ECS gene has been reported. Overexpression of oxidative stress enzymes such as ACC aminase resulted in transgenic plants which accumulated a variety of metals (Ezaki et al. 2000). Selected examples of transgenic plants developed for phytoremediation are shown in Table 9.

**Table 9.** Selected examples of transgenic plants for phytoremediation

| Gene transferred | Plant | Effect |
|---|---|---|
| MT-1 gene from human | Tobacco, Seed rape | Cd toletrance |
| CUP-1 gene from yeast | Cauliflower | Cd accumulation |
| γ-glutamyl cysteine synthetase gene from rice | Indian mustard | Cd acumulation |
| At MTP-1 from Thlaspi goesingense | Arabidopsis | Zn accumulation |
| Arsenate reductase γ-glutamyl cysteine synthetase from *E.coli* | Indian mustard | As tolerance |
| Mer A and Mer B gene | Arabidopsis, Yellow poplar | Phytovolatilization of Hg |

### 9.2 Field Testing of Transgenics and Risk Assessment

Transgenic mustard overexpressing phytochelatins were used for greenhouse studies in Leadville, Colarado such plants were shown to accumulate significant levels of Zn and Cd (Bennett et al. 2003). Some of the possible risks associated with the transgenics are enhanced exposure risk to wild life and humans. Suitable fencing off of the area and use of non-palatable species will prevent grazing/ingestion by wild animals/birds. No transgenic has been commercially used currently for phytoremediation, although mercury volatilizing plants pose no risk (Lin et al. 2002). The risk of escape of genes from transgenic plants is also negligible (Meagher et al. 2000).

## 10. Companies Developing Phytoremediation

In the last few years, several commercial companies on phytoremediation have started springing up in US and Europe and is similar to microbial bioremediation industries as listed in Table 10.

Dedicated companies exclusively working on phytoremediation are developing plants for remediation of metals and radionuclides from soil and water. Phytotech Inc., for example, has used *Brassica* species to remove lead from soil and sunflower to remove uranium and cesium from aqueous waste streams while Phytoworks Inc. is focusing on remediation of organics and mercury by introducing transgenic plants which metabolize mercury. Another company, Earthcare Inc., is working on phytoremediation of organic contaminants using different plants. Similarly, phytokinetics is using grasses to stimulate rhizospheric biodegradation of organics. A number of large industrial companies, principally the oil ad chemical industry, are also conducting or supporting phytoremediation. Phytoremediation is expected to have a large market in future as reflected in Table 11 for USA alone.

**Table 10.** Companies conducting Phytoremediation

| | |
|---|---|
| 1. | Applied Natural Science (USA) |
| 2. | Aquaphyte Remediation (Canada) |
| 3. | BioPlanta (Germany) |
| 4. | Consulagri (Italy) |
| 5. | Earthcare (USA) |
| 6. | Ecolotree (USA) |
| 7. | OEEL (UK) |
| 8. | Piccoplant (Germany) |
| 9. | Phytotech (USA) |
| 10. | PhytoWorks (USA) |
| 11. | Plantechno (Italy) |
| 12. | Slater (UK) |
| 13. | Thomas Consultants (USA) |
| 14. | Verdant Technologies (USA) |
| 15. | Viridian Resources (USA) |

**Table 11.** US Phytoremediation markets (2005) in millions of US Dollars*

| | |
|---|---|
| Metals from soil | 70-100 |
| Metals from groundwater | 1-3 |
| Metals from wastewater | 1-2 |
| Radionuclides | 40-80 |
| Organics from groundwater | 35-70 |
| Others | 65-115 |
| Total | 214-370 |

* Taken from Glass Associates Inc.

## 11. Regulatory Acceptance and Public Acceptance

Phytoremediation's ability to make further strides will depend on how quickly the regulators become convinced of the efficacy of the technology. The regulatory agencies by nature are conservative and tend to have more confidence in technologies longest known to them. The use of plants is generally considered to be aesthetically pleasing means of remediating contaminated sites and is preferable than excavation and other remedial activities, which may involve environmental disruption, noise and frequent worker activity.

## 12. Conclusion

Phytoremediation is an emerging technology for contaminated sites and is attractive due to its low cost, high public acceptance and environmental friendliness nature. It is not a panacea for all waste problems, but a supplement to the existing technologies. The technology has been demonstrated, but not yet commercially exploited. More research background for development of plant tailored for remediation needs use of genetic engineering. The concept of manipulating plant genes for toxic metal uptake is today a cutting edge research area. The likelihood of public acceptance of genetically engineered plants for phytoremediation will be welcomed, since it will clean up the environment of toxic metals. No doubt phytoremediation technology has attracted a great deal of attention in recent years and it is expected that phytoremediation will capture a significant share of the environmental market in the coming years.

Acknowledgements. Shraddha Singh is grateful for financial assistance to Board of Research in Nuclear Sciences, Department of Atomic Energy, Govt. of India in form of Dr. K.S. Krishnan Research Associateship.

## References

Arthur WJ (1982) Radionuclide concentration in vegetation at a solid radioactive waste disposal area in southeastern Idaho. J Environ Qual 11(3):394-399

Baker AJM, Brooks RR (1989) Terrestrial higher plants which hyperaccumulate metallic elements- A review of their distribution, ecology and phytochemistry. Biorecovery 1:81-126

Banuelos GS, Cardon G, Mackey B, Ben-Asher J, Wu L, Beuselinck P, Akohoue S, Zambrzuski S (1993) Plant and Environment Interactions. Boron and selenium removal in boronladen soils by four sprinkler irrigated plant species. J Environ Qual 22:786-792.

Banuelos GS, Ajwa HA, Mackey B, Wu LL, Cook C, Akohoue S, Zambrzuski S (1997) Evaluation of different plant species used for phytoremediation of high soil selenium. J Environ Qual 26:639-646

Barry SAS, Clark SC (1978) Problems of interpreting the relationship between the amount of lead and zinc in plants and soil in metalliferous waste. New Phytol 81:773-783

Begonia GB, Davis CD, Begonia MFT, Gray CN (1998) Growth responses of Indian Mustard [*Brassica juncea* (L.) Czern.] and its phytoextraction of lead from a contaminated soil. Bull Environ Contam Toxicol 61:38-43

Bennett LK, Burkhead JL, Hale KL, Terry N, Pilon M, Pilon-Smits AH (2003) Analysis of transgenic Indian mustard plants for phytoremediation of metal contaminated mine tailings. J Environ Qual 32:432-440

Bhainsa KC, D'Souza SF (1999) Biosorption of uranium (VI) by *Aspergillus fumigatus.* 13:695-699

Bizily SP, Rugh CL, Meagher, R.B. (2002). Phytoextraction of hazardous organo mercurials by geneticallyengineered plants. Nature Biotechnol 18:213-217

Blaylock MJ, Salt DE, Dushenkov S, Zakharova O, Gussman C, Kapulnik Y, Ensley BD, Raskin I (1997) Enhanced accumulation of Pb in Indian mustard by soil-applied chelating agents. Environ Sci Technol 3:860-865

Blanchfield LA, Hoffman LG (1984) Environmental surveillance for the INEL radioactive waste management complex and other areas. Annual Report 1983. EG & G 2312, INEL

Broadley MR, Willey NJ (1997) Difference in root uptake of radiocesium by 30 plant taxa. Environ Pollut 97(1):2-11

Brooks RR, Morrison RS, Reeves RD, Dudley TR, Akman Y (1979) Hyperaccumulation of nickel by *Alyssum linnaeua* (Cruciferae). Proc Soc Lond Biol Sci 203:387-403

Brooks RR, Reeves RD, Morrison RS, Malaisse F (1980). Hyperaccumulation of copper and cobalt - a review. Bull Soc Roy Bot Belg 113:166-172

Clint GM, Dighton J (1992) Uptake and accumulation of radiocesium by mycorrhizal and non-mycorrizal heather plants. New Phytol 121:555-561

Cornish JE, Goldberg RS, Levine RS, Benemann JR (1995) Phytoremediation of soils contaminated with toxic elements and radionuclides. In: Hinchee RE, Means JL, Burris DR (eds) Bioremediation of Inorganics, Battelle Press, Columbus, OH, pp 55-63

Cornish JM, Fuhrmann L, Kochian LV, D Page (1997) Phytoextraction treatability study: removal of $^{137}Cs$ from soils at Brookhaven National Laboratory's Hazardous Waste Management Facility Site. In: Progress Report. U.S. Department of Energy, February 1997

Christopher CG, David RP, Christopher A, Yiqiang Z (2003) Soil selenium uptake and root system development in plant taxa differing in Se-accumulating capability. New Phytologist 159(2):391-402

Cunningham SD, Berti WR, Huang JWW (1995) Phytoremediation of contaminated soils. Trends Biotechnol 13:393-397

Curie J, Alonso JM, Le JM, Ecker JR, Briat JF (2000) Involvement of Nramp from Arabidopsis thaliana in iron transport. Biochem J 347:749-755

De la Fuenta JM, Ramirez-Rodriguez Y, Cabrera-Ponce JL, Herrera Estrella L (1997) Aluminium tolerance in transgenic plants by alteration of citrate synthesis. Science 276:1566-1568

Dhankher OP, Li Y, Rosen BP, Shi J, Salt D, Senecoff JF, Sashti Na, Meagher RB (2002) Engineering tolerance and hyperaccumulation of arsenic in plants by combinig arsenic reductase and $\gamma$-glutamyl cysteine synthetase expression. Nature Biotechnol 20:1140-1145

Doksopulo EP (1961) Nickel in rocks, soils, water and plants adjacent to tail deposits of Alyssum pintoclasilvae. T.R. Dudley sp. Nov. Feddes Reporter 97:135-138

Dushenkov V, Kumar PBAN, Motto R, Raskin I (1995) Rhizifiltration : The use of plants to remove heavy metals from aqueous streams. Environ Sci Technol 29:1239-1245

Dushenkov S, Vasudev D, Kapulnik Y, Gleba D, Fleisher D, Ting KC, Ensley B (1997) Removal of uranium from water using terrestrial plants. Environ Sci Technol 31:3468-3474

Dushenkov S, Mikheev A, Prokhnevsky A, Ruchko M, Sorochinsky B (1999) Phytoremediation of radiocesium-contaminated soil in the vicinity of Chernobyl, Ukraine. Environ Sci Technol 33:469-475

Dushenkov S (2003) Trends in phytoremediation of radionuclides. Plant Soil 249:167-175

Eapen S, Suseelan KN, Tivarekar S, Kotwal SA, Mitra R (2003) Potential for rhizofiltration of uranium using hairy root cultures of Brassica juncea and Chenopodium amaranticolor. Environ Res 91(2):127-33

Ebbs SD, Lasat MM, Brady DJ, Cornish J, Gordon R, Kochian LV (1997) Phytoextraction of cadmium and zinc from a contaminated soil. J Environ Qual 26:1424-1430

Ebbs SD, Kochian LV (1997) Toxicity of zinc and copper to *Brassica* species: Implications for phytoremediation. J Environ Qual 26:776-781

Ebbs SD, Kochian LV (1998) Phytoextraction of zinc by oat (*Avena sativa*), barley (*Hordeum vulgare*), and Indian mustard (*Brassica juncea*). Environ Sci Technol 32(6):802-806

Ebbs SD, Brady DJ, Kochian LV (1998) Role of uranium speciation in the uptake and translocation of uranium in plants. J Exp Bot 49(324):1183-1190

Entry JA, Rygiewicz PT, Emmingham WH (1993) Accumulation of cesium$^{137}$ and strontium$^{90}$ in Ponderosa pine and Monterey pine seedlings. J Environ Qual 22:742-745

Entry JA, Watrud LS, Manasse RS, Vance NC (1997) Phytoremediation and reclamation of soils contaminated with radionuclides. In: Kruger EL, Anderson TA, Coats, JR (eds) Phytoremediation of Soil and Water Contaminants, ACS Symposium Series No. 664. American Chemical Society, Washington, DC

Evans KM, Gatehouse JA, Lindsay WP, Shi J, Tommey AM, Robinson NJ (1992) Expression of pea metallothionin like gene Ps MTA in *Escherichia coli* and *Arabidopsis thaliana* and analysis of trace metal ion accumulators: Implications of Ps MTA function. Plant Mol Biol 20:1019-1028

Ezaki B, Gardner RC, Ezaki Y, Matsumuto H (2000) Expression of aluminum -induced genes in transgenic *Arabidopsis* plants can ameliorate aluminum stress and/or oxidative stress. Plant Physiol 122:657-666

Garland TR, Cataldo DA, Wildung RE (1987) Factors affecting uptake and distribution of plutonium in barley and soybean plants. In: Vaughan BE (ed) Pacific Northwest laboratory Annual Report for 1974 to the US atomic energy commission. Division of Biomedical and Environmental Sciences. Part 2. Ecological Sciences, BNWL-1950 PT-2 UC-48

Hasegawa I, Terada E, Sunair M, Wakita H, Shinmachi F, Noguchi A, Nakajima M, Yazaki J (1997) Genetic improvement of heavy metal tolerance in plants by transfer of the yeast metallothionin (CUPI). Plant Soil 106:277-281

Hse W (1996) Metals soil pollution and vegetative remediation by using poplar trees at two heavy metal contaminated sites. MS Thesis. Univ Iowa, Iowa City, USA

Huang JWW, Chen JJ, Berti WR, Cunningham SD (1997) Phytoremediation of lead contaminated soils: Role of synthetic chelates in lead phytoextraction. Environ Sci Technol 31:800-805

Huang JW, Blaylock MJ, Kapulnik Y, Ensley BD (1998) Phytoremediation of uranium contaminated soils: Role of organic acids in triggering uranium hyperaccumulation in plants. Environ Sci Technol 32(13):2004-2008

IAEA (1981) Tritium in some typical ecosystems. Technical Reports Series no. 207, Vienna

IDAHO Department of Health and Welfare (1998) Proposal plan for waste area group 9. Argonne National Laboratory-West, IDAHO National Engineering and Environmental Lab

Jayaraman AP, Prabhakar S (1982) The water hyacinth uptake of Cs and Sr and its decontamination potential as an approach to the zero release concept. In: Proc. International Symp of Migration in the Terrestrial environment of long lived radionuclides from the nuclear fuel cycle. Knoxville. T.N. International Atomic Energy Agency, Vienna

Kumar PBAN, Dushenkov V, Ensley BD, Chet I, Raskin I (1995) Phytoremediation: A novel strategy for the removal of toxic metals from environment using plants. Biotechnol 13:1232-1238

Lasat MM, Fuhrmann M, Ebbs SD, Cornish JE, Kochain LV (1998) Phytoremediation of a radiocesium-contaminated soil: Evaluation of Cesium-137 bioaccumulation in the shoots of three plant species. J Environ Qual 27(1):165-169

Lasat MM, Norvell WA, Kochian LV (1997) Potential for phytoextraction of $^{137}$Cs from a contaminated soil. Plant, Soil 195(1):99-106

Langmuir D (1978) Uranium solution mineral equilibria at low temperature with applications to sedimentary ore deposits. Geochimica Et Cosmochimica Acta 42:547-560

Lin Z-Q, Souza MDe, Pickering LJ, Terry N (2002) Evaluation of the macroalga, muskgrass for the phytoremediation of selenium contaminated agricultural drainage water by microcosms. J Environ Qual 31:2104-2110

Ma LQ, Komar KMM, Tu C, Zhang W, Cai Y, Kennelley ED (2001) A fern that hyperaccumulates arsenic. Nature (London) 409:579

Meagher RB, Rugh CL, Kandasamy MK, Gragson G, Wang NJ (2000) Engineered phytoremediation of mercury pollution in soil and waters using bacterial genes. In: Terry N, Banuelos G (eds) Phytoremediation of contaminated soil and water, Leurs Boca Raton, Florida, pp 201-221

Melo JS, D'Souza SF (2003) Removal of chromium by mucilaginous seeds of *Ocimum basilicum*. Bioresourse Technol 92(2):151-155

Minguzzi C, Vergano O (1948) Il continuto di nichel nelle ceneri di Alyssum bertolonii Desv. Att Soc Toscana Sci Nat Mem Serie 55:49-77

Negri CM, Hinchman RR (2000) The use of plants for the treatment of radionuclides. In: Raskin I, Ensley BD (eds) Phytoremediation of toxic metals: using plants to clean up the environment, Chapter 8, 2000, Wiley-Interscience, New York

Norwell WA (1984) Comparison of chelating agents as extractants for metals in diverse soil materials. Soil Sci Soc Am J 48:1285-1292

Norwell WA (1991) Reactions of metal chelates in soils and nutrient solution. In: Mortvedt JJ (ed) Micronutrients in agriculture, Soil Sci. Soc. of America. Madison. Wl, pp 187-227

Noshkin VE (1972) Ecological aspects of plutonium dissemination in aquatic environments. Health Phys 22:537-549

Pan A, Yang M, Tie F, Li L, Chen Z, Ru B (1994) Expression of mouse metallothionin-1 gene confers cadmium resistance in transgenic tobacco plants. Plant Mol Biol 24:341-351

Rascio N (1977) Metal accumulation by some plants growing in zinc mine deposits. Oikos 29:250-253

Raskin I (1996) Plant genetic engineering may help with environmental cleanup. In: Proc. of the National Academy of Sciences of the United States of America, pp 3164-3166

Salt DE, Blaylock M, Kumar PBAN, Dushenkov V, Ensley BD, Chet I, Raskin I (1995) Phytoremediation: a novel strategy for the removal of toxic metals from the environment using plants. Biotechnology 13:468-474

Sar P, D'Souza SF (2001) Biosorptive uranium uptake by a Pseudomonas strain characterization and equilibration studies. J Chem Tech Biotech 76:1286-1294

Sar P, D'Souza SF (2002) Biosorption of thorium (VI) by a Pseudomonas strain. Biotech Letters 24:239-243

Severne BC, Brooks RR (1972) A nickel accumulating plant from Western Australia. Planta 103:91-94

Singh S, Sinha S, Saxena R, Pandey K, Bhatt K (2004) Translocation of metals and its effects in the tomato plants grown on various amendments of tannery waste: evidence for involvement of antioxidants. Chemosphere 57(2):91-99

Sinha S, Saxena R, Singh S (2002) Comparative studies on accumulation of Cr from metal solution and tannery effluent under repeated metal exposure by aquatic plants: Its toxic effects. Environ Monito Assess 80(1):17-31

Suresh B, Ravishankar (2004) Phytoremediation- a novel and promising approach to environmental clean up. Critical Rev Biotechnol 24(2):1-28

Wallace A, Romney EM (1972) Radioecology and ecophysiology of desert plants at the Nevada test site. Environmental Radiation Division. Los Angeles Soil Science and Agricultural Engineering. Univ. California, Riversida TID-25954

Wild H (1974) Indigenous plants and chromium in Rhodesia. Kirkia 9:233-241

Zehnder HJ (1995) Uptake and transport of radioactive cesium and strontium into grapevines after leaf contamination. Radiat Phys Chem 46(1):61-69

Zhu Y, Pilon-Smits Eah, Jouanin L, Terry N (1999a) Overexpression of glutathione synthetase in *Brassica juncea* enhances cadmium tolerance and accumulation. Plant Physiol 119:73-79

Zhu Y, Pilon-Smits Eah, Tarun A, Weber SU, Jouanin L, Terry N (1999b) Cadmium tolerance and accumulation in Indian mustard is enhanced by overexpressing $\gamma$-glutamylcysteine synthetase. Plant Physiol 121:1169-1177

Zhu YG, Shaw G (2000) Soil contamination with radionuclides and potential remediation. Chemosphere 41:121-128

9

# Nanotechnology for Bioremediation of Heavy Metals

P. Rajendran[1] and P. Gunasekaran[2]

[1]Department of Zoology, Vivekananda College, Tiruvedakam West, Madurai 625 217, INDIA, [2]Department of Microbial Technology, Centre for Excellence in Genomic Sciences, School of Biological Sciences, Madurai Kamaraj University, Madurai 625 021, INDIA, Email: pguna@eth.net

## 1. Introduction

Nanotechnology, a highly promising discipline in science and technology is the emerging and novel trend that will redesign the future of several existing know-how, which will change every aspect of our lives and lead to the generation of uniqueness in all the streams of technology. The current revolution in nanoscience is brought about by the concomitant development of several advances in technology. Nanotechnology applies the techniques and processes of microfabrication to build devices for studying bio-systems and has a wide range of applications in variety of fields from space science to deep oceanic research (Vincent 2003). Noria Taniguchi used the term 'Nanotechnology' while measuring precise machining tolerances of materials in the range of 0.1-100 nanometer (Bhat 2003).

Biological synthesis of metal nanoparticles using microbes, such as bacteria, yeasts, algae, actinnomycetes and fungi, is gaining momentum due to the eco-friendly nature of the organisms which reduce toxic chemicals (Muralisastry et al. 2003). Metal-microbe interaction is very important in several biotechnological applications, including in the fields of biomineralization, bioremediation, bioleaching, and microbial corrosion (Joerger et al. 2001). Nano materials, besides providing new research challenges, form the basis of a new class of atomically engineered materials. Confluence of environmental biotechnology and nanotechnology will lead to the most exciting progress in the development of nano-devices having bio-capabilities in novel metal remediation strategies.

## 2. Nanotechnology - A New Scientific Frontier

The parentage of the modern subject of nanoparticles derives from the work of Michel Faraday, who carried out studies on nanoscale gold particles in aqueous solution and established the first scientific basis (Thomas and Kulkarni 2003). Nanotechnology is an *enabling technology* that leads to generation of new capabilities, new products and new markets. Multiple events have converged to provide a persuasive argument for supporting a focus in nanotechnology: i) historical trends and a projected end of this trend in the absence of new scientific principles ii) new research trends to explain relatively unknown frontiers iii) discovery of new phenomena iv) superior products designed by nature v) advanced computational methods coupled with massive computational capabilities and vi) possibility of new high-performance products (Tolles and Rath 2003). Nanomaterials research is now concentrating on the development of materials that can be designed to have desired properties by manipulating and attaching atoms in different ways.

## 3. Unique Properties of Nanoparticles

Nanocrystals cover a size range 1-100 nm and are intermediate to the molecular size regime on one hand and the macroscopic bulk on the other. The significance of nanophase particle is that the behavior is completely different from the commonly accepted and familiar properties of the macro particles. The physical, chemical and electronic properties of nanoparticles depend strongly on the number, kind of atoms that make up the particle, interaction of crystal atoms and atoms in the grain boundaries. Laws relating to physical, chemical, biological, electrical, magnetic and other properties at the nano-scale are different from those that apply to macro matter. Van der Waal's forces, electron resistance and magnetism are the more important governing forces of nanoparticles instead of forces, such as gravity or inertia (Bhat 2003). The unusual physicochemical and optoelectronic properties of nanoparticles are due to confinement of electrons within particles of dimensions smaller than the bulk electron delocalization length, termed quantum confinement. Because of the special properties of the nanophase materials, there is great deal of interest in the cost-effective synthesis.

## 4. Synthesis of Nanophase Materials

Many important nanostructures are composed of the group IV elements Si or Ge, type III-V semiconducting compounds, such as GaAs or type II-VI semiconducting materials such as CdS (Poole and Owens 2003). The materials used to form various types of nanostructures generally have bulk properties. However, it is modified when their sizes are reduced to nanorange. Mechanical, ferroelectric and ferromagnetic properties of materials change when measurements are made in micrometer or nanometer range.

Nanophase materials can be synthesized by low temperature and high temperature methods (Komarneni 1995). Low temperature method includes precipitation of solutions from room temperature to 100°C, hydrothermal synthesis (>100°C and > 1 atmosphere pressure), inverse micelle method and sol gel synthesis. The high temperature nanophase material synthesis includes gas condensation, wire explosion and liquid aerosol thermolysis. Hydrothermal, microwave-hydrothermal and microwave solvothermal are the conventional techniques used for the preparation of nanophase materials of different sizes and shapes (Komarneni 2003).

Fabrication of nanopowder/colloidal particles includes i) extensive ball milling ii) condensation or precipitation iii) drawing glassy materials iv) self assembly that includes biological fabrication v) forming materials around/within templates, and vi) growth of a second material on a crystalline lattice in which the lattice parameters don't match.

Widely used method for the fabrication of nanostructures is lithography, which makes use of a radiation-sensitive layer to form well-defined pattern on a surface. Molecular-beam epitaxy and the growth of one crystalline material on the surface of another, is a second technique that has been perfected. There are also chemical methods: the utilization of self-assembly and the spontaneous aggregation of molecular groups (Poole and Owens 2003). Gedanken (2003) reported that 20 kHz, ultra sound radiation could rupture chemical bonds and explained the role of few parameters in determining the yield of reaction and the unique products that were obtained in the form of amorphous nanoparticles in material science. These methods are cheaper because of less energy consumption and are ideally suited for precise control of size and shape of nanophases. However the main drawback with these techniques is the cost and chemical contamination.

## 5. Instrumentation for Nanotechnology

Nanotechnology revolution is due to the improvement of old and the introduction of new instrumentation systems for evaluating and characterizing nanostructures. Research in this vast area has been possible only because of the development of tools and instruments that are effective at nano levels. Many of the systems are very large and expensive, often requiring specialists to operate them. Whan (1986), in his review, described the instruments for determining the position of atoms in materials, instruments for observing and characterizing the surface of the structures, and various spectroscopic devices for obtaining information of the properties of nanostructures. Electron beams provide crystallographic information about nanoparticle surfaces and also produce images of the surface.

In a transmission electron microscope (TEM), the electrons from source, such as electron gun, enter the sample, are scattered as they pass through it, are focused by an objective lens, are amplified by a magnifying (projector) lens, and finally produce the desired image. Field ion microscopy is another technique in which the resolution approach is interatomic. The scanning transmission electron microscope (SEM), the scanning tunneling microscope (STM) and the atomic force microscope (AFM) are the efficient instrumentation systems to obtain images of the surface of a specimen by scanning the surface with an electron beam in a raster pattern.

Nanomaterials can be investigated and characterized using spectroscopic techniques in the infrared and Raman region of the spectrum (frequencies from $10^{12}$ to $4 \times 10^{14}$ Hz, wavelength $\lambda$ from 300 to 1μm), as well as visible and ultraviolet spectroscopy (frequencies from $4 \times 10^{14}$ to $1.5 \times 10^{15}$, $\lambda$ from 0.8 to 0.2 μm). Emission spectroscopy can be studied by varying the frequency of the incident light, by studying the frequency distribution of the emitted light, or by combining both techniques (Poole and Owen 2003).

Photoluminescence excitation (PLE) is a standard one for obtaining information on the nature of nanostructures, such as quantum dot. This technique involves scanning the frequency of the excitation signal, and recording the emission within a very narrow spectral range. Thermoluminescence is another spectral technique that can provide information on surface states, detrapping, and other processes involved in light emission from nanoparticles. In this technique, the emission of light is brought about by heating.

## 6. Application and Current Status of Nanotechnology

Nanotechnology is concerned with materials and systems whose structure and components exhibit significantly improved physical, chemical and biological properties and that enable the exploitation of novel phenomenon and processes due to their nanoscale size. The unique chemical, electrical, magnetic, optical and other properties of nanoscale particles have already led to their evaluation and use in a broad range of industries, including biotechnology, catalysis, data storage, energy storage, microelectronics and others. The possibility to modify existing materials through technology has become a recipe for the preparation of advanced materials (Komarneni 2003). The domain of this technology is not restricted to only the realm of materials and applications, but also extends to life sciences.

## 7. Metal Pollution and its Impact

Contamination of heavy metals in the environment is a major global concern because of their toxicity and threat to human life and environment (Ceribasi 2001). Urbanization, industrialization and modern agriculture activities are the main reasons for heavy metal pollution. The group of heavy metals are about 65 and are defined in a number of criteria, such as their cationic-hydroxide formation, specific gravity greater than 5 g/ml, complex formation, hard-soft acids and bases, and, more recently, association with eutrophication and environmental toxicity. Metal concentration has been linked to birth defects, cancer, skin lesions, retardation leading to disabilities, liver and kidney damage and a host of other maladies (ATSOR 2001). Wastewater from various industries, such as electroplating, cement, paint etc., discharge heavy metals, such as cadmium, copper, lead, mercury, nickel, zinc and arsenic which are highly toxic to living systems. Persistence and non-biodegradability of toxic heavy metals with their hazardous effect cause serious threat to living organisms. Changes in trace element profile of the soil cause physiological and genetic changes in various life, such as plants, aquatic and benthic fauna, insects, earthworms, fish, birds and mammals as evidenced by recent research work (Mudakavi et al. 1998).

## 8. Current Strategies for Metal Remediation

Technologies involving physical, chemical or biological agents are available for the remediation of heavy metal contaminated effluents and sludge (Table 1). Microbe based technology presents an economic alternative for today's mining, mineral and waste water treatment industries. In the past few decades, new metal treatment and recovery techniques, based on biosorption, have been explored using both dead and living microbial biomass with remarkable efficiency. Biological approach for metal detoxification offers high potential for selective removal of toxic metals. It has an advantage of operation flexibility and easy adaptability for *in-situ* and *ex-situ* application in a range of bioreactors (Lloyd and Lovley 2001).

## 9. Bioremediation through Nanotechnology

Researchers in the field of nanoparticle synthesis and assembly have turned to biological systems, since they have potential to control the shape, which is not possible in conventional chemical synthesis. Muralisastry et al. (2004) reported that an amalgamation of curiosity, environmental compulsions, and conviction, that nature has evolved the best process for synthesis of inorganic materials on nano and macro-length scales, has contributed to the development of a relatively new and largely unexplored area of research based on the use of microbes in the biosynthesis of nanomaterials. Organisms, synthesizing inorganic materials, include magnetotactic bacteria, sillceous material synthesizing diatoms and S- layer bacteria which produce gypsum and calcium carbonate layers (Joerger et al. 2001). Advancement in nanoscience will achieve the control of matter via controlled molecular assembly.

**Table 1.** Comparison of conventional and bioremediation metal clean up strategies

| Strategy | Methods | Disadvantage | Remarks |
|---|---|---|---|
| **Conventional:** | | | |
| Evaporation | Single/multi stage or vapor compression evaporator | Scaling or fouling | High/ commercial |
| Distillation | Packed column with heating and concentration device | Scaling or fouling | Medium/ commercial |
| Solvent extraction | Standard process | Required for the processing | Moderately high/ commercial |
| Adsorption | Batch or continuous Adsorption beds | Limited to low concentration | Medium/ commercial |
| Ion exchange | Synthetic product | Require pretreatment | High/ commercial |
| Membrane process | Standard manufacture units | Separation is imperfect | Medium/ commercial |
| Electrochemical process | DC power and plating apparatus | Impurity upsets the process | Medium/ commercial |
| Starch xanthate process | Synthetic process | Preparation is tedious | Medium/ experimental |
| **Bioremediation:** | | | |
| Bioaccumulation | Live microbes/ideal for genetic manipulations. | Emerging technology | Lab level |
| Biosorption | Live or dead microorganism | Emerging technology | Low cost/ commercial |
| Phytoremediation | Live or dead plant biomass | Emerging technology | Low cost/*ex-situ* / *in-situ* remediation |
| Plant microbe interaction | Plant and microorganisms. | Emerging technology | Low cost/*ex-situ* remediation |

Material scientists are viewing the uses of microbes in toxic heavy metal bioremediation with interest for nanofabrication of environmentally useful submicron scale particles. If we could build it in microbes, it is possible to use them as eco-friendly and effective nanofactories for heavy metal remediation. Formation of inorganic particles within microorganisms might become a central discipline in biometric and bioengineering applications. Biological systems provide many examples of specifically tailored, nanostructured molecules with highly optimized properties and characteristics. Thus biological materials are considered as a nanophase system in its own right and as the starting point for producing other novel nanophase systems (Table 2). The fungal and actinomycete-mediated green chemistry approach towards the synthesis of nanoparticles has many advantages, such as ease with which the process can be scaled up, economic viability and possibility of easily covering large surface areas by suitable growth of the mycelia, etc (Muralisastry et al. 2003).

**Table 2.** Microorganisms in nanoparticles synthesis

| Organism | Nanoparticle | Mechanism | Size (nm) | Reference |
|---|---|---|---|---|
| *Pseudomonas stutzeri* AG259 | Silver | Intracellular | 200 | Joerger et al. (2001) |
| *Verticillium* sp | Gold / Silver | Intracellular | 2-20 | Muralisastry et al. (2003) |
| *Thermomonospora* sp | Gold / Silver | Extracellular | - | Muralisastry et al. (2003) |
| *Lactobacillus* | Gold / Silver | Intracellular | - | Nair & Pradeep (2002) |
| *Torulla* sp | Lead | Intracellular | - | Kowshick et al. (2002) |
| *Schizosaccharomyces pombe* | Cadmium | Intracellular | - | Kowshick et al. (2002) |
| *Fusarium oxysporium* | Gold / Silver | Extracellular | 2-50 | Mukherjee (2001) |
| Magnetotactic bacterium | Magneite / Greigite | Intracellular/ Extracellular | 35-120 | Joerger et al. (2001) |
| Diatoms | Siliceous | Intracellular/ Extracellular | - | Joerger et al. (2001) |
| *Rhodococcus* sp | Gold | Intracellular | 5-15 | Ahmad et al. (2003) |

## 10. Case Studies

Joerger et al. (2001) have shown that the bacteria *Pseudomonas stutzeri* AG259 isolated from silver mine, when placed in a concentrated aqueous solution of $AgNo_3$, resulted in the reduction of the $Ag^+$ ions and the formation of silver nanoparticles of well defined size and distinct morphology within the periplasmic space of bacteria. Ahmad et al. (2003) reported an alkalotolerent actinnomycetes (*Rhodococcus* sp) capable of synthesizing gold nanoparticles of the dimension 5-15 nm with good monodispersity formed on the cell wall as well as on the cytoplasmic membrane. However, the particles are more concentrated on the cytoplasmic membrane than on the cell wall, possibly due to reduction of the metal ions by the enzymes present in the cell wall and on the cytoplasmic membrane. An acidophilus fungus, *Verticillium* sp isolated from the *Taxus* plant when challenged with $Ag^+$ and $AuCl^-_4$ ions, led to their reduction and accumulation as silver and gold nanoparticles. The growth of the silver nanoparticles occurred within the fungal biomass and the possible mechanism could be the extracellular reduction of the $Ag^+$ ions in the solution, followed by precipitation onto the cells (Muralisastry et al. 2003). A novel alkalothermophilic (extremophilic) actinomycete, *Thermomonospora* sp., isolated from self-heating compost exposed to $AuCl_2$, completely reduced it to $AuCl_4$ ions producing gold nanoparticles, indicating that it secretes four distinct proteins of molecular masses between 80 and 10 kDa.

## 11. Magnetotactic Bacteria

Alivisatos (2001) reported the presence of inorganic crystals in magnetotactic (magnetic sensing) bacteria. The bacterium has fixed within it a chain of about 20 magnetic crystals with the size between 35 and 120 nm diameter. The chain of magnetic crystals (magnetosomes) is visible in electron microscope and imparts the bacterium with a magnetic dipole movement along its length. These crystals constitute a miniature compass and it is a marvel of natural nanoscale engineering. It is madeup of the perfect material-either magnetite or greigite, both highly magnetic iron materials. The crystals align the bacteria with the external magnetic field. In nature, this enables the bacteria to navigate with respect to the earth's magnetic field towards their ideal environment in the upper micro-aerobic sediments of ponds and streams (magnetotaxis). The magnetic separation of heavy metals and radionuclides in conjugation with microbial accumulation by magnetotactic bacteria, can be applied to mineral processing and environmental management of wastes. Magnetotactic bacteria immobilize heavy metals from a surrounding solution and applying a low intensity, focusing magnetic field and can easily separate them. This principle can be extended to develop a treatment process for the removal of metals from wastewater.

## 12. Comparison of Current Strategies with Nanotechnology

Material scientists have been viewing microbes as an eco-friendly nanofactories for metal remediation though biotechnological applications employing microbes, such as bacteria, yeast, algae, diatoms and actinnomycetes. However, compared to bacteria, fungi and actinnomycetes are known to secrete much higher amounts of proteins, thereby significantly increasing nanoparticles by biosynthetic approach. Nanomaterial *in vivo* biosynthesis is the best option for metal bioremediation, since biologically controlled mineralization process produces materials with well-defined characteristics. The biominerals are composite materials and consist of an inorganic component and a special organic matrix; the organic matrix has a vital influence on the morphology of the inorganic compound.

Metal nanoparticles bring about halocarbon mineralization efficiently, economically and eco-friendly. The reaction, studied with silver and gold nanoparticles, results in the catalytic destruction of halocarbons forming silver halide (silver chloride) and amorphous carbon.The reaction is more efficient with silver nanoparticles in the size range of 2-150 nm (Nair and Pradeep 2003). Many hydrocarbons are toxic, mutagenic and resistant to microbial degradation. However, they can be catalytically destroyed by metal nanoparticles. Application of this reaction in detection, extraction, and degradation of environmentally significant halocarbons in general and pesticides in particular, will be a promising and novel technology.

## 13. Future Prospects

The impact from advances emerging from nanotechnology developed over the next 15-20 years has been estimated by National Science Foundation to be approximately $ 1 trillion. In anticipation of this economic impact, nanotechnology research programme in several countries has increased substantially in recent years (Tolles and Rath 2003). Technological merits of nanoparticles provide a vision for transmitting new discoveries into products. It is possible to produce synthetic macroscopic 'living-like' organisms made of nanoparticles that would remediate hazardous heavy metals from contaminated environment. Attempts are being made to develop nano-thick particulate coatings onto macroscopic and microscopic structures using a novel pulse laser deposition technique. There have been other concerted efforts of integrating microelectronics and molecular biology into a platform technology with a number of potential commercial applications (Bhat 2003). Surface study of the biogenic nanoparticles (i.e. nature of capping surfactants/peptides/proteins) would lead to the possibility of genetically engineered microbes to overexpress specific reducing molecules and capping agents and there by, control the size. The rational use of constrained environment within cells, such as periplasmic space and cytoplasmic vesicular compartments (e.g. magnetosomes) to modulate nanoparticles size and shape, is an exciting possibility yet to be explored (Muralisastry et al. 2003). Traditional metallurgical research, organic matter, optical property optimization, biological materials and function are the vital areas in nanotechnology that could be the inspiration to make eco-friendly nanomaterials to remediate heavy metal pollution in the environment.

## 14. Conclusion

In future, modification and adaptation of nanotechnology will extend the quality and length of life. The breath of anticipated opportunities, cross-disciplinary nature, potential for innovation, historical track records and the impact of the potential gains of nanotechnology research have led to the recognization of this area with special emphasis. The social benefits are significant from nanomaterials and the new products are applicable to information technology, medicine, energy, and environment. An important challenge in nanotechnology is to tailor optical, electric and electronic properties of nanoparticles by controlling the size and shape. Utilization of microbe for intracellular/extracellular synthesis of nanoparticles with different chemical composition, size/shapes and controlled monodispersity can be a novel, economically viable and eco-friendly strategy that can reduce toxic chemicals in the conventional protocol.

## Reference


Ahmad A, Senapati S, Khan MI, Rajivkumar, Ramani R, Srinivas V, Muralisastry (2003) Intracellular synthesis of gold nanoparticles by a novel alkalotolerent actinomycete, *Rhodococcous* species. Nanotechnology 14:824-828

Alivisatos AP (2001) Less is more in medicine. Scientific American 9:59-65

ATSDR (2001) Agency for toxic substances and disease registry. CERLLA Comprehensive environmental response, compensation and liability act, Priority list of hazardous substances. http:// www.atsdr.cdc.gov/clist.html

Bhat JSA (2003) Hearlding a new future-Nanobiotechnology? Curr Sci 85(2):147-154

Ceribasi IH, Yetis U (2001) Biosorption of Ni (ii) and Pb (ii) by *Phanaerochaete chysosporium* from a binary metal system- kinetics. Water SA 27(1):15-19

Gedanken A (2003) Sonochemistry and its application to nanochemistry. Curr Sci 85(12):1730-1734

Joerger K, Joerger R, Olsson E, Granqvist CG (2001) Bacteria as workers in the living factory: Metal accumulating-bacteria and their potential for material science. Tibtech 19:15-20

Komarnei S (1995) Nanophase materials In: McGraw-Hill Year book of Science and Technology, McGraw-Hill, New York, pp 285-288

Komarnei S (2003) Nanophase materials by hydrothermal. Microwave-hydrothermal and microwave-solvothermal methods. Curr Sci 85(12):1730-1734

Kowshik M, Vogel W, Urban J, Kulkarni SK, Paknikar KM (2002) Microbial synthesis of semiconductor CdS nanoparticles, their characterization, and their use in the fabrication of an ideal diode. Biotechnol Bioeng 78:583-588

Lloyd JR, Lovley DR (2001) Microbial detoxification of metals and radionuclides. Curr Opi Biotech 12:248-253

Mudakavi JR, Narayana (1998) Toxic heavy metal contamination of the soil and biota: Part II Environmental implications. IJEP 18(2):101-108

Mukherjee P, Ahmad A, Mandal D, Senapati S, Sainkar SR, Khan MI, Parischa R, Ajayakumar PV, Alam M, Kumar R, Muralisatry (2001) Fungus-mediated synthesis of silver nanoparticles and their immobilization in the mycelial matrix: A novel biological approach to nanoparticle synthesis. Nano Lett 1:515-519

Muralisastry, Ahmad A, Khan MI, Rajivkumar (2003) Biosynthesis of metal nanoparticles using fungi and actinnomycetes. Curr Sci 85(2):162-170

Muralisastry, Ahmad A, Khan MI, Rajivkumar (2004) Microbial nanoparticles production. In: Niemeyer C, Mirkin C (eds) Nanobiotechnology, WILEY-VCH Verlag GmbH& Co.KgaA, Weinherim, pp 126-135

Nair AS, Pradeep T (2003) Halocarbon mineralization and catalytic destruction by metal nanoparticles. Curr Sci 84(12):1560-1564

Nair B, Pradeep T (2002) Coalescence of nanoclusters and the formation of sub-micron crystallites assisted by Lactobacillus strains. Cryst Growth Des 2:293-298

Poole CP Jr, Owens FJ (2003) Introduction to nanotechnology. Wiley interscience, A John Wiley & Sons, Inc, Publication, Hoboken, New Jersey

Thomas JP, Kulkarni GU (2003) From colloids to nanotechnology: Investigations on magic nuclearity palladium nanocrystals. Curr Sci 85:(12)1760-1766

Tolles WM, Rath BB (2003) Nanotechnology, a stimulus for innovation. Curr Sci 85(12):1746-1759

Vincent SGP (2003) Nanotechnology. In: Selvin J (ed) Biotechnology Emerging Trends, Biotech books, New Delhi, pp 4-7

Whan RE (1986) Materials characterization Vol 10 of Metals hand book, American Society for metals, Metals Park OH

# 10

# Biotechnological Approaches to Improve Phytoremediation Efficiency for Environment Contaminants

Rana P. Singh, Geeta Dhania, Asha Sharma and Pawan K. Jaiwal

Department of Bioscience, M.D. University, Rohtak 124 001, INDIA, Email: rana_psingh@rediffmail.com

## 1. Introduction

The realization, that plants serve the mankind by cleanup of the toxic contaminants, is quite old, but the problems of the contaminated land sites, water bodies and ground water and spoiled air worldwide have increased many folds due to anthropogenic activities during second half of the $20^{th}$ century and hence deserve special attention. The environmental concerns of government and non-government agencies and the people at large have increased enormously, which have paved the way for the establishment of a large number of research institutes and commercial groups to develop new techniques and technologies for rapid cleanup of the contaminants from the sites identified for alarming contaminations. Phytoremediation, as a sustainable, cost effective and potential cleanup technology over the conventional methods, has emerged very fast as an alternative technology in the last decade (see Cunningham et al. 1995; Cunningham and Ow 1996; Salt et al. 1998; Saxena et al. 1999; Macek et al. 2000; Baker et al. 2000; Morikawa and Takahashi 2000; Singh et al. 2001; Morikawa et al. 2002; Kassal et al. 2002; Dhankhar et al. 2002; Maiti et al. 2004; Prasad 2004; Datta and Sarkar 2004; Schwitzguébel 2004; Pan et al. 2005).

Phytoremediation technology can be implemented *in situ* or *ex-situ* to cleanup a variety of the organic contaminants e.g.. petroleum hydrocarbons, gas condensates, crude oil, chlorinated compounds, pesticides, herbicides, explosive compounds as well as typical inorganic toxicants, such as heavy metals, metalloids, radionuclides, etc. (Morikawa and Takahashi 2000). Air pollutants like nitrogen and sulfur oxides, ozone and suspended particulate matters (SPMs) can also be ameliorated by growing efficient naturally occurring plants as well as more efficient genenetically modified plants (see Wellburn 1990; Morikawa and Takahashi 2000; Takahashi et al. 2001; Schwitzguébel 2004; Morikawa et al. 2005). Phytoremediation is considered as an aesthetically pleasing and solar energy driven cleanup technology, which causes minimal environmental disruption and *in situ* treatment preserves the topsoil (Morikawa and Takahashi 2000). It is inexpensive (60-80% or even less costly than conventional physio-chemical methods) and useful for treating a broad range of the environmental contaminants, especially at sites with shallow or low levels of contaminants. Possibly due to their static (non-mobile) nature, plants had to evolve their survival modes even in odd environments including sites contaminated with the xenobiotic substances, which are non-essential or even harmful for them. The natural adaptations and genetic mutations have evolved a wide range of preferential or general tolerance to the toxic substances in plants. Naturally occurring tolerance to plants is based on the mechanisms like phytostabilization, rhizodegradation, phytoaccumulation, phytodegradation, phytovolatization and evapotranspiration etc. which facilitate plants various means to avoid, escape, partition or remove the toxic contaminants as an adaptation measure. Such naturally evolved potential of plants, on the other hand, can be used for cleanup purposes. Bioprospecting of the suitable plant species and genotypes having higher tolerance, agroclimatic fitness, higher biomass and faster growth cycle is needed for various kinds of the contaminants.

In addition, to commercially exploite those naturally occurring plants selected for the remediation of the pollutants, some biotechnological approaches such as rhizosphere manipulations to increase bioavailability or biodegradation of the contaminants for higher uptake and rapid removal by the phytoremediator (Vassil et al. 1998; Chaudhary et al. 1998; de Souza et al. 1999; Singh et al. 2003; Saxena et al. 1999; Morikawa and Takashashi 2000; Geebelen et al. 2002; Piechalak et al. 2003; Thangavel and Subburaam 2004) and genetic engineering of plants to increase uptake, transport, partitioning, tolerance, *in situ* degradation, volatization or evaporation etc (Rugh et al. 1998; Zhu et al. 1999,a,b; Pilon Smits et al. 1999; Gleba et al. 1999; Zaal et al. 1999; Saxena et al. 1999; Morikawa and Takahashi 2000; Hirschi et al. 2000; Bizily et al. 2000; Hannink et al. 2001; Singh et al. 2001; Takahashi et al. 2001; Dhanker et al. 2002; Lee et al. 2003a,b; Pilon et al. 2003; Singh and Jaiwal 2003; Maiti et al. 2004; Datta and Sarkar 2004; Marikawa et al. 2002 2005; Pan et al. 2005) have been persued to increase the phtoremediation efficiency.

Such biotechnological efforts are also made to resolve the specific problems for the improvement of a phytoremediator to suit to the specific contaminant(s) and site(s) to make it commercially successful. This review is an attempt to analyse such approaches and efforts in the light of the present challenges towards the alarming contaminations of toxic heavy metals, major gaseous pollutants like nitrogen oxides, sulfur oxides and organic pollutants of agrochemicals and industrial origin (Fig. 1). We have confined our discussions largely on the higher plants and focused on the need to understand the key regulatory steps and mechanisms to produce superhyperaccumualtors of commercial grade by gene technologies. We have also discussed the needs of rhizosphere manipulations of plants for their better performance.

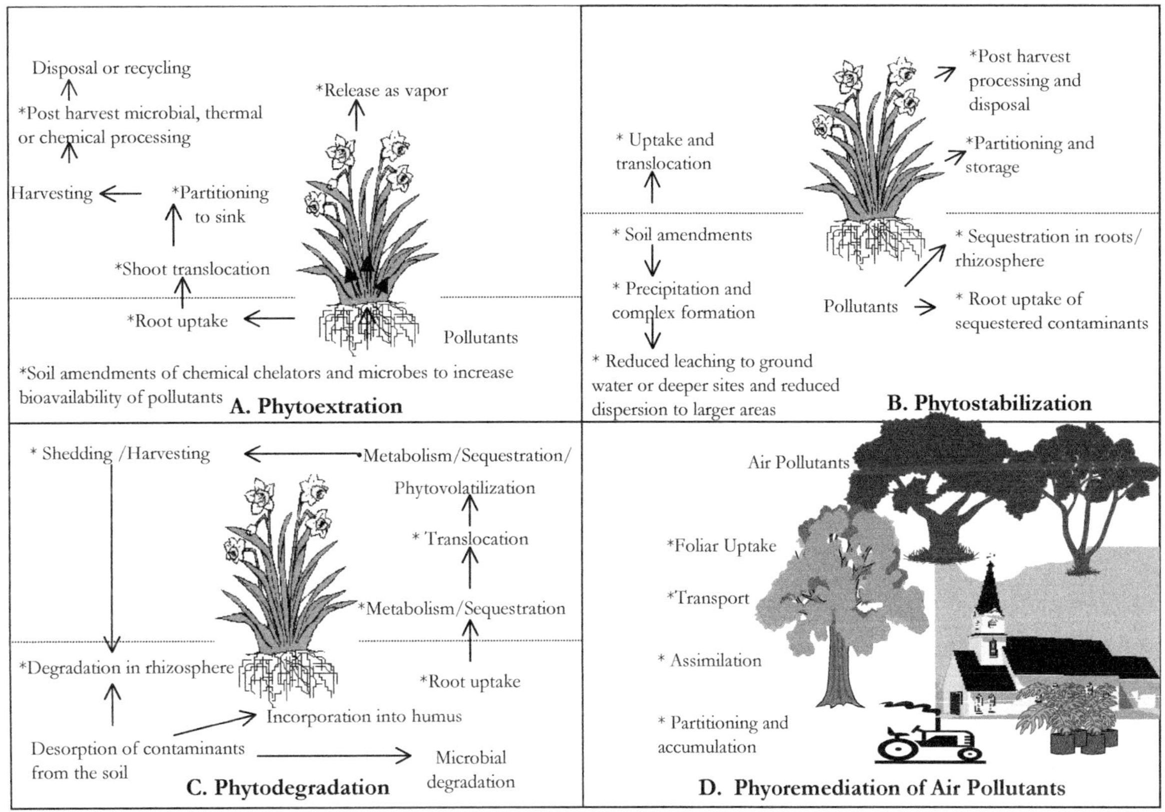


**Fig. 1.** Phytoremediation types and postulated sites for biotechnological interventions A. Phytoextraction / Phytovolatilization, B. Phytostabilization, C. Phytodegradation, D. Phytoremediation of air pollutants; *Sites for Biotechnological input

## 2. Phytoremediation: The Processes, Potentials and Limitations

Phytoremediation is based on the fact that a living plant can be considered as a solar driven pump, which can extract, concentrate, degrade, volatize or vaporize soluble toxic substances from the soil, water or air through their natural water and mineral uptake, transport, partitioning, assimilation and transpiration systems. In addition, plants need to survive in several odd environments, and hence they posses more flexible metabolic systems evolved genetically or adopted physiologically to avoid, partition, degrade, store or exclude various undesired and toxic substances. They have developed various specific and general adaptation mechanisms to protect them from the abiotic and biotic stresses. The biotechnological approaches focus to exploit these evolved potentialities of the plants and other associated organisms and to modify their characteristics with some needed alterations in favour of the human needs. Cleanup of the toxic substances from the contaminated sites using the principles of phytoremediation can be achieved in many ways (see Table 1). The details of these processes have been discussed in many past and recent reviews (Brooks et al. 1979; Baker and Brooks 1989; Raskin et al. 1997; Salt et al. 1998; Saxena et al. 1999; Baker et al. 2000; Maiti et al. 2001; Raskin and Ensley 2000; Morikawa and Takahashi 2000; Prasad 2004; Thangavel and Subburaam 2004; Schwitzguébel 2004) and also in this chapter. The popularity of this technology is increasing with increase in the awareness for a need of sustainable environment around us. The remediation of soil pollution may involve a cost of 300 billion of dollars (Raskin et al. 1997; Maiti et al. 2004). Phytoremediation and other bioremediation techniques are not only significantly cost effective over the physical and chemical means of the soil, water or air remediations, they also reduce the risk from exposure to the hazardous constituents at waste and spill sites (Salt and Rauser 1995; Salt et al. 1995; Salt 2001; Raskin et al. 1994; Cunningham and Ow 1996).

The efforts to understand the physiological and molecular mechanisms involved in the processes of the phytoremediation by plants have come to the focus of attention more precisely with a view point to apply these *in situ* processes to enhance the phytoremediation potentials using biological and engineering strategies designed to optimise and improve the process (Schwitzguébel 2004). Several plant species have been explored and the treatment systems for decontamination of the toxicants from sites have been set up, but, most of them were used without exact understanding of the mechanisms involved. Certain woody plant species, shrubs, other perennials, and annual herbs including crop plants have been found suitable for the phytoremediation techniques (Table 2).

In addition to pulling out the toxic contaminants from the soil to metabolize, concentrate or evaporate, the phytoremediation techniques involve extensive pull out and evaporation of water from the plant covered sites. This high consumption

**Table 1.** Biotechnological approaches for the various modes of phytoremediation enhancement.

| Mode | Meaning | Target | Possible Phytoremediation Enhancement strateies |
|---|---|---|---|
| Phytoextraction | The extraction of pollutants from soil, water or air and its higher accumulation and compartmentaion in harvestable plant parts | Toxic metals | Overexpression or insertion of uptake, transport, partitioning storage and binding related genes (including regulatory transcription factors and organ specific promoters) |
| Phytoaccumulation | The uptake and concentration of the contaminants within the roots or aboveground portions of the plants | -do- | -do- |
| Phytodegradation (Phytotrans-formation) | The partial or total degradation of complex organic molecules within the plants | Organic pollutants | Overexpression or insertion of uptake, transport, degradation and metabolism related genes and transcription factors. |
| Phytovolatilization | The uptake, transport and volatilization of volatile organics through stomata | Volatile pollutants or pollutants producing volatile products on catabolism | Insertion and overexpression of uptake, transport, degradation, metabolism and volatilization related genes and transcription. |
| Evapotranspiration | The uptake, transport and evaporation of pollutants through the transpiration pathways. | Contaminants reached to deeper sites or at wet, marshy sites | Gene manipulation to increase water uptake and transpiration rate |
| Phytostabilization | The reducing mobility of pollutants towards ground water or its dispersion in soil or water by enhancing precipitation or sequestering to the roots | To avoid leaching or dispersal and to concentrate pollutants in the rhizosphere of plants | Amendments of binders/sequesters and microbial population suitable for the purpose |

| | | | |
|---|---|---|---|
| Tree as pump | The use of trees to evaporate water and to extract pollutants from soil | Deep rooted pollutants from wasteland not expected to be used shortly | Genetic engineering for higher water uptake and enhanced transpiration rates |
| Phytostimulation (Rhizodegration) | The release of plant exudates/enzymes into the rhizosphere which stimulates the microbial and fungal degradations of organic pollutants | *Ex-situ* degradation of organic pollutants in rhizosphere of plants | Over expression /insertion of genes producing such microbial stimulants |
| Rhizofiltartion | The use of plant roots to absorb or adsorb pollutants from water and aqueous waste stream | Clean-up of shallow waterlogged areas or for municipal waste water treatment | Manipulation for desired and extensive root systems and higher uptake of the pollutants |

and recycling of water can also prevents pollutant wash out and slows down the possible migration of toxic compounds through the soil and into the groundwater. In many cases, associated microflora play a important, if not the decisive, role in the treatment of the polluted sites (Siciliano and Germida 1998; Schwitzguébel 2004).

Though several plants have been identified from the natural plant populations as hyperaccumulators of toxic heavy metals (Prasad 2004 for a recent review), oxides of nitrogen (Morikawa et al. 2002, 2005) and organic pollutants (see Schwitzguébel 2004), bioprospecting for the natural phytoremediators has not been done adequately. For example, phytodiversity and the polluted sites are enormous in India, and many other developing countries, but-there have not been adequate work on biodiversity prospecting for the exploration of minerals and other natural resources and for the environmental cleanup (see Prasad 2004). Most of the knowledge generated on the different kinds of phytoremediation, improvements in phytoremediation potentials by engineering and biotechnological approaches and its commercialization, belongs to the countries which are more planned and environmentally careful, though many of them posses less plant diversity. Bioprospecting of the natural plant diversity for the environmental cleanup potentials will not only provide insights to use more appropriate phytoremediators, which are cheapest, sustainable and most acceptable in the public domain, but it will also provide very significant informations for gene pool available to produce superior quality genetically manipulated plants, more suitable for the commercial viability as phytoremediation systems. Generally fast growing plants with high biomass and different kinds of root system suitable to be used to cleanup the pollutants at different depth are considered as ideal phytoremediators. However, they should be tolerant enough for the target

**Table 2.** Some case studies and commercial phytoremediation field project based on websites (http:/www.mobot.org/jwcross/phytoremediation/phytorem-sponsors-corp.htm; Saxena et al. 1999; Morikawa and Takahashi 2000; Schwitzguébel 2004)

| Contaminant | Plant species and technique used | Institution/Industry/Company | Site name /Location |
|---|---|---|---|
| Removal of nitrogen | Poplar tree planting | CH2M HILL, Potland, OR, USA | Mill Greek, USA |
| Treatment of oily waste through land application | Rhizosphere amendants with rotation of grass, grains and clover crops on the sites two times each year .The crops are seasonally plowed into the soil with the applied waste to provide a stabilizing "green manure" nutrient source | -do- | Texaco, Anacortes, Washington, USA |
| Remediation of diesel contaminated soil | Cultivation of grass and clover and rhizosphere bioremediation | -do- | Daishowa paper Mill, Port Angeles,Washington,USA |
| Remediation of wood preservative wastes through plant cultivation (contaminants included pentachlorophenol (PCP) and PAH s) | Planting of native cottonwood, willow, alfalfa and several grasses in 1999 to 2001 added with rhizosphere bioremediation | -do- | Union Pacific Railroad, Laramine, WyominG, USA (140 Acre site) |
| Soil and ground water contamination with petroleum related organics, PAHs and chlorinated organics released by accidental spills in year 2000 | Hybrid poplar trees, buried upto 10 feet below the surface and a sub-surface aeration system (to encourage deep rooting into ground water) | Ecolotree, Inc., Iowa city, IO, USA (Ecolotree (r) cap (Ecap) and Ecolotree (r) Buffer (EBuffer) | Milwaukee, Wisconsin, USA |
| Fertilizer and pesticide | 440,12-18 feet tall bare root hybrid poplar were planted into 6'deep trenches | -do- | Illinois, USA (April, 1999) |
| Treated 80,000 gallons per day of municiple sewage contaning | South Burlington's Living Machine | Living Technologies, Taos, NM, USA (Living Machines®) | Lake Champlain, USA (1995) |

| | | | |
|---|---|---|---|
| Trichloroethanol | Hybrid poplar | Occidental Petroleum Corp., Los Angles, CA USA& University of Washington; USA | Various sites in USA |
| Heavy metals | Indian mustard and sunflowers (the patented plants can take up heavy metals more than 3.5% of their dry weight) | Edenspace system carporation, Reston, VA, USA | Various sites in USA |
| Uranium soil contamination 47mg/kg | Sunflower (Accumulation in plants 764 mg/kg-1669mg/kg) | -do- | US Army site in Aberdeen, Maryland, USA |
| Arsenic | Fern *P. vittata* (brake fern). Phytoextraction in above ground part by more than upto 200 fold higher than other plants | .-do- | 1.5 Acre site in New Jersey, North Carolina, USA (2001) |
| $^{89/90}$ Sr (radionuclide) | Specially selected Indian mustard ($^{89/90}$ Sr in plants was more than 10-15 fold higher that than in soil); Phyto-extraction +soil amendments | -do- | Fort Greely, Alaska, USA |
| $^{137}$Cs (radionuclide) | -do- | -do- | Chernobyl Nuclear Power Plant accident in 1986 in Ukraine |
| Organic pollutants including dichloro-benzidine (a human carcinogen) | Mixed native species e.g. Willows and Poplars (13,000 trees) | Phytokinetics, Inc. North Logan, UT | Bofors-Nobel Superfund site, USA 20 Acre site) |
| Ground water treatment of chlorinated volatile organic | Poplar & willow trees (1000); 'Pump and treat' system (Evapotranspiration of contaminated water)+ Enhanced rhizosphere degradation | Solvent Recovery Services of New England (SRSNE) | Superfund site in Southington, Connecticut, USA (1998) |
| PATHs, heavy metals | Various | Stockholm University | Old gasworks site (Husarviken, Sweden) |
| Contaminations with wood presser-vatives including pentachlorophenol & polyaro-matic hydrocarbons | Perennial rye grass (*Lolium perenne*) | | Mc Cormick and Baxter Superfund site, USA (1996-1998) |

| | | | |
|---|---|---|---|
| Cadmium, zinc, lead | Alpine pennycress (*Thlaspi caerulescens*) Take up Zn@125Kg/ha per year and Cd @ 2Kg/ha per year with optimum growth condition; Phytoextraction | Dr Chaney and coworkers | Pig's Eye landfill site in St Paul, Minnesota, USA |
| $^{137}$Cs and $^{90}$ Sr | Indian mustard and redroot pigweed (*Amaranthus retroflexus);* Phytoextration | - | Brookhaver National Lab New Jersy and in Ashtabula Ohio, USA |
| Lead and Cadmium | Indian mustard (*Brassica juncea*) | Phytotech, Florida State University, IETU | Czechowice oil refinery (Katowice, Poland) |
| Zinc and Cadmium | *Salix viminalis* (willow) | Swiss Federal Institute of Technology | Former landfill (Switzerland) |
| Nickel,copper,zinc ,cadmium | *Salix* species | University of Glasgow | Sewage disposal site (United Kingdom) |
| Zinc | H.annuus, Z.mays, C.halleri | International Graduate School Zittau | Zinc waste landfill (Hlemyzdi, Czech Republic) |
| Copper, zinc, cadmium | Improved tobacco | Several institutes | Zinc/Copper (Dornach, Switzerland) |
| Zinc, copper, lead, cadmium | Grasses for phytostabilization | Limburgs University | Zinc smelter site (Lommel, Belgium) |
| Zinc, copper, lead, cadmium | Grasses for phytostabilization | Limburgs University | Contaminated playing ground (Overpelt, Belgium) |
| Zinc, copper, lead, cadmium | *B. napus* for phytoextraction | Limburgs University | Zinc / Cadmium contaminated soil (Balen, Belgium) |
| Lead, cadmium, zinc, copper, Ti, Sb, As | Various plants | Several institutes | Guadiamar river area, Donana National Park (Aznalcollar mine, Spain) |

| | | | |
|---|---|---|---|
| Lead | Successive crops of sunflower and indian mustard planted in 24" deep *ex-situ* treatment cell on an impermeable concrete base .The single season phytoremediation treatment achieved the regulatory goal of 900 mg/kg. Total cost of phytoremediation treatment was less than $50 per cubic yard, which saved more that $1.1 million comparedto the estimated cost of excavation and disposal. | -do- | Daimler Chysler's Detroit Forge Site, USA. (4300 cubic younds of soil with $Pb^{+2}$ ranging from 75-3,450 mg/kg soil) in 1998 |
| Lead | Sunflower and indian mustard were planted. A combined phytoextraction and Phytostabilization treatment for three years costed less than $40 per cubic yard of treated soil | -do- | Industrial facility in Connecticut, USA (1997-2000) |
| Lead | Indian mustard, Phytoextraction + rhizosphere amendments with EDTA | -do- | A Site at Trenton, NJ, USA(1996-1997) |
| BTEX | *Populus x Canadensis* (poplar) | Limburgs University | BTEX contaminated groundwater (Genk, Belgium) |
| Chlorinated organics | Various | Stockholm University | Eka Chemicals site, (Bohus, mercury Sweden) |
| Gasoline and diesel compounds | Poplars and willow | Technical University of Denmark | Old gas filling station (Axelved Denmark) |
| Cyanide, BTEX, PAHs and oil | Poplars and willow | Technical University of Denmark | Former municipal gasworks site |
| Pesticides | Poplars | Polish Academy of Sciences, Kornik ISTEA-CNR Bologna | Resort pollution by pesticides stored in bunkers (Niedwiady, Poland) |

toxicant(s) to survive with prosperous vegetative growth on the contaminated site(s) and should be suitable for the agro-climatic conditions of the area under cleanup. It will be best to search out a naturally evolved phytoremediator with all such positive characters during the phytoprospecting, but it is likely that one may need to incorporate one or more character(s) artificially by genetic manipulations to achieve such goals.

## 3. Commercial Viability of Phytoremediation Projects

Phytoremediation has been carried out commercially or demonstrated at pilot scale at nearly 200 sites in USA involving all the contaminant categories (Glass 1999; Shekhar et al. 2004). A growing concern over the safe and sustainable environment has created a huge space globally for such eco-friendly techniques within a viable commercial set up. Several universities, research institutes, government bodies and private companies are collaborating to develop large scale economically viable projects for cleanup of the notorios toxicants contaminating various sites accidentally or slowly (Table 2). Such efforts and practices are, however, confined to developed countries which are getting better public perception and pressure for the sustainable eco-friendly developmental projects. Other parts of the world including most of the developing countries are yet to be adequately sensitized to the cause of the environmental cleanup and a central focus on the sustainable development which is a task ahead. It is evident, that phytoremediation, as a technology, will gain momentum throughout the world, as we don't have better options to treat the contaminated water, air and land sites which are creating a high risk health hazards to human and live stocks and damaging green cover and plant productivity enormously.

Large scale phytoremediation of the contaminated sites has been achieved for heavy metals, organic xenobiotics and radionuclides (Table 2. Glass 1999, Dietz and Schnoor 2001; Schwitzguébel et al. 2002; Schwitzguébel 2004). Developing a commercial phytoremediation strategy needs attention to both pre-harvest (e.g. contaminant level monitoring, plant selection, decontamination rates, agro-climatic suitability of phytoremediator, groundwater capture zone, transpiration rate and required cleanup time etc.) and post harvest processing (e.g. harvestable biomass collection, leftovers and underground residues disposal and treatment removal of the contaminated plant materials etc.) steps. With minimal environmental disturbances, the phytoremediation techniques can be applied to a broad range of toxicants, which generate less secondary air or water waste as compared to other traditional methods. The organic pollutants may ideally be degraded to $CO_2$ and $H_2O$, reducing environmental toxicity. It is always beneficial for treating large volumes of water, air or land having low to moderate concentration of the contaminants. During land reclamation using phytoremediation, the topsoil is left in usable condition and may be developed for agricultural use as the soil remains intact at the site after contaminants are removed in contrast to conventional methods.

Rhizosphere amendments with chelators, bacteria and mycorrhizae have been used to enhance bioavailability of the contaminants to the remediating plants for large scale remediation strategies (Table 2. Chaudhary et al. 1998; Khan et al. 2000; Thangavel and Subburaam 2004; Schwitzguébel 2004). Rhizosphere manipulations to deal with various layers/depth of the contaminants and to provide sub-surface aeration etc. have been provided in some systems developed by companies dealing with this technology. Though hybrid poplar willows (*Salix* sp.), clover, alpine pennycress (*Thlaspi* sp.), grasses, Indian mustard, sunflower, geraniums, fern (*Pteris vittala*), perennial ryegrass, redroot pigweed etc. have been plants of choice for many commercial phytoremediation systems (Table 2), several new plants with higher efficiency and better suitability for phytoremediation can be searched out with the extensive phytoprospecting of new sites. In addition, genetically modified superior quality phytoremediators can be developed to handle specific situations. A large number of large scale demonstration/ treatment projects have established the commercial viability of phytoremediation as a sustainable and viable cleanup technology of present and for the future.

## 4. Rhizosphere Manipulations for Enhanced Bioavailability of the Toxic Substances

Amongst the major factors that can make a phytoremediation successful and commercial, rhizosphere manipulations for increased bioavailability of toxic substances have been a focus of attention in the recent past. In addition to genetic ability of the phytoremediating species /cultivars, optimal agronomic (soil and crop management) practices can increase the efficiency of the system (Li et al. 2000; Khan et al. 2000; Thangavel and Subburaam 2004; Datta and Sarkar 2004).

Heavy metals are one very significant category of the industrial contaminants, which are unique being selectively toxic, persistent and non-biodegradable (Baker and Brooks 1989; Bharti and Singh 1993, 1994; Kumar et al. 1993; Singh et al. 1994a,b, 1996, 1997a,b,c, 2001, 2003; Dabas et al. 1995; Bharti et al. 1996). The United States Environmental Protection Agency (USEPA) has indicated recently that the sites polluted with toxic heavy metals should receive priority for cleanup during the next few years (Eccles 1998). The contaminated land sites may consist of a heterogeneous mixture of different minerals, organic, organomineral substance and other solid components. The binding mechanisms of the heavy metals are, therefore, complex and vary with the composition of soil, soil acidity and redox conditions (Thangavel and Subburaam 2004). The bioavailability and mobility of heavy metals in soils is dependent upon the redistribution processes between solution and solid phases and among solid phase components. The rates of redistribution of metals and their binding intensity in soils were affected by the metal species, loading levels, ageing and soil properties (Eccles 1998; Han et al. 2003). The slow desorption of heavy metals in soil has been a major impediment to the successful phytoextraction of the metal contaminated sites (Thangavel and Subburaam 2004). Generally, only a fraction of soil metal is readily available (bioavaialable) for the plant uptake. The bulk of the metal in soil is commonly found as insoluble compounds unavailable for transport into roots from the aqueous phase. Cadmium and zinc are considered as easily mobile heavy metals as they occur primarily as soluble or exchangeable, readily bioavailable forms (Thangavel and Subburaam 2004). Copper, molybdenum and chromium are mainly bound in silicates and thus are slightly mobile. Lead occurs as insoluble precipiate (phosphates, carbonate and hydroxy-oxides), which are largely unavailable for plant uptake (Pitchel et al. 1999). It appears, therefore, that soluble, exchangeable and chelated species of trace elements are the most mobile in soils and these properties of the metals govern their migration and phytoavilability (Kabata-Pendias 1997). Binding and immobilization of the toxic metals within the soil matrix can significantly restrict their uptake and removal from the site. The bioavailability of the metals and other toxic substances, however, can be enhanced by manipulating the rhizosphere of the potential remediator plants by changing soil pH (lowering of pH is recommended to increase the bioavailability of heavy metals), adding chelating agents, using appropriate fertilizers (ammonium containing fertilizers), altering soil ion composition, adding adequate consortia of soil microbes and phytosiderophores and soil exudates managements (Table 3. Singh et al. 1996, 1999, 2001; Chaudhary et al. 1998; Khan et al. 2000; Thangavel and Subburaam 2004; Schwitzguébel 2004; Datta and Sarkar 2004).

Amendments of soil with ammonium containing fertilizers, organic and inorganic acids and elemental sulfur, $HNO_3$ and $CaCO_3$ lower the soil pH and enhance phytoaccumualtion of the toxic metals (Huang et al. 1997; Cristofaro et al. 1998; Chaney et al. 2000; Gao et al. 2003; Thangavel and Subburaam 2004), however, contrary reports are also available (Singh et al. 1996; Khan et al. 2000). Therefore, more precise and focused studies are needed to evaluate the independent effect of soil pH and soil amendments on hyperaccumulators yield and metal removal efficiency.

Artificial chelates e.g.. EDTA has been studied to enhance the heavy metal bioavailability and subsequent uptake and translocation to the shoots (Table 3. Fuentes 1997; Huang et al. 1997; Khan et al. 2000; Kayser et al. 2000). The chelates may be added at once a few days before harvest or gradually during the entire growth period. The uptake of Fe, Mn and Cu by maize plants was increased when EDTA or DTPA (1g/kg soil) was added in the soil prior to planting (Fuentes Bolomey 1997). Biosurfactants have also been shown to enhance the metal bioavailability in contaminated soil and sediments (Mulligan et al. 2001).

**Table 3.** Changes in bioavailability of the environmental contaminants especially heavy metals in rhizosphere and their uptake and accumulation by plants leading to altered phytoremediation efficiency due to rhizosphere amendments

| The toxic contaminant | Rhizosphere Amendments | Plant | Response | Reference |
|---|---|---|---|---|
| Cadmium | Iron | *Thlaspi caerulescens* | Decrease uptake by 3 folds | Lombi et al. (2002) |
| Cadmium and Zinc | Root exudates by the plant (organic legands) | *Thlaspi caerulescens* | Enhanced metal accumulation | Zhao et al. (2001) |
| Cadmium, Iron and Manganese | *Bacillus* sp., *Pseudomonas* sp. (Exude organic compounds) | *Brassica juncea* | Enhanced metal accumulation | Salt et al. (1995); Shekhar et al. (2004) |
| Iron, Manganese and Copper | EDTA | *Zea mays* | Enhance metal uptake | Fuentes (1997) |
| Iron, Manganese and Copper | Phytosiderophores | *Graminaceous species* | Enhance metal accumulation | Khan et al. (2000); Treeby et al. (1989); Ma and Nomoto (1996) |
| Lead | EDTA (0.5-1 mM) | *Pisum sativum* | 2 fold increase in accumulation | Piechalak et al. (2003) |
| Lead | EDTA (0.25 mM) | *Brassica juncea* | 75 fold higher Pb in plants than in hydroponics solution | Vassil et al. (1998) |
| Lead | EDTA (1 g/Kg soil) | *Garcinia cambogia* | Increased accumulation by 1.5 fold | Sekhar et al. (2004) |
| Lead | NaCl (6-12 EC) | *Vigna radiata* | Decreased accumulation by 3.5 to 5 fold | Singh et al. (2003 |
| Lead | $K_2HPO_4$ (10 mM), $CaCl_2$ (10 mM), $KNO_3$ (10 mM) | *Vigna radiata* | Decreased metal accumulation in roots and leaves | Singh et al. (1994b) |
| Nickel | NPK fertilizers | *Alyssum bertolonii, Thlaspi caerulescens, Streptanthus polygaloids* | Enhanced biomass with same concentration of nickel in aerial parts | Bennett et al. (1998) |

| | | | | |
|---|---|---|---|---|
| Selenium | Rhizosphere bacteria | *Brassica juncea* | 4-5 fold higher Se accumulation and volatilization | de Souza et al. (1999) |
| Trace metals and organic pollutants | Mycorrhizae | Many plants | Enhance uptake, phytostabilization and Biodegradation of contaminants | Chaudhary et al. (1998); Schwitzguéb el, (2004) |
| Zinc | Lime stone, cattle manure and poultry litter | *Zea mays* | Reduced bioavailability | Pierzynski and Schwab (1993) |
| Zinc | Phytosiderophores | *Triticum aestivum* | Increased uptake | Zhang et al. (1991) |

Another approach to enhance the rate of phytoremediation relates to the better agronomical management, which may yields an enhanced harvestable biomass of the remediating plants. Application of N-fertilizers (Bennett et al. 1998) to *Alyssum bertolonii, Streptanthus polygaloides* and *Thlaspi careulescens* have been shown to increase biomass very significantly without reducing the shoot nickel concentration. Addition of phosphate to soil may also help extract ion of Cr, Se and As by competing for the binding sites and thereby increasing bioavailability of the metals (Thangavel and Subburaam 2004).

Soil microbes have been found suitable to enhance the bioavailability and phytoremediation potential by complimenting the processes in many ways. Microbial activity in the rhizosphere of plants is several folds higher than in the bulk soil. Chemolithotrophic bacteria have been shown to enhance metal availability (Kelley and Tuovinen 1988). Several strains of *Bacillus* and *Pseudomonas* have been reported to increase cadmium accumulation by *Brassica juncea* (Salt et al. 1995). Naturally occurring rhizobacteria were found to promote Se and Hg accumulation in plants growing in wetland (de Souza et al. 1999). These microbes can grow more well, if organic manures are added to the soil. The mechanisms by which they increase the bioavailability and uptake of the heavy metals is not adequately elucidated yet, however, the possible mechanisms might include soil acidification and changes in the solubility of the metal complexes through their exudates (organic compounds exude from soil bacteria). The soil microbes may degrade organic pollutants and supply nutrients to plants for enhanced phytoremediation of the site.

It is generally considered that the majority of plants growing under natural conditions have symbiosis with mycorrhizae in roots, which result in increase in root surface area and nutrient acquisition (Khan et al. 2000). Mycorrhizal fungi have been reported in plants growing on heavy metal contaminated sites indicating its heavy metal tolerance and a potential role in the heavy metal phytoremediation (Table 3. Shetty et al. 1995; Weissenhorn and Leyval 1995;

Pawlawska et al. 1996; Chaudhary et al. 1998; Khan et al. 2000; Schwitzguébel 2004). Mycorrhizal fungal taxa, such as species like *Glomus, Gigaspora* and *Entrophospora*, have been reported to be associated with most of the plants growing in the heavy metal polluted habitats (Khan et al. 1990). The transport of the toxic metals absorbed by the mycorrhizal surface to the aerial part of the remediating plants is an obvious mechanism, which can enhance the total uptake and transport of the toxic metals in a defined period due to an increased surface area of the rhizosphere by the mycorrhizal associations.

Phytosiderophores (a class of organic compounds e.g. mugineic and avenic acids) exudated by roots of the many plants especially graminaceous species have been reported to enhance bioavailability of soil metals e.g.. Fe, Cu, Zn and Mn etc (Treeby et al. 1989; Thangavel and Subburaam 2004). Other kinds of root exudates can also reduce the rhizosphere soil pH and thus modulate the metal availability for uptake by the plants (Thangavel and Subburaam 2004), however, no direct evidence that indicates the involvement of root exudates in the phytoremediation has been documented.

## 5. Molecular Mechanisms of Uptake, Detoxification, Transport and Accumulation of Toxic Substances by Plants and Genetic Engineering for Enhanced Phytoremediation

Uptake of the toxic substances by the remediating plants is a pre-requisite for the phytoremediation. Following its bioavilability in the rhizosphere, their enhanced uptake and transport to the sink or metabolism sites can increase the efficiency of the phytoremediation of a selected plant. Transport proteins and intracellular high-affinity binding sites mediate the uptake of the metals and other substances across the plasma membrane. Many metal transporters genes have been cloned recently (Table 4. Datta and Sarkar 2004). Maser et al. (2001) have cloned genes of ZIP (Zn-regulated transporter/Fe-regulated transporter like proteins) family e.g.. *ZNT1* and *ZNT2,* from *Thlaspi careulescens*, which are highly expressed in roots of the accumulator plants, but their expression are not responsive to Zn status of the plants. Through functional complementation in yeast, it has been shown, however, that *ZNT1* protein mediates high affinity uptake of Zn and low-affinity uptake of $Zn^{+2}$ and $Cd^{+2}$(Pence et al. 2000). The transcription (factors) activators, such as Zn-responsive element, have been suggested to play an important role in Zn hyperaccumulation in *T. careulescens* (Pence et al. 2000). An increased uptake of Cd by *T. careulescens* and *A. thaliana* by enhanced expression of *IRT1* gene, which is essential for Fe uptake has been demonstrated (Lombi et al. 2002; Vert et al. 2002; Connolly et al. 2002; Datta and Sarkar 2004).

**Table 4.** Strategies for genetic engineering of plants to produce superior transgenic plants for phytoremediation of the environmental contaminants

| Plant genotype | Foreign gene introduced | Promoter | Vector | Response obtained | Phytoremediation efficiency of transformed plants | References |
|---|---|---|---|---|---|---|
| *Brassica juncea* L. cv 173874 | *Arabidopsis APSl* encoding ATP-sulfurylase | CaMV, 35S | *Agrobacterium tumefaciens* | Overexpression of plastidic ATP sulfurylase | 2-3 fold higher Se accumulation in shoots and 1.5 fold higher Se in roots as compared to the wild type plants | Pilon-Smits et al. (1999) |
| ,, | *E. coli gshII* encoding glutathione synthetase (GS) | ,, | ,, | Overexpression of cytosolic glutathione synthetase | 3 fold high Cd accumulation in transformed plants | Zhu et al. (1999a) |
| ,, | *E. coli gshI* encoding $\gamma$-glutamyl cysteinethione synthetase (GS) | ,, | ,, | Overexpression of $\gamma$-glutamyl cysteine synthetase targeted to the plastids | Increased tolerance to Cd, higher accumulation of phytochelatins, glutathione and total non-protein thiols, and accumulated more Cd (40-90% higher) in shoot than wild plants | Zhu et al. (1999b) |
| ,, | *E. coli gor* gene encoding gluthione reductase (GR) | ,, | ,, | Overexpression of gluthione reductase targeted to the plastids (cpGR) as well as cytosol (cystGR) | Reduced Cd uptake and /or translocation: Cd levels in shoots of (cpGR) plants were half as high as those in wild type shoots. Two times higher root glutathione levels in transformed (cpGR) plants than in wild type | Pilon-Smits et al. (2000) |

| | | | | | | |
|---|---|---|---|---|---|---|
| *Nicotiana tabaccum* and *B. juncea* | Mouse metallothionein1 (*MT1*) and human *MT2* genes | CaMV, 35S | ,, | Increased cadmium tolerance and accumulation | Increased cadmium tolerance 10µM to 200 µM | Pan et al. (1994), Misra and Gedamu (1989) |
| *Arabidopsis thaliana* ecotype c-24 | Chimeric plasmid pSNIRH containing spinach *NiR* cDNA and hygromycin phosphotransferase *hph* gene | CaMV, 35S and *NOS* polyadenylation signal | *A. tumefaciens* | Overexpression of *NiR* cDNA in transgenic plants | 40% increase was observed in $NO_2$ assimilation | Takahashi et al. (2001)<br>Morikawa et al. (2002) |
| *B. oleracea* var. botrytis | Yeast *cup1* gene | - | ,, | ,, | 16 fold higher Cd tolerance and accumulation | Hasegawa et al. (1997) |
| *B. juncea* | γ- glutamyl cysteine synthetase (γ-ECS) and glutathione synthetase (GS) | - | - | Increased cadmium and zinc uptake | Accumulated 1.5 fold more Cd and 2 fold more Zn in green house experiments based on the field contaminated soils compared to the wild type Indian mustard | Bennett et al. (2003) |
| *Arabidopsis thaliana* | *Arabidopsis* PC synthase (*At PCS1*) gene | -do- | ,, | Increased phytochelatin synthesis and higher tolerance to Cd | Increased cd phytoremediation | Lee et al. (2003b) |
| *A. thaliana* | Zinc transporter (*ZNT*), a putative vacuolar transporter, which encode a Pb-II/CdII/Zn II pump | -do- | ,, | Increase Zn, Pb, and Cd tolerance | Two fold higher Zn accumulation in roots | van der Zaal et al. (1999) |
| *A. thaliana* | -do- | -do- | *A. tumefaciens* | Lower accumulation | Accumulated more Cd in | Lee et al. |

| | | | | | | |
|---|---|---|---|---|---|---|
| | | | | of Cd in shoot protoplast | vaculoes | (2003a) |
| *N. tabaccum* | Calcium vacuolar transporter *Arabidopsis antiporter CAX2* | - | - | Higher tolerance to $Mn^{+2}$ levels | Accumulated more $Ca^{+2}$, $Cd^{+2}$ and $Mn^{+2}$ | Hirachi et al. (2000) |
| *N. tabaccum* and *A. thaliana* | Bacterial genes *ars c* from *E.coli* | - | ,, | Higher cadmium tolerance than wild type | Higher cadmium tolerance than wild type | Dhankher et al. (2003) |
| *A. thaliana* | Bacterial genes (*E.coli*) arsenate reductase (*ars c*) and γ- glutamyl cysteine synthetase (*γ-ECS*) together | Two different promoters; SRS1p and CaMV, 35S | ,, | Higher arsenic tolerance | 4-17 fold higher arsenic hyperaccumulation in shoots | Dhankher et al. (2002) |
| *A. thaliana* | Mouse Se-cysteine lyase (*pSLY*) and *pSCH*) | CaMV, 35S | ,, | Expression in the cystol or chloroplast of *Arabidopsis* resulted 2 fold (cytosolic lines) or 6 fold (chloroplastic lines) higher SL activities in transgenic plants than wild type and enhanced tolerance to Se | Higher Se- volatilization than wild type | Pilon et al. (2003) |
| *B. juncea* | Cystathionine-gamma-synthase (*CGS*) gene from *A. thaliana* | - | *A. tumefaciens* | Higher Se tolerance than wild type | 2-3 fold higher Se volatilization than wild type | van Huysen et al. (2003) |
| *Lycopersicon* | 1-aminocyclopropane-1- | | ,, | Produces lower | Higher metal accumulation | GrichKo et al. |

| | | | | | | |
|---|---|---|---|---|---|---|
| *esculeutum* and *B. juncea* | carboxylate (ACC) deaminase gene | | | levels of ethylene and protects from deleterious effects of six metals e.g. $Cd^{+2}$, $Co^{+2}$, $Cu^{+2}$, $Mg^{+2}$, $Ni^{+2}$, $Pb^{+2}$ or $Zn^{+2}$ | than the wild types | (2000), Nie et al. (2002) |
| *N. tabaccum* | Murine monoclonal antibody *IgG1* gene | - | - | Higher metal uptake | Higher level of phytoremediation | Drake et al. (2002) |
| *A. thaliana* | Yeast vacuole transporter *YCF1* gene | - | - | Enhance tolerance to $Pb^{+2}$ and $Cd^{+2}$ | Higher accumulation in transgenic plants | Song et al. (2003) |
| *A. thaliana* | Mercuric ion reductase *merA* genes from *E.coli* | | ,, | Transgenic plants expressing *merA* gene was more tolerant to $HgCl_2$ and $Au^{+3}$ and volatilized elemental mercury | Higher mercury phytoremediation | Rugh et al. (1996) |
| *A. thaliana* | Mercuric ion reductase *merB* genes from *E.coli* | | ,, | Transgenic plants were more tolerant to methyl mercury and other organomercurials | Phytoremedation of organomercurials; can grow on 10 fold higher methyl mercury than wild type | Bizily et al. (1999) |
| *A. thaliana* | Both of the above genes | | *A. tumefaciens* | -do- | Transgenic plants can grow on 50 fold higher methyl mercury than the wild type plants | Bizily et al. (2000) |

| | | | | | | |
|---|---|---|---|---|---|---|
| *A. thaliana* | Both, *merA* and *merB* genes of bacteria origin | | ,, | Transfer to the chloroplast genome resulted in high levels of tolerance to the organomercurials compound, phenylmercuric acetate (PMA) and increased biomass | The use of chloroplast transformation to enhance Hg phytoremediation is particularly beneficial because it prevents the escape of transgenes via pollen to the related weeds or crops and there is no need for codon optimization to improve transgene expression | Ruiz et al. (2003) |
| *N. tabaccum* | Bacterial nitroreductase gene (*pNITRED3*) | CaMV, 35S promoter modified *nfsI* and *nos* termination sequences | ,, | Increased tolerance to 2,4,6-trinitrotoluene (TNT) | Remediation/Detoxification of TNT (recalcitrant military explosive) | Hannink et al. (2001) |

The metal transporters e.g. metal (or CPx-type) ATPases, that are involved in the overall metal ion homeostasis and tolerance in plants and natural resistance associated macrophase (Nramp) family of proteins and cation diffusion facilitator (CDF) family of proteins have been characterized in a wide range of organisms including plants (Belouchi et al. 1997; Tabata et al. 1997; Alonso et al. 1999; Guerinot and Eide 1999; Thomine et al. 1999; van der Zaal et al. 1999; Williams et al. 2000; Datta and Sarkar 2004). CPx-type metal ATPases have been implicated in the transport of essential as well potentially toxic metals like Cu, Zn, Cd and Pb etc across the cell membranes (Williams et al. 2000). They share a common feature of a conserved intra-membranous cystein-proline-cystein, cystein-proline-histidine or cystein-proline-serine(CPx) motif, which is thought to function in the metal transduction. These transporters use ATP to pump a variety of charged substrates across the cell membranes and are distinguished by the formation of a charged intermediate during the reaction cycle. *Arabidopsis* P-type ATPases (PAA1) was the first CPx-ATPases reported in the higher plants (Tabata et al. 1997; Datta and Sarkar 2004).

Though the physiological role of the metal ATPases in higher plants is not precisely demonstrated, most CPx –type ATPases identified have been involved in the Cu or Cd transport. Since *Arabidopsis* CPx-ATPases show fairly low similarities to each other, they are specific for transporting different substrates (Datta and Sarkar 2004). The ATPases located in plasma membrane may function as efflux pumps removing potentially toxic metals from the cytoplasm, or may also be present at the various intracellular membranes and be responsible for the compartmentalization of the metals, e.g. sequestration in the vacuoles, golgi or endoplasmic recticulum (Datta and Sarkar 2004). To control the intracellular levels of the metals, regulation of transporters, which could occur in higher plants, similarly as has been observed in the bacteria and yeast, at the transcriptional level (control on initiation rates, mRNA stability, differential mRNA splicing) or at the post translational level (control on targeting and/or stability) have been postulated, though the precise mechanisms for the regulation of the metal transport by CPx-ATPases in higher plants is not known (Williams et al. 2000; Datta and Sarkar 2004).

Another divalent metal ion transporters of Nramp family, encoded by *Nramp* genes, have been identified in rice and *Arabidopsis* (Belouchi et al. 1997; Alonso et al. 1999). Cation diffusion facilitator (CDF) proteins have also been involved in the transport of Zn, Co, Cu and Cd in bacteria and plants e.g.. poplar (Blaudez et al. 2003). Related Zn transporters *ZAT1*, which may have a role in Zn sequestration in plants, have been reported in *Arabidopsis* (van der Zaal et al. 1999). Enhanced Zn resistance has been demonstrated in transgenic plants overexpressing *ZAT1*. constitutively throughout. Zinc transporter (ZIP) proteins have also been found to be involved in Zn and Fe uptake (Guerinot and Eide 1999). The metal uptake which may lead to an enhanced phytoremediation efficiency can be increased by increasing number of uptake sites, specific transporters and regulators of the transport system, intracellular high affinity binding sites by incorporating/over-expressing the target genes in the plants by genetic engineering (Table 4). However, a comprehensive understanding of the metal transport processes in plants is essential for formulating the effective strategies to develop genetically engineered plants that can be used commercially for rapid cleanup of the contaminated sites.

The toxic heavy metal detoxification mechanisms involve chelation of metals by a ligand, followed by the sequestration of the metal-ligand complexes into the vacuoles. Intracellular metal complex formations have been reported with peptide and protein legands, such as metallothioneins (MTs) and phytochelatins (PCs). Metallothineins are first identified in mammalian tissues as Cd –binding peptides and subsequently in the plants (Murphy and Taiz 1995; Foley et al. 1997; de Borne et al. 1998; Garcià-Hernández et al. 1998; Salt et al. 1998; Datta and Sarkar 2004). Phytochelatins are a family of sulfur rich peptides, first identified in yeast and subsequently in a wide variety of plant species including angiosperms (both monocots and dicots), gymnosperms, algae, fungi and marine diatoms but not in animals (Rauser 1995; Cobett 2000; Vatamaniuk et al. 2002; Datta and Sarkar 2004 and references therein). Molecular-genetic studies on yeast and *Arabidopsis* PCs have revealed significant insights during the last decade (Rauser 1995; Cobbett 2000). PCs are induced rapidly in cells and tissues on exposure to a range of metal ions (cations), such as Cd, Ni, Cu, Zn, Ag, Hg and Pb and anions, such as arsenate and selenite (Rauser 1995, 1999; Friederich et al. 1998; Ha et al. 1999; Leopold et al. 1999; Cobbett 2000; Hartley-Whitaker et al. 2001; Cosio et al. 2004; Hussain et al. 2004; Küpper et al. 2004; Raab et al. 2004; Song et al. 2004; Datta and Sarkar 2004).

The PCs are synthesized from glutathione by adding a terminal glycine (gly) into the dipeptides $(\gamma\text{-Glu-Cys})_n$ by the action of enzyme phytochelatin synthase. PCs form a family of structures with increasing repetitions of the γ-Glu-Cys dipeptide, followed by a terminal Gly;$(\gamma\text{-Glu-Cys})_n$-Gly, where $_n$ has been reported as being as high as 11, but is generally in a range of 2-5 (Cobbett 2000).

It has been demonstrated that GSH deficient mutants of *Arabidopsis* are deficient in PCs and are found Cd-sensitive (Cobbett et al. 1998). Metal ion induced and GSH dependent PC synthase activity that related to the metal tolerance has been shown in *Silene cucubalis* (Grill et al. 1989), tomato (Chen et al. 1997), pea (Klapheck et al. 1995) and *Arabidopsis* (Howden et al. 1995a,b). In Azuki beans (*Vigna angularis*), an essentiality of PC synthase for Cd tolerance has been demonstrated (Inouhe et al. 2000). PC synthase genes *AtPCS1* in *Arabidopsis*, whose expression mediated an increased Cd accumulation (Vatamaniuk et al. 1999) and *TaPCS1* in wheat that increased Cd-resistance and accumulation (Clemens et al. 1999) were reported simultaneously. Both *AtPCS1* and *TaPCS1* mediated Cd tolerance has been found GSH dependent and function in vacuole-deficient mutants, suggesting a cytosolic localization. These genes mediate *in vivo* PC biosynthesis in yeast

(Datta and Sarkar 2004). The role of GSH and PCs in hyperaccumulation of the heavy metals in plants has been demonstrated using transgenic approach in few plants (Table 4).

The availability of amino acids, especially that of sulfur amino acids and regulation of PC synthase activity, is considered as the most important regulatory mechanism of the PC biosynthetic pathways. Another important molecular event that regulates hyperaccumulation of toxic heavy metals in plants relates to sequestration of the metals in the vacuoles. The PC-metal complexes are driven by various membrane transporters (Cobbett 2000; Blaudez et al. 2003; Küpper et al. 2004; Cosio et al. 2004; Raab et al. 2004: Datta and Sarkar 2004). These membrane transporters include CPx-type ATPases, Nramp family of proteins and CDF family proteins as discussed earlier. More detailed insights on the characterization, isolation, cloning and regulation of transport of the PC-metal complexes from source to sink are needed to achieve better phytoremediation efficiency of the heavy metals using biotechnological approaches. Free histidine (His) has been reported to be Ni-chelator in *Alyssum lesbiacum* and *Brassica juncea* and it has been found to enhance release of Ni into the xylem during its transport to aerial parts (Kerkeb and Krämmer 2003). However, Persans et al. (1999) have reported that the Ni-hyperaccumulation phenotype in *Thlaspi goesingense* could not be related to the overproduction of His in response to nickel.

The phytoremediation mechanisms for most of the heavy metals thus seem to be governed by the ion transport and hyperaccumulation in the vacuolar sinks of the tolerant plants. Phytodegradation and phytovolatilization are the preferred mechanisms for the cleanup of organic xenobiotics (Morikawa and Takahashi 2000; Schwitzguébel 2004). These processes, however, also rely on the movement of the pollutants into plant roots and subsequent translocation into other tissues and parts of the plants, where the detoxification and metabolization take place (Schroeder et al. 2002). Higher plants have evolved many genes and enzymes, which have potentials to metabolize or degrade different kinds of xenobiotic compounds. Xenobiotic metabolism in plant cells proceeds through different partially linked stages (Schwitzguébel 2004 for a recent review). The reductive, oxidative and hydrolytic enzymes introduce functional groups (-OH, -$NH_2$, -SH) into lipophilic substrates in phase I reactions. Hydrolytic reactions, catalysed by esterases or amidases, are quite common and the multiple isoforms of substrate inducible enzymes have been reported. The oxidation reactions (epoxidation, O- or N-dealkylation, aryl- or alkylhydroxylation, N-or S-oxidation) appear to be catalysed by the cytochrome P450 mono-oxygenases. This process seems to be the most important in xenobiotics, phytoremediation. These enzymes are microsomal in localization and have been characterized well in mammalian systems. In plants, they are induced by wounding, pathogenesis and chemical stresses e.g. organic xenobiotic compounds. The wide range of transferases catalyze removal of glucosyl moieties, amino acids, malonic acid or glutathione residues in Phase II reaction. The herbicide and other xenobiotics metabolites containing these residues can be deposited as "bound residues" in the extracellular matrix/cell wall, or stored as water soluble metabolites in the vacuoles (Phase III) (Schwitzguébel 2004).

One of the major limitations in the phytoremediation of the organic pollutants, especially for the soil contaminants, has been realized as the poor understanding of the soil chemistry of these pollutants, their mobilization in the rhizosphere, their uptake and the transport within the plants (Cunningham et al. 1996; Sicilano and Germida 1998; Trapp and Karlson 2001; Mehmannavaz et al. 2002; Campanella et al. 2002; Harvey et al. 2002; Muratova et al. 2003; Schwitzguébel 2004). Rhizosphere microbes can play an important role in enhancing the bioavailability of the organic pollutants for the plant uptake. Uptake of hydrophobic xenobiotics of larger size can be facilitated by the primary microbial biodegradations in the rhizosphere. The hydrophobic persistent organic pollutants like polychlororinated dibenzo-p-dioxins and furans (PCDD/Fs) and 2,3,7,8-tetrachlorodibenzo-p-dioxin (TCDD) with present log $K_{ow}$ value above 4 are taken up by roots and transported to shoots by the transpiration stream of plants like Zucchini (*Cucurbita pepo*) (Campanella and Paul 2000). A proteinaceous molecule able to increase apparent aqueous solubility and binding during transport of such organic compounds have been found in the xylem sap and leaf extracts (Campanella and Paul 2000; Campanella et al. 2002).Hybrid poplar (*Populus* species) have also been demonstrated to remediate organic pollutants including trichloroethylene (TCE), a potential carcinogen commonly found in ground water and the contaminated sites (Kassel et al. 2002). Although many organic pollutants are metabolized or degraded to less toxic substances and accumulated in the phytoremediating plants, certain volatile organic chlorinated compounds e.g. BTEX (benzene, toluene, ethylbenzene, xylene), and MTBE (methyl tert-butyl ether) etc. can be released to the atmosphere. However, volatilization undermines the merits for phytoremediation for these applications (Schwitzguébel 2004). For such problems, rhizodegradation is usually attempted as a solution. However, large root absorption area, big root tip mass, high biomass with high enzymatic capabilities can make plants as ideal cleaning system of soil-based organic pollutants too, if bioavailability, uptake, transport and its metabolism can be regulated upto the desired extent. Though some success have been achieved to develop large scale commercial phytoremediation projects for cleanup of the sites or groundwater contaminated with organic xenobiotics (Table 2. Glass 1999; Trapp and Karlson 2001; Schwitzguébel 2004), but still this area needs more attention in the future.

Isolation, characterization and cloning of most appropriate genes from the organisms across the taxonomic boundaries, adequate promoters and regulatory genes (e.g. transcription factors), efficient genetic transformation and *in vitro* regeneration protocols can be seen as biotechnological approaches to resolve such problems of persistent organic pollutant, contamination. Plant genetic engineering has emerged as a technology which can create new potential character in a plant from a distantly related organism (beyond taxonomic boundaries) or even using synthetic genes and promoters. Many appropriate genes of foreign origin have been transferred in the plants like *Arabidopsis thaliana, Nicotiana tabaccum, Brassica juncea, Brassica oleracea* var *botrytis, Lycopersicon esculeutum* etc. to enhance the phytoremediation efficiency of these plants (Table 4. Raskin 1996; Rugh et al. 1996; Arazi et al. 1999; Arisi et al. 2000; Meagher 2000; Nedelkoska and Doran 2000; Assuncao et al. 2001). The genes of choice are related to the regulatory genes of sulfur metabolism, glutathione biosynthesis for the synthesis of binding peptide and proteins, uptake and transport proteins for the partitioning, targeting and metabolizing proteins/enzymes etc. which have enhanced significantly the potential of the phytoremediation using transgenic plants. Transgenic plants so far have been developed for the hyperaccumulation of toxic heavy metals e.g. Hg, As, Pb, Cd, Co, Ni, Zn, Cu etc. air pollutants e.g. $NO_2$ and $SO_2$ and organic pollutants e.g. 2,4,6-trinitrotoluene and organomercurials etc. The literature available on the genetic engineering of plants for phytoremediation indicate clearly that this technology can be used successfully to enhance rhizosphere degradation, bioavailbility, uptake, transport, targeting, partitioning, storage and hyperaccumulation of toxic pollutants of various kinds and also to resolve the problems associated with post harvest, management and recycling of the contaminated phytomass. It can combine the various characters of ideal phytoremediation in one plant which has fast growth, higher biomass, suitability for easy post harvest, agroclimatic adaptations and desired root size and root depth alongwith high efficiency to remediate specific contaminants as well as mixture of many contaminants. Rhizosphere management can also be enhanced by introducing genes for required plant exudates and microbial strains for better potential for supplementing phytoremediation by enhancing bioavailbility and solubility of the pollutants.

A lot of challenges are to be addressed, by the biotechnologists to meet out the commercial needs and to utilize an optimal potential of this technology. The major limitations of plant genetic engineering as a technology have been the availability of most appropriate genes based on wider prospecting of huge biodiversity, novel promoters and transcription regulators (transcription factors regulating larger metabolic pathways), genes for factors regulating post translation modification, targeting and transport proteins and peptides and the factors for the storage management of the metabolites etc. In addition, removal of non-required or deleterious associated genes (e.g. selectable and visible markers) and avoidance of pollen mediated flow of foreign gene e.g. chloroplast transformation etc. will be a focal attention in the recent future. Addressing these challenges environmental safer, free from any health hazards, high potentials economic phytoremediation can be developed using extensive bioprospecting and genetic engineering in recent future.

## 6. Conclusion

Phytoremediation is an eco-friendly cost-effective technology, as compared to classical physical, chemical and even to the microorganisms-based bioremediation techniques. It is useful for the remediation of sites contaminated with non-biodegradable toxic heavy metals, hazardous air pollutants like oxides of nitrogen and sulfur, and photoxidants like ozone, recalcitrant organic pollutants, like chlorinated pesticides, organophosphate, insecticides, petroleum hydrocarbons, polynuclear aromatic hydrocarbons (PAHs), sulphonated biphenyl (PCBs) and chlororinated solvents (TCE, PCE) etc.

Amongst the major limitations of the technique, tolerance level of plants to high contamination zones, treatment of only bioavailable fraction of the contaminants and remediation of the contaminants largely from within a meter of the surface of the soil and within a few meters of the surface of the groundwater can be counted. The agro-climatic and hydrological conditions may also limit the plant growth on the treatment site and chances of entering of the contaminants in food chain through animals /insects that eat plant material containing the contaminants need to be attended while advocating for this technology. Plant biomass and agricultural vegetable wastes can also be used as adsorbant systems for the remediation of waterbodies from organic and inorganic pollutant's contaminations. Due to the low cost of the technique, the low disturbance in the *in situ* treatments, a higher probability for the public acceptance and an easy handling, this technology indicates a strong potential as a natural, or improved, solar energy driven remediation approach for the treatments of the various kinds of the pollutants.

## References

Alonso JM, Hirayama T, Roman G, Nourizadesh S, Ecker JR (1999) EIN2, a bifunctional transductor of ethylene and stress response in *Arabidopsis*. Science 284:2148-2152

Arazi T, Sunkar R, Kaplan B, Fromm H (1999) A tobacco plasma membrane calmodulin-binding transporter confers $Ni^{+2}$ tolerance and $Pb^{+2}$ hypersensitivity in transgenic plants. Plant J 20:171-182

Arisi ACM, Mocquot B, Lagriffoul A, Mench M, Foyer CH, Jouanin L (2000) Response to cadmium in leaves of transformed poplars over expressing γ-glutamyl-cysteine synthetase. Physiol Plant 109:143-149

Assuncaáo AGL, Martins PDC, Folter SD, Vooijs R, Schat H, Aatrss MGM (2001) Elevated expression of metal transporter genes in three accessions of the metal hyperaccumulator *Thlaspi caerulescens*. Plant Cell Environ 24:217-226

Baker AJM, Brooks RP (1989) Terrestrial higher plants which hyperaccumulate metal element: A review of their distribution, ecology and phytochemistry. Biorecovery 1:81-126

Baker AJM, McGrath SP, Reeves RD, Smith J (2000) Metal hyperaccumulator plants: a review of the ecology and physiology of a biological resource for phytoremediation of metal-polluted soils. In: Terry N, Banuelos GS (eds) Phytoremediation of Contaminated Soils and Water, CRC Press Inc., FL, pp 85-107

Belouchi A, Kwan T, Gros P (1997) Cloning and characterization of the OsNramp family from *Oryza sativa*, a new family of membrane proteins possibly implicated in the transport of metal ions. Plant Mol Biol 33:1085-1092

Bennett FA, Tyler EK, Brooks RR, Gregg PEH, Stewart RB (1998) Fertilization of hyperaccumulators to enhance their potential for phytoremediation and phytomining. In: Brooks RR (ed) Plants that Hyperaccumulates Heavy Metals, CAB International, UK, pp 249-259

Bennett LE, Burkhead JL, Hale KL, Terry N, Pilon M, Pilon-Smits EAH (2003) Analysis of transgenic Indian mustard plants for phytoremediation of metal contaminated mine tailings. J Environ Qual 32:432-440

Bharti N, Singh RP (1993) Growth and nitrate reduction by *Sesamum indicum* cv. PB-1 respond differentially to lead. Phytochemistry 33:531-534

Bharti N, Singh RP (1994) Antagonistic effect of sodium chloride to differential heavy metal toxicity regarding biomass accumulation and nitrate assimilation in *Sesamum indicum* seedlings in a lead enriched environment. Phytochemistry 35:1157-1161

Bharti N, Singh RP, Sinha SK (1996) Effect of calcium chloride on heavy metal induced alterations in growth and nitrate assimilation of *Sesamum indicum* seedlings. Phytochemistry 41:105-109

Bizily SD, Rugh CL, Meagher RB (2000) Phytodetoxification of hazardous organomercurials by genetically engineered plants. Nature Biotech 18:213-217

Bizily SP, Rugh CL, Summers AO, Meagher RB (1999) Phytoremediation of methylmercury pollution :*mer* B expression in *Arabidopsis thaliana* confers resistance to organomercurials. Proc Natl Acad Sci USA 96:6806-6813

Blaudez D, Kohler A, Martin F, Sanders D, Chalot M (2003) Poplar metal tolerance protein 1 confers zinc tolerance and is an oligomeric vacuolar zinc transporter with an essential leucine zipper motif. Plant Cell 15:2911-2928

Brooks PR, Morison RS, Reeves RD, Dudey TP, Akman Y (1979) Hyperaccumulation of nickel by *Alyssum linneaus* (cruciferae). Proc Roy Soc Lond Biol Sci 203:387-403

Campanella B, Paul R (2000) Presence, in the rhizosphere and leaf extracts of Zucchini (*Cucurbita pepo* L.) and Melon (*Cucumis melo* L.), of molecules capable of increasing the apparent aqueous solubility of hydrophobic pollutants. Int J Phytorem 2:145-158

Campanella B, Bock C, Schroeder P (2002) Phytoremediation to increase the degradation of PCBs and PCDD/Fs potential and limitations. Environ Sci Pollut Res 9:73-85

Chaney RL, Li YM, Brown SL, Homer FA, Malik M, Angle JS, Baker AJM, Reeves RD, Chin M (2000) Improving metal hyperaccumulator wild plants to develop commercial Phytoextration system: Approaches and progress. In: Terry N, Banuelos G, Vangronsveld J (eds) Phytoremediation of Contaminated Water and Soil, CRC Press, pp 129-158

Chaudhary TM, Hayes WJ, Khan AG, Khoo CS (1998) Phytoremediation-focusing on accumulator plants that remediate metal-contaminated soils. Aust J Ecotoxic 4:37-51

Chen J, Zhou J, Goldsbrough PB (1997) Characterization of phytochelatin synthase from tomato. Physiol Plant 101:165-172

Clemens S, Kims EJ, Neumann D, Schroeder JI (1999) Tolerance of toxic heavy metal by a gene family of phytochetin synthase from plants and yeast. EMBO J 18:3325-3333

Cobbett CS, May MJ, Howden R, Rolls B (1998) The glutathione-deficient, cadmium-sensitive mutant, *cad2*-1 of *Arabidopsis thaliana* is deficient in γ-glutamylcysteine synthetase. Plant J 16:73-78

Cobbett CS (2000) Phytochelatins and their roles in heavy metal detoxification. Plant Physiol 123:825-832

Companella B, Paul R (2000) Presence, in the rhizosphere and leaf extracts of *Zucchini* (*Cucurbita pepo* L.) and Melon (*Cucumis melo* L.) of molecules capable of increasing the apparent aqueous solubility of hydrophobic pollutants. Int J Phytoremed 2:145-158

Connoly EL, Fett JP, Guerinot ML (2002) Expression of the *IRT1* metal transporter is controlled by metals at the levels of transcript and protein accumulation. Plant Cell 14:1347-1357

Cosio C, Martinoia E, Keller C (2004) Hyperaccumulation of cadmium and zinc in *Thlaspi careulescens* and *Arabidopsis halleri* at the leaf celluar level. Plant Physiol 134:716-725

Cristofaro AD, Zhon DH, He JZ, Violante A (1998) Comparison between oxalate and humate on copper adsorption on goethite. Fresenius Environ Bull 7:570-576

Cunningham SD, Ow DW (1996) Promises and prospects of phytoremediation Plant Physiol 110:715-719

Cunningham SD, Berti WR, Huang JW (1995) Phytoremediation of contaminated soils. Trends Biotechnol 13:393-397

Cunningham SD, Anderson TA, Schwab AP, Hsu FC (1996) Phytoremediation of soils contaminated with organic pollutants. In: Sparks DL (ed) Advances in Agronomy, vol 56, Academic Press London, pp 55-114

Dabas S, Singh RP, Sawheny V (1995) Nitrogen fixation and ammonia assimilation in mungbean nodules during lead contamination. Physiol Mol Biol Plants 1:135-140

Datta R, Sarkar D (2004) Biotechnology in phytoremediation of metal-contaminated soils. Proc Indian Natn Sci Acad, B70:99-108

de Borne FD, Elmayan T, de Roton C, de Hys L, Tempfer M (1998) Cadmium partitioning in transgenic tobacco plants expressing a mammalian metallothionein gene. Mol Breed 4:83-90

de Souza MP, Chu D, Zhao M, Zayed AM, Ruzin SE, Schichenes D, Terry N (1999a) Rhizosphere bacteria enhance selenium accumulation and volatilization by Indian mustard. Plant Physiol 119:565-573

de Souza MP, Huang CP, Chee N, Terry N (1999b) Rhizosphere bacteria enhance the accumulation of selenium and mercury in wetland plants. Planta 209:259-263

Dhankher OP, Shasti NA, Rosen BP, Fuhrmann M, Meagher RB (2003) Increased cadmium tolerance and accumulation by pants by expressing bacterial arsenate reducatse. New Phytol 159:431-441

Dhankher OP, Li Y, Rosen BP, Shi J, Salt D, Senecoff JF, Sashti NA, Meagher RB (2002) Engineering tolerance and hyperaccumulation of arsenic in plants by combining arsenate reductase and γ-glutamylcysteine synthetase expression. Nature Biotech 20:1140-1145

Dietz AC, Schnoor JL (2001) Advances in phytoremediation, Environ. Heath Perspect 109:163-168

Drake PMW, Chargelegue D, Vine ND, van Dollewerd CJ Obregon P, Ma JKC (2002) Transgenic plants expressing antibodies: a model for phytoremediation. FASEB J 16:1855-1860

Eccles H (1998) Metal contaminated soil-Is natural attention acceptable? Biochem Soc Tran 26:657-661

Foley RC, Liang ZM, Singh KB (1997) Analysis of type 1 metallothionein cDNA in *Vicia faba*. Plant Mol Biol 33:583-591

Francova K, Macek, T, Demnerove K, Mackova M (2001) Transgenic plants- a potential tool for decontamination of environmental pollutants. Chemicke Listy 95:630-637

Friederich M, Kneer R, Zenk MH (1998) Enzymic synthesis of phytochelatins in gram quantities. Phytochemistry 49:2323-2329

Fuentes Bolomy HD (1997) The influence of selected soil parameters on DTPA extractable metals and the uptake of *Zea mays* (corn), Honours Thesis, University of Western Sydney, Macarthur

Gao Y, He J, Ling W, Hu H, Liu F (2003) Effects of organic acids on copper and cadmium desorption from contaminated soils. Environ Int 29:613-618

Garcià-Hernàndez M, Murphy A, Taiz L (1998) Methallothioneins 1 and 2 have distinct but overlapping expression patterns in *Arabidopsis*. Plant Physiol 118:387-397

Geebelen W, Vangronsveld J, Adriano DC, Poucke LCV, Clijsters H (2002) Effects of Pb-EDTA and EDTA on oxidative stress reaction and mineral uptake in *Phaseolus vulgaris*. Physiol Plant 115:377-384

Glass JD (1999) US and International Markets for Phytoremediation 1999-2000, Dglass Associates, Inc pp 4

Gleba D, Borisjuk NV, Borisjuk LG, Kneer R, Poulev A, Skarzhinskaya M, Dushenkov S, Logendra S, Glebs YY, Raskin I (1999) Use of plant roots for phytoremediation and molecular farming. Proc Natl Acad Sci USA 96:5973-5977

Grichko VP, Filby B, Glick BR (2000) Increased ability of transgenic plants expressing the bacterial enzyme ACC deaminase to accumulate Cd, Co, Cu, Ni, Pb, and Zn. J Biotech 81:45-53

Grill E, Loffler S, Winnacker E-L, Zenk MH (1989) Phytochelatins, the heavy-metal-binding peptides of plants, are synthesized from glutathione by a specific r-glutamylcysteine dipeptidyl transpeptidase (phytochelatin synthase). Proc Natl Acad Sci USA 86:6838-6842

Guerinot ML, Eide D (1999) Zeroing in on zinc uptake in yeast and plants. Curr Opin Plant Biol 2:244-249

Ha S-B, Smith AP, Howden R, Dietrich WM, Bugg S, O'Connell J, Goldsbrough PB, Cobbett CS (1999) Phytochelatin synthase genes from *Arabidopsis* and the yeast *Schizosaccharomyces prombe*. Plant Cell 11:1153-1163

Hall JL (2002) Cellular mechanism for heavy metal detoxification and tolerance. J Exp Bot 53:1-11

Han FX, Banin A, Kingery WL, Triplrtt GB, Zhou LX, Zheng SL, Ding WX (2003) New approach to studies of heavy metal redistribution in soil. Adv Env Res 8:113-120

Hannink N, Rosser SJ, French CE, Barran A, Murray JAH, Nicklin S, Bruce NC (2001) Phytodetoxification of TNT by transgenic plants expressing a bacterial nitroreductase. Nature Biotech 19:1168-1172

Hartley-Whitaker J, Ainsworth G, Vooij R, Bookum WT, Schat H, Meharg AA (2001) Phytochelatins are involved in differential arsenate tolerance in *Holcus lanatus*. Plant Physiol 126:299-306

Harvey PJ, Campanella BF, Castro PML, Harms H, Lichtfouse E, Schaeffner AR, Smrcek S, Werck-Reichhart D (2002) Phytoremediation of polyaromatic hydrocarbons, anilines and phenols. Environ Sci Pollut Res 9:29-47

Hasegawa I, Terada E, Sunairi M, Wakita H, Shinmachi F, Noguchi A, Nakajima M, Yazaki J (1997) Genetic improvement of heavy metal tolerance in plants by transfer of the yeast metallothionein gene *cup*1. Plant Soil 196:277-281

Hirschi KD, Korenkov VD, Wilganowski NL, Wagner GL (2000) Expression of *Arabidopsis CAX*2 in tobacco; Altered metal accumulation and increased manganese tolerance. Plant Physiol 124:125-133

Howden R, Goldsbrough PB, Andersen CR, Cobbett CS (1995a) Cadmium sensitive, cad1 mutants of *Arabidopsis thaliana* are phytochelatin deficient. Plant Physiol 107:1059-1066

Howden R, Andersen CR, Goldsbrough PB, Cobbett CS (1995b) A cadmium sensitive, glutathione-deficient mutant of *Arabidopsis thaliana*. Plant Physiol 107:1067-1073

Huang JW, Chen J, Berti WR, Cunningham SD (1997) Phytoremediation of lead contaminated soils: Role of synthetic chelates in lead Phytoextration. Environ Sci Technol 31:800-805

Hussain D, Haydon MJ, Wang Y, Wong E, Sherson SM, Young J, Camakaris J, Harper JF, Cobbett CS (2004) P-type ATPase heavy metal transporters with roles in essential Zn homeostasis in *Arabidopsis*. Plant Cell 16:1327-1339

Inouhe M, Ito R, Ito S, Sasada N, Tohoyama H, Joho M (2000) Azuki bean cells are hypersensitive to cadmium and do not synthesize phytochelatins. Plant Physiol 123:1029-1036

Kabata-Pendias A (1997) Trace metal balances in soil- a current problem in agriculture. In: Adriano DC, Chen ZS, Yang SS, Iskander IK (eds) Biogeochemistry of Trace Metals, Science Reviews Lewis Publishers, New York, pp 139-167

Kassal AG, Ghosal D, Goyal A (2002) Phytoremediation of trichloroethylene using hybrid poplar. Physiol Mol Biol Plant 8:3-10

Kayser A, Wenger K, Keller A, Attinger W, Felix HR, Gupta SK, Schulin R (2000) Enhancement of Phytoextration of Zn, Cd and Cu from calcareous soil: the use of NTA and sulfur amendments. Environ Sci Technol 34:1778-1783

Kelley BC, Tuovinen OH (1988) Microbial oxidation of minerals in mine tailings. In: Solomons W, Foerstner V (eds) Chemistry and Biology of Solid Waste, Springer Verlag, Berlin, pp 33-53

Kerkeb L, Krämer U (2003) The role of free histidine in xylem loading of nickel in *Alyssum lesbiacum* and *Brassica juncea*. Plant Physiol 131:716-724

Khan AG, Kuek C, Chaudhary TM, Khaoo CS, Hayes WJ (2000) Role of plants, mycorrhizae and phytochelators in heavy metal contaminated land remediation. Chemosphere 21:197-207

Klapheck R, Schlunz S, Bergmann L (1995) Synthesis of phytochelatins and homo-phytochelatins in *Pisum sativum* L. Plant Physiol 107:515-521

Kumar G, Singh RP, Sushila (1993) Nitrate assimilation and biomass production in *Sesamum indicum* L. seedling in a lead enriched environment. Water Air Soil Poll 66:163-171

Küpper H, Mijovilvovich A, Meyer-Klaucke W, Kroneck PMH (2004) Tissue and age – dependent differences in the complexation of cadmium and zinc in the cadmium/zinc hyperaccumulators *Thlaspi careulscens* (Ganges ecotype) revealed by x-ray absorption spectroscopy. Plant Physiol 134:748-757

Lee S, Moon JS, Ko T-S, Petro D, Goldsbrough PB, Korban SS (2003a) Overexpression of *Arabidopsis* phytochelatin synthase paradoxically leads to hypersensitivity to cadmium stress. Plant Physiol 131:656-663

Lee L, Bae H, Jeong J, Lee JY, Yang YY, Hwang I, Martinoia E, Lee Y (2003b) Functional expressions of a bacterial heavy metal transporter in *Arabidopsis* enhance resistance to and decrease uptake of heavy metals. Plant Physiol 133:589-596

Leopold I, Günther D, Schmidt J, Neumann D (1999) Phytochelatins and heavy metal tolerance. Phytochemistry 50:1323-1328

Li YM, Chanery RL, Angle JS, Baker AJM (2000) Phytoremediation of heavy metal contaminated soils, In: Wise DL et al (eds) Bioremediation of Contaminated Soils, Marcel Dekker, New York, pp 837-884

Lombi E, Tearall KL, Howarth JR, Zhao F-J, Hawkesford MJ, McGrath SP (2002) Influence of iron status on cadmium and zinc uptake by different ecotype of the hyperaccumulator *Thlaspi caerulescens*. Plant Physiol, 128:1359-1367

Ma JE, Namoto K (1996) Effective regulation of iron acquisition in graminaceous plants. The role of mugeneic acid as phytosiderophores. Physiol Plant 97:609-617

Macek T, Mackova M, Kas J (2000) Exploitation of plants for removal of organics in environmental remediation. Biotechnol Advanc 18:23-34

Maiti RK, Piñero JLH, Oreja JAG, Santiago DL (2004) Plant based bioremediation and mechanisms of heavy metal tolerance of plants: a review. Proc Indian Natn Sci Acad B70:1-12

Marser P, Thomine S, Schroeder JI, Ward JM, Hirschi H, Sze IN, Talke A, Amtmann F, Maathuis MJ, Sanders D, Harper JF, Tchieu J, Gribskov M, Persans MW, Salt DE, Kim SA, Guerinot ML (2001) Phylogenetic relationship within cation transporter families of *Arabidopsis*. Plant Physiol 126:1646-1667

Meagher RB (2000) Phytoremediation of toxic elements and organic pollutants. Curr Opin Plant Bio 3:153-162

Mehmannavaz R, Prasher SO, Ahmad D (2002) Rhizospheric effects of alfalfa on biotransformation of polychlorinated biphenyls in a contaminated soil augmented with *Sinorhizobium melilotii*. Process Biochem 37:955-963

Misra S, Gedamu L (1989) Heavy metal tolerant transgenic *Brassica Napus* L. and *Nicotiana tabacum* L. plants. Thero Appl Genet 78:161-168

Morikawa H, Takahashi M (2000) Remediation of soil, water and air by naturally occurring and transgenic plants. In: Proc Gamma Field Symposia (No. 39) Institute of Radiation Breeding NIAR, MAFF, Japan, pp 81-104

Morikawa H, Takahashi M, KawamuraY (2002) Metabolism of nitrogen dioxide in plants-assimilation, dissimilation and novel nitrogen metabolites. Physiol Mol Biol Plants 8:19-29

Morikawa H, Takahashi M, Hakata M, Matsubara T, Sakamoto A (2005) Higher plants and metabolism of oxides of nitrogen, In: Singh RP, Jaiwal PK (eds) Molecular Strategies to Improve Nitrogen Use Efficiency in Plants, Studium Press, LLC, Honston, Texas, USA (In Press)

Mulligan CN, Yong RN, Gibbs BF (2001) Remediation technologies for metal contaminated soil and ground water: an evaluation. Eng Geol 60:193-207

Muratova A, Hübner T, Tischer S, Turkovskaya O, Möder M, Kuschk P (2003) Plant-rhizosphere-microflora association during phytoremediation of PAH-contaminated soil. Int J Phytoremed 5:137-151

Murphy A, Taiz L (1995) Comparison of metallothionein gene expression and nonprotein thiols in ten *Arabidopsis* ecotype: correlation with copper tolerance. Plant Physiol 109:945-954

Nedelkoska TV, Doran PM (2000) Hyperaccumulation of cadmium by hairy roots of *Thlaspi caerulescens*. Biotech Bioeng 67:607-615

Nie L, Shah S, Rashid A, Burd GI, Dixon G, Glick BR (2002) Phytoremediation of arsenate contaminated soil by transgenic canola and the plant growth – promoting bacterium *Enterobacter cloacae* CAL2. Plant Physiol Biochem 40:355-361

Noij M, Saito M, Nakamura M, Aono M, Saji H, Saito K (2001) Cysteine synthase overexpression in tobacco confers tolerance to sulfur containing environmental pollutants. Plant Physiol 126:973-980

Pan AH, Yang MZ, Tie F, Li LG, Chen ZL, Ru B (1994) Expression of mouse metallothionein-I gene confers cadmium resistance in transgenic tobacco plants. Plant Mol Biol 24:341-351

Pan X, Zhang B, Cobb GP (2005) Transgenic plants: environmental benefits and risks. Physiol Mol Biol Plants (In Press)

Pawlowska TE, Blaszkowski A, Ruhling A (1996) The mycorrhizal status of plants colonizing a calamine spoil mound in southern Poland. Mycorrhiza 6:499-505

Pence NS, Larsen PB, Ebbs SD, Letham DLD, Lasat MM, Garvin DF, Eide D, Kochian A (2000) The molecular physiology of heavy metal transport in the Zn/Cd hyperaccumulator, *Thlaspi caerulescence*. Proc Natl Acad Sci USA 97:4956-4960

Persans MW, Yan X, Patnoe ML, Krämer U, Salt DE (1999) Molecular dissection of the role of histidine in nickel hypertaccumulation in *Thlaspi goesingense* (Hālacsy). Plant Physiol 121:1117-1126

Piechalak A, Tomaszewska B, Baralkiewicz D (2003) Enhancing phytoremediative ability of *Pisum sativum* by EDTA application. Phytochemistry 64:1239-1251

Pierzynski GM, Schwab AP (1993) Bioavailability of zinc, cadmium and lead in metal contaminated alluvial soil. J Environ Qual 22:247-254

Pilon M, Ower JD, Garifullina GF, Kurihara T, Mihara H, Easki N, Pioln-Smits EAH (2003) Enhanced selenium tolerance and accumulation in transgenic Arabidopsis expressing a mouse selenocysteine lyase. Plant Physiol 131:1250-1257

Pioln-Smits EAH, Hwang S, Lytle CM, Zhu Y, Tai JC, Bravo RC, Chen Y, Leustek T, Terry N (1999) Overexpression of ATP-sulfurylase in Indian mustard leads to increased selenate uptake, reduction and tolerance. Plant Physiol 119:123-132

Pilon-Smits EAH, Zhu YL, Sears T, Terry N (2000) Overexpression of glutathione reductase in *Brassica juncea*: Effects of cadmium accumulation and tolerance. Physiol Plant 110:445-460

Pitchel J, Kuroiwa K, Sawyer HT (1999) Distribution of Pb, Cd and Ba in soils and plants of two contaminated soils. Environ. Pollut 110:171-178

Prasad MNV (2004) Phytoremediation of metals in the environment for sustainable development. Proc Indian Natn Sci Acad B70:71-98

Raab A, Feldmann J, Meharg AA (2004) The nature of arsenic phytochelatin complexes in *Holcus lanatus* and *Pteris cretica*. Plant Physiol 134:1113-1122

Raskin I, Kumar PBAN, Dushennov S, Salt D (1994) Bioconcentration of heavy metals by plants. Curr Opin Biotech 5:285-290

Raskin I, Smith RD, Salt DE (1997) Phytoremediation of metals: using plants to remove pollutants from environment. Curr Opin Biotechnol 8:221-226

Raskin I, Ensely B (2000) Phytoremediation of toxic metals using plants, Wiley & Sons, Inc., Canada, ISBN.471: 19254-19256

Raskin I (1996) Plant genetic engineering may help with environmental cleanup. Proc Natl Acad Sci USA 93:3164-3166

Rauser WE (1995) Phytochelatins and related peptides, structure, biosynthesis and function. Plant Physiol 109:1141-1149

Rauser WE (1999) Structure and function of metal chelators produced by plants; the case for organic acids, phytin and metallothioneins. Cell Biochem Biophys 31:19-48

Rosser SJ, French CE, Bruce NC (2001) Engineering plants for the phytodetoxification of explosives. In vitro Cell Dev Biol Plant 37:330-333

Rugh CL, Senecoff JE, Meagher RB, Merkle SA (1998) Development of transgenic yellow poplar for mercury phytoremediation. Nature Biotech 16:925-930

Rugh CL, Wilde HD, Stacks NM, Thompson DM, Summers AO, Meagher RB (1996) Mercuric ion reduction and resistance in transgenic *Arabidopsis thaliana* plants expressing a modified bacteria *mer*A gene. Proc Natl Acad Sci USA 93:3182-3187

Ruiz ON, Hussein HS, Terry N, Daniell H (2003) Phytoremediation of organomercurial compounds via chloroplast genetic engineering. Plant Physiol 132:1344-1352

Salt D, Rauser W (1995) Mg-ATP-dependent transport of phytochelatins across the tonoplast of oat roots. Plant Physiol 107:1293-1301

Salt DE, Blaylock M, Kumar PBAN, Dushenkov V, Ensley BD, Chet L, Raskin L (1995) Phytoremediation: a novel strategy for the removal of toxic metals from the environment using plants. Biotechnology 13:468-474

Salt DE (2001) Nickel hyperaccumulation in *Thlaspi goesigense*: A scientific travelogue. Cellular Dev Biol Plant 37:326-329

Salt DE, Smith RD, Raskin I (1998) Phytoremediation. Annu Rev Plant Physiol Mol Biol 49:643-668

Saxena PK, Krishna Raj S, Dan T, Perras MR, Vettakkorumakankav NN (1999) Phytoremedaition of heavy metal contaminated and polluted soils. In: Prasad MNV, Hayemeyer J (eds) Heavy Metal Stress in Plants: From Molecules to Ecosystems. Springer-Verlag Berlin Heidelberg, pp 305-329

Schroeder P, Harvey PJ, Schwitzguébel JP (2002) Prospects for the phytoremediation of organic pollutants in Europe. Environ Sci Pollut Res 9:1-13

Schwitzguébel J-P (2004) Potential of phytoremediation, an emerging green technology: European trends and outlook. Proc Indian Natn Sci Acad B70:131-152

Schwitzguébel J-P, van der Lelie D, Baket A, Glass DJ, Vangronsveld J (2002) Phytoremediation: European and American trends, success, obstacles and needs. J Soil Sediments 2:91-99

Sekhar KC, Chary NS, Kamala CT, Anjaneyulu Y (2004) Utilization of plant metal interaction for environmental management: from a general disbelief to universal acceptance. Proc Indian Natn Sci Acad B70:13-30

Shetty KG, Banks MK, Hetrick BA, Schwab AP (1995) Effects of mycorrhizae and fertilizer amendments on zinc tolerance of plants. Enviro Pollut 88:307-314

Siciliana SD, Germida JJ (1998) Mechanisms of phytoremediations: biochemical and ecological interactions between plants and bacteria. Environ Rev 6:65-79

Singh RP, Bharti N, Kumar G (1994a) Differential toxicity of heavy metals to growth and nitrate reductase activity of *Sesamum indicum* seedling. Phytochemistry 35:1153-1156

Singh RP, Maheshwari R, Sinha SK (1994b) Recovery of lead caused decrease in biomass accumulation of mungbean (*Vigna radiata* L.) seedlings by $K_2HPO_4$ and $CaCl_2$. Indian J Exp Biol 32:507-510

Singh RP, Dabas S, Chaudhary A (1996) Recovery of $Pb^{+2}$ caused inhibition of chlorophyll biosynthesis in leaves of *Vigna radiata* (L) Wilczek by inorganic salts. Indian J Exp Biol 34:1129-1132

Singh RP, Chaudhary A, Gulati A, Dahiya HC, Jaiwal PK, Sengar RS (1997b) Response of plants to salinity in interaction with other abiotic and biotic factors. In: Jaiwal PK, Singh RP, Gulati A (eds) Strategies for Improving Salt Tolerance in Higher Plants. Oxford & IBH Pub Co Pvt Ltd Delhi, Calcutta (India) and Enfield (USA), pp 25-54

Singh RP, Dabas S, Chaudhary A, Masheswari R (1997/1998c) Effect of lead on nitrate reductase activity and alleviation of lead toxicity by inorganic salts and 6-benzylaminopurine. Biol Plant 40:399-404

Singh RP, Jaiwal PK (2003) Arsenic phytoremetaion: new hopes for old problem. Physiol Mol Biol Plants 9:1-3

Singh RP, Tripathi RD, Sinha SK, Maheshwari R, Srivastava HS (1997a) Response of higher plants to lead contaminated environment. Chemosphere 34:2467-2493

Singh RP, Singh HP, Sharma A, Rizvi SMH, Jaiwal PK (2001) Indian mustard: a potential phytoremediator of heavy metal contaminated soil. Brassica 3:31-39

Singh RP, Tripathi RD, Dabas S, Rizvi SMH, Ali MB, Sinha SK, Gupta DK, Mishra S, Rai UN (2003) Effect of lead on growth and nitrate assimilation of *Vigna radiata* (L.) Wilczek seedling in a salt affected environment. Chemosphere 52:1245-1250

Smith RD, Salt DE (1997) Phytoremediation of metals: using plants to remove pollutants from environments, Curr Opin Biotech 8:221-226

Song WY, Sohn EJ, Mortinoia E, Lee YJ, Yang YY, Jasinski M, Forestier C, Hwang I, Lee Y (2003) Engineering tolerance and accumulation of lead and cadmium in transgenic plants. Nature Biotech 21:914-919

Song W-Y, Martinoia E, Lee J, Kim D, Kim D-Y, Vogt E, Shim D, Choi KS, Hwang I, Lee Y (2004) A novel family of cys-rich membrane proteins mediates cadmium resistance in *Arabidopsis*. Plant Physiol 135:1027-1039

Tabata K, Kashiwagi S, Mori H, Ueguchi C, Mizuno T (1997) Cloning of cDNA encoding a putative metal transporting P-type ATPase from *Arabidopsis thaliana*. Biochim Biophys Acta 1326:1-6

Takahashi M, Sasaki Y, Ida S, Morikawa H (2001) Nitrate reductase gene enrichment improves assimilation of nitrogen dioxide in *Arabidopsis*. Plant Physiol 126:731-741

Thangavel P, Subbhuraam CV (2004) Phytoextraction; Role of hyperaccumulators in metal contaminated soils. Proc Indian Natn Sci Acad B70:109-130

Trapp S, Karlson U (2001) Aspects of phytoremediation of organic pollutants. J Soils Sediments 1:1-7

Treeby M, Marschner H, Romheld V (1989) Mobilization of iron and other micronutrient cations from a calcareous soil by plant borne, microbial and synthetic metal chelators. Plant Soil 114:217-226

van der Zaal BJ, Neuteboom LW, Pinas JE, Chardonnes AN, Schat H, Verleij JAC, Hooykass PJJ (1999) Overexpression of a novel *Arabidopsis* gene related to putative zinc transporter genes from animals can lead to enhanced zinc resistance and accumulation. Plant Physiol 19:1047-1055

van Huysen T, Abdel-Ghany S, Hale KL, Leduc D, Terry N, Pilon-Smits EAH (2003) Overexpression of cystathionine- $\gamma$-synthase enhances selenium volatilization in *Brassica juncea*. Planta 218:71-78

Vassil AD, Kapulink Y, Raskin I, Salt DE (1998) The role of EDTA in lead transport and accumulation by Indian mustard. Plant Physiol 117: 447-453

Vatamaniuk OK, Mari S, Lu Y-P, Rea PA (1999) AtPCS1, a phytochelatin synthase from *Arabidopsis* :isolation and *in vitro* reconstitution. Proc Natl Acad Sci USA 96:7110-7115

Vert G, Grotz N, Dedaldechamp F, Gaymard F, Guerinot ML, Briata FM, Curie C (2002) *IRT1*, an *Arabidopsis* transporter essential for iron uptake from the soil and for plant growth. Plant Cell 14:1223-1233

Weissenhorn I, Leyval C (1995) Root colonization of maize by a cd-sensitive and a Cd-tolerant *Glomus mosseae* and cadmium uptake in sand culture. Plant Soil 175:233-238

Wellburn AR (1990) Why are atmospheric oxides of nitrogen usually phytotoxic and not alterative fertilizers? New Phytol 115:395-429

Wellburn FAM, Greissen GP, Lake JA, Mullineaux PM, Wellburn AR (1998) Tolerance to atmospheric ozone in transgenic tobacco over-expressing glutathione synthetase in plastids. Physiol Plant 104:623-629

Williams LE, Pittman JK, Hall JL (2000) Emerging mechanisms for heavy metal transport in plants. Biochim Biophys Acta 1465:104-126

Zhang FS, Romheld V, Marschner H (1991) Diurnal rhythm of release of phytosiderophores and uptake rate of zinc in iron-deficient wheat. Plant Nutrit 37:671-678

Zhao FJ, Hamon RE, Mc Laughlin MJ (2001) Root exudates of the hyperaccumulator *Thlaspi caerulescens* enhance metal mobilization. New Phytol 151:613-620

Zhu YL, Pilon-Smits EAH, Jouanin L, Terry N (1999a) Overexpression of glutathione synthetase in Indian mustard enhances cadmium accumulation and tolerance. Plant Physiol 119: 73-79

Zhu YL, Pilon-Smits EAH, Tarun AS, Weber SU, Jouanin L, Terry N (1999b) Cadmium tolerance and accumulation in Indian mustard is enhanced by overexpressing $\gamma$-glutamylcysteine synthetase. Plant Physiol 121:1169-1177

# 11

# Aquatic Plants for Phytotechnology

M.N.V. Prasad

Department of Plant Sciences, University of Hyderabad, Hyderabad-500 046, INDIA, Email: mnvsl@uohyd.ernet.in

## 1. Introduction

Aquatic macrophytes are represented by a variety of algal and macrophytic species that occur in many types of habitats. Members of Cyperaceae, Potamogetonaceae, Ranunculaceae, Typhaceae, Haloragaceae, Hydrocharitaceae, Najadaceae, Juncaceae, Pontederiaceae, Zosterophyllaceae, Lemnaceae, mainly represent aquatic plants. These plants are either emergent, submerged, or free floating. Some non-vascular plants, like macro algae, are rootless and capable of growing with their thalli in the water. Aquatic macrophytes are extremely important components of an aquatic ecosystem for primary productivity and nutrient cycling (Aksorn and Visoottiviseth 2004; Prasad et al. 2001, 2005). They also provide habitat, food and refuge for a variety of other organisms. The aquatic plants have been reported for long to detoxify environmental pollutants. The notable environmental contaminants are radionuclides as well as inorganic and organic pollutants which can be phytoremedaited in various ways (Fig. 1). The efficacy of the detoxification or remediation function of the of aquatic plants depends on a) sediment geochemistry, b) water physico-chemistry (Adriano et al. 2004), c) plant physiology (Prasad 2004) d) plant genotype and e) nature of the contaminant or pollutant (Pilon-Smits 2005).

## 2. Phytotechnologies

Environmental protection strategies involving plants are called "Phytotechnologies". Phytotechnologies employ plants to remediate, stabilize or control contaminated or polluted sites. Phytoremediation is one of the approaches in phytotechnologies (COST action 837).

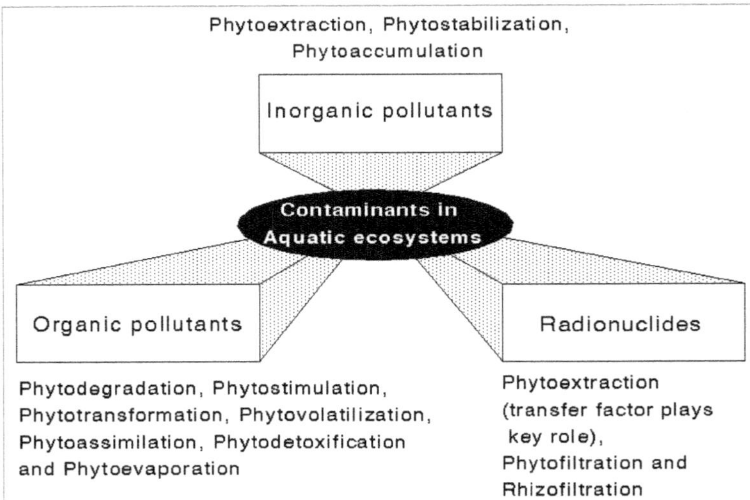


**Fig. 1.** Different sub-sets of phytoremediation for removal/detoxification of both organic and inorganic contaminants

## 2.1 Pollutants and Contaminants in Aquatic Ecosystem

### *2.1.1 Inorganic Pollutants*

These include nitrate, phosphate, per chlorate, cyanide etc; trace elements essential to plants when present in excess viz., B, Cu, Fe, Mn, Mo and Zn; trace elements essential for animal nutrition when present in excess i.e. As, Co, Fe, Mn, Zn, Cr, F, Ni, Se, Sn and V and the most toxic trace elements like Cd, Hg and Pb which are not required by any organisms. Trace elements essential for human nutrition are identical to animal nutrition with the exception of As and V. Aquatic plants and constructed wetlands have been designed and used for the treatment of a wide range of inorganic pollutants and mine drainage, salt water and removal of radionuclides (Tables 1 and 2). Methods of phytoremediation have been demonstrated in Figures 2-4.

**Phytodetoxification of cyanide.** Cyanide is the leach reagent of choice for gold and silver extraction, but also a toxic chemical that may cause severe environmental pollution problems. Vascular plants possess an enzyme system that detoxifies cyanide by converting it to the amino acid asparagine. The phytotoxicity of cyanide is directly connected to the efficiency of this enzyme system. Plants only survive cyanide exposure up to a dosage they can eliminate. Cyanide elimination with plants seems to be a feasible option for gold and silver mine waste and wastewater. During several metabolic reactions, plants are confronted with cyanide as byproduct, e.g., during the ethylene synthesis of mature tissue (Manning 1988), where hydrogen cyanide is formed as a by- product.

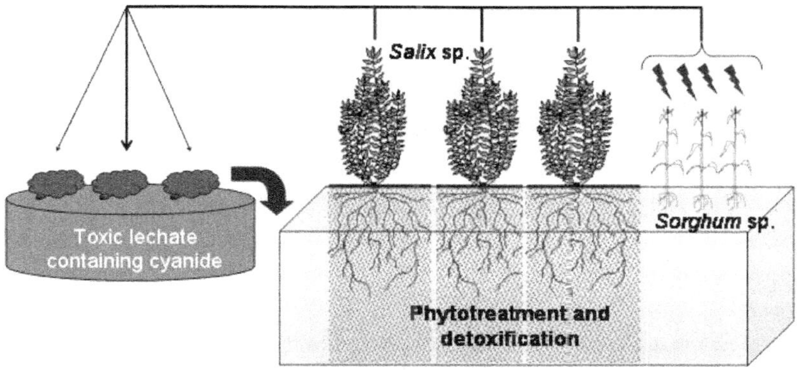


**Fig. 2.** Cyanide (CN-) is the leach reagent (ammonium thiocynate) of choice for gold and silver extraction. *Salix* sp. and *Sorghum* sp. contain the enzyme β-cyanoalanine synthase which is involved in its detoxification. The final product in this treatment is asparagines, a non-toxic essential amino acid

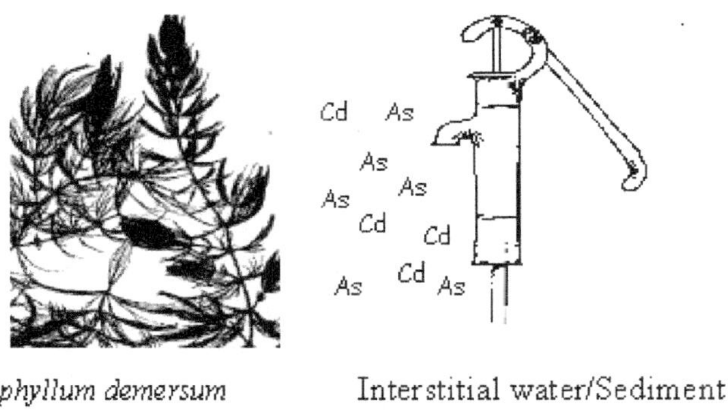


**Fig. 3.** *Ceratophyllum demersum*, a freshwater free floating species could serve a biofilter of toxic metals. It is reported to accumulate arsenic with a 20,000-fold concentration factor (Weis and Weis 2004)

**Fig. 4.** *Talinum cuneifolium* (Portulacaceae), an ideal experimental system developed for rhizofiltration of environmental contaminants in hydroponic system

**Table 1.** Aquatic plants for biomonitoring of toxic trace elements in a wide range of toxicity bioassays (Prasad et al. 2001, 2005)

| Plant species | Metal |
|---|---|
| *Azolla fililiculioides* | Cr, Ni, Zn, Fe, Cu, Pb |
| *A. pinnata* | Cd, Cr, Zn |
| *Bacopa monnieri* | Hg, Cr, Cu, Cd |
| *Carex juncell* | Cu, Pb, Zn, Co, Ni, Cr, Mo, U |
| *Carex rostrata* | Cu, Pb, Zn, Co, Ni, Cr, Mo, U |
| *Carex Sp.* | Cd, Fe, Pb.,Mn |
| *Ceratophyllum demersum* | Cd, Cu, Cr, Pb, Hg, Fe, Mn. Zn, Ni, Co and radionuclides |
| *Cyperus eragrostis* | Cd, Cu, Pb, Zn |
| *Distichlis spicata* | Cd, Fe, Pb, Mn |
| *Elodea densa* | Hg, methyl-Hg |
| *E. nuttallia* | Cu |
| *E. sptangulare* | Hg, Pb, Cd, Cu and Fe |
| *Eichhornia crassipes* | As, Cd, Co, Cr Cu, Al, Ni, Pb, Zn, Hg, P, Pt, Pd, Os, Ru, Ir, Rh |
| *Elodea canadensis* | Cu, Pb, Cd, Zn, Cr, Ni |
| *Eriocaulon septangulare* | Hg, Pb, Cd, Fe |
| *Euryale ferox* | Cd, Cr, Pb, Cu |
| *Hydrilla verticillata* | Hg, Fe, Ni, Pb |
| *Hygrophila onogaria* | Hg, methyl-Hg |
| *Isoetes lacustris* | Cu. Pb |
| *Lemna minor* | Mn, Pb, Ba, B, Cd, Cu, Cr, Ni, Se, Zn, Fe |
| *L. trisulca* | Cu, Cd |
| *L. gibba* | Cu, Cd |
| *L. palustris* | Zn, Cu, Fe, Hg |
| *L. paucicostata* | Cd, Zn, EDTA, Cu, Ca |
| *L. perpusilla* | Cd |
| *L. polyrrhiza* | Cd |
| *L. valdivinia* | Cd, Cu |
| *Littorella uniflora* | Cu, Pb |
| *Ludwigia natans* | Hg, methyl-Hg |
| *Lysimachia nummularia* | Hg, methyl-Hg |
| *Myriophyllum spicatum* | Cd, Cu, Zn, Pb, Ni, Cr |
| *M. alterniflorum* | Cu, Pb |
| *M. exalbescens* | Zn, Pb |
| *M. aquaticum* | Zn, Cu, Fe, Hg, Cd, Pb |

| | |
|---|---|
| *Melilotus indica* | Se |
| *Mentha aquatica* | Cd, Zn, Cu, Fe, Hg |
| *Najas marina* | Cd, Fe, Pb, Mn |
| *Nasturtium officinale* | Cd |
| *Nuphar lutea* | Cu, Ni, Cr, Co, Zn, Mn, Pb, Cd, Hg, Fe |
| *N. variegatum* | Cu, Zn |
| *Nymphaea alba* | Ni, Cr, Co, Zn, Mn, Pb, Cd, Cu, Hg, Fe |
| *Nymphoides germinate* | Cd, Cu, Pb, Zn |
| *Potamogeton attenuatum* | Cd, Cu. Pb. Zn |
| *P. communis* | Ni, Cr, Co, Zn, Mn, Pb, Cd, Cu, Hg, Fe |
| *P. crispus* | Cu, Pb, Mn, Fe, Cd |
| *P. filiformis* | Cd, Fe, Pb, Mn |
| *P. lapathifoilum* | Cd, Cu, Pb, Zn |
| *P. orientalis* | Cd, Cu, Pb, Zn |
| *P. pectinatus* | Mn. Pb, Cd, Cu, Cr, Zn, Ni, As, Se |
| *P. perfoliatus* | Cu, Pb, Cd, Zn, Ni, Cr |
| *P. richardsonii* | Cd, Cr, Cu, Ni, Zn, Pb |
| *P. subsessiles* | Cd, Cu. Pb. Zn |
| *Phragmites karka* | Cr |
| *Pistia stratoites* | Cu, Al, Cr, P, Hg |
| *Ranunculus aquatilis* | Mn, Pb, Cd, Fe, Pb, |
| *R. baudotii* | Cd, Cu, Cr, Zn, Ni, Pb |
| *Ruppia maritima* | Mn, Pb, Cd, Pb, Fe, Se |
| *Salvinia acutes* | Mn, Pb |
| *S. maritimus* | Cd, Fe, Pb, Mn |
| *S. natans* | Pb, Cr |
| *S. undulata* | Pb |
| *S. molesta* | Hg |
| *Scapania uliginosa* | B, Ba, Cd, Co, Cr, Cu, Li, Mn, Mo, Ni, Pb, Sr, V, Zn |
| *Schoenoplectus lacustris* | Ni, Cr, Co, Zn, Mn, Pb, Cd, Cu, Hg, Fe |
| *Scirpus lacustris* | Cr |
| *Spirodela polyrhiza* | Cr |
| *Typha domingensis* | Cd, Cu, Pb, Zn |
| *T. latifolia* | Ni, Cr, Co, Zn, Mn, Pb, Cd, Cu, Hg, Fe |
| *Vallisneria americana* | Cd, Cr, Cu, Ni, Pb, Zn |
| *V. spiralis* | Hg |
| *Wolffia globosa* | Cd, Cr |

**Table 2.** Aquatic macrophyts a for phytotechnologies to treat inorganic pollutants, acid mine drainage, salt water, regulation of water and removal of radionuclides (Prasad et al. 2001, 2005; McCutcheon and Schnoor 2003, COST action 837, 2003 and COST action 859, 2005, Kamal et al. 2004; Peles et al. 2002; Hattink et al. 2000; Sheppard and Motycka 1997)

| Plant name | Common name | Phytormediation function |
|---|---|---|
| *Azolla filiculoides* | Water fern | Metal hyperaccumulation |
| *Bacopa monnieri* | Water hyssop | Metal accumulation |
| *Canna flaccida* | | HM removal in constructed wetland |
| *Carex pedula* | | HM removal in constructed wetland |
| *Chara, Nitella, Mougeotia, Ulothrix* | Algae | If they could be induced to grow in mining effluents, they would provide a simple, long-term solution remove U and other radionuclides |
| *Cladium Jamaicense* | Sawgrass | Brine concentration |
| *Eichhornia crassipes* | Water hyacinth | Metal accumulation and biosorption |
| *Elodea canadensis* | | Phytofiltration of storm water and removal of zinc |
| *Eriophorum angustifolium* | | Phytostabilization of metal rich mine tailings |
| *Eriophorum scheuchzeri* | | Phytostabilization of metal rich mine tailings |
| *Glyceria fluitans* | | Phytostabilization of mine tailings, treatment of acid mine drainage |
| *Hydrilla verticillata* | - | TNT transformation and metals accumulation |
| *Hydrocotyle umbellata* | Pennywort | Biosorption of toxic metals |
| *Ipomea aquatica* | Water spinach | Metal accumulation |
| *Juncus articulatus* | | Phytostabilization of mine tailings |
| *Lemna minor* | Duck weed | Concentrates technetium-99 |
| *Lemna, Spirodela* and *Wolffia* | Duckweeds | Biosorbents of inorganic and organic pollutants and metals accumulation |
| *Miscanthus floridulus* | | HM removal in constructed wetland |
| *Miscanthus sacchariflorus* | | HM removal in constructed wetland |
| *Nymphaea violacea* | Waterlily | Uranium and thorium series radionuclides |
| *Pistia stratiotes* | Water lettuce | Metal accumulation |
| *Polygonum punctatum* | | radiocesium ($^{137}Cs$) |
| *Potamogeton natans* | | Phytofiltration of storm water and removal of zinc |
| *Potamogeton natans* | - | Metals uptake |

| | | |
|---|---|---|
| *Sagittaria latifolia* | - | Radiocesium ($^{137}Cs$) |
| *Salvinia molesta* | Kariba weed | Metals accumulation |
| *Scirpus* spp | Bulrush | Used in constructed wetland |
| *Scirpus validus* | - | Brine concentration |
| *Spartina alterniflora* | Cordgrass | Saltwater, brine concentration |
| *Spirodela oligorrhiza* | Giant duckweed | Metal accumulation |
| *Sporobolus virginicus* | Coastal dropseed | Brine concentration |
| *Tamarix* spp. | Salt cedar | Hydraulic control of arsenic |
| *Vallisneria spiralis* | Eel grass | Metal hyperaccumulation |
| *Zizania aquatica* | Wild rice | Uptake of $^{129}I$. |

Consequently, plants have evolved effective detoxifying strategies. The detoxifying enzyme system (beta-cyanoalanine synthase) connects free cyanide and cysteine to cyanoalanine. The final metabolite is asparagine, a non-toxic essential amino acid (Manning 1988; Trapp et al. 2003).

The fact, that plants can remove high amounts of cyanide in waste and wastewaters from gold mining, was demonstrated in constructed wetlands or artificial ponds with aquatic plants, if the concentration is low, or by planting selected crops or trees in an area and irrigated with cyanide containing wastewater (at higher concentrations). The chemistry of the cyanide leaching follows the equation:

$$Au + 2NaCN + \frac{1}{2} H_2O \rightarrow Na[Au(CN)_2] + NaOH$$

Gold dissolves as negatively charged complex. The same reaction occurs with silver. In gold and silver mining, a diluted sodium cyanide solution (0.05%) is sprayed on gold-containing crushed ore, placed in heaps. The cyanide readily forms a water-soluble complex with the gold from which the gold can be recovered. Since its commercial introduction in New Zealand over a century ago, cyanide has been used worldwide in the extraction of gold and silver. Although chemical replacements for cyanide have been investigated for decades, it still remains the exclusive leaching reagent of choice due to a combination of availability, effectiveness and economics. This technique to leach silver and gold from low-grade ores and old mining wastes, has been increasingly used since the 1980s. In the US alone, more than 150 heap-leach operations were active in the 1990s. Currently, there are about 875 gold and silver operations throughout the world, of which about 460 utilize cyanide, using 347 000 tons of sodium cyanide per year. The cyanide heap leach process is an environmental hazard.

### *2.1.2 Organic Pollutants*

A number of aquatic plants work well also for remediation of organic pollutants. Sediments contaminated with organics can be cleaned with the plant enzymes (e. g. dehalogenase, laccase, peroxidase, nitrilase and nitrate reductase). Enzymes excreted from plant roots into the rhizosphere can degrade the organic molecules (Tables 3 and 4). In the case of PAHs, there is evidence for both uptake and metabolism by plants (McCutcheon and Schnoor 2003). However, the uptake of large molecules by plant cells is difficult depending on the narrow "channels" in the structure of the cell wall system, especially when they are lipophilic.

Oxygenation is an important initial mode of attack and this step serves to increase water solubility and provides an opportunity for conjugation via glycosidic bond formation. Cytochrome P450, peroxygenases, and peroxidases are involved in plant oxidation of xenobiotics. Other enzyme classes like gluthathione S-transferases, carboxylesterases, o-glucosyltransferases o-malonyltransferases, N-glucosyltransferases, and N-malonyltransferases are associated with xenobiotic metabolism in plant cells, transport of intermediates, and compartmentation processes (Macek et al. 2000). In addition, the plant roots serve as a habitat for biodegrading microbes and these microbes thrive much better and degrade organics much faster in the rhizosphere of specific plant species. Remediation of water contaminated with chlorinated alkanes and other organic chemicals has been shown with aquatic plants (Fig. 5). Phytotransformation of perchlorate using parrot-feather (*Myriophyllum aquaticum*) was described by Susarla et al. (1999). This plant has already been tested for successful remediation of soils contaminated with TNT as well as other contaminants (e.g. TCE, PCP). There are numerous defence disposal sites across the USA with explosives contaminated groundwater. The U.S. Army Environmental Centre is developing technologies to effectively clean up groundwater contaminated with residues of explosives like TNT, RDX, octahydro-1,3,5,7-tetranitro-1,3,5,7- tetraazocine (HMX), and DNT. Current groundwater cleanup technologies, such as granular activated carbon and advanced oxidation, are cost prohibitive. One potential treatment alternative is phytoremediation using constructed wetlands (Betts 1997).

**Table 3.** Plant enzymes implicated in phytoremediation of organics (Sandermann 1994; Macek et al. 2000)

| Enzyme | Contaminants degraded/ transformed into less toxic forms |
|---|---|
| Phosphatase | Organophospates |
| Aromic dehalogenase | Chlorinated aromatics (DDT, PCBs etc.) |
| O-demethylase | Alachlor, metoalchor |
| Cytochrome P450, Peroxygenases, Peroxidases | PCBs |
| Glutathione s-transferase, carbooxylesterases, o-glucosyltransferases, o-malonyltransferases, N-glucosyltransferases, N-malonyltransferases | Xenobiotics |

**Table 4.** Aquatic macrophytes for phytotechnologies to treat organic pollutants (Prasad et al. 2001, 2005; COST action 837, 2003; McCutcheon and Schnoor 2003; COST action 859, 2005)

| Plant name | Common name | Phytormediation function |
|---|---|---|
| *Alisma subcordatum* | Water-plantain | Uptake of explosives |
| *Apium graveolens* | Celery | Removes sulphonated anthraquinones in textile wastewater |
| *Carex gracilis* | | Degrades Trinitrotoluene (TNT) |
| *Ceratophyllum demersum* | Coontail | Degradation of organics |
| *Chara, Nitella, Mougeotia, Ulothrix* | Algae | If they could be induced to grow mining effluents, they would provide a simple, long-term solution remove U and other radionuclides |
| *Eleocharis obtusa* | Blunt spike | Transformation of TNT (explosive) |
| *Eleocharis tuberosa* | Water chestnut | Transformation of TNT |
| *Iris pseudocorus* | Swamp/yellow iris | Methyl bromide and TNT transformation, |
| *Juncus glaucus* | | Degrades TNT |
| *Myriophyllum aquaticum* | Parrot feather | Explosives sensitivity to and transformation, halocarbon metabolism, halogenated organics transformation, hormesis, organophosphorus degradation, perchlorate degradation |
| *Myriophyllum spicatum* | Milfoil | TNT monitoring and transformation |
| *Nelumbo nucifera* | Indian lotus | TNT transformation |
| *Nymphaea odorata* | fragrant water lily | TNT transformation |
| *Phragmites australis* | | Degrades TNT |
| *Phragmites* sp | Reed | Methyl iodide volatilization, integral component in wetlands |
| *Phragmites australis* | Common reed | Treatment of dairy wastes in constructed wetland |
| *Potamogeton nodosus* | Pondweed | Explosives degradation |
| *Potamogeton pectinatus* | - | Transformation of explosives, uptake of metals |
| *Potamogeton pusilus* | - | Used in free-surface wetlands, |
| *Rumex hydrolapatum* | - | Removes sulphonated anthraquinones in textile wastewater |
| *Sagittaria latifolia* | Arrowhead | Explosives degradation |
| *Salicornia virginica* | Perennial glasswort | Perchlorate tolerance, Brine concentrator |

| | | |
|---|---|---|
| *Salvinia rotundifolia* | Floating moss | TNT transformation |
| *Scirpus* spp | Bulrush | Used in constructed wetland |
| *Spirodela oligorrhiza* | giant duckweed | Organic degradation and metals accumulation |
| *Trifolium pratense* | Red clover | Rhizodegradation |
| *Typha angustifolia* | Cattail | Degradation of explosives |
| *Typha latifolia* | Cattail | Degrades TNT, Biosorption and perchlorate degradation |
| *Vallisneria americana* | tape grass | TCE transformation and metals accumulation |
| *Vallisneria spiralis* | Eel grass | Metals hyperaccumulation |

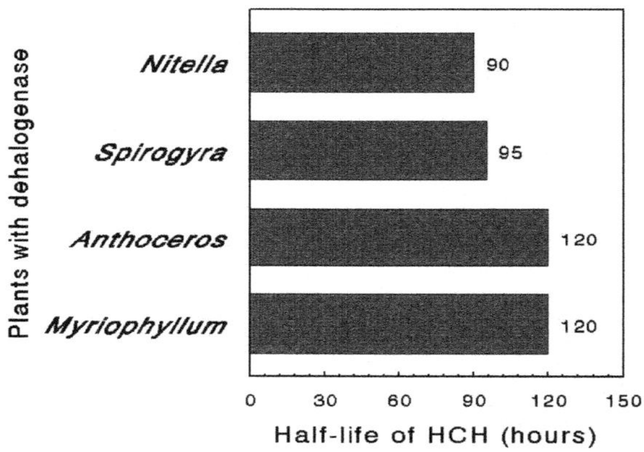


**Fig. 5.** Phytodegradation of pesticides by plant dehalogenases

### *2.1.3 Radionuclides*

Major sources of radioactive contamination and major radionuclides released with long-term impact are (Sheppard and Motycka 1997; Hattink et al. 2000; Peles et al. 2002; Kalin et al. 2004):

a) The nuclear weapon testing (release of mainly $^{14}C$, $^{137}Cs$, $^{90}Sr$ and $^{95}Zr$)
b) Nuclear weapon production (release of mainly $^{137}Cs$, $^{106}Ru$, $^{95}Zr$)
c) Nuclear power production
   i. During the mining operation, the main radionuclide discharged is $^{222}Rn$; the environment of the mining and milling sites is contaminated through dispersion of $^{238}U$ (and daughters: e.g. $^{226}Ra$, $^{210}Pb$, $^{210}Po$ and $^{232}Th$).
   ii. During the operational phase small amounts of radionuclides are routinely released mainly $^{14}C$.
   iii. Nuclear accidents can involve only small local contamination (coctail of $^{37}Cs$, $^{90}Sr$, $^{131}I$, $^{210}Po$, $^{95}Zr$, $^{144}Ce$).

iv. Natural sources of contamination. Others, e.g. zircon and rare earths, the concentration of $^{238}U$ and $^{232}Th$ may be considerably elevated.

The flux of an element from aquatic environment to plant is often referred as plant uptake or removal. The removal of the radionuclide from water with the harvested part of the plant (in Bq / area or volume), is the product of the concentration in the plant (Cplant, in Bq / kg) and the yield of the harvested biomass (kg per unit area/volume of water):

Plant Removal = Yield x Cplant
Transfer Factor (TF, dimensionless) TF = Cplant/Cwater or sediment

$$TF = \frac{\text{Activity in plant (Bq / kg dry or fresh weight)}}{\text{Activity in background water (Bq / kg volume)}}$$

A translocation factor (TLF) which is similar to TF is also used.

$$TLF = \frac{\text{Activity in plant (Bq / kg dry or fresh weight)}}{\text{Activity in background water (Bq surface area per m}^2\text{)}}$$

or

$$TLF = \frac{\text{Activity in plant (Bq / kg dry or fresh weight)}}{\text{Activity in soil for wetland or marshy plants (Bq / m}^2\text{)}}$$

Hence, the transfer factor is an important parameter for determining the potential of phytoextraction of aquatic macrophytes. *Alternanthera philoxeroides* is one classic example that was widely used for removal of radionuclides (Prasad 2001).

## 2.2 Aquatic Plants as Biomonitors of Contaminants and Pollutants

The use of aquatic plants in water quality assessment has been in practice for years as *in-situ* biomonitors and bioremediators (St-Cyr and Campbell 1994; St-Cyr et al. 1994; Kamal et al. 2004; Lytle et al. 1994, 1998). The occurrence of aquatic macrophytes has been found related to water chemistry and using these plant species or communities as indicators or biomonitors has been an objective for surveying water quality (Wang and Freemark 1995). Aquatic plants have also been used frequently in waste water treatment to remove suspended solids, nutrients, heavy metals, toxic organics from acid mine drainage, agricultural landfill and urban storm-water runoff. However, the response of an organism to deficient or excess levels of metal (i.e. bioassays) can be used to estimate metal impact. Such studies performed under defined experimental conditions can provide results that can be extrapolated to natural environment. There are multifold advantages in using an aquatic macrophyte as a study material. Macrophytes are cost-effective universally available aquatic plants and with their ability to survive adverse conditions and high colonization rates, are excellent tools for the study of phytoremediation. Rooted macrophytes especially play an important role in metal bioavailability through rhizosphere secretions and exchange processes. This naturally facilitates metal uptake by floating and emergent macrophytes. Macrophytes readily take up metals in their reduced form from sediments, which exist in anaerobic situations due to lack of oxygen and oxidize them in the plant tissues making them immobile and bioconcentrate them to a great extent. Metal concentrated macrophytes are available to be eaten by fishes. These may also be available for herbivorous and detrivorous invertebrates. This may be a major route for incorporating metals in the aquatic food chain. It is, therefore, of interest to assess the level of heavy metals in macrophytes due to their importance in ecological processes. The immobile nature of macrophytes makes them an effective bio-indicator of metal pollution, as they represent real level of metals present at that site. Data on phytotoxicity studies are also considered in the development of water quality criteria to protect aquatic life, in the toxicity evaluation of municipal and industrial effluents. In addition, aquatic plants have been used to assess the toxicity of contaminated sediment and hazardous waste leachates.

In the past, researches on macrophytes have focused mainly to find out effective eradication techniques for several aquatic species, such as *Elodea canadensis, Eichhornia crassipes, Ceratophyllum demersum etc.* Scientific literature indicates use of a wide diversity of macrophytes in toxicity tests designed to evaluate the hazard of potential pollutants, but the test species and methods used are quite scattered. Estuarine and marine plant species are being used considerably less than freshwater species in the toxicity tests conducted for regulatory reasons (Mohan and Hosetti 1999). The suitability of a test species is usually based on the specimen bioavailability, sensitivity to toxicant, reported data etc. The sensitivity of various plants to metals was found to be species and metal specific, differing in the uptake as well as toxicity of metals. Many submersed plants have been used as test species, but there is no single species being widely used. In a literature survey, only 7% of 528 reported phytotoxicity tests used macrophytic species. Their use in microcosm and mesocosm studies is even rarer and although it has been highly recommended. Several plant species, like *Lemna, Myriophyllum, Potamogeton* have been exhaustively used in phytotoxicity assessment, but several others have given less importance as a bioassay tool. Duckweeds have received the highest attention for toxicity tests as they are relevant to many aquatic environments, including lakes, streams, effluents. Duckweeds comprise *Spirodela, Wolfiella, Lemna* and *Wolffia*, of which *Lemna* has been widely studied.

## 2.3 Constructed Wetlands

The most important role of plants in wetlands is that they increase the residence time of water and thereby increase the sedimentation of particles and associated pollutants. Thus, they are indirectly involved in water cleaning. Plants also add oxygen to the roots generating favourable conditions for microbes and bioremediation. For efficient removal of pollutants, a high biomass per volume of water of the submerged plants is necessary.

Uptake of metals in emergent plants only accounts for 5% or less of the total removal capacity in wetlands (Sobolewski 1999). Not many studies have been performed on submerged plants, however, higher concentration of metals in submerged than emergent plants has been found (Fritioff and Greger 2003) and in microcosm wetland, the removal by *Elodea canadensis* and *Potamogeton natans* showed up to 69 % removal of Zn.

## 2.4 Potential Role of Aquatic Plants in Phytotechnology

Phytoremediation is defined as the use of plants for environmental cleanup. Aquatic macrophytes have paramount significance in the monitoring of metals in aquatic ecosystems (eg: *Lemna minor, Eichhornia crassipes, Azolla pinnata*) (Mohan and Hosetti 1999). Aquatic plants are represented by a variety of macrophytes including algal species that occur in various habitats. They are important in nutrient cycling, control of water quality, sediment stabilization and provision of habitat for aquatic organisms. The use of aquatic macrophytes in water quality assessment has been a common practice for biomonitoring.

The submerged aquatic macrophytes have very thin cuticle and therefore, readily take up metals from water through the entire surface. Hence, the integrated amounts of bioavailable metals in water and sediment can be indicated to some extent by using macrophytes. Macrophytes with their ability to survive adverse conditions and high colonization rate are the excellent tools for phytoremediation. Further, they redistribute metals from sediments to water and finally take up in the plant tissues and hence maintain circulation. Benthic rooted macrophytes (both submerged and emergent) play an important role in metal bioavailability from sediments through rhizosphere (Mohan and Hosetti 1999; Prasad et al. 2001) exchanges and other carrier chelates. This naturally facilitates metal uptake by other floating and emergent forms. Macrophytes readily take up metals in their reduced form from sediments and oxidize them in the plant tissues making them immobile and hence bioconcentrate them to a high extent.

Constructed wetlands are man-made wetlands designed to intercept and remove a wide range of contaminants from water. These wetlands can save the time and money by using natural mechanisms to treat non-point source pollutants such as oils, nutrients, suspended solids, and other substances before it reaches our lakes, rivers, and oceans. Conventional wastewater treatment plants can also effectively remove non-point source pollution, but are very expensive to build and operate.

**Treatment mechanisms**

These include:

- Filtration and uptake of contaminants
- Settling of suspended solids
- Water velocity and trapping action of plants, leaves, and stems
- Precipitation, adsorption, and sequestration of metals
- Microbial decomposition of petroleum hydrocarbons and other organics.

**Benefits**

- Cost-effective treatment of non-point source pollution.
- Compliance with water quality goals.
- Reduction of operation and maintenance costs relative to conventional water treatment plants.
- Conservation of natural resources.
- Reduction of flood hazard and erosion.
- Creation of wildlife habitat and aesthetic resource.

Plants may reduce element leakage from submerged mine tailings by phytostabilisation. However, high shoot concentrations of elements might disperse them and could be harmful to grazing animals. Plants that are tolerant to elements of high concentrations have been found useful for reclamation of dry mine tailings containing elevated levels of metals and other elements. Mine tailings rich in sulphides, e.g. pyrite, can form acid mine drainage (AMD) which may also promote the release of metals and metalloids such as As. To prevent AMD formation, mine tailings rich in sulphides may be saturated with water to reduce the penetration of atmospheric oxygen. An organic layer with plants on top of the mine tailings would consume oxygen, as would plant roots through respiration.

Some plant species seem to have an inherent tolerance to heavy metals. Since, some wetland plant species have been found with inherent metal tolerance, for example *Thypha latifolia*, *Glyceria fluitans* and *Phragmites australis*, wetland communities may easily establish on submerged mine tailings. Some plant species have mechanisms that make it possible to cope with high external levels of elements. Low-accumulators are plants that can reduce the uptake when the substrate has high element concentrations, or have a high net efflux of the element in question. Thus the plant tissue concentration of the element is low even though the concentration in the substrate is high.

## 3. Conclusion

Surface flow constructed wetlands are being designed for the treatment of municipal waste waters in developed nations. However, use of constructed wetlands is not gaining momentum in tropical nations due to water scarcity and high surface evapotranspitration. But, in there countries for the bioremediation mine drainage, agricultural waste waters and flood water there is considerable scope as they have rich plant diversity.

## References

Adriano DC, Wenzel WW, Vangronsveld J, Bolan NS (2004) Role of assisted natural remediation in environmental cleanup. Geoderma 122:121-142

Aksorn E, Visoottiviseth P (2004) Selection of Suitable Emergent Plants for Removal of Arsenic from Arsenic Contaminated Water. Science Asia 30:105-113

Betts KS (1997) Native aquatic plants remove explosives. Environ Sci Technol 31:304A

COST Action 837 View (2003) In: Vanec T, Schwitzguebel J-P (eds) UOCHB AVCR, Prague, pp 41

COST Action 859 (2005) In: Schwitzguebel J-P (ed) Phytotechnologies to promote sustainable land use and improve food safety, 2005

Fritioff Å, Greger M (2003) Aquatic and terrestrial plant species with potential to remove heavy metals from stormwater. Intl J Phytorem 5:211-224

Hattink J, de Goeij JJM, Wolterbeek HTH (2000) Uptake kinetics of $^{99}$Tc in common duckweed. Environ Exp Bot 44:9-22

Kamal M, Ghaly AE, Mahmoud N, Côté R (2004) Phytoaccumulation of heavy metals by aquatic plants. Environ Intl 29:1029-1039

Lytle CM, Lytle FW, Yang N, Qion JH, Hansen D, Zayed A, Terry N (1998) Reduction of Cr (VI) to CR (III) by wetland plants: Potential for *in situ* heavy metal detoxification. Environ. Sci Technol 32:3087-3093

Lytle CM, Smith BN, McKinnon CZ (1994) Manganese accumulation along Utah roadways: a possible indication of motor vehicle exhaust pollution. Sci Tot Environ 162:105

Macek T, Mackova M, Kas J (2000) Exploitation of plants for the removal of organics in environmental remediation. Biotechnol Adv 18:23-34

Manning K (1988) Detoxification of cyanide by plants and hormone action. In: Cyanide compounds in biology, ed. Ciba foundation, 1988, John Wiley & Sons, Chichester, UK

McCutcheon SC, Schnoor JL (eds) (2003) Phytoremediation – transformation and control of contaminants. Wiley Interscience, pp 985

Mohan BS, Hosetti BB (1999) Aquatic plants for toxicity assessment. Environ Res 81:259-274

Peles JD, Smith MH, Brisbin IH Jr (2002) Ecological half-life of $^{137}$Cs in plants associated with a contaminated stream. J Environ Radioactivity 59:169-178

Prasad MNV, Greger M, Aravind P (2005) Biogeochemical cycling of trace elements by aquatic and wetland plants: relevance to phytoremediation. In: Prasad MNV,

Sajwan KS, Naidu R (eds), Trace elements in the environment: Biogeochemistry, Biotechnology and Bioremediation. CRC Press, Florida, USA, Now Taylor and Francis, Chap 24, pp 443-474

Prasad MNV (2001) Bioremediation Potential of Amaranthaceae In: Leeson A, Foote EA, Banks MK, Magar VS (eds), Phytoremediation, Wetlands, and Sediments, Vol. 6(5), Proc 6th Int In Situ and On-Site Bioremediation Symposium, Battelle Press, Columbus, OH, pp 165-172

Prasad MNV, Greger M, Smith BN (2001) Aquatic macrophytes, in Metals in the Environment: Analysis by biodiversity. In: Prasad MNV (ed) Marcel Dekker Inc., New York, 259.

Prasad MNV (2004) Heavy metal stress in plants: from biomolecules to ecosystems, Narosa Publishing House, New Delhi, 2nd Ed. Pp 462+XIV

Sandermann H (1994) Higher plant metabolism of xenobiotics: the ‘green liver’ concept. Pharmacogenetics 4:225-241

Sheppard SC, Motycka M (1997) Is the akagare phenomenon important to iodine uptake by wild rice (*Zizania aquatica*)? J Environ Radioactivity 37:339-353

Sobolewski A (1999) A review of processes responsible for metal removal in wetlands treating contaminated mine drainage. Int J Phytorem 1:19-51

St-Cyr L, Campbell PGC (1994) Trace metals in submerged plants of St. Lawrence river. Can J Bot 72:429

St-Cyr L, Campbell PGC, Guertin K (1994) Evaluation of the role of submerged plant beds in the metal budget of fluvial lake. Hydrobiologia 291:141

Susarla S, Bacchus TS, Wolfe NL, McCutcheon CS (1999) Phytotransformation of perchlorate using parrot feather. Soil Groundwater Cleanup 2:20-23

Trapp S, Larsen M, Pirandello S, Danquah-Boakye J (2003) Feasibility of cyanide elimination using plants. Europ J Min Porc Environ Prot 3(1):128-137

Wang W, Freemark K (1995) The use of plants for environmental monitoring and assessment. Ecotoxicol Environ Safety 30:289-301

Weis JS, Weis P (2004) Metal uptake, transport and release by wetland plants:implications for phytoremediation and restoration. Environ Intl 30:685-700

# 12

# Phytomonitoring of Air Pollutants for Environmental Quality Management

Jeetendra K. Upadhyay and Nobuyuki Kobayashi

Wind Engineering Research Center, Tokyo Polytechnic University, 1583, Iiyama, Atsugi, Kanagawa 243-0297, JAPAN, Email: upadhyay@arch.t-kougei.ac.jp

## 1. Introduction

Presently rapid changes in the Earth system are an issue of prime importance for the sustainability of the biosphere. The physical and chemical features of the Earth are intimately tied to the organisms and the activities required for their sustenance. Anthropogenic disturbances, such as growing population and its consequent increasing needs, rapid industrialization, increased energy consumption and exploitation of the natural resources, have led to a number of negative effects, appearing in the form of pollution and general degradation of the ecology and environment. The biosphere and human organism can cope to a certain extent with these adverse changes, but the level of pollutants and accompanying phenomena, that nature and man can endure without damage, is often exceeded today in a number of developed and developing countries. The pollution has attained such unacceptable levels that vast forest areas have been damaged, agricultural production lowered, and the health of the whole population endangered.

One of the major environmental concerns of today is the excessive pollution of air. Air is a resource not confined by political or geographical boundaries. The human body requires ~50lb of air a day for its oxygen needs (Perkins 1974). If one assumes an average daily consumption of food ~1.5 kg per person, the intake of air is ~15 to 20 times the amount of food. This explains why air quality, which is characterized by the nature of pollutants and their concentrations, is a serious public health and environmental problem.

The pollutants in the atmospheric air may be in solid, liquid and gaseous form e.g., wind blown dust, volcanic dust and gases, sea-spray, oxides of nitrogen and sulfur, hydrocarbons, hydrogen sulfide from decaying organic matters etc. They are transported to the terrestrial and marine surfaces from their sources of origin by wind and turbulence. The mean wind speed in the atmospheric boundary layer varies typically in the range ~5-10 m/sec among regions, thus the horizontal transport of pollutants over a day is typically ~500-1000 km. During transport process, these pollutants may undergo change of form, such as secondary gaseous pollutants and aerosols through chemical reactions under a set of different meteorological conditions in the atmosphere. The transformation of physical and chemical form greatly influences rates of removal of the pollutants from the atmosphere by direct deposition as gases or aerosols to the terrestrial surfaces or marine layers, known as *dry deposition* and by precipitation as *wet deposition*. Atmospheric pollutants may also interact with short-wave and long-wave terrestrial radiation through scattering and absorption processes and thus may cause perturbations in the radiation energy balance of the earth atmospheric system. This may lead to climatic changes which may have local, regional and global repercussions in terms of temperature, rainfall, soil moisture and food production. Excellent reviews of many historical aspects and sources of pollutants, atmospheric transport and transformations of pollutants, and issues of global change are provided in the book by Bell and Treshow (2002). It is, thus, clear that atmospheric pollution has serious consequences not only for human health, but the planet life itself.

In order to mitigate environmental pollutants and to protect the biosphere from the adverse effects of pollution, four important issues should be highlighted explicitly. These issues include changing lifestyle to control or decrease the emissions of pollutants, developing technologies to avoid or mitigate emissions, making rules and regulations to reduce or cut emissions and decontamination of existing pollutants in the environment. Gaseous pollutants and particulates, once released in the atmosphere, disperse rapidly. Mechanical treatment processes in such situation are very energy-intensive and costly; while plants are driven by solar energy, self-reproducing and concentrate and detoxify pollutants. The ability of a plant to clean up dispersed ambient pollutants has been confirmed in a number of studies (Hill 1971; Okano et al. 1988; Simonich and Hites 1994, 1995; Weber et al. 1995; Yunus et al. 1996; Salt et al. 1998; Pacala et al. 2001). Thus, plant is a natural monitor and detoxifier "device" of toxic pollutants in our ambient environment while adding value to our buildings, landscapes, and communities.

Air pollution has both direct and indirect impacts on plant life. It has been known for several decades that air pollution can adversely affect plant health. Many studies have been conducted on the responses of plants to air pollution (Treshow 1984; Posthumus 1985; Hutchinson and Meema 1987; Heck et al. 1988; Treshow and Anderson 1991; Alscher and Wellburn 1994; Alfani et al. 1996; DeKok and Stulen 1998). Studies have also demonstrated a relationship between trace gas emissions and agricultural crops with respect to $CH_4$ and $N_2O$ in particular (Singh 2000). Amongst these, a number of studies were carried out under controlled exposure conditions inside the chamber. The results from chamber studies are valuable and can provide casual links between pollution and onset of injuries to plants; nevertheless field survey reveals the integrated effects of pollutants on the plants over longer duration under different pollutant mixtures and set of environmental conditions (Lee et al. 2004).

Plant injury symptoms by air pollutants are most common near large cities, smelters, refineries, electric power plants, airports, highways, refuse dumps, pulp and paper mills, and coal-, gas- or petroleum-burning furnaces. Damage to plants and vegetations in isolated areas also occurs when pollutants are spread long distances by the wind under different climatic conditions. Damage to vast forested areas in Europe and North America is a good example of long-range transport of pollution (Bell and Treshow 2002). Injuries to plants due to air pollution include mottled foliage, "burning" at leaf tips or margins, twig dieback, stunted growth, premature leaf drop, delayed maturity, abortion or early drop of blossoms, and reduced yield or quality. In general, the visible injuries to plants are of three types: (1) collapse of leaf tissue with the development of necrotic patterns, (2) yellowing or other color changes, and (3) alterations in growth or premature loss of foliage.

The transport of gaseous pollutants and aerosols from the atmosphere to vegetation is by the turbulent wind field, generated by the frictional drag by the vegetation surfaces on the wind. It is this turbulent wind field that drives exchange of scalar concentrations between vegetation and the atmosphere. The aerodynamically rough surfaces like, forests and vegetation, generate much greater frictional drag on airflow than flat terrain and as a consequence, the rates of transport of pollutants from free atmosphere to the surface are much greater over forests and vegetation than over short vegetation or flat terrain. Thus, the nature of surface strongly affects the rate of transfer. This turbulent transfer of pollutants to the vegetation surface, together with processes at the surface, determines the uptake of gases and capture of aerosols by plants (Fig. 1).

Plants are very sensitive to the surrounding habitats. Alteration in normal environmental conditions, such as temperature, wind, light, soil water content, nutrients and air pollutants, directly affects the physiology of plant functioning like, developing injuries, abnormal symptoms or growth. Injury is often evident on plants before it can affect human being and other animals. The appearance of such abnormal symptoms/injuries or growth is a good indicator of the danger of environmental pollution to human beings. Some plants are relatively tolerant to air pollutants, and so can accumulate pollutants. The possible use of plants as passive monitors/indicators was early recognized (Bleasedale 1973; Harward and Treshow 1975; Roose et al. 1982). Phillips (1980) outlined the criteria for suitable bioindicator species that include relative tolerance to pollution exposure; abundant presence; sedentary habit; ease of laboratory holding and testing; and the ability to accumulate some pollutants and hence show dose-response relationship. Canas et al. (1997) further categorized the plants, used for biomonitoring of air pollutants into two: (1) sensitive species in which visible injuries indicate damage, and (2) tolerant species that can accumulate pollutants and demonstrate dose-response relationships. In a more recent study,

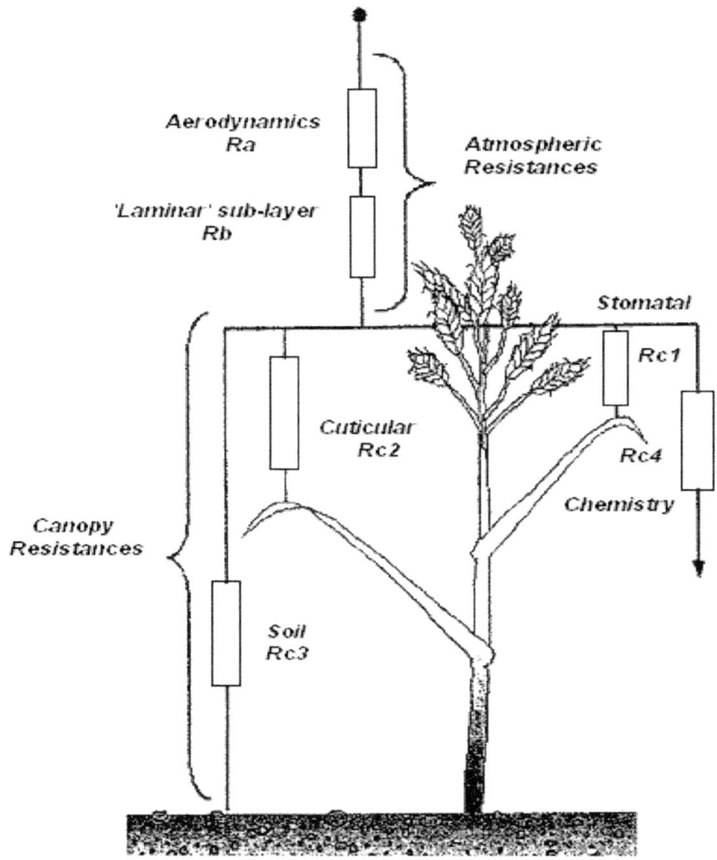


**Fig. 1.** Resistance diagram to show the effects of atmospheric and surface processes on pollutant deposition to terrestrial surfaces

Lee et al. (2004) demonstrated the use of tolerant plants to restore a coastal forest ecosystem severely damaged by air pollutants discharged from an industrial complex in two industrial cities of Korea. Further, results from transplant tests indicated that a field survey was the most reasonable method for the selection of tolerant plants to restore a pollution-damaged ecosystem. There are many plant species which fulfill these criteria and are useful ecosystem indicators. Any alteration in them has implications for the whole ecosystem. Accordingly, other studies have also acknowledged the possibility of using plants as an indicator to monitor air quality (Angold 1997; Loppi et al. 1997; Beckett et al. 1998; Roy and Sharma 1998; Freer et al. 2004; Santitoro et al. 2004). Hence, use of plant, as an indicator "device" to provide information on the toxicity of pollutants, is an inexpensive method, and can act as an early-warning indicator of deteriorating air quality.

A number of air pollutants, such as sulfur dioxide ($SO_2$), nitrogen oxides ($NO_x$), ozone ($O_3$), peroxyacetyl nitrate (PAN), halogens and acid rain can onset early visible damage on plants. Hence, plants offer an excellent alarm system for detecting the presence of excessive concentrations of these pollutants and often provide the very first evidence on polluted air. Plant responses, characteristic visible foliar symptoms in particular, have long been used as indicators of air pollutants. In additions, the amount of metal accumulation has also been used as a bioindicator. This chapter considers the potential of plants as a phytoindicator/phytomonitor for management of air quality. A section of this chapter also outlines the role of plants in fighting indoor air pollution.

## 2. Plants as Bioindicators of Air Pollutants

### 2.1 Bioindicators for Sulfur Dioxide ($SO_2$)

Sulfur dioxide ($SO_2$) is a major pollutant in the atmosphere, especially in developing countries. Common sources of $SO_2$ include power plants, fossil-fuel furnaces, oil refineries, copper and iron smelters. The exposure of succulent, broad-leaved plants to $SO_2$ and its by-product sulfuric acid ($H_2SO_4$) usually results in dry, papery blotches colored tan, straw or even white, and turn to interveinal browning or necrosis. However, the leaf veins remain green. Young and mid-aged plants and leaves are more sensitive. Exposure to 0.5 ppm for 4 hours or 0.25 ppm for 8 hours may be injurious to some crops which may show symptoms as far as 50 km from its source. Plants are more sensitive to $SO_2$ during periods of bright sun, high relative humidity, and adequate plant moisture during the late spring and early summer.

Many plants are known to be injured by $SO_2$ under natural and experimental exposure conditions (Fig. 2). If $SO_2$ injury is suspected, one can check nearby, more sensitive crops, such as alfalfa, beans, beets, buckwheat, soybean, and sunflower, or sensitive weeds, such as pigweeds, ragweed and morning glory. In the National Monitoring Network of The Netherlands, alfalfa (*Medicago sativa*) and buckwheat (*Fagopyrum esculentum*) were used for monitoring the effects of $SO_2$ (Posthumus 1984). DeSloover and LeBlanc (1968) developed an Index of Atmospheric Purity (IAP), based on mathematical formula that correlated the lichen and bryophyte vegetation of an area with the air quality around urban areas or point sources of $SO_2$.

### 2.2 Bioindicators for Fluorides

Fluorides are compounds containing the elemental fluorine (F). Fluorides are produced by glass, aluminum, pottery, brick and ceramic industries and by refineries, metal ore smelters, and phosphate fertilizer plants. The typical injury by gaseous or particulate fluorides is either a yellowish mottle to a wavy, red-

**Fig. 2.** Effect of $SO_2$ on several species, under controlled exposure of $SO_2$ (Source: University of Newcastle, UK)

dish or tan "scorching" at the margin or tips of the broad-leaved plants or a "tipburn" of grasses and conifers. Accumulated leaf-fluoride concentrations of 20 to 150 ppm often injure sensitive plants, although resistant varieties and species of plants will tolerate leaf concentrations of 500 to 4,000 ppm or more without any visible injury. Gladiolus (*Gladiolus hortulanus*) is the most widely used plant for biomonitoring fluoride (Manning and Feder 1980). A 4-week exposure of susceptible *Gladiolus hortulanus* to an air concentration of 0.0001 ppm, or less than 24 hours at 10 ppb, produces leaf concentrations of 150 ppm and definite tissue necrosis.

### 2.3 Bioindicators for Chlorides

Like fluorides, chlorides are compounds containing the elemental chlorine (Cl). Hydrogen chloride (HCl) and chlorine ($Cl_2$) are emitted from the stacks of glassmaking industries and refineries. These can be also produced by incineration and spillage, such as chlorine tanker storage tanks. Injury caused by chlorine is similar to that caused by $SO_2$ and fluorides, in that it is marginal and interveinal. On broad-leaved plants, necrotic, bleached, or tan to brown areas tend to be near the leaf margins, tips, and between the principal veins. Middle-aged or older ones are more susceptible that the young ones. Conifers may show tipburn on the current seasons. Susceptible plants, when exposed for 2 hours or more at concentrations of chlorine ranging from 0.1 to 4.67 ppm, show injury symptoms. Chlorine-injured vegetation is often observed near swimming pools, water-purification plants, and sewage-disposal facilities. Grasso et al. (1999) reported the capacity of lichens to accumulate atmospheric contaminators like, halides and particulate matters linked to volcanic activity in Italy: Mount Etna and Vulcano Island.

### 2.4 Bioindicators for Ethylene (Ethene)

Ethylene ($H_2C$-$CH_2$) is a known and important plant-toxic air pollutants. Ethylene is one of the many products of auto, truck, and bus exhaust. Ethylene also results from the incomplete combustion of coal, gas and, oil for heating and is a by-product of polyethylene manufacture. Ethylene ($H_2C$-$CH_2$) modifies the activities of plant hormones and growth regulators, which affect developing tissues and normal organ development, without causing leaf-tissue collapse and necrosis (Abeles and Heggestad 1973). Injury to broad-leaved plants occurs as a downward curling of the leaves and shoot (epinasty), followed by a stunting of growth. Posthumus (1983) suggested the use of petunia (*Petunia axilliaris hybrida*) as a bioindicator plant for $H_2C$-$CH_2$ in The Netherlands. Pleijel et al. (1994) used potted petunia (*Petunia hybrida*), placed at distances 10, 20, 40, 80 and 120 m from a motorway with approximately 30,000 vehicles/day, as an indicator for ethylene in Sweden in 1989. The result showed that the petunia flowers were significantly smaller on plants closer to the motorway that those at distance. Furthermore, the abortion rate of flower buds of plants closer to motorway was more frequent and the ripening of fruits was also high near motorway. Thus, the authors inferred from the survey that ethylene ($H_2C$-$CH_2$) concentrations were high enough to influence the petunia reproductive structures, close to the motorway.

### 2.5 Bioindicators for Ozone ($O_3$)

Ozone, a molecule ($O_3$), formed by three atoms of oxygen, is a photochemical oxidant that disrupts photosynthetic and metabolic functions. It is probably the most important phytotoxic air pollutant in the troposphere. Ozone is brought down to ground level by vertical winds from the stratosphere during electrical storms. But the most important mechanism of ozone formation in the tropospheric atmosphere is reaction of $NO_x$ and hydrocarbons (HC) in presence of sunlight. $O_3$ is a widespread air pollutant in the industrialized countries (Stockwell et al. 1997). Leaf symptoms to ozone exposure are termed "stippling" or "speckling" characterized by numerous tiny dots on the upper leaf surface. On the other hand, long-term exposure to near-ambient ozone levels may lead to chlorotic symptoms or may reduce photosynthesis and crop yield without visible injury (Heath and Taylor 1997; Pell et al. 1997). Injury occurs mostly in the afternoon and the least at night.

The ozone sensitivity of plant species and cultivars varies greatly. There are some excellent bioindicator plant species that have been used widely to detect $O_3$ in the lower atmosphere. For example, the tobacco (*Nicotiana tabacum*) cultivars Bel-W3 (super-sensitive to ozone) and Bel-B (ozone-tolerant) have been used as ozone biomonitor and control, respectively, for three decades. This has greatly contributed to the awareness of people to recognize ozone as a pollutant (Heggestad 1991). Susceptible tobacco plants are injured when concentrations of ozone exceed 0.04 ppm. Further detail on tobacco, as an indicator plant for ozone, is considered later as an example. Morning glory (*Ipomoea violacea*) in Japan (Nouchi and Aoki 1979) and clover in Sweden (Karlsson et al. 1995) have also been reported as indicator plants for $O_3$. Reduction in growth of radish (*Raphanus sativus*) has been also observed as an indicator of ozone in Japan and Egypt (Izuta et al. 1993; Hassan et al. 1995). Several other plant species are also known as bioindicators of ozone exposures. Observations of symptoms from an open-top exposure chamber investigation in central Pennsylvania have confirmed that black cherry, yellow poplar, white ash, common milkweed, spreading dogbane, and blackberry were sensitive to ambient ozone exposures (Skelly 2000).

## 2.6 Tobacco

Ozone injury to tobacco is called weather fleck (Fig. 3). This symptom was first observed in 1959 (Heggestad and Middleton 1959). Tobacco (*Nicotiana tabacum*) is known to be particularly sensitive to ozone and the ozone-sensitive tobacco cultivar Bel-W3 has been widely used as biomonitor of tropospheric ozone (Heggestad 1991). Furthermore, they observed that the cultivar Bel-W3, developed from progeny of two plants, showed pergament-like lesions two to three times larger than those typically associated with ozone injury in cigar wrapper tobacco (Heggestad 1991). In contrast, the genetically related cultivar Bel-B was visibly unaffected by ambient ozone levels (Heggestad 1991; Langebartels et al. 1991). The "classical" ozone symptoms in tobacco cultivar Bel-W3 plants occur as sharply defined dot-like lesions on the adaxial side of the leaf resulting from the death of a group of palisade cells (Loreto et al. 2001). In a recent study, Nali et al. (2004) surveyed the use of vascular plants for the bioindication of tropospheric ozone in the area of Pisa (Tuscany, Central Italy). They observed that with the exposure of photochemical ozone surpassing 100 ppb (maximum hourly means) during the warm season, supersensitive tobacco cultivar Bel-W3 confirmed the value of detailed, cost-effective, monitoring surveys. Trials with clover clones demonstrated that sensitive plants underwent severe biomass reduction in the current ozone regime. Therefore, a set of tobacco plant species: Bel-W3 and Bel-B, as sensitive and tolerant cultivars, would be highly recommended for bioindication of ozone.

**Fig. 3.** Necrotic lesions on tobacco BEL-W3 leaves after growth at ambient ozone concentrations (Source: NCSU, Raleigh)

### 2.7 Bioindicators for Peroxyacetyl Nitrate (PAN)

Another photochemical oxidant is peroxyacetyl nitrate (PAN). After ozone, it is the most phytotoxic air pollutant. Like ozone, PAN is produced when sunlight reacts with various exhaust gases. PAN causes leaves to develop bands, blotches, bronzed and silvery areas. In some plants, such as petunia, pinto bean, tomato, and tobacco, the collapse may be through the entire thickness of the leaf blade. Pre-mature senescence and defoliation may also occur. PAN is most toxic to small plants and young leaves. Exposure to 0.01 to 0.05 ppm for one hour induces symptoms in susceptible plants. In the early 1940s, in Los Angeles basin, plants, such as romaine lettuce (*Lactuca sativa*), Swiss chard (*Beta chilensis*) and annual blue grass (*Pao annua*) were identified as bioindicators of PAN even when PAN had not yet been chemically identified (Manning and Feder 1980). Petunia plants are also known to be highly sensitive to PAN. But the sensitivity of petunia varies among cultivars and, in general, cultivars with white flower are more sensitive to PAN than those of blue or red flowers.

## 3. Phytoremediation and Urban Air Quality Management

Natural and planted vegetation are an efficient sink for various air pollutants including nitrogen oxides ($NO_x$) (Yunus et al. 1996), carbon dioxide ($CO_2$) (Pacala et al. 2001) and polycyclic aromatic hydrocarbons (PAHs) (Simonich and Hites 1994, 1995). Several other investigations too proposed that the plants should be utilized to reduce pollutant concentrations in the atmosphere (Hill 1971; Okano et al. 1988; Simonich and Hites 1994, 1995; Weber et al. 1995; Salt et al. 1998). Poor air quality has brought attention to trees as air pollution remedies since trees/plants directly absorb carbon dioxide in their life-dependent process, photosynthesis.

Plants play an important role in the mitigation of highly polluted atmosphere and extreme climates in urban and semi-urban areas. Pollutants in urban areas are of myriad types and distributed unevenly, as shown by some studies (Pfeffer 1994; Raaschou et al. 1995). Street/park trees in urban areas can be very helpful in mitigation of harmful pollutants and chemicals, including heavy metals from the environment (Pfeffer 1994; Raaschou et al. 1995). Removal of airborne pollutants is done by the process of respiration. During photosynthesis, plant intakes $CO_2$ simultaneously with several other pollutants, such as nitrogen oxides ($NO_x$), airborne ammonia, sulfur dioxide ($SO_2$), and ozone ($O_3$), that is also a part of the smog and greenhouse gases, through its stomata (Bergmann et al. 1995; Singh et al. 1995; Lea 1998; Morikawa et al. 1998; Wellburn 1998). Once inside the leaf, gases diffuse into the spaces between the cells of the leaf to be absorbed by water films or chemically altered by the plant tissues. Street trees in the urban areas are particularly important for this due to their close proximity to vehicles, the major source of air pollutants.

Plants also reduce air pollution by intercepting particulate matter (PM), and aerosols and retaining them on the leaf surface by process of *dry deposition*. Leaf surfaces are most efficient at removing pollutants that are water-soluble including $SO_2$, $NO_2$ and $O_3$. Pollutant removal rates are the highest when vegetative surfaces are wet or damp; these conditions can increase removal rates ten-fold because the entire tree surface is available for the pollutant uptake. A number of field measurements have suggested that the vegetation can significantly reduce their adverse effects through their ability to capture pollutant particles (Nasarullah et al. 1994; Beckett et al. 1998; Roy and Sharma 1998). In a more recent study, Freer et al. (2004) presented relative deposition velocities and capture efficiencies of five species used widely in woodland of urban and sub-urban areas of Europe i.e. oak (*Quercus petraea*), alder (*Alnus glutinosa*), ash (*Fraxinus excelsior*), sycamore (*Acer pseudo-platanus*) and Douglas fir (*Pseudotsuga menziesii*), and two species being used increasingly in semi-arid regions, i.e. weeping fig (*Ficus nitida*) and Eucalyptus (*Eucalyptus globulus*). The measurements were made at three wind speeds, and deposition velocities and capture efficiencies were compared with those published for other tree species. It was found that the values of deposition velocity ranging from 0.1 to 0.3 cm/s at a wind speed of 3 m/s to maximum values of 2.9 cm/s at 9 m/s wind speed. Further, the authors noticed that species with more complex stem structure and smaller leaves had greater deposition velocities. Such data sets can be used in the models to guide species choice and planting design in order to maximize particle removal from the urban air. It is also clear that species choice, planting design and location relative to pollution source are critical in determining the effectiveness of particle capture by trees.

Plants remove (sequester) carbon from the atmosphere through photosynthesis, extracting carbon dioxide from the air, separating the carbon atom from oxygen, and returning oxygen to the atmosphere (Pacala et al. 2001). Plant's ability to offset carbon emissions is determined by average size, canopy cover, health, and age, but larger tress can help in lowering annual carbon emissions by 2 to 3% in the atmosphere. Generally, trees are comprised of 45% carbon, 50% water, and 5% minerals, but these constituents vary from species to species.

Higher urban temperatures also accelerate the production of smog, of which ozone is a major component causing respiratory and other health problems. One of the major causes of smog is "heat-island effect", caused by the internal build-up of heat in cities from incoming solar energy absorbed onto concrete and asphalt, such as roads, parking lots, and buildings (Voogt and Oke 1989). This is further compounded by emissions from vehicles, houses and heating. Vegetation in urban areas helps to mitigate air quality problem by reducing temperature-dependent production of air pollutants, such as, ozone, VOCs and others (Taha 1997). Tree species strategically planted to shade homes can generate about 10 to 50% savings in cooling expenses depending upon tree type, location, and climatic variation. This not only reduces the amount of carbon-based fuels used, but also attenuate emissions that reduce air quality. Improvement in the air quality can be expected, if trees can absorb more air pollutants close to the pollutant sources and thus the number of exceedance days can be reduced. Nevertheless, species choices, planting design and location relative to pollution sources are necessary requirements for the phytoremediation of urban air quality. Mixed plantings should be planned, with the more susceptible plants acting as bioindicators for early-warning of deteriorating air quality and tolerant ones for amelioration of pollution level.

## 4. Phytoremediation and Indoor Air Quality (IAQ)

Air pollution is not confined to outdoor environment in cities, urban areas and industrial sites only. Now one's home itself could be a potent source of potentially harmful chemicals. "Energy crisis" of seventies, resulted in growing demand of airtight and insulated buildings to conserve energy. An unintended effect of this improved energy efficiency was poor indoor air quality (IAQ) because of airtight buildings hampering the circulation of airflow. Most buildings use recirculated air and mix it with minimum amount of fresh air being brought into the buildings through an outside duct for building ventilation. As a result, more and more buildings have indoor air quality (IAQ) problems due to building-up of hazardous pollutants and chemical compounds released from building materials and furnishings. This chemically polluted indoor environment has been related to symptoms of illness, known as the "sick house syndrome". The pollutants most widely present in indoor environment are: carbon monoxide (CO), nitrogen oxides ($NO_x$), undesirable products of burning tobacco and wood, formaldehyde, volatile organic compounds (VOCs), including chemicals like, toluene, xylene, ethylbenzene and chloropyrifos. Indoor air pollution has become a serious public health concern. This has fuelled growing demand for healthier indoor air, to which health professionals, architects, researchers and housing industry have made beginning to respond.

It is well acknowledged that plants are known for their ability to remove air pollutants from outdoor environment. They absorb carbon dioxide ($CO_2$) and significant amounts of harmful gases from the air and release oxygen as a part of photosynthetic process. Over the past few years, studies have shown that house plants have been able to reduce levels of some chemicals in the laboratory experiments. Many common house plants and blooming potted plants help fight against indoor air pollution (Wolverton et al. 1984; Wolverton et al. 1985). "Indoor" potted-plants can remove airborne contaminants, such as volatile organic compounds (VOCs), over 300 of which have been identified for indoor air pollution. Studies have shown that many house plants can absorb benzene, formaldehyde, trichloroethylene and other VOCs, (Wolverton and Wolverton 1993; Wolverton 1997; Orwell et al. 2004). The foliage of indoor plants is also capable of extracting particulate matters (PM) from the air. In an experiment, Lohr and Pearson (1996) reported that the presence of foliage plants in interior spaces changed particulate matter (PM) accumulation: accumaltion was lower in both rooms when plants were present than when plants were absent. In particular, vegetation with rough surfaces with fine hairs or raised veins is more effective in intercepting PM than smooth vegetation. Plant roots can also absorb some pollutants and render them harmless in the soil.

In a study sponsored by National Aeronautics and Space Administration (NASA), spider plants (*Chlorphytum elatum*) were placed in closed chambers with 120 ppm of CO or 50 ppm of $NO_2$ (Wolverton et al. 1985). After 24 hours, spider plants removed 96% of CO and 99% of $NO_2$. Experiments with Golden pothos plants (*Epipremnum aureum*) showed that 75% of CO was removed after 24 hours. Another study, conducted jointly through NASA and the Associated Landscape Contractors of America (ALCA), investigated the use of common indoor plants to provide a natural way to combat "Sick Building Syndrome" (Wolverton et al. 1989). The chemicals screened for the removal were benzene, formaldehyde and trichloroethylene. The results of these tests suggested:

- Low-light-requiring house plants with activated carbon plant filters have potential for improving IAQ.
- The plant root-zone is an effective area for removing VOCs. (maximum air exposure to plant root-soil area for best filtration).
- Use of activated carbon filter should be part of the house plant/air-cleaning plan.

However, NASA studies were conducted in a closed chamber climate controlled environment with activated carbon, air blown through the soil and single contaminant release. The purpose of their studies was to see if plants could be used for space habitation; nevertheless, the results provided impetus to use foliage plants in offices and other workplaces to improve the quality of indoor air.

Plants need sunlight in order to convert $CO_2$ into oxygen by the process of photosynthesis. From the perspective of indoor environment, it would be very helpful to study some common house plants that need less light or no light for photosynthesis process. Raza et al. (1995) evaluated the status of indoor air quality of a hospital using several plants that do not need light. They found that *Apicra deltoidea* is the most effective, followed by *Sedum pachyphyllum*, in converting carbon dioxide into oxygen at night when there is no sunlight.

Below is the list of most effective plants with large leaf surface area to be used in removing pollutants like, formaldehyde (Source: UF/IAS):

- Heart-leaf philodendron (*Philodendron scandens*)
- Elephant ear philodendron (*Philodendron domesticum*)
- Green spider plant (*Chlorphytum elatum*)
- Lacy tree philodendron (*Philodendron selloum*)
- Golden pothos (*Epipremnum aureum*)
- Chinese evergreen (*Aglonema modestum*)
- Mini-Schefflera (*Bassaia arboricola*)
- Peperomia (*Peperomia obtusifolia*)
- Peace lily (*Spathiphyllum clevelandii*)
- Corn plant (*Dracaena fragrans 'massangeana'*)
- Snake plant (*Sansevieria traifasciata*)

To some extent, these plants can also be used against pollutants like, benzene and trichloroethylene. Most of the house plants listed above are commonly found in tropical and sub-tropical forests, where they received light filtered through the branches of taller trees. Because of this, their leaf photosynthesizes efficiently under relatively low light conditions, which, in turn, allows them to process gasses in the air efficiently.

However, careful selection of indoor plants is necessary, if anyone suffers from exposure to molds, pollen, odors or dust. House plants also add moisture to the indoor environment. Molds can grow in the soil of the plant and release spores into the air. This can have negative effects on comfort and health of the occupants. Wolverton (1997) has detailed the role of house plants in fighting indoor air pollution in his book.

## 5. Conclusion

Air pollution has both direct and indirect impacts on the plant life. Some plants are very sensitive to the air pollution. If there is any injury caused by air pollution, the plant shows an appropriate response. The early recognition of pollutant damage to plants, notably characteristic visible foliar symptoms, acts as an alarm for toxic dangers to humans and their environment. Hence, the bioindicator method indicates directly whether the ambient concentration of a pollutant is harmful to biological tissues, and reveals the synergetic and antagonistic effects of multiple pollutants of the environment. A suitable bioindicator plant must be sensitive to a specific pollutant and respond proportionally to the pollutant or dose; be native or adaptable to the region and abundant presence; and be tolerant to pests and diseases. Bioindicator/biomonitoring method provides a relatively low-cost and easy method of environmental surveillance compared to high tech measuring methods.

Despite being novel technology for environmental monitoring, the great potential of bioindicators is often confronted with difficult questions of methodology how to use "living measuring instruments". The effects of environmental load can not always be clearly differentiated from natural stress factors. Lack of practical experience with certain bioindicators makes interpretation of findings very difficult, especially if, no comparable pollutant measurements are available. Hence, efforts should be made to develop standardized indicator species that will show known, reliable dose-response relationships with any gaseous pollutants and mixture under various environmental conditions.

It can be concluded that a more integrated and detailed approach, a combination of physical and chemical methods together with indicator plants, is most reliable means of monitoring air quality for protecting human health and the environment. Phytoremediation of air pollutants using street/park trees with abundant foliage helps to a greater extent in improving urban air quality. They are capable of removing pollutants, like gases and particulate matters; reduce energy expenditures and lower air temperatures. Similarly, many common house plants and blooming potted plants help fight against pollution in indoor environment. They scrub significant amount of toxic pollutants and chemical compounds from air and render them harmless. Systematic studies of responses of plants in indoor and outdoor environment would greatly increase our understanding of plants as biological indicators of air quality. Bioindicator method provides a novel and cost-effective technology to visualize and monitor environmental air pollution keeping public health in mind.

Acknowledgements. Authors would like to thank the Ministry of Education, Culture, Sports, Science and Technology, Japan for providing financial support in form of “Academic Frontier” Research Fellowship Project, 2000-2004. We also thank Mr. N. Pillai for going through the manuscript.

## References

Abeles FB, Heggestad HE (1973) Ethylene: an urban air pollutant. J Air Pollut Contr Assoc 23:517-521

Alfani A, Maisto G, Iovieno P, Rutigliano FA, Bartoli G (1996) Leaf contamination by atmospheric pollutants as assessed by elemental analysis of leaf tissue, leaf surface deposit and soil. J Plant Physiol 148:243-248

Alscher RG, Wellburn AR (1994) Plant responses to the gaseous Environment. Chapman & Hall, London

Angold PG (1997) The impact of a road upon adjacent vegetation: effects on plant species composition. J Appl Ecol 34:409-417

Beckett KP, Freer PH-S, Taylor G (1998) Urban woodlands: the role in reducing effects of particulate pollution. Environ Poll 99:347-360

Bell JNB, Treshow M (2002) Air Pollution and Plant Life, 2nd Edition. John Wiley and Sons, Ltd

Bergmann E, Bender J, Weigel HJ (1995) Growth responses and foliar sensitivities of native herbaceous species to ozone exposures. Water Air Soil Poll 85:1437-1442

Bleasedale JKA (1973) Effects of coal-smoke pollution gases on the growth of ryegrass (*Lolium perenne* L.). Environ Poll 5:275-285

Canas MS, Carreras HA, Orellana L, Pignata ML (1997) Correlation between environmental conditions and foliar chemical parameters in *Ligustrum lucidum* Ait. Exposed to urban pollutants. J Environ Manag 49:167-181

DeKok LJ, Stulen I (1998) Responses of Plant metabolism to Air Pollution and Global Change. Backhuys, Leiden, The Netherlands

DeSloover J, LeBlanc F (1968) Mapping of atmospheric pollution on the basis of lichen sensitivity. In: Misra R, Gopal B (eds) Proc Symposium on Recent Advances on Tropical Ecology, pp 42-56

Freer PH-S, El AA-K, Taylor G (2004) Capture of Particulate Pollution by Trees: A Comparison of Species Typical of Semi-Arid Areas (*Ficus nitida* and *Eucalyptus globulus*) with European and North American Species. Water Air Soil Pollut 155:173-187

Grasso MF, Clocchiatti R, Carrot F, Deschamps C, Vurro F (1999) Lichens as bioindicators in volcanic areas: Mt. Etna and Vulcano Island (Italy). Environ Geol 37:207-217

Harward M, Treshow M (1975) Impact of ozone on the growth and reproduction of understorey plants in the aspen zone of western USA. Environ Conserv 2:17:23

Hassan IA, Ashmore MR, Bell JNB (1995) Effect of ozone on radish and turnip under Egyptian field conditions. Environ Poll 89:107-114

Heath RL, Taylor GE (1997) Physiological processes and plant responses to ozone exposure. In: Sandermann H, Wellburn AR, Heath RL (eds) Forest Decline and Ozone: a Comparison of Controlled Chamber and Field Experiments, Ecological Studies, 127, Springer, Berlin, pp 317-368

Heck WW, Taylor OC, Tingey DT (1988) Assessment of Crop Loss from Air Pollutants. Elsevier Applied Science, London and New York

Heggestad HE (1991) Origin of Bel-W3, Bel-C and Bel-B tobacco varieties and their use as indicators of ozone. Environ Poll 74:264-291

Heggestad HE, Middleton JT (1959) Ozone in high concentrations as cause of tobacco leaf injury. Science 129:208-210

Hill AC (1971) Vegetation: a sink for atmospheric pollutants. J Air Pollut Contr Assoc 21:341-346

Hutchinson TC, Meema GHM (1987) Effects of atmospheric pollutants on forests, wetlands and agricultural systems. Springer-Verlag, Berlin

Izuta T, Miyake H, Totsuka T (1993) Evaluation of air-polluted environment based on growth of radish plants cultivated in small-sized open-top chambers. Environ Sci 2:25-37

Karlsson GP, Sellden G, Skarby L, Pleijel H (1995) Colver as an indicator plant for phyotoxic ozone concentrations: visible injury in relation to species, leaf age and exposure dynamics. New Phytol 129:355-365

Langebartels C, Kerner K, Leonardi S, Schraudner M, Trost M, Heller W, Sandermann H (1991) Biochemical plant responses to ozone: I. Differential induction of polyamine and ethylene biosynthesis in tobacco. Plant Physiol 95:882-889

Lea PJ (1998) Oxides of nitrogen and ozone: can our plants survive? New Phytol 139:25-26

Lee CS, Lee KS, Hwangbo JK, You YH, Kim JH (2004) Selection of Tolerant Plants and Their Arrangement to Restore a Forest Ecosystem Damaged by Air Pollution. Water Air Soil Poll 156:251-273

Lohr VI, Pearson C-M (1996) Particulate matter accumulation on horizontal surfaces in interiors: Influence of foliage plants. Atmos Environ 30:2565-2568

Loppi S, Nelli L, Ancora S, Bargagli R (1997) Passive monitoring of trace elements by means of tree leaves, epiphytic lichens and bark substrate. Environ Monitor Assess 45:81-88

Loreto F, Mannozzi M, Maris C, Nascetti P, Ferranti F, Pasqualini S (2001) Ozone quenching properties of isoprene and its antioxidant role in leaves. Plant Physiol 126:993-1000

Manning WJ, Feder WA (1980) Biomonitoring air pollutants with plants. Applied Science Publishers, London

Morikawa H, Higaki A, Nohno M, Takahashi M, Kamada M, Nakata M, Toyohara G, Okamura Y, Matsui K, Kitani S, Fujita K, Irifune K, Goshima N (1998) More than a 600-fold variation in nitrogen dioxide assimilation among 217 plant taxa. Plant Cell Environ 21:180-190

Nali C, Crocicchi L, Lorenzini G (2004) Plants as indicators of urban air pollution (ozone and trace elements) in Pisa, Italy. J Environ Monitor 6:636-645

Nasrullah M, Tatsumoto H, Misawa A (1994) Effect of roadside planting on suspended particulate matter concentration near road. Environ Techno 15:293-298

Nouchi I, Aoki K (1979) Morning glory as a photochemical oxidant indicator. Environ Poll 18:289-303

Okano K, Machida T, Totsuka T (1988) Absorption of atmospheric $NO_2$ by several herbaceous species: estimation by the $^{15}N$ dilution method. New Phytol 109:203-210

Orwell RL, Ronald L, Wood RL, Tarran J, Torpy F, Burchett MD (2004) Removal of Benzene by the Indoor Plant/Substrate Microcosm and Implications for Air Quality. Water Air Soil Poll 157:193-207

Pacala SW, Hurtt GC, Baker D, Peylin P, Houghton RA, Birdsey RA, Heath L, Sundquist GT, Stallard RF, Ciais P, Moorcroft P, Caspersen JP, Shevliakova E, Moore B, Kohlmaier G, Holland E, Gloor M, Harmon ME, Fan SM, Sarmiento JL, Goodale CL, Schimel D, Field CB (2001) Consistent land-and atmosphere-based U. S. carbon sink estimates. Science 292:2316-2320

Pell EJ, Schlagnhaufer CD, Arteca RN (1997) Ozone-induced oxidative stress: Mechanisms of action and reaction. Physiol Plant 100:264 273

Perkins HC (1974) Air Pollution. McGraw-Hill Book Co., New York

Pfeffer HU (1994) Ambient air concentrations of pollutants at traffic-related sites in urban areas of North Rhine-Westphalia, Germany. Sci Tot Environ 146/147:263-274

Phillips DH (1980) Quantitative aquatic biological indicators. Applied Science, London

Pleijel H, Ahlfors A, Skarby L, Pihl G, Sellden G, Sjodin A (1994) Effects of air pollutant emissions from a rural motorway on *Petunia* and *Trifolium*. Sci Tot Environ 146/147:117-123

Posthumus AC (1983) Higher plants as indicators and accumulators of gaseous air pollution. Environ Monitor Assess 3:263-272

Posthumus AC (1984) Monitoring levels and effects of air pollutants. In: Treshow M (ed) Air Pollution and Plant Life, John Wiley and Sons, Chichester, pp 73-95

Posthumus AC (1985) Plants as indicators for atmospheric pollution. In: Nurnberg HW (ed) Pollutants and their ecotoxicological significance, Wiley, London, pp 55-65

Raaschou O-N, Nielsen M-L, Gehl J (1995) Traffic-related air pollution: exposure and health effects in Copenhagen street cleaners and cemetery workers. Arch Environ Health 50:207-213

Raza RH, Shylaja G, Gopal BV (1995) Different abilities of certain succulent plants in removing $CO_2$ from the indoor environment of a hospital. Environ Intern 21:465-469

Roose ML, Bradshaw AD, Roberts TM (1982) Evolution of resistance of gaseous air pollutants. In: Unsworth MH, Ormond DP (eds) Effects of gaseous pollutants in Agriculture and Horticulture, Butterworth, London, pp 379-409

Roy RK, Sharma SC (1998) Bioremediation of urban pollution by orientation of landscaping. Ind J Environ Health 40:203-208

Salt DE, Smith RD, Raskin I (1998) Phytoremediation. Annu Rev Plant Mol Biol 49:643-668

Santitoro A, Aprile GG, Baldantoni D, Bartoli G, Alfani A (2004) Trace Element Analyses in an Epiphytic Lichen and its Bark Substrate to Assess Suitability for Air Biomonitoring. Environ Monitor Assess 98:59-67

Simonich SL, Hites RA (1994) Importance of vegetation in removing polycyclic aromatic hydrocarbons from the atmosphere. Nature 370:49-51

Simonich SL, Hites RA (1995) Organic pollutant accumulation in vegetation. Environ Sci Techno 29:2905-2914

Singh N, Yunus M, Srivastava K, Singh SN, Pandey V, Misra J, Ahmad KJ (1995) Monitoring of auto exhaust pollution by roadside plants. Environ Monitor Assess 34:13-25

Singh SN (2000) Trace Gas Emissions and Plants. Kluwer Academic Publishers, The Netherlands

Skelly JM (2000) Tropospheric ozone and its importance to forests and natural plant communities of the Northeastern United States. Northeastern Naturalist 7:221-236

Stockwell WR, Kramm G, Scheel H-E, Mohnen VA, Seiler W (1997) Ozone formation, destruction and exposure in Europe and the United States. In: Sandermann H, Wellburn AR, Heath RL (eds) Forest Decline and Ozone: a Comparison of Controlled Chamber and Field Experiments, Ecological Studies, 127, Springer, Berlin, pp 1-38

Taha H (1997) Urban Climates and Heat Islands: Albedo, Evapotranspiration, and Anthropogenic Heat. Energ Build - Special Issue on Urban Heat Islands 25:99-103

Treshow M (1984) Air Pollution and Plant Life. John Wiley and Sons, Chichester

Treshow M, Anderson FK (1991) Plant Stress from Air Pollution. John Wiley and Sons, Chichester

Voogt JA, Oke TR (1989) Complete Urban Surface Temperatures. J Appl Meteor 36:1119-1132

Weber P, Nubbaum S, Fuhrer J, Gfeller H, Schlungger U, Brunold P, Rennenberg H (1995) Uptake of atmospheric $^{15}NO_2$ and its incorporation into free amino acids in wheat (*Teitcium aestivum*). Physiol Plant 94:395-429

Wellburn AR (1998) Atmospheric nitrogenous compounds and ozone-Is $NO_x$ fixation by plants a possible solution? New Phytol 139:5-9

Wolverton BC (1997) How to Grow Fresh Air: 50 Houseplants that Purify Your Home or Office. Penguin Books, New York. First published in UK as *Eco-Friendly Houseplants.* Weidenfeld & Nicolson, London, 1996

Wolverton BC, Wolverton JD (1993) Plants and Soil Microorganisms - Removal of Formaldehyde, Xylene and Ammonia from the Indoor Environment. J Missi Acad Sci 38:11-15

Wolverton BC, Johnson A, Bounds K (1989) Interior Landscape Plants for Indoor Air Pollution Abatement. NASA/ALCA Final Report, Plants for Clean Air Council, Davidsonville, Maryland.

Wolverton BC, McDonald RC, Mesick HH (1985) Foliage Plants for the Indoor Removal of the Primary Combustion Gases Carbon Monoxide and Nitrogen Oxides. J Missi Acad Sci 30:1-8

Wolverton BC, McDonald RC, Watkins EA Jr (1984) Foliage Plants for Removing Indoor Air Pollution from Energy-Efficient Homes. Econo Botan 38:224-228

Yunus M, Singh N, Iqbal M (1996) Global status of air pollution: an overview. In: Yunus M, Iqbal M (eds) Plant Response to Air Pollution, John Wiley, UK, pp 1-34

# 13

# Phytoremediation of Air Pollutants: A Review

S.N. Singh and Amitosh Verma

Environmental Science Division, National Botanical Research Institute, Lucknow 226 001, INDIA, Email: drsn_s@rediffmail.com

## 1. Introduction

Industrialization and urbanization are vital for the economic development of a nation. In fact, these are the indicators of prosperity and progress today. However, unplanned industrialization and urbanization may lead to multi-faceted environmental problems. Hence, in a developing nation like India, future industrialization and urbanization should be carefully planned to avoid any irreparable damage to the environment. As far as possible, air, being the lifeline, should be protected from the evils of pollution, as its quality depletion beyond a threshold limit may lead to serious health hazards to both man and livestock. Even, a possibility of change in the genetic make-up cannot be ruled out in case of spill over of nuclear materials. Despite this fact, a progressing country can avoid neither industrialization nor urbanization for its economic growth, which is triggered even at the cost of ecological imbalance. Therefore, the best way out appears today is to be an all-out effort to use such measures (engineering or biological), which can help in mitigation of ambient pollution.

In this endeavour, various laws and policies have been framed to direct the industries and vehicle manufacturers to adopt new technologies to reduce the emission of pollutants from both stationary and mobile sources. In order to curb the menace of air pollution, mechanical collectors, fabric collectors, wet scrubbers, electromagnetic precipitators and fume incineration are often used to attenuate pollutant emission at source. But, as these devices are of mechanical nature, the possibility of their occasional failure cannot be ruled out. Moreover, engineering devices are very expensive, demanding a big budget which is itself a major problem in many cash-strapped industries. Alternatively, less efficient systems are, at times, used by many medium and small-scale industries, leading to the discharge of an alarming proportion of pollutants in the industrial pockets of our country. This situation demands our adequate attention for immediate remedial recourse.

In fact, there is no device available, either mechanical or chemical, which can completely check the emission of pollutants at the source. Once the pollutants are released to the atmosphere, only the plants are the hope, which can mop up the pollutants by adsorbing, absorbing and metabolizing them from the atmosphere. Therefore, the plant's role in the air pollution abatement has been increasingly recognized in recent years.

Phytoremediation is an emerging eco-friendly technology, dealing with degradation, mitigation or stabilization of air, soil or water pollutants by plants (Macek et al. 2000; Garbisu et al. 2002 and Lasat et al. 2002). The main advantages of phytoremediation technology include; (1) it is an aesthetically pleasing and solar energy-driven cleanup technology (2) chances of environmental degradation, because of *in situ* application, are very less (3) a variety of environmental contaminants may be treated simultaneously and (4) it is a cost-effective technology, as cost involved in phytoremediation technology is 60-80% less than the conventional physico-chemical or mechanical technologies (Schnoor 1997). Despite of several advantages, this technology has also got some limitations like (1) it is a time consuming approach, as it may take several growing seasons to cleanup a site (2) plants after phytoremediation got loaded with toxic heavy metals or persistent chemicals that may pose a risk to wildlife or contaminate a food chain, and (3) accumulation of organic or inorganic compounds may lead to the formation of number of cytotoxic intermediates into plants or animals. Therefore, to prove the applicability of phytoremediation, there is a need to analyse the mass balance and metabolic fate of pollutants in the plant system (Morikawa and Erkin 2003).

Most common air pollutants in the urban environment are sulfur oxides (SOx), nitrogen oxides (NOx), carbon monoxide (CO), suspended particulate matter (SPM) hydrocarbons (HC) and ozone ($O_3$) (D'Amato 1999). Out of which, SPM is of the greatest concern, as it contributes 50% to total air pollution (Fuller 1974) and causes respiratory disorders in human beings on prolonged exposure (Freer-Smith et al. 2004). Joshi (1998) carried out the monitoring of respirable suspended particulate matter (RSPM) and total suspended particulate matter (TSPM) in the core city area of Indore (Madhya Pradesh, India) and found higher RSPM and TSPM levels at about all the selected road intersections as compared to the prescribed standards of CPCB, New Delhi. $NO_2$ and $SO_2$ are the gases that contribute to acidic deposition in terrestrial ecosystem as dry deposited gases or in dissolved form in precipitation (Cox 2003). In aerosol form, they also impact visibility. $NO_2$ is of particular nature, as it is a precursor of the formation of the photochemical oxidants, which directly impact health of human beings. Gaseous pollutants, such as $SO_2$, $NO_2$ and $O_3$, have pernicious effects of varying magnitude on wheat, mustard, mung and palak plants, depending upon individual pollutant concentration, in combination, plant species and seasons (Rhode et al. 2002; Agrawal et al. 2003).

## 2. Phytotoxicity of Air Pollutants

Pollutant emissions are likely to continue to increase in developing countries with their worsening impacts on vegetation (Emberson et al. 2002). The phytotoxicity of major primary pollutants, like $SO_2$, $NO_2$ and SPM, alone or in combination, emitted mainly through the fossil fuel burning, to the plant tissues when adsorbed, absorbed and assimilated by the avenue trees, is discussed as under:

### 2.1 Sulphur Dioxide

About 95% of the $SO_2$ in the atmosphere arises from the anthropogenic sources, of which fossil fuel combustion is responsible for nearly 80%. Out of all the air pollutants in the atmosphere, $SO_2$ is thought to be the unique one, as at low concentration, it is beneficial to plants, while its high concentration causes phyto-toxicity (Winner et al 1985). $SO_2$ has the potential to reduce both yield and nutritional quality of the crop plants (Jäger et al. 1993; Ashmore and Marshall 1999). Chapekar (2000) concluded that $SO_2$ might cause severe damage to the mango crop yield. $SO_2$ inflicts injury to plants both in visible and invisible form. Therefore, there is a possible correlation between the $SO_2$ absorption and injury symptoms. (Furuhawa et al. 1980; Moraes et al. 2002). It has been reported that favourable environmental condition facilitates stomatal opening and then entry of more $SO_2$ deposes plants (Darral 1989). Gaseous $SO_2$ is highly soluble in water and is ionized to form the hydrogen ($H^+$), sulfite ($SO_3^{2-}$) and bisulfite ($HSO_3^-$) ions depending upon the pH (Giordano et al. 2005). The toxic effects of resultant ions depend upon the capacity of the plant tissue to convert them into non-toxic forms. Free radicals, produced during $SO_3^{2-}$ oxidation, have been known to destroy many physiologically important compounds, such as amino acids, plant harmone IAA, chlorophyll and β carotene (Arora et al. 2002). The physiological and biochemical imbalance may lead to shunted plant growth, morphological alterations and yield reductions (Mass et al. 1987; Heck 1989). Laboratory experiments have shown significant reductions in non-structural carbohydrates and proteins and nitrogen contents of seeds fruits and vegetables when exposed to $SO_2$ (Agrawal et al. 1984; Pell et al. 1997).

### 2.2 Nitrogen Dioxide

$NO_2$ is thought to be the less disruptive for plants, however, its prolonged exposure may lead to development of toxicity symptoms (Hicks et al. 2000). $NO_2$, after entering the leaves, dissolves into the intracellular fluid to form the nitrous acid which further dissociates into the toxic nitrites and $H^+$ ions. Normally, majority of nitrite accumulated is converted to ammonia by nitrite reductase and consequently incorporated in the formation of amino acids and proteins, thus alleviating the toxicity on one hand and benefiting the plants on the other hand. Beneficial effects of lower concentration of $NO_2$ on the plant growth and development have been demonstrated in a number of plant species (Sabratnam et al. 1988; Sandhu and Gupta 1989; Pandey and Agrawal 1995). Low $NO_2$ concentrations have been found to induce the chlorophyll production (Prasad and Rao 1979), but at higher concentrations, reduction in photosynthetic pigments was observed (Pandey and Agrawal 1994). The exact mechanism of the $NO_2$-induced chlorophyll reduction, however, is not known. Patterns of carbohydrate allocation are directly influenced by the excess N in tissue by altering the partitioning between root and shoot (Waring 1987; Moraes et al. 2002). The reduced carbohydrate allocation to roots may enhance the plant's susceptibility to pathogens (Matson and Waring 1984).

Combination of primary pollutants, $SO_2$ and $NO_2$, has been chosen for study by several researchers because of their common source i.e. fossil fuel burning. The interactive effects of $SO_2$ and $NO_2$ may be important in two situations; i) in or near urban areas or close to the emission source, where short episodes of high gaseous concentrations may occur and ii) in rural locations situated near urban areas where prolonged exposures to low concentrations of both gases occur. Greater additive effects of $SO_2$ and $NO_2$, in combination, were indicated by a series of long-term fumigations of grasses (whitemore and Mansfield 1983). Agrawal et al. (2003) concluded that $SO_2$ and $NO_2$ in combinations were more detrimental in causing yield losses. Treatment with $NO_2$ alone often stimulated growth, but the same concentration of $NO_2$ combined with $SO_2$ caused greater damage to plants than $SO_2$ alone. Similarly, field-grown soybean showed a significant decrease in the yield with combined pollutant treatment even when there was no effect of either $NO_2$ or $SO_2$ alone (Miller et al. 1980; Amundson and Maclean 1982). The foliar N-level was also induced by the other pollutants ($SO_2$, $O_3$) along with $NO_2$.

## 2.3 Suspended Particulate Matter (SPM)

SPM may cause ultrastructural and physiological disturbances in plants (Dixit 1988; Bacic 1999). Wax crystals, which are the barriers between the plant and the environment (Bermadinger-Stabentheiner 1994) fuse and flatten with age, but in presence of particulate matter, erosion rate of wax structure increases (Huttunen 1994). At worst, the epistomatal chamber of the leaf/needle surface may be plugged totally by the withered and fused wax, inhibiting transpiration which could have far-reaching physiological consequences, such as prevention of gas exchange and photosynthesis (Sauter and Voß 1986; Sauter et al. 1987). Encrustation or dust deposition on leaf cuticle due to particulate penetration into the epicuticular wax may reduce the intensity of incident light, leading to a reduction in net photosynthesis. As the result, unconsumed $CO_2$ into the substomatal cavities might induce stomatal closure. The clogging of stomata by the occluded particulate matter may lead to inhibition or reduction in photosynthetic process of plants through interference with gaseous exchange and also impairing of thermal balance of the leaf.

## 3. Absorption and Assimilation of Pollutants

The adsorption, absorption or assimilation of phytotoxic pollutants by the plants has been described below:

### 3.1 $SO_2$ Absorption and Assimilation by Plants

$SO_2$, from the atmosphere, finds its entry mainly through stomatal openings into the plant tissues. The plants can utilize absorbed $SO_2$ in a reductive sulfur cycle as illustrated in the Scheme 1 for their growth and development.

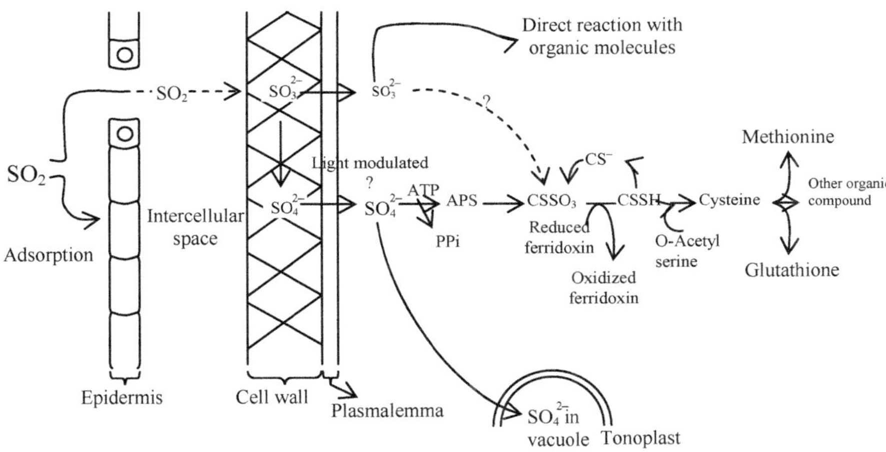


**Scheme 1.** Reduction of sulfur dioxide at cellular level (after Schiff and Hodson 1973)

| | | |
|---|---|---|
| APS | – Adenyl-5-phosphosulphate | Intermediates in the sulphate reduction pathway |
| $CS^-$ | – Carrier protein | |
| $CSSO_3$ | – Carrier protein with bound sulphite | |
| $CSS^-$ | – Carrier protein with bound sulphide | |

In polluted area, the plants are reported to have higher sulfate content (Reddy and Dubey 2000; Krupa and Legge 2001). Manninen and Huttunen (2000) also observed increased $SO_4^{2-}$ concentration in both Scotspine and Norway spruce needles when exposed to $SO_2$ gas. This may be partly due to foliar absorption of $SO_2$ gas and partly due to increased uptake of sulfur and nitrate from the soil (Roberts 1974; Pandey and Rao 1980; Pawar 1981; Kumar and Dubey 1998). In another study, Singh et al. (1995), while working on effects of automobile exhaust pollution on roadside plants, also reported higher sulfate content in the foliar tissues at polluted sites. Murthy et al. (1988) also observed that plants growing in $SO_2$ polluted environments had high sulfate content. The investigation near two thermal power plants also proved the higher accumulation capacity of mango for sulfur rich compounds (Agrawal and Singh 2000). Because of its pan-tropical distribution and high sensitivity, mango tree is considered a promising bio-indicator species in tropical and sub-tropical regions (Chapekar 2000). It has been further explained that nitrogen was also very important, as its balance might help in assimilating excess sulfate accumulated in the tree due to $SO_2$ exposure and plants would finally be left over with a certain degree of sulfate accumulation after its utilization in the plant metabolism. This sulfate is named as "Residual sulfate" (S). The residual sulfate will be toxic enough to induce chlorosis and nutritional imbalance (Keller and Jager 1980; Giodano et al. 2005). Therefore, residual sulfate has been considered as a final toxicant of $SO_2$ metabolism in plants.

### 3.2 $NO_2$ Absorption and Assimilation by Plants

Nitrogen dioxide may reach plant system either directly through its foliar deposition or indirectly through rainwater or soil deposition. The surface deposition of the gas on the foliage is governed by a variety of plant and environmental factors, including pubescence, cuticular reactivity, foliar hydration states and temperature (Taylor et al. 1988). Viskari et al. (2000), working on the effect of $NO_2$ exhaust exposure on ultrastructural changes and stomatal behaviour of spruce seedlings (*Picea abies*), concluded that even relatively short term exposure to realistic concentration of exhaust gas in the atmosphere could induce severe injuries to the needle surface structures as well as ultrastructure at the cellular level. The leaf penetration of $NO_2$ is through open stomata and is governed by various factors including the plant species (Okano et al. 1989), plant age (Srivastava et al. 1975), $NO_2$ concentration and a number of environmental and nutritional factors (Thoene et al. 1991). Direct evidence for the foliar absorption of $NO_2$ has been obtained by using $^{15}N$ isotope of nitrogen. Among the sevenal species examined by Okano et al. (1989), the maximum $NO_2$ absorption was by three cultivars of *Populus* hybrids, i.e. 0.3 mg $N/dm^2/d$. The uptake rate generally increases in a concentration dependent manner, as observed in bean in the concentration range of 100-400 ppb (Roger et al. 1979), in potato in a concentration range of 120-430 ppb (Sinn and Pell 1984) and in sunflower and maize in a concentration range of 200-1000 ppb (Okano et al. 1986). After its entry into leaf, $NO_2$ is rapidly translocated to all other parts of the plants (Yoneyama et al. 1980).

Nitrogen dioxide may also reach plants through rainwater in the form of $HNO_3$ and $HNO_2$. Because of high solubility of this gas in the atmospheric water, its residence time in the atmosphere is only about one week (Derwent and Steward 1973). Soils, specially the alkaline soil, may directly absorb $NO_2$ from the atmosphere (Parther et al. 1973; Ghiorse and Alexander 1976). There is every possibility that the $NO_2$ absorbed by the soil is taken up by the plants through roots.

By using $^{15}NO_2$, Roger et al. (1979) have shown that about 65% of the absorbed $NO_2$ is incorporated into organic nitrogen during a three-hour exposure period in bean. Several factors influence the incorporation of $NO_2$ into organic nitrogen. In most studies, the contribution of $NO_2$ to total organic nitrogen is higher in those plants, which are raised at deficient or sub-optimum levels of soil nitrogen (Srivastava and Ormrod 1984; Rowland et al. 1987). The increase in organic nitrogen content generally increases with the increase in $NO_2$ concentrations up to a certain level (Srivastava and Ormrod 1986). However, at very high concentration, the organic nitrogen content may decrease as well (Sabaratnam and Gupta 1988). In sunflower, 300 ppb $NO_2$ exerted a nutritional effect on plant growing on nitrogen-deficient soil, while 2000 ppb $NO_2$ was phytotoxic at all (0, 5 and 15 mM) nutrient nitrate levels (Okano and Totsuka 1986).

Related enzymic determinations have indicated that assimilation of $NO_2$ follows the established route of inorganic nitrogen assimilation. Nitrate and nitrite produced by the dissolution of $NO_2$ in the cell sap are reduced by the activities of nitrate and nitrite reductases, respectively, to generate ultimately $NH_4^+$, which is then assimilated to glutamate, preferentially through GS-GOGAT pathway. Increase in nitrate reductase (NR) activity following $NO_2$ exposure has been demonstrated by several workers (Murray and Wellburn 1985; Srivastava and Ormrod 1989; Wellburn 1990; Weber et al. 1995;). An increase in nitrite reductase activity (NiR) has been also demonstrated in barley shoots (Rowland et al. 1987), spinach (Yu et al. 1988) and tomato (Murray and Wellburn 1985).

Extrapolating the data from laboratory experiments with $NO_2$ assimilation, Hanson et al. (1989) have calculated annual nitrogen inputs between 0.08 and 1.9 Kg/ha/yr for forest canopies in urban environments at prevailing $NO_2$ levels. Based on annual nitrogen requirement for forest trees, this would supplement about 3% of the tree's annual needs in natural forests and about 10% in the urban forests. The nutritive effect of low levels of $NO_2$ is reflected in increased growth of plant or plant parts in some species (Whitmore and Mansfield 1983; Freer-Smith 1985). The stimulation of growth is more apparent at lower soil nitrogen levels than at high levels (Rowland 1986; Rowland et al. 1987), indicating thereby that $NO_2$ may serve as an alternative source of nutrient nitrogen of plants growing in nitrogen-deficient soils (Singh et al. 2005).

## 4. Phytofiltration of Particulate Matter

The exposed surface of plants, such as leaves and bark, form a natural sink for particulate matter, as they offer site for gravity or wind-blown settlement of particulates (Romney et al. 1963; Dochinger 1980; Lindberg and Lovett 1985; Kovar and Mejstrik 1987; Vora and Bhatnagar 1987; Smardon 1988; Varshney and Mitra 1993). The ability of trees to take up particulates has been characterized through the measurement of relative deposition velocities and trapping and capture efficiencies using wind tunnel and through the measurement of relative deposition velocities using micro-meteorological techniques in the field (Gallagher et al. 1997).

The use of vegetation in filtering out the dust, shoot and particulates from the polluted atmosphere has been accepted in many developed countries. Meetham (1964) noted 27% reduction in dust particles in London (Hyde Park) by a green cover of only one square mile (2.5 $km^2$). The planting of trees and shrubs was recommended as a way to combat dust pollution in Russian cities by Novoderzhikina et al. (1966), who reported a 2-3 times reduction in dust fall by planting a 8 m wide green belt between the roads and buildings. Dochinger (1980), who examined the ability of plants to abate particulate pollutants, reported a reduction of up to 42% in overall dust fall by a canopy of coniferous plants in the urban areas of Ohio, USA. El-Khatib and El-Swaf (2001) reported that foliar SPM contamination was particularly severe at roadsides in urban and in suburban cities of countries in transition. Bach (1972), who studied the dust collecting potential of some plants, observed interesting relationship between certain leaf surface parameters and their dust trapping potential. Freer-Smith et al. (2004) carried out modeling experiments related to the capture of pollution by trees and suggested their usefulness in selection of tree species and planting design to filter particulate pollutants. As a result of their large leaf areas and the turbulent air movements created by their structures, trees take up more pollutants, including particulates (PM10), than shorter vegetation (Fowler et al. 1989; Beckett 2000; Freer-Smith and Taylor 2000a).

In India, some preliminary studies have been carried out by a few workers. Shetey and Chaphekar (1978) used plants for biomonitoring of dust load in different localities of Bombay and based on this study, they have prepared a pollution map of the city. Das et al. (1981), working at Calcutta, made a comparative study of the dust filtering potential of some common Indian avenue trees. Varsney and Mitra (1993), while working at New Delhi, assessed the particulate abatement capacity (PAC) of three commonly grown hedge species, *Bougainvillea spectabilis* Willd, *Duranta plumieri* Jacq. and *Nerium indicum* Mill. The PAC of the species was found in the order of *D. plumieri*>*B. spectabilis*>*N. indicum*. They also concluded that the row of roadside hedges trapped nearly 40% of particulate matter, most of which arises from the traffic movement. In a study carried out at Lucknow city by Khan et al. (1989), dust trapping potential of 10 plant species growing along with road side in the polluted atmosphere was examined, and among all these species, the maximum dust load was found on the leaves of *Nyctanthes arbortristis* L. while the lowest dust load was observed on *Tabernaemontana coronaria* Willd. The dust filtering ability of the plant species was correlated with foliar surface characteristics. The morphological characteristics which alone or in combination play a significant role in the interception of dust load from the ambience are: orientation of leaf on the main axis, size (leaf area in square centimeters) and shape, surface nature (smooth/striate), the presence or absence of trichomes and wax deposition etc. (Verma 2003).

## 5. Plant Tolerance to Ambient Pollutants

Inspite of adverse effects of these pollutants, there are a few reports on pollution tolerant plants (Singh et al. 1995; Varshney and Mitra 1995; Singh and Rao 1983; Nivane et al. 2001), which can adsorb, absorb, detoxify, metabolise and accumulate the pollutants to act as a living filter for the air pollution (Varshney 1985; Singh et al. 1995). Yang et al. (2005), during their study on role of urban forest in air pollution reduction, concluded that there was 1261.4 Tons of pollutant reduction by the forest cover of Beijing. On the basis of foliar biochemical features, Nivane et al. (2001) classified the plants into sensitive and tolerant species. In a study carried out at Warangal city (Andhra Pradesh), Kalyani and Singaracharya (1995) screened out a list of plants on the basis of their tolerance levels. Joshi et al. (1997), during their study on urban air pollution effects on two species of *Cassia*, observed that the *Cassia siamea* was more tolerant to urban pollution than *Cassia fistula*. Rangarajan et al. (1995) discussed the relative tolerance of a few ornamental plant species to automobile exhaust pollution. Farooq et al. (1988) exposed 12 common Indian tree species to varying concentration of $SO_2$ to determine their tolerance level and an order of sensitivity was emerged as *Tamarindis indica* > *Pithecolobium dulce* > *Mangifera indica* > *Ficus rumphie* > *Holoptelea integrifolia* > *Bambax ceiba* > *Ficus bengalensis* > *Azadirachta indica* > *Ficus religiosa* > *Syzygium cuminii* > *Psidium guajava* > *Ficus recemosa*. Agrawal et al. (1991) observed the effect of $SO_2$ pollution on different plants and categorized them into mitigator and bioindicator of $SO_2$ pollution.

However, the ability of each plant species to absorb and adsorb pollutants by their foliar surface varies greatly and depends on several biochemical, physiological and morphological characteristics. As vegetation provides a site for absorption and adsorption of air pollutants, planting of trees along the city roads to mitigate the urban pollution is greatly stressed now-a-days. Therefore, the selection of plant species for the abatement of pollution is to be based on certain scientific criteria, which attribute to the efficacy of plant species in pollution mitigation. Generally, there are four classes of plant species on the basis of their tolerance and absorbance of air pollution.

1. *Less absorption, strong tolerance*: Plants of this type have strongest resistance, so can be grown in the heavily polluted area.
2. *Less absorption, weak tolerance*: Plants of this type have weak resistance and cannot be grown in the polluted area.

3. *More absorption, strong tolerance*: These types of plants are the most suited for use as mitigators of air pollution.
4. *More absorption, weak tolerance*: Such plants have little resistance, so they could not be grown in polluted area for mitigation, but may be used as indicator species.

Therefore, the degree of plant's resistance to air pollutants is determined by the relation or balance between absorption and tolerance. The contributions of absorption and tolerance to plant resistance against air pollutants are related to their concentrations. Most of the plants, when exposed to higher concentration of air pollutants, tend to restrict the entry of pollutants. Mansfield and Freer-Smith (1981) have shown that there is a linear relationship between stomatal conductance and net sulfur flux into leaves. Some plants, like peanut and tomato, close their stomata quickly during $SO_2$ exposure and hence decrease $SO_2$ entry. These plants are resistant to $SO_2$, especially in the case of exposure to high concentration of $SO_2$. Mansfield and Freer-Smith (1984) have found that stomatal closure in silver birch in the presence of $SO_2$, operated as an effective avoidance mechanism. They showed that 0.07 ppm $SO_2$ caused a depression in net photosynthesis amounting to about 19% and a loss of 46% in transpiration. The main influence of 0.07 ppm $SO_2$ was on the stomatal movement with little effect on the internal resistance to $SO_2$ intake. So they suggested that the stomatal closure induced by $SO_2$ might represent a mechanism for avoiding $SO_2$ stress without any major interference with photosynthesis.

## 6. Factors Controlling Plant Tolerance

Morphological characters of plants are very important in determining plant's resistance to air pollution. Characteristics, such as sunken stomata, thick cuticle, small and dense cells and suberised cell wall and so on, are in favour of reducing air pollutant entry into leaves and cells. Pollutants may also cause erosion of epicuticular wax, which protects the entry of pollutants through leaf cuticle by serving as a barrier. Therefore, the structural resistance of cuticular wax to the erosion effect of air pollutants would be an important factor in providing overall resistance of plants to air pollution (Dixit 1988; Huttunen 1994; Bacic 1999).

The cause of air pollutant (particularly $SO_2$ and $NO_2$) injury has been ascribed partially to its acidic property. Cytoplasmic pH, however, is relatively insensitive to moderate external pH changes (Smith and Raven 1979). The apparent resistance is interpreted as a mechanism of internal pH regulation. However, gaseous air pollutants with acidic properties can alter both the intracellular pH and buffering capacity (Nieboer et al. 1984). Buffering capacity of cells is a function of the total buffer concentration, the dissociation constant of week acid and the value of pH. The buffer components include inorganic salts, organic phosphates, proteins, several amino acids, such as histidine, cysteine and cystine etc (Priebe et al. 1978). Skye (1968) investigated the relation between air pollution level and buffering capacity of lichens growing in the vicinity of Stockholm, and found that the species with the lowest buffering capacity for acid substances disappeared first when one proceeded from the normal area towards the "lichen desert" area. Spedding et al. (1980) have shown the importance of pH in modifying the toxicity of $SO_2$. Chinese scientists carried out a great deal of research work to demonstrate the correlation between the resistance of plants to $SO_2$ and cell sap pH. Plants with lower pH were found more susceptible, while those with pH around 7 were found to be more tolerant (Wu et al. 1975). Farooq et al. (1988) have found a strong correlation between the pH values of leaf-extract and tolerance level of plants of Indian origin. The activity of ascorbic acid is also pH controlled, being more at higher and less at lower pH. Hence, the leaf-extract pH, on the higher side, gave tolerance to plants against air pollution (Agrawal 1986). Chlorophyll is the main triggering molecule of green plants and its significance, while assessing resistance of a plant against stress, can never be underestimated (Verma 2003). Depletion in chlorophyll immediately causes a decrease in productivity of plant and subsequently plant exhibits poor vigour. Therefore, plants maintaining their chlorophyll even under polluted environment are said to be tolerant ones (Joshi 1998). Another parameter that may decide the tolerance of plant to air pollution is ascorbic acid content, which is also called vitamin C (Singh et al. 2005). It plays a significant role in light reaction of photosynthesis (Noctor and Foyer 1998), activates defense mechanism (Sakaki et al. 1983; Arora et al. 2002), and under stress condition, it can replace water from light reaction II (Sigurd et al. 1988). The response of various antioxidants to automobile exhaust pollutants was studied and it was concluded that *Amaranthus spinosus* L. and *Cephalandra indica* Naud. were equipped with a very good scavenging system to combat air pollution (Mandal and Mukerji 2001). Due to its multiple role in metabolism and defense of plants, ascorbic acid is used as a very reliable parameter to denote tolerance level of plants against stress, especially the pollution stress (Kumar and Dubey 1998). Pollution often increases their phytotoxicity by impinging a decrease in the ascorbic acid content of plants, which results in increased susceptibility of plants to pollution. While working on $SO_2$ and ascorbic acid interaction, Malviya (1986) has reported that ascorbic acid has the potential of mitigating the $SO_2$-induced injury in plants.

Water is crucial prerequisite for plant life, the shortage of water may cause severe stress to terrestrial plants. A suit of physiological, anatomical, morphological and life history adaptations ensure that plants are able to maintain a water status suitable for survival and reproduction even under stress conditions (Riederer and Schreiber 2001). Under pollution stress, the transpiration rate remains very high, which may lead to desiccation. Therefore, maintenance of relative water content by the plant may decide the relative tolerance of plants to air pollution (Verma 2003).

Nearly all higher plants can use nitrate ($NO_3^-$) as a source of nitrogen (N) and the majority of species are capable of reducing nitrate in both roots and shoots (Runge 1983). Nitrate acquired by the plants, has to be reduced and assimilated ultimately into amino acids (Yamasaki 2005). Rowland et al. (1987), working on the effect of $NO_2$ on nitrate uptake in barley (*Hordeum vulgare* L.), concluded that 300 nl/L $NO_2$ for 9 days was either ineffective or inhibitory on nitrate uptake by roots. Nitrate reductase is the enzyme, which is used by the plants to reduce accumulated nitrate into nitrite. Increase in nitrate reductase (NR) activity, following $NO_2$ exposure, has been demonstrated in several systems, such as barley shoots (Rowland et al. 1987), tomato (Murray and Wellburn 1985), bean (Srivastava and Oremrod 1984), pea (Zeevart 1974), and *Picea rubers* (Murray and Wellburn 1985) leaves and in the needles of *Pinus sylvestris* (Wingsle et al. 1987). Norby et al. (1989), Wellburn (1990) and Srivastava et al. (1995) have also found an increase in NR activity in response to $NO_2$ exposure. The increase in NR activity in $NO_2$-exposed plants is considered to be associated with accumulation of nitrate from the dissolution of $NO_2$ in apoplastic or symplastic water (Srivastava and Ormrod 1984, 1989, Yamasaki 2004). In barley, an increase in enzyme activity has been seen even after termination of exposure with 500 nl/L $NO_2$ for 3 days (Srivastava et al. 1994). Apparently, the barley leaves store nitrate in storage pools during exposure, which is released subsequently to the metabolic pool during post exposure growth of the plant. The reduction of nitrate to nitrite, therefore, is a rate-limiting step (Guerrero et al. 1981), and the activity of this enzyme may be an appropriate marker for determining whether trees assimilate foliar-absorbed $NO_2$. Plants with higher NR activity in $NO_2$ polluted environment are said to be the tolerant ones (Norby 1989).

Today "Wonder Plants" produced by the genetic manipulations are of great demand to attenuate the toxic air pollutants from the atmosphere. A number of such type of plants like *Arabidopsis* (Takahashi et al. 2001), *Pittosporum tobire* (Kondo et al. 2002), *Raphiolepis umbellate* (Irkin et al. 2003) etc. are now available in market to serve as sink for air pollutants. Key enzymes, helpful in metabolizing $NO_2$ into plants, include nitrate reductase (NR), nitrite reductase (NiR), glutamine synthase (GS), while $SO_2$ is metabolized into plant tissues with the help of sulfite oxidase, sulfate oxidase. Therefore, over-expression of genes of these enzymes may play a key role in developing transgenic $NO_2$ or $SO_2$-philic plants.

## 7. A Case Study

In order to evaluate the role of plants in mitigation of air pollutants, an intensive study was carried out in the Lucknow city, an emerging metro of tomorrow. The whole city was divided into three regions i.e. trans-Gomti, central and southern. In each region, different road intersections representing low to high traffic density were selected for the purpose of air monitoring in different seasons and collection of plant samples to test their efficacy in minimization of urban pollution. In order to select the plant species for study, a survey of avenue trees was made at different selected sites in each zone of the city, which was earlier known as the city of gardens. Based on this survey, 15 tree/shrub species *Ficus religiosa* L., *Zizyphus jujuba* Mill., *Bougainvillea spectabilis* Willd., *Saraca indica* L., *Callistemon lanceolatus* L., *Delonix regia* L., *Nerium odorum* Mill., *Syzygium cumini* L., *Cassia siamea* Lam., *Tamarindus indica* L., *Dalbergia sissoo* Roxb., *Azadirachta indica* A. Juss., *Bauhinia variegata* L., *Thevetia nerifolia* L. and *Mangifera indica* L., which were of common occurrence at most of the sites, except a few sites located in the main city, were identified for investigation.

The fresh leaf samples were analysed for chlorophyll, pH, nitrate reductase activity, relative water content and ascorbic acid content, while dry samples were were used for nitrate and sulfate content estimation.

By using the data obtained from detailed biochemical estimations of plant samples, air pollution tolerance index (APTI), sulfur dioxide tolerance index (STI) and nitrogen dioxide tolerance index (NTI) were calculated as per formula given by Singh and Rao (1983) for APTI, Murthy et al., (1988) for STI and Verma (2003) for NTI as follows:

$$\text{APTI} = \frac{A\,(T + P) + R}{10}$$

Where:
A = Foliar ascorbic acid content (mg/g DW)
T = Total chlorophyll content (mg/g FW)
P = pH of leaf extract
R = Relative water content (%)
Total sum was divided by 10 to obtain a manageable figure.

$$\text{STI} = \left[\frac{A\,(T+P) + R}{10 \times S}\right]$$

Where:
A = Ascorbic acid content of leaf (mg/g DW)
T = Total chlorophyll content (mg/g FW)
P = pH of leaf extract
R = Relative water content (%)
S = Sulfate content of leaf (%)

$$\text{NTI} = \left[A\,(T+P) + R\right] \times \left[\frac{NR}{N}\right] \times 10$$

Where:
A = Ascorbic acid content in plant leaves (mg/g DW)

T = Total chlorophyll content (mg/g FW)
P = pH of leaf extract
R = Foliar relative water content (%)
NR = Nitrate reductase activity of leaf (μ moles $NO_2^-$/h/g FW)
N = Nitrate content of leaves (mg/g dw)

Degradation of chlorophyll in the plants under air pollution stress is directly related to the cell pH under two regimes i.e. below and above 3.5 (Rao and LeBlanc 1966; Yu 1980). Therefore, in the physiological sum [A(T+P)+R], addition of chlorophyll and leaf-extract pH values (T+P) were included, as they are strongly correlated with each other and a plant has to maintain their levels to tolerate pollution stress. The multiplication of ascorbic acid content with (T+P) measures the plant's detoxification ability. A correlation of ascorbic acid with the chlorophyll and cell pH is well known. At pH more than 3.5, superoxide radicals are dismutated into hydrogen peroxide ($H_2O_2$) by superoxide dismutase (SOD). Ascorbic acid plays an important role in protection of chlorophyll from $H_2O_2$-induced damage. Therefore, a high level of ascorbic acid is required by a plant to acquire resistance against pollution. As the chlorophyll synthesis is also mediated by ascorbic acid, a reduction in ascorbic acid may hamper the chlorophyll synthesis in green parts of the plants (Saran et al. 1988). Thus, together with cell pH, ascorbic acid also plays a significant role in determining the air pollution sensitivity of plants (Agrawal et al. 1991). Since reducing power of ascorbic acid protects the chloroplasts from pollutants in a pH-dependent manner, introduction of ascorbic acid as a multiplicant in the formula as [A(T+P)], represents the capacity of chloroplast in mitigation of pollutants after their entry inside the plant cells. The introduction of relative water content to [A(T+P)] shows the potential of cell membrane in maintenance of cell integrity under polluted or stressed condition. Thus, under field conditions, a combination of these four parameters represents the best index to determine the tolerance level of plants to air pollution.

To the 'physiological sum' of the formula, "residual sulfate", as discussed earlier, occupies a place on the denominator side to calculate the sulfur dioxide tolerance index (STI) of a particular plant to ameliorate $SO_2$ in the polluted environment. Increased accumulation of sulfate indicates decreased tolerance of a plant species to $SO_2$ pollution.

During this study, the nitrate content and nitrate reductase activity in the plant foliage increased in the presence of $NO_2$ pollution, as reported by Norby (1989). All the plants use nitrate ($NO_3^-$) as a source of nitrogen, which gets reduced to amino acids through GS/GOGAT pathway (Yamasaki 2004). The increase in NR activity, in presence of $NO_2$ pollution from auto-exhaust, facilitates the removal of $NO_2$ from the atmosphere and its metabolism in protein building. Therefore, higher NR activity confers tolerance to plants growing in $NO_2$ polluted environment (Yamada et al. 2001). In view of interdependence of nitrate and NR activity, NR/N was separately multiplied with physiological sum to find out nitrogen dioxide tolerance index (NTI), which would help us to screen out the tolerant species for mitigation of $NO_2$ from the polluted environment.

The yearly mean values of air pollution tolerance index (APTI), $SO_2$ tolerance index (STI) and $NO_2$ tolerance index (NTI) for a particular plant species were calculated by taking the average of its region-wise values to categorize it into different groups depending upon the tolerance level. Based on APTI values, *Bougainvillea spectabilis* was found to be the tolerant one; *A. indica, Z. jujuba, C. lanceolatus, T. nerifolia, C. siamea, M. indica* and *T. indica* plants fell under intermediate category; whereas *D. sissoo, F. religiosa, B. variegata, S. cumini, D. regia, N. odorum* and *S. indica* were placed under the sensitive class. Likewise, on the basis of STI values, *C. lanceolatus, T. nerifolia* and *T. indica* were placed under susceptible class, *B. spectabilis, B. variegata, D. regia* and *M. indica* under sensitive category, *D. sissoo, F. religiosa, S. cumini, C. siamea* and *N. odorum* under moderately tolerant category; while *A. indica, Z. jujuba* and *S. indica* were kept under tolerant category, and based on NTI values, *T. indica* plant was placed under sensitive category; *D. sissoo, F. religiosa, A. indica, Z. jujuba, S. cumini, D. regia* and *S. indica* under moderately tolerant class; whereas *B. spectabilis, B. variegata, C. lanceolatus, T. nerifolia, C. siamea, N. odorum* and *M. indica*, were categorized under tolerant class (Table 1).

**Table 1:** Yearly mean values of APTI, STI and NTI of different plant species growing along roadsides in Lucknow city

| Plants | APTI | STI | NTI |
|---|---|---|---|
| *Dalbergia sissoo* | 10.48 | 17.30 | 27.04 |
| *Ficus religiosa* | 10.87 | 28.52 | 19.49 |
| *Azadirachta indica* | 13.39 | 36.09 | 20.74 |
| *Zizyphus jujuba* | 15.02 | 85.60 | 26.35 |
| *Bougainvillea spectabilis* | 17.93 | 14.76 | 50.72 |
| *Bouhinia variegata* | 8.96 | 12.97 | 31.24 |
| *Syzygium cumini* | 9.87 | 25.88 | 26.28 |
| *Callistemon lanceolatus* | 14.03 | 8.04 | 35.30 |
| *Delonix regia* | 8.72 | 14.81 | 18.80 |
| *Thevetia nerifolia* | 11.43 | 6.54 | 66.78 |
| *Cassia siamea* | 11.46 | 19.18 | 118.00 |
| *Nerium odorum* | 10.64 | 21.65 | 51.48 |
| *Mangifera indica* | 11.92 | 12.08 | 124.19 |
| *Saraca indica* | 9.62 | 54.34 | 26.44 |
| *Tamarindus indica* | 11.12 | 8.82 | 12.39 |

APTI $\leq$ 11, sensitive; 11-16, intermediate; $\geq$ 16, tolerant

STI $\leq$ 12, susceptible; 12-16, sensitive; 16-30, moderately tolerant; $\geq$ 30, tolerant

NT I $\leq$ 12, susceptible; 12-16, sensitive; 16-30, moderately tolerant; $\geq$ 30, tolerant

The APTI, STI and NTI values, obtained for different plant species growing along the roadside of the Lucknow city, were compared with the sensitivity or tolerance of the same plant species, as determined under laboratory and field conditions (Table 2). It was found that plants with high index value were generally tolerant to air pollutants and *vice versa*. Although our results were, by and large, in consistency with the findings of the other workers, but some exceptional cases were also observed, where our observations did not match with the results of earlier workers. The disagreement might be attributed to the

**Table 2:** Comparison of plant responses based on APTI, STI and NTI values with experimental and field observations of other workers

| Plant species | Present study | | | Experimental/field observations | References |
|---|---|---|---|---|---|
| | Evaluated response | | | Response | |
| | APTI | STI | NTI | | |
| *Dalbergia sissoo* | S | MT | MT | S | Rao etal. (1983) |
| *Ficus religiosa* | S | MT | MT | T | NBRI Annual Report (1983) |
| *Azadirachta indica* | I | T | MT | T | Yunus and Ahmad (1979) |
| *Zizyphus jujuba* | I | T | MT | T | Pandey (1983) |
| *Bougainvillea spectabilis* | T | S | T | T | Varshney (1985) |
| *Bouhinia variegata* | S | S | T | MT | NBRI Annual Report (1983) |
| *Syzygium cumini* | S | MT | MT | S | Rao etal. (1983) |
| *Callistemon lanceolatus* | MT | Sus | T | MT | Singh et al. (1995) |
| *Delonix regia* | S | S | MT | S | Varshney (1985) |
| *Thevetia nerifolia* | MT | Sus | T | - | - |
| *Cassia siamea* | MT | MT | T | T | Joshi et al. (1997) |
| *Nerium odorum* | S | MT | T | T | Chaphekar (1972) |
| *Mangifera indica* | MT | S | T | MT<br>S | Farooq et al. (1988), Pandey (1983) |
| *Saraca indica* | S | T | MT | MT<br>T | Prasad et al (1979), NBRI Annual Report (1983) |
| *Tamarindus indica* | I | Sus | S | S | Farooq and Beg (1980) |

T=tolerant, MT=moderately tolerant, I=intermediate, S=Sensitive, Sus=susceptible differences in methodologies and/or criteria used for screening sensitivity or tolerance level of plants. In laboratory conditions, plants are generally exposed to one or two pollutants, while under field conditions, air-shed is polluted with a number of pollutants, which could modulate the response of plants in a different way. In addition, the tolerance level of plants, exposed to high concentration of pollutants for a short duration, may differ from that of its exposure at a lower concentration for a longer duration.

## 8. Conclusion

Thus, there are several plant, edaphic and environmental factors which regulate plant resistance to air pollution. Suitability of plants for the pollution abatement depends on how fast they are able to absorb pollutants from the atmosphere and metabolise or detoxify them at cellular levels. However, the plants with pollutant avoidance mechanism may not be recommended for mitigating air pollution level in urban or industrial areas. This makes crystal clear that effectiveness of avenue trees in urban areas, and greenbelts in and around industrial units largely depends on the selection of suitable plant species and its number.

## References

Agrawal M, Singh J (2000) Impact of coal power plant emissions on foliar elemental concentration in plants in a law rainfall tropical region. Environ Monitor Assess 60:261-282

Agarwal M, Singh SK, Singh J, Rao DN (1991) Biomonitoring of air pollution around industrial sites. J Environ Biol 211-222

Agrawal M, Singh B, Rajput M, Marshall F, Bell JNB (2003) Effect of air pollution of peri-urban agriculture, a case study. Environ Pollut 126:323-329

Agarwal SK (1986) A new distributional function of foliar phenol concentration in the evaluation of plants for their air pollution tolerance index. Acta Ecol 8:2

Ahmad KJ, Yunus M, Singh SN, Srivastava K, Singh N, Kulshreshtha K (1988) Survey of Indian plants in relation to atmospheric pollutants: A research project. In: Perspectives in Environ Bot 2:283-306, Today and Tomorrow's Printers and Publishers, New Delhi 110 005 (India)

Arora A, Sairam RK, Srivastava GC (2002) Oxidative stress and antioxidative system in plants. Curr Sci 82(10):1227-1238

Ashmore MR, Marshall FM (1999) ozone impacts on agriculture; an issue of global concern. Adv Bot Res 29:32-49

Bach W (1972) Atmospheric pollution. New York. Mc Graw Hill

Bacic T, Lynch AH, Cutler D (1999) Reactions of cement factory dust contamination by *Pinus halepensis* needles. Environ Exp Bot 41:155-166

Bermadinger-Stabentheiner E (1994) Problems in interpreting affects of air pollutants on spruce epicuticular wax. In: Perey KE, Cape JN, Jugels R, Simpson CJ (eds) Air pollutants and the leaf cuticle (NATO ASI Series Vol. G36). Springer-Verlag Berlin, pp 321-327

Chapekar SB (2000) Phytomonitoring in industrial areas. In: Agrawal SB, Agarwal M (eds) Environmental Pollution and Plant Responses. Lewis Publishers, Boca Raton, USA. pp-329-342

Cox RM (2003) The use of passive sampling to monitor forest exposure to $O_3$, $NO_2$ and $SO_2$, a review and some case studies. Environ Pollut 126:301-311

Darral NM (1989) The effects of air pollutants on physiological processes in plants. Plant Cell Environ 12:1-30

Das TM, Bhaumik A, Chakravarty A (1981) Trees as dust filters. Science Today 15(12):19-21

Derwent RC, Stewart HNM (1973) Atmos Environ 7:385

Dixit AB (1988) Effects of particulate pollutants on plants at ultrastructural and cellular levels. Ann Bot 62:643-651

Dochinger LS (1980) Interception of air borne particulates by tree planting. J Environ Qual 9:265-268

Farooq M, Saxena RP, Beg MU (1988) Sulfur dioxide resistance of Indian trees. Water, Air, Soil Pollut 40:307-316

Freer-Smith PH (1985) The influence of $SO_2$ and $NO_2$ on the growth development and gas exchange of *Betula pendula* Roth. New Phytol 99:417-430

Fuller EC (1974) Chemistry and man's environment. Houghton Mifflin Company, Boston, pp 502

Ghiorse WC, Alexander M (1976) J Environ Qual 5:227

Giordano M, Norici A, Hell R (2005) Sulfur and phytoplankton: acquisition, metabolism and impact on the environment. New Phytol. 166:371-382

Guerrero MG, Vega JM, Losada M (1981) The assimilatory nitrate reducing system and its regulation. Ann Rev Plant Physiol 32:169-204

Hanson PJ, Lott K, Taylor GE Jr, Gunderson CA, Lindberg SE, Ross-Toad BM (1989) Atmos Environ 23:1783

Huttunen S (1994) Effects of air pollutants of epicuticlular wax structure. In: Perey KE, Cape JN, Jagels R, Simpson CJ (eds) Air Pollutants and leaf cuticle (NATO ASI Series Vol. G36) Springer-Verlag, Verlag, pp 81-96

Ito O, Yoneyama T, Kumazawa K (1978) Amino acid metabolism in plant leaf, IV: The effect of light on ammonium assimilation and glutamine metabolism in the cells isolated from Spinach leaves. Plant and Cell Physiol 19:1109-19

Jäger HJ, Unsworth MH, DeTimmerman L, Mathy P (Eds) (1993) Effects of air pollution on agricultural crops in Europe. Air Pollution research report 46, Commission of the European Communities, Brussels, Belgium

Joshi G (1998) Ambient air quality at road side of an urban area with special reference to respirable dust and total suspended particulate matter. Pollut Res 17(1):79-81

Joshi OP, Wagela DK, Pawar K (1997) Urban air pollution effects on two species on *Cassia*. Poll Res 16(1):1-3

Kaji M, Yoneyama T, Totsuka T, Iwaki (1980) Absorption of atmospheric $NO_2$ by plants and soils VI. Transformation of $NO_2$ absorbed in the leaves and transfer of nitrogen through the plants. Res Rep Natl Inst Environ Stud Japan 11:51-58

Kalyani Y, Singaracharya MA (1995) Biomonitoring of air pollution in Warangal city, Andhra Pradesh. Acta Botanica Indica 23:21-24

Keller T, Jager HJ (1980) Der eunflurs boden burtiga sulfation auf den schwefchalf sulfur dioxide-begaster. Assimilation usorgane von wald aumarten. Augewante. Botanik. 54:74-89

Khan AM, Pandey V, Yunus M, Ahmad KJ (1989) Plants as dust scavengers. A case study. The Indian Foresters 115(9):670-672

Krupa SV, Legge AH (1999) Foliar injury symptoms of Saskatoon service berry (*Amelanchier alnifolia* Nutt.) as a biological indicator of ambient sulfur dioxide exposures. Environ Pollut 106:449-454

Kumar GS, Dubey PS (1998) Differential response and detoxifying mechanism of *Cassia siamea* Lam. and *Dalbergia sissoo* Roxb. of different ages to $SO_2$ treatment. J Environ Biol 9(3):243-249

Lewis S (1976) Vitamin C: In Molecular Biology and Medicinal Potential, London, Academic Press

Malviya NK (1986) Sulfur dioxide and ascorbic acid interaction bioassay, Dissertation Thesis, School of Studies in Botany, Vikram University, Ujjain (India)

Mandal M, Mukherji S (2001) A study on the activities of few free radicals scavenging enzymes present in fine road side plants. J Environ Biol 22(4):301-305

Manninen S, Huttunen S (2000) Response of needle sulfur and nitrogen concentration of Scots pine versus Norway spruce to $SO_2$ and $NO_2$. Environ Pollut 107:421-436

Mansfield TA, Freer-Smith PH (1981) Effects of urban air pollution on plant growth. Biol Rev 56:343-368

Mansfield TA, Freer-Smith PH (1984) The role of stomata in resistance mechanisms. In: Kozoil MJ, Whatly FR (eds) Gaseous air pollutants and plant metabolism 131-146, London Butter Worths

Matsushima J (1972) Bull Fac Agric Mic Univ Tsu 44:131

Meetham AR (Ed) (1964) Atmospheric Pollution: Its Origin and Prevention. Oxford: Pergamon Press

Murray AJS, Wellburn AR (1985) Differences in nitrogen metabolism between cultivars of tomato and pepper during exposure to glass-house atmosphere containing oxides of nitrogen. Environ Pollut 39:303-316

Murthy MSH, Raza SH, Adeel A (1988) A new method in evaluation of $SO_2$ tolerance of certain trees. Air Pollut. & Forest Decline. In: Bucher JB, Bucher I, Wall N, (eds) Proc 14th Int. Meeting for specialists in air pollution effects on Forest. Ecosystem, IUFRO $P_2O_5$, Interlaken, Switzerland, Oct. 2-8, 1988. Birmensdorf, 1989, pp 486-488

Nieboer E, Mac Farlane LD, Richardson DHS (1984) Modification of plant cell buffering capacities by gaseous air pollutants. In: Koziol MJ, Whatley FR (eds) Gaseous air pollutants and plant metabolism 313-330, London, Butter Worths

Nivane SY, Chaudhari PR, Gajghate DG, Tarar JL (2001) Foliar biochemical features of plants as indicators of air pollution. Bull Environ Contam Toxicol 67:133-140

Noctor G, Foyer CH (1998) Ascorbate and glutathione: keeping active oxygen under control. Ann Rev Plant Physiol Plant Mol Biol 49:249-279

Norby RJ (1989) Foliar nitrate reductase: A marker for assimilation of atmospheric nitrogen oxides. In: Grosblatt N (ed) Biological Markers of Air-Pollution Stress and Damage in Forests. National Academy Press, Washington DC, pp 245-250

Novoderzhikina Yu G, Andrianova LA, Zheldakkova GG (1966) Effect of plantings on the sanitary and hygienic conditions of densely polluted settlement. In: Nuttonson M (ed) AICE Survey of USSR, Vol. 2, silver spring Md. American Institute of crop ecology, pp 25-31

Okano K, Totsuka T (1986) Absorption of nitrogen dioxide by sunflower plants grown at various levels of nitrate. New Phytol 102:551-556

Okano K, Fukuzawa T, Tazaki T, Totsuka T (1986) $^{15}N$ dilution method for estimating the absorption of atmospheric $NO_2$ by plants. New Phytol 102:73-84

Okano K, Machida T, Totsuka T (1989) Differences in ability of NO2 absorption in various broad leaved tree species. Environ Pollut 58:1-18

Pandey SN (1983) Impact of thermal power emission on vegetation and soil. Water, Air, Soil Pollut 19:87-100

Pandey SN, Rao DN (1980) Effect of coal smoke sulfur dioxide pollution on the accumulation of certain minerals and chlorophyll content of wheat plant. Tropical Ecol 19(2):155-162

Parther RJ, Miyamoto S, Bohn HL (1973) Soil Sci Soc Am J 37:914

Pawar K (1981) Pollution studies in Nagda area due to Birla Industrial Discharges, Ph.D. Thesis, School of Studies in Botany, Vikram University, Ujjain, M.P. (India)

Priebe A, Klein H, Jager HJ (1978) Role of polyamines in $SO_2$-polluted pea plants. J Exp Bot 29:1045-1050

Rangarajan TN, Arjunan MC, Ponnammal NR (1995) Effect of automobile pollution on few ornamental plants. Ecol Env Cons 1:1-4

Reddy BM, Dubey PS (2000) Scavenging potential of trees to $SO_2$ and $NO_2$ under experimental condition. Intl J Ecol Environ Sci 26:99-106

Riederer, Lukas-Schreiber (2001) Protecting against water loss: analysis of the barrier properties of plant cuticle. J Exp Bot (Special issue on plants under stress). 52 (363):2023-2032

Roberts BR (1974) Folia sorption of atmospheric $SO_2$ by higher plants. Environ Pollut 7:133-140

Rodhe H, Dentener F, Schulz M (2002) The global distribution of acidifying wet deposition. Environ Sci Technol 36:4382-4388

Roger HH, Campbell JC, Volk RJ (1979) Nitrogen-15 dioxide uptake and incorporation by *Phaseolus vulgaris* (L.). Science 206:333-335

Romney EM, Lindberg RG, Hawthorne HA Bystrom BG, Larson KH (1963) Contamination of plant foliage with radioactive fallout. Ecology 44:343-349

Rowland AJ (1986) Plant Soil 91:53

Rowland AJ, Drew MC, Wellburn AR (1987) Foliar entry and incorporation of atmospheric nitrogen dioxide into barley plants of different nitrogen status. New Phytol 107:357-371

Runge M (1983) Physiology and ecology of nitrogen nutrition. In: Lange OL, Nobel PS, Osmond CB, Ziegler H (eds) Encyclopedia of Plant Physiology, New Series, vol. 12C. Springer-Verlag, Berlin, pp 163-200

Sabaratnam S, Gupta G (1988) Effect of $NO_2$ on leaf chlorophyll and nitrogen content of soybean. Environ Pollut 51:113-120

Sauter JJ, Kammerbauer H, Panber L, Hock B (1987) Evidence for the accelerated micro morphological degradation of *Epistomatal wax* in Norway spruce by motor vehicle emission. European J Forest Pathol 17:444-448

Sauter JJ, Voß JU (1986) SEM-observations on the structural degradation of epistomatal waxes in *Picea abies* (L.) Karst. – and its possible role in the "Fichtensterben". European J Forest Pathol 16:408-423

Shetey RP, Chephekar SB (1978) Some estimations dust fall in the city of Bombay, using plants. Proc. Seminar on Recent Advances in Ecology, New Delhi, Today and Tomorrow, pp 61-70

Sigurd S-H, Nancy MD, Lutz WB, Wellburn AR (1988) Air pollution and plant metabolism. Elsevier Applied Science, London and New York

Singh N, Yunus M, Srivastava K, Singh SN, Pandey V, Mishra J, Ahmad KJ (1995) Monitoring of auto exhaust pollution by road side plants. Environ Monitor Assess (USA) 34:13-25

Singh SK, Rao DN (1983) Evaluation of plants for their tolerance to air pollution. In: Mathur HB, Pal K (eds) Proc Sym Air Pollut Control, Delhi IIT, pp 218-224

Sinn JP, Pell EJ (1984) Uptake rate of nitrogen dioxide by potato plants. J Air Pollut Control Assoc 34:668

Smardon RC (1988) Perception and aesthetics of urban environment: review of the role of vegetation. Land scape urban Plann 15:85-106

Smith PA, Raven JA (1979) Intercellular pH and its regulation. Ann Rev Plant Physiol 30:289-311

Spedding DJ Ziegler I, Hampp R, Ziegler H (1980) Effect of pH on the uptake of $S^{35}$ sulfur from sulfate, sulfite and sulfide by isolated spinach chlroplast. Z Pflanzenphysiol 96:351-364

Srivastava HS, Oremrod DD (1984) Effects of $NO_2$ and nitrate nutrition on growth and nitrate assimilation in bean leaves. Plant Phsyiol 76:418-423

Srivastava HS, Ormrod DP (1989) Nitrogen dioxide and nitrate nutrition effects of nitrate reductase activity and nitrate content of bean leaves. Environ Exp Bot 29:433-439

Srivastava HS, Jolliffe PA, Runeckles VC (1975) Inhibition of gas exchange in bean leaves by $NO_2$. Can J Bot 53:466-474

Srivastava HS, Ormrod DP, Hale BA (1994) Responses of greening bean seedling leaves to nitrogen dioxide and nitrate supply. Environ Pollut 86:37-42

Srivastava HS, Ormrod DP, Hale BA (1995) Assimilation of nitrogen dioxide by plants and its effects or nitrogen metabolism. In: Nitrogen nutrition in Higher plants (Srivastava HS, Singh RP, Eds), pp 417-430. Associated Publishers Co., New Delhi, India

Takeuchi Y, Nihira J, Kondo N, Tezuka T (1985) Plant Cell Physiol. 26: 1027

Taylor GE Jr, Hanson PJ, Baldochi DD (1988) In: Heck WW, Taylor OC, Tingey DT (eds) Assessment of crop loss by air pollutants, Elsevier Pub. New York, pp 227

Thoene B, Schrader P, Papen H, Egger A, Rennenberg H (1991) New Phytol 117:575

Varshney CK (1985) Effects of sulphur dioxide on plants Final Technical Report, DoEn, Ministry of Environment and Forest, Govt. of India

Varshney CK, Mitra I (1993) Importance of hedges in improving urban air quality. Landscape and Urban Planning 25:75-83

Varshney CK, Mitra I (1995) Response of tropical stress of sulphur dioxide stress and recovery. Intl J Environ Studies 49:13-21.

Verma A (2003) Attenuation of automobile generated air pollution by higher plants. Ph.D. Thesis, University of Lucknow, Lucknow, India

Viskari E-L, Kossi S, Holopainen JK (2000) Norway spruce and spruce shoot aphid as indicators of traffic pollution. Environ Pollut 107:305-314

Vora AB, Bhatnagar AR (1987) Comparative study of dust fall on the leaves in high pollution and low pollution area in Ahmedabad. V. Caused foliar injury. J Environ Biol 8(4):339-346

Wellburn AR (1990) Why are atmospheric oxides of nitrogen usually phytotoxic and not alternative fertilizers. New Phytol 115:395-429

Wellburn AR, Capron TM, Chan HS, Horsman DC (1980) In: Mansfield TA (ed) Effects of Air Pollutants on Plants, Cambridge University Press, Cambridge, pp 105

Whitmore ME, Mansfield TA (1983) Environ Pollut 31:217

Wingsle GO, Nasholm T, Lundmark T, Erricson A (1987) Induction of nitrate reductase in Scots pine by NOx and $NO_3^-$. Physiol Plant 70:399-403

Wu Y, Hau J, Li W, Fu L (2002) Calculating emissions of exhaust particulate matter from motor vehicles with PART 5 model, Huan Jing Ke Xue 23(1):6-10

Yamasaki H (2004) The NO world for plants: Achieving balance in an open system. Plant Cell Environ 28:78-84

Yamasaki H (2005) Nitric oxide research in plant biology: its past and future. In: Magalhaes JR, Singh RP, Passos LP (eds) Nitric oxide signaling in higher plants, Stadium press, Houston (in press)

Yoneyama T, Arai K, Totsuka I (1980) Plant Cell Physiol 21:1367

Yu S-W, Li L, Shimazaki KI (1988) Environ Pollut 55:1

Zeevart AJ (1974) Induction of nitrate reductase by $NO_2$. Acta Bot Neert 23:345-346

Zeevart AJ (1976) Some effects of fumigating plants for short periods with $NO_2$. Environ Pollut 11:97-108

14

# Phytoremediation: Role of Plants in Contaminated Site Management

Rajiv K. Sinha[1], Sunil Herat[1] and P.K. Tandon[2]

[1]School of Environmental Engineering, Griffith University, Nathan Campus, Brisbane, Queensland – 4111, AUSTRALIA; [2]Department of Botany, Lucknow University, Lucknow, INDIA

## 1. Introduction

Bioengineering is a new branch of civil engineering which integrates live materials, mainly plants and microorganisms, to address the problems of environmental management and sustainable development. The technology originated in Germany in the 1930s, but gained importance in the 1980s, when researches in environmental biotechnology discovered the environmental virtues of some specially adapted plants and microbes. Bioengineering is the 'green' or 'soft' cheaper alternative to the 'hard' and costly civil engineering works for environmental reconstruction.

Phytoremediation (Greek: *phyton* = plant; Latin: *remediare* = remedy) is emerging 'green bioengineering technology' that uses plants to remediate environmental problems. A number of green plants- trees, herbs, grasses and shrubs, both aquatic and terrestrial, have been discovered to have been endowed with the wonderful properties of environmental restoration, such as decontamination of polluted soil and water, stabilization of engineered slopes and embankments on highways, railways, bridges and dams, and prevention of soil erosion. They are aesthetically pleasing, passive, solar-energy driven and pollution abating nature's (green) technology meeting the same objectives of fossil-fuel driven and polluting conventional technology. They thrive in very harsh environmental conditions of soil and water; absorb, tolerate, transfer, assimilate, degrade and stabilise highly toxic materials (heavy metals and organics such as solvents, crude oil, pesticides, explosives and polyaromatic hydrocarbons) from the polluted soil and water; and firmly holds the soil in place by their extensive root network to prevent any erosion. The plants act both as 'accumulators', and 'excluders'. Accumulators survive despite concentrating contaminants in their aerial tissues. They biodegrade or biotransform the contaminants into inert forms in their tissues. The excluders restrict contaminant uptake into their biomass. The plant biomass eventually becomes valuable biological source for the community or for the plant-based industries.

## 2. Plant Species Involved in Phytoremediation

Several plants are being identified and trialed to be used in phytoremediation task. The most versatile plant species, both terrestrial and aquatic that have been identified after rigorous laboratory and field experiments are as listed below:

| | | | |
|---|---|---|---|
| 1. | Vetiver grass (*Vetiveria zizanioides*); | 2. | Barmuda grass (*Cynodon dactylon*); |
| 3. | Bahia grass (*Paspalaum notatum*); | 4. | Sunflower oil plant (*Helianthus annus*); |
| 5. | Poplar tree (*Populus spp.*); | 6. | Mustard oil plant (*Brassica juncea*); |
| 7. | Periwinkle (*Catheranthus roseus*); | 8. | Cumbungi (*Typha angustifolia*); |
| 9. | Water hyacinth (*Eichhornia cressipes*); | 10. | Duck Weed (*Lemna minor*); |
| 11. | Red Mulberry (*Morus rubra*); | 12. | Kochia (*Kochia scoparia*); |
| 13. | Foxtail barley (*Hordeum jubatum*); | 14. | Switch grass (*Panicum variegatum*); |
| 15. | Musk thistle (*Carduus nutans*); | 16. | White raddish (*Raphanus sativus*); |
| 17. | Catnip (*Nepeta cataria*); | 18. | Big bluestem (*Andropogan gerardii*); |
| 19. | Indian grass (*Sorghastrum nutans*); | 20. | Canada wild rye (*Elymus candensis*); |
| 21. | Nightshade (*Solanum nigrum*); | 22. | Wheat grass (*Agropyron cristatum*); |
| 23. | Alfa-alfa (*Medicago sativa*); | 24. | Tall Fescue (*Festuca anundinacea*); |
| 25. | Lambsquarters (*Chenopodium berlandieri*); | 26. | Reed grass (*Phragmites australis*); |
| 27. | Tall wheat grass (*Thynopyron elongatum*); | 28. | Rhodes grass (*Chloris guyana*); |
| 29. | Flatpea (*Lathyrus sylvestris*); | 30. | Carrot (*Daucus carota*); |

Other species are *Elodea canadensis, Ceratophyllum demersum, Potamogeton spp., Myriophyllum spp, Spartina alterniflora, Pinus sylvestris, Poa alpine, Bouteloua gracilis* (Rice et al. 1995; Watanabe 1997). A number of them are still wild, while others have been domesticated for their food value. They are highly salt and toxicity tolerant, have extensive root binding system and were tried in the rehabilitation works. A number of them readily absorb, volatilise and / or metabolise compounds such as tetrachloroethane, trichloroethylene, metachlor, atrazine, nitrotoluenes, anilines, dioxins and various petroleum hydrocarbons. Ideal species for the job are members of the grass family Gramineae and Cyperaceae and the members of families

Brassicaceae (in particular the genera Brassica, Alyssum and Thalapsi), and Salicaceae (in particular willow and poplar trees). Grasses such as the vetiver, clover and rye grass, Bermuda grass, tall fescue etc. have been particularly effective in the remediation of soils contaminated by heavy metals and crude oil (Kim 1996).

Large scale plantation of sunflower plants (*Helianthus annus*) have been made around Chernobyl (erstwhile USSR), where nuclear disaster in 1985 spewed vast amount of radioactive materials into the environment. The land and soil in the area was badly contaminated. Sunflower is reported to absorb radionuclides from soil and decontaminate it. This phytoremediation technology costs $ 2 per hectare for decontamination of soil which might have costed million of dollars by other means.

Duckweeds can 'absorb' and 'adsorb' all the dissolved gases and substances, including the heavy metals, from the wastewater. Within 2 to 3 weeks, the quality of wastewater improves significantly in terms of BOD and DO values, heavy metals and suspended solids and becomes useful for irrigation, industrial uses and aquaculture. It purifies the wastewater rich in phosphorus, nitrate and potassium until the water is crystal clear with phosphorus and nitrogen contents coming down to 0.5 mg/litre within 20 days. Water hyacinths harbor a large number of microorganisms in symbiotic relationships on their roots which feed off upon minerals and organic chemicals (contaminants) from the effluents. Water hyacinth can remove heavy metals by 20-100%. In just 24 hours, the weed can extract more than 75% of lead from contaminated water. It also absorbs cadmium, nickel, chromium, zinc, copper, iron and pesticides and several toxic chemicals from the sewage. In just 7 days of exposure, it can lower BOD by 97% and remove over 90% of nitrates and phosphates. It can also remove radioactive substances.

The current paper focuses mainly about the phytoremediation techniques of the vetiver grass (*Vetiveria zizanioides*).

## 3. Phytoremediation: The Biophysical and Biochemical Mechanisms

Remediation of organic and inorganic contaminants involves either physical removal of compounds or their bioconversion (biodegradation or biotransformation) into biologically inert forms. The conversion of metals into inert forms can be enhanced by raising the pH (e.g. through liming), or by addition of organic matter (e.g. sewage sludge, compost etc.), inorganic anions (e.g. phosphates) and metallic oxides and hydroxides (e.g. iron oxides). The plants themselves can play a role here by altering soil redox conditions and releasing anions and /or lignins. Phytoremediation technology works mainly through:

### 3.1 Phytoextraction and Phytoaccumulation

Plant roots uptake (extract) metal contaminants from the soil, polluted and the wastewater, and accumulate them in their roots. Plant roots absorb both organics and inorganics. The bioavailability of a given compound depends upon the lipophilicity and the soil or water conditions e.g. pH and clay content. Considerable amount of the contaminants may be translocated above through the xylem and accumulated in the shoots and leaves. The roots, shoots and leaves are collected (harvested) and incinerated to decompose the contaminants.

### 3.2 Phytostabilisation

Certain plant species immobilise contaminants in the soil and groundwater through absorption by and adsorption on to roots or precipitation within the root zone (rhizosphere).

### 3.3 Phytodegradation

Some plant species breakdown the contaminants after absorbing them. This they do through enzyme-catalyzed metabolic process within their root or shoot cells. Others breakdown the contaminants in the substrate itself by secreting enzymes and chemical compounds. The enzymes secreted are usually dehydrogenases, oxygenases and reductases. The biodegraded constituents are converted into insoluble and inert materials that are stored in the lignin or released as exudates (Watanabe 1997). Some plants biodegrade contaminants with the aid of microbes which live in symbiotic association on their roots.

### 3.4 Phytotransformation

Several inorganic and organic contaminants once absorbed inside the root, may become biochemically bound to cellular tissues (biotransformed), in the forms that are biologically inert or less active (Watanabe 1997).

### 3.5 Phytovolatilisation

Plants absorb and transpire the impurities from soil and water through their aerial organs. Some contaminants like selenium (Se), mercury (Hg) and volatile organic compounds (VOCs), can be released through the leaves into the atmosphere (Cunningham and Ow 1996).

### 3.6 Rhizofiltration

It is based on a combination of principle of phytoextraction and phytostabilization specially suited to remove metals and radionuclides from polluted water. Contaminants are absorbed and concentrated by plant roots, then precipitated as their carbonates and phosphates (Salt et al. 1995). Hydroponically grown terrestrial plants like vetiver (*Vetiveria zizanioides*) and sunflower (*Helianthus annuus*) which have large root systems and greater biomass, are specially suitable. Species that do not readily transfer contaminants from the roots to stem are preferred, since the accumulated metals and radionuclides can be removed by simply harvesting the roots. Rhizofiltration works in the effecient removal of organics such as tetrachloroethane, trichloroethylene, metachlor, atrazine, nitrotoluenesanilines, dioxins and various petroleum hydrocarbons (Rice et al. 1997).

### 3.7 Plant - Assisted Microbial Degradation

Certain plant roots release substances that are nutrients for microorganisms like bacteria and fungi. This results in increased biological activity of the microbes in the area immediately surrounding the root zone (rhizosphere). By encouraging a microbiologically active rhizosphere, the plants facilitate accelerated digestion (biodegradation) of wide variety of organic contaminants in the upper soil layers and / or wastewater / polluted water (Anderson et al. 1993). Many organic compounds are degraded by microorganisms located in the rhizospheres (on the roots) of plants. The enhanced rhizosphere biodegradation results from the ability of certain plants to provide favourable habitats for soil microbes to act (Cunningham and Ow 1996). Mackova et al. (1997) reported effective degradation of PCBs (Polychlorinated Biphenyls) by cells of *Solanum nigrum* that were infected with bacterial strains of *Agrobacterium tumefaciens* and *A. rhizogenes.* The water hyacinths (*Eichhornia cressipes*) works on the same biological principle. It harbours several microbes in its root zone which perform the task of biodegradation of heavy metals in polluted water and also helps in absorption and adsorption of chemical impurities.

Certain metals, such as mercury (Hg) and selenium (Se), can be phytovolatalised usually through plant-microbe interactions (Cunningham and Ow 1996). Genes for synthesizing the enzyme 'bacterial mercuric ion reductase' has been engineered into *Arabidopsis thaliana* and the resulting transformant transgenic plant is capable of tolerating and volatalising mercuric ions. The toxic cation is absorbed by the root and reduced to volatile Hg (O) by the introduced mercuric ion reductase (Rugh et al. 1996).

## 4. The Vetiver Grass Technology (VGT)

Worldwide use of vetiver grass, for soil and water conservation and to protect the farmlands from soil erosion, started in the 1980s following its promotion by the $US 100 million World Bank Watershed Management Project in India (Sinha 1996).

Major research works are being done in India, China, Thailand and Australia on this grass for its uses in environmental management. A global network with 4000 members over 100 countries, and a regional network have been established in Latin America, Europe, China, the Pacific Rim and the Oceania. U.S., France, Italy, Spain, Soviet Russia, China, India, Sri Lanka, Malaysia, Fiji and Thailand are using the grass extensively for protection of their lands and water bodies (Greenfield 1989).

Australia has also taken great initiative towards the use of this wonder grass for various environmental purposes including decontamination and rehabilitation of contaminated lands (sites) and water bodies, stabilization of mining overburdens, sediment control and soil conservation (Sinha et al. 2003). It was introduced into Australia by the Indian settlers in Fiji early in the 1900s. All the researches and its environmental applications conducted in Australia are based on the genotype 'Monto' (Truong and Loch 2000).

### 4.1 Biological Diversity in the Wonder Grass Vetiver

The 'wonder grass' vetiver, also sometimes referred as the 'miracle grass' is native of India, and has been used for land protection as well as, soil and moisture conservation for centuries. Two genotypes of *V. zizanioides* viz. the wild and fertile north Indian and the sterile south Indian genotype exist and are being mostly used in Asia. The sterile one is preferred globally, because it does not pose the threat of becoming a weed. Two other species used for soil conservation are *V. nigratana* (native of Thailand) and *V. nemoralis* (native of southern Africa).

Australia selected the genotype 'Monto' after its rigorous test for sterility. This is genetically similar to the majority of sterile south Indian genotype of *V. zizanioides* used in other countries. The 'Monto' genotype is highly palatable and readily grazed by cattle, dairy cows, sheep and horses as well as some native animals in Australia (Truong and Baker 1998a).

### 4.2 Morphological Character and Ecological Adaptations of Vetiver

i. Vetiver grows very rapidly and becomes effective for environmental restoration works in only 4-5 months as compared to 2-3 years taken by trees and shrubs for the same job.

ii. It has stiff and erect stem and finely structured network of 'deep and spongy root system' often reaching 3- 4 meters in the very first year of growth. When buried under sediment, vetiver root will establish from the nodes thus continuing to grow with the new soil level. New shoots emerge from the base helping it to withstand heavy traffic and heavy grazing pressure.
iii. It is also non-invasive, has no runners or rhizomes, and only spread by tillering.
iv. It is highly resistant to pest, diseases and fire and tolerant to prolonged drought, flood, frost and submergence. It is difficult to burn vetiver even in dry and frosted conditions. Vetiver not only survived but continued to grow through the worst drought in Australia early in the 1990s. It can re-grow very quickly after being affected by adverse environmental conditions.
v. Vetiver's survival and growth is significantly increased (2 ton / ha) by mulching and application of fertilizer di-amonium phosphate (DAP).
vi. It can withstand extreme temperatures from –15°C to 48°C in Australia and even higher in India and South Africa (over 55°C).
vii. It can grow in regions where annual rainfall vary from 200 mm to 3000 mm. In Sri Lanka, it has been shown to survive where rainfall is as much as 5000 mm per annum.
viii. It can tolerate very high acidity and alkalinity conditions (pH from 3.0 to 10.5); high soil salinity (EC = 8 dScm), sodicity (ESP = 33%) and magnesium;
ix. It can tolerate very high levels of heavy metals Al, Mn, Mg, As, Cd, Cr, Ni, Cu, Pb, Hg, Se, Zn and the herbicides and pesticides in soils (Table 1).

**Table 1.** Tolerance and toxicity levels of Vetiver and other plants to heavy metals in soil

| Heavy metals | Other plants (mg/kg) | Vetiver (mg/kg) |
|---|---|---|
| Arsenic (As) | 20 | 100-250 |
| Cadmium (Cd) | 1.5 | 20- 60 |
| Nickel (Ni) | <60 | 100-200 |
| Selenium (Se) | 2-14 | >74 |
| Zinc (Zn) | 200 | > 750 |
| Manganese (Mn) | 500 | 578 |
| Copper (Cu) | 35-60 | 50-100 |
| Chromium (Cr) | 50 | 200-600 |
| Lead (Pb) | 300 | > 1500 |
| Mercury (Hg) | 1 | >6 |

Source: Truong & Baker (1998a): *Vetiver Grass System for Environmental Protection*

x. Vetiver is highly sensitive to shading and can even disappear. This property of Vetiver is of great advantage in rehabilitation of a disturbed waste land.

Vetiver would first stabilise the eroded ground, improve the micro-environment of the habitat for the local and native species (trees and shrubs) to grow and eventually give up after shading. Experience has shown that within two years, native species can reduce vetiver growth and dominate the area. Vetiver is thus, a very suitable species for land rehabilitation which eventually makes way for the native species to flourish.

### 4.3 Propagation and Planting of Vetiver

Although vetiver can be planted as bare root slips by splitting up older plants, a better establishment rate is obtained by raising young plants first. Young vetiver plant is broken into planting slips of two to three tillers with intact root and stem. The top of slips is cut to 200 mm and the roots to 50 mm. Each slip is planted in a pot with sandy loam soil fertilized by 5 gm of di-amonium phosphate (DAP). Pots are watered everyday and kept in full sun. Vetiver becomes ready for planting on site when at least two new shoots appear.

The rooted vetiver slips can be directly planted on ground (site) at 150 mm apart to ensure a close hedge. Roots are covered with 20-30 mm of soil and firmly compacted. DAP is added @ of 50 g/meter length. Water is given every second day and twice a week after it is established. Trimming the young plants stimulates early tillering and the hedge closes up faster. Mature hedge requires no further fertilizer or water.

### 4.4 The Bioengineering of Vetiver Action

Vetiver works as a 'biological sieve' in preventing the movement of soil (and the attached pollutants), by conserving and 'cleaning' water, and by strengthening, through its root system, the soil profiles, thus preventing water induced slippage and collapse and subsequent damage to life and property. It can stabilize engineering structures such as river banks, small dams, and levees which require hard engineering solutions (of stones, gabions, mattresses) to strengthen all these structures and thus help prevent catastrophic events due to structural failures.

VGT is a 'biological' or 'soft engineering' method that is responsive to serious environmental mitigation needs over a broad range of ecological conditions for wide applications that normally require 'hard engineering' solutions. In Malaysia, shear tests done on vetiver roots showed that the tensile strength of the roots was at 75 Mpa (one third of the strength of mild steel reinforcement) is as strong as, or even stronger than that of many hardwood species which have been proven positive for 'root reinforcement' in steep slopes. The US Corps of Engineers Construction Engineering Research Laboratory have been using vetiver grass for bioengineering solutions in borrow pits, abandoned strip mines, stream banks and embankments and gully heads. It has been found to reduce soil loss by 90% and rainfall run-off by 70%, thus improving groundwater recharge; remove excess agrochemicals from the farm soil and increase crop yield by as much as 40%; improve tree seedling growth (15%) and survival rate (95%); rehabilitate wastelands (gullies, mined areas, degraded lands) and improve polluted sites (landfills). It can even prevent or at least significantly reduce natural disasters caused by hurricanes, landslides and massive floods (Grimshaw 2000).

## 5. Role of VGT in Environmental Management

### 5.1 Erosion Control and Sediment Trapping by VGT

Vetiver is a 'living wall'. The massive root systems of vetiver bind the soil firmly and make it very difficult to be dislodged and eroded under high velocity of wind or water flows. Stems also stand up to relatively deep water flow and when planted close together, form dense hedges which reduce water flow velocity and work as an effective 'sediment filter' (for both coarse and fine sediment) trapping the silt from the run-off water behind the hedge. Chemical pollutants in run-off water are often adsorbed by these sediments. Vetiver filter strips are extensively used in Queensland Australia, to trap sediments in both agricultural and industrial lands. At working quarries, vetiver hedges are planted across waterways and drainage lines. This significantly reduced erosion and trapped the silts thus lessening the sediments in the dam water.

In Louisiana, US, the vetiver grass was very successfully used for 'gully erosion' control. Three scenic streams were getting filled with silt. Check dam was built to control the problem but it failed. Vetiver was planted near the check dams, on the sides and slopes. Within 8 weeks, the hedges grew to 2m and trapped the silt and mud that was going into the stream (Truong and Baker 1998b).

### 5.2 Decontamination of Polluted Soils by VGT

Vetiver roots can absorb and accumulate several times of some of the heavy metals present in the soil and water (Truong and Baker 1998a). Studies further indicated that very little (1 to 5%) of the arsenic (As), cadmium (Cd), chromium (Cr) and mercury (Hg) and very moderate amount (16 to 33%) of copper (Cu), lead (Pb), nickel (Ni) and selenium (Se) absorbed were translocated to the shoots (Table 2). Hence, its green shoots can be harvested for mulch. Vetiver can be disposed off safely elsewhere, thus gradually reducing the contamination levels.

**Table 2.** Absorption and distribution of heavy metals in Vetiver shoot and root

| Metals | Soil (mg/kg) | Shoot (mg/kg) | Root (mg/kg) |
|---|---|---|---|
| Arsenic (As) | 959.00 | 9.6 | 185.00 |
| Cadmium (Cd) | 1.60 | 0.31 | 14.20 |
| Chromium (Cr) | 600.00 | 18.00 | 1750.00 |
| Copper (Cu) | 50.00 | 13.00 | 68.00 |
| Lead (Pb) | 1500.00 | 72.30 | 74.50 |
| Mercury (Hg) | 6.17 | 0.12 | 10.80 |
| Nickel (Ni) | 300.00 | 448.00 | 1040.00 |
| Selenium (Se) | 23.60 | 8.40 | 12.70 |
| Zinc (Zn) | 750.00 | 880.00 | 1030.00 |

Source: Truong, Paul (1999): *Vetiver Grass Technology for Mine Rehabilitation*

In a study made at Griffith University, we found that vetiver removed nearly 30% of cadmium (Cd) from the contaminated soil in just 5 weeks.

### 5.3 Farm Soil Decontamination

With the heavy use of agro-chemicals in the wake of green-revolution, most farmlands in world today are badly polluted. Vetiver has high capacity to absorb and remove agro-chemicals like carbofuran, monocrotophos and anachlor from soil thus preventing them from contaminating and accumulating in the crop plants. At the Scott Lumber Company site in Missouri, U.S., 16,000 tonnes of soils, contaminated with polyaromatic hydrocarbons (PAHs), were biologically treated with VGT. The PAH concentration was effectively reduced by 70% (Pinthong et al. 1998).

## 6. Stabilization and Rehabilitation of Mining Overburdens

### 6.1 Some Case Studies from Australia

Vetiver Grass Technology (VGT) is now being successfully used in Australia to stabilize mining overburdens. It is currently being used to stabilize a very large dam wall of a bauxite mine in Northern Territory and a bentonite mine, coal and gold mines in Queensland. It is also being used for a large-scale application to control dust storm and wind erosion on a 300 ha tailings dam.

#### *6.1.1 Bentonite Mine Tailings*

Commercial Minerals Limited, operates a large bentonite mine and processing plant in Queensland Australia. The mine spoils were extremely erodible, as they had high sodium content, high sulphate, very little moisture and extremely low in nutritional value. The major ecological concern of the mining operation was the run-off of sediment laden stormwater from the disturbed areas to the surrounding catchment areas. Vetiver grass was grown as hedges on the highly sodic bentonite spoils to arrest the run-off and also for erosion and sediment control. Mulching and fertilization was done and within 10 months of planting, excellent results were seen. Shoot growth was on an average 3 cm per week over the first three weeks and root growth was also extensive. The hedges supported 100% soil moisture within a 3.4m arc along the rows. When the hedges were complete (with no gaps), it trapped up to 200 mm deep sediment. This sediment now hosts several annual and perennial native species. Samples of runoff water was collected upstream and downstream of the vetiver hedges which indicated that vetiver was able to remove most of the solids and pollutants from the clay contaminated stormwater. Heavy rains inundated the vetiver rows and some plants remained submerged for over 2 weeks and yet in healthy conditions.

### *6.1.2 Coal Mine Tailings*

The overburden of open cut coal mine in central Queensland, Australia is generally highly erodible. These soils are usually highly sodic (ESP 33%), saline, acidic (pH 3.5) and alkaline (pH 9.5), and extremely low in nitrogen (1.3 mg/kg) and phosphorus (13 mg/kg) and high in soluble sulphur (6.1 mg/kg), magnesium (2400 mg/kg), calcium (1200 mg/kg) and sodium (2760 mg/kg). Plant available copper, zinc, magnesium, and iron are also high. Soil with exchangeable sodium percentage (ESP) higher than 15 is considered to be strongly sodic. Moreover, the sodicity of coal tailings is further exacerbated by the very high levels of magnesium compared to calcium.

To rehabilitate an old coal mine tailings dam with a surface area of 23 ha and capacity of 3.5 million cubic meter, vetiver grass was grown on these mining spoils with 20% slopes. Mulching and fertilization was done with DAP application. Within 2-3 months vetiver established firmly and stabilized the slope of spoil dump. The microenvironment also became receptive for the growth of native species (Radloff et al, 1995; Truong and Baker 1996; Truong 1999).

### *6.1.3 Gold Mine Tailings*

Fresh gold mine tailings in Australia are typically alkaline (pH 8-9), low in plant nutrients, and very high in free sulphate (830 mg/kg), sodium and total sulphur (1-4%) and high in arsenic. Vetiver established on such spoils even without fertilizers, but growth was improved with application of 500 kg/ha of DAP.

Due to high sulphur content, old gold mine tailings are often extremely acidic (pH 2.5-3.5), high in heavy metals and low in plant nutrients. Arsenic is

1120 mg/kg, chromium 55 mg/kg, copper 156 mg/kg, manganese 2000 mg/kg, lead 353 mg/kg, strontium 335 mg/kg, and zinc 283 mg/kg. These tailings are source of contaminants, both above ground and underground to the local environment. Field trials were conducted on two 8 years old gold mine, one with soft surface (pH 3.6; sulphate 0.37%) and the other with hard crust (pH 2.7; sulphate 0.85). Excellent growth of vetiver was observed when supplied with DAP at 300 kg/ha (Truong 1999).

#### *6.1.4 Bauxite Mines Tailings*

Bauxite mine tailings are highly caustic (alkaline) with pH as high as 12. Vetiver is successfully growing on these aluminum tailings in Northern Territory of Australia and has stabilised a very large dam wall of bauxite spoils (Truong 1999).

### 6.2 Case Study from China

Environmental rehabilitation works were carried out with *Vetiveria zizanioides*, *Cynodon dactylon, Paspalaum notatum* and *Imperata cylndraca* var. *major* at the Lechang lead (Pb) and zinc (Zn) mines in Guangdong Province which covered an area of 1.5 square kilometer producing approximately 30,000 tons of tailings annually, with a dumping area of 60,000 square meter. The tailings contained very high content of heavy metals lead (Pb), cadmium (Cd), zinc (Zn) and copper (Cu). It had very low levels of nutrients nitrogen (N), potassium (K) and phosphorus (P) and organic matter. The tailings were amended with 10 cm of domestic refuse + complex fertilizer (NKP). *V. zizanioides* was the best species to revegetate the mine tailings.

### 6.3 Case Study from South Africa

In South Africa vetiver has been very successfully used to stabilise / rehabilitate ‘slime dams’ (tailings) at de Beers Diamond Mines where surface temperature was 55°C (Knoll 1997).

## 7. Rehabilitation of Waste Landfills: Leachate Retention and Purification

Municipal and industrial waste landfills and industrial waste sites are usually contaminated with heavy metals such as arsenic (As), cadmium (Cd), chromium (Cr), nickel (Ni), copper (Cu), lead (Pb) and mercury (Hg) which are highly toxic to both plants and humans. Works done in Queensland have shown that vetiver can stabilise the highly erodible slopes and drainage lines and also suck up the leachate substantially from the contaminated landfill sites. Leachate from a landfill near Judy Holt Park, at Wellington point in Australia, was polluting a nearby watercourse. A biological barrier of vetiver was laid and today the area is ecologically restored with no sign of toxic leachate and the native species have come up in the area. Vetiver is successfully being used for checking landfill seepage problems by the Redland Shire Council in Brisbane. It is proving its worth in Brisbane valley, preventing run-off into local waterways from the effluent of landfills and acid sulphate soils that might otherwise leach into the Lake Somerset. The massive root system is removing extensive nitrogen and phosphorus build up from the effluent at Church Youth Camp, just 200 meters from the lake (Truong and Baker, 1998b).

At a major landfill in Bangkok, where 5000 tons of garbage was dumped everyday, a section was marked for vetiver plantation in July 1999. After four months, it was found that vetiver was able to survive fairly well despite the presence of leachate and toxicity normally present at all waste dump sites. Work done in China showed that vetiver could also purify and cleanse the urban garbage leachate. Small-scale planting of vetiver was carried out on a garbage dump in Guangzhou city and it was found that the grass could not only survive well, but also eliminate some of the foul odor from the dump site. Of all, the ammoniac nitrogen was the best cleansed, and its purification rate was between 83 – 92% indicating that vetiver can strongly absorb ammoniac nitrogen dissolved in water. Phosphorus was removed by 74% (Xia 1998).

## 8. Removal of Nutrients and Heavy Metals and Prevention of Eutrophication in Streams and Lakes by VGT

Because vetiver grass can withstand prolonged submergence in water, it also behaves as a wetland plant. It can efficiently absorb dissolved nitrogen (N), phosphorus (P), mercury (Hg), cadmium (Cd), lead (Pb) and all other heavy metals from the polluted streams, ponds and lakes and its efficiency increase with age. Works done in China have confirmed that vetiver can effectively remove dissolved nutrients, specially the N and P from wastewater and reduce the growth of blue green algae (which cause eutrophication) within two days under experimental conditions. Phosphorus (P) is removed up to 99% after 3 weeks and nitrogen (N) 74% after 5 weeks (Zheng et al. 1998).

Vetiver has the potential of removing up to 102 tonnes of nitrogen and 54 tonnes of phosphorus / year / hectare of vetiver. This can be achieved by both planting vetiver on the edges of the streams or on the shallow parts of the lakes where usually high concentrations of soluble N and P occur. An innovative idea is to grow vetiver hydroponically on floating platforms which could be moved from one place to the other, and to the worst affected parts of the lakes and ponds. The advantage of the platform technology is that the top portions of the grass can be harvested easily for stock feed or mulch and the roots can also be removed for oil production.

## 9. Wastewater / Storm water Treatment by VGT in Constructed Wetlands

Constructed wetland technology (CWT) using aquatic and wetland plants in artificially created wetlands for municipal wastewater / storm water treatment and purification are also considered as a part of phytoremediation technology. Vetiver can easily thrive in wetlands and can be used in the constructed wetlands for removal of nitrogen (N) and phosphorus (P) and heavy metals from the polluted storm water, municipal and industrial wastewater, and effluents from abattoirs, feedlots, piggeries and other intensive livestock industries. Works done in Thailand show that VGT can also effectively remove substantial quantities of cadmium (Cd), mercury (Hg), chromium (Cr), arsenic (As) and lead (Pb) from municipal wastewater. Chinese study also revealed successful use of vetiver as a wetland plant to remediate animal waste from a piggery (Hengchaovanich, 2003). Vetiver roots can accumulate several times of some of the heavy metals present in the wastewater (Truong and Baker, 1998a).

### 9.1 Environmental - Economics of VGT

Environmental - economics works highly in favor of VGT. It can reduce point source erosion from highways and building sites at much reduced costs, often less than 90% of the cost of the ‘hard engineering’ solutions. The cost-benefit analysis of VGT done in China (developing country), where labor cost is cheaper, indicates that the soft engineering solution costs approximately 10% of the corresponding hard engineering solution for environmental problems. In Australia (a developed nation), where the labor cost is higher, the VGT would cost between 27 to 40% of the hard engineering solution (Hengchaovanich 2003). In the U.S., VGT costs around one-tenth to one-third of conventional engineering technology and its use is likely to increase by more than 10 fold in future.

### 9.2 Economic Importance of Vetiver Grass

The root of vetiver produces an essential oil called ‘vetiver oil’ which is used in perfumery industry. The south Indian genotype is specially useful in the oil production. The Department of Natural Resources in Australia is producing a world class perfume ‘Guerlain’. Vetiver oil is also an ‘insect repellant’. Vetiver grass also has herbicide / weedicide properties. Methanol extracts of ground stem and root were found to be very effective in preventing the germination of a number of monocot and dicot weed species (Techapinyawat et al. 1996).

## 10. Conclusion

Phytoremediation by VGT is a low cost technology as compared to conventional (engineering) methods for site remediation. It is also virtually maintenance free, the grasses regrow very quickly and its efficiency improves with age (Truong 1999). Social acceptance of a particular technology in remediation of contaminated lands and water bodies has also become an important issue, as it directly affects the life of community. Biological technologies based on the use of plants are more acceptable to people, as it creates a green and aesthetic view and also provides some useful materials. Several plants are being identified and trialed to be used in the phytoremediation task. Important among them are other grasses like the Bermuda grass (*Cynodon dactylon*), Bahia grass (*Paspalum notatum*), Rhodes grass (*Chloris guyana*), the tall wheat grass (*Thynopyron elongatum*), common reed grass (*Phragmites australis*), the munj grass (*Sachharum munja*) and *Imperata cylindrica*. Other plants are the marine couch (*Sporobolus virginicus*), cumbungi (*Typha domingensis*) and *Sarcocrina* spp. They are highly salt and toxicity tolerant and have extensive root binding system. They were tried in the rehabilitation works, but none succeeded so well as vetiver. There is need to educate the society, the general people and the planner about the ecological and economic value of this 'wonder grass'.

## References

Anderson TA, Guthrie EA, Walton BT (1993) Bioremediation in the rhizosphere. Plant roots and associated microbes clean contaminated soils. Environ Sci Technol 27:2630-2636

Cunningham SD, Ow DW (1996) Promises and prospects of phytoremediation. Plant Physiol 110:715-719

Greenfield JC (1989) Vetiver grass: The ideal plant for vegetative soil and moisture conservation, The World Bank, Washington D.C., USA

Grimshaw Dick (2000) Vetiver and the Environment-The Future, Paper presented at the 2nd International Conference on Vetiver, Thailand, January 19-22, 2000

Hengchaovanich Diti (2003) VGT: A Bioengineering and Phytoremediation Option for the New Mellennium, APT Consult Company Limited, Bangkok, Thailand

Knoll C (1997) Rehabilitation with Vetiver, Journal of African Mining, Vol. 2(2), pp 43

Kim I (1996) Harnessing the green clean. J Chem Eng 103:39-41

Mackova M, Macek T, Ocenaskova J, Burkhard J, Demnerova K, Pazlarova J (1996) Selection of the potential plant degraders of PCB's. Chemistry Listy 90:712-713

Pinthong J, Impithuksa S, Ramlee A (1998) The capability of vetiver hedgerows in decontamination of agrochemical residues: A case study on the production of cabbage at Nong Hoi Development Centre, Proceedings of 1st International Conference on Vetiver, Chiang Rai, Thailand, Feb 4-8, 1996, pp 91-98

Radloff B, Walsh K, Melzer A (1995) Direct Revegetation of Coal Tailings at BHP Saraji Mine, Australian Mining Council Environment Workshop, Darwin, Australia

Rice PJ, Anderson TA Anhalt JC, Coats JR (1997) Phytoremediation of atrazine and metachlor contaminated water with submerged and floating aquatic plants, Proc. Of the 12th Annual Conference on Hazardous Waste Research, Kansas City, Missouri, May 19-22, 1997; pp 52

Rugh CL, Wilde HD, Stack NM, Thompson DM, Summers AO, Meagher RB (1996) Mercuric ion reduction and resistance in transgenic *Aribidopsis thaliana* plants expressing a modified merA gene; Proceeding of the National Academy of Science; U.S.A.; Vol. 93, pp 3182-3187

Sinha RK (1996) Global Biodiversity; INA Shree Publications, Jaipur, India; ISBN 81-86653-04-X, pp 396

Sinha RK, Herat S and Tandon PK (2003) A Review of Phytoremediation as a Cost-effective, Ecologically Sustainable and Socially Acceptable Bioengineering Technology, Proceedings of National Environment Conference (NEC), Brisbane, June 18-20, 2003

Tamaya (2003) Removal of cadmium (Cd) from contaminated soil by Vetiver (*Vetiveria zizanioides*), 20 CP Project Report, School of Environmental Engineering, Griffith University, Brisbane, Australia (Supervised by Sinha RK).

Techapinyawat S, Sripen S, Komkris T (1996) Allelopathic effects of vetiver grass on weeds, Paper presented at the 1st International Conference on Vetiver; Chiang Rai, Thailand, Feb. 4-8, 1996

Truong, P, Baker, D (1996) Vetiver grass for the stabilisation and rehabilitation of acid sulfate soils, Proceedings of 2nd National Conference on Acid Sulfate Soils; Coff's Harbour, NSW, Australia, pp 196-198

Truong P, Baker D (1998a) Vetiver Grass System for Environmental Protection, Technical Bulletin No. 1998/1; Pacific Rim Vetiver Network, Bangkok, Thailand

Truong P, Baker D (1998b) The role of vetiver grass for the stabilisation and rehabilitation of toxic and contaminated lands in Australia, Proceedings of International Vetiver Workshop, Fuzhou, China, Oct. 21-26, 1997

Truong P (1999) Vetiver Grass Technology for Mine Rehabilitation, Technical Bull. No. 1999/2; Pacific Rim Vetiver Network; Bangkok, Thailand

Watanabe M (1997) Phytoremediation on the brink of commercialization. Environ Sci Technol 31:182A-186A

Xia H, Ao H, Lui S, He D (1998) A preliminary study on vetiver's purification for garbage leachate, Proceedings of International Vetiver Workshop, Fuzhou, China, Oct. 21-26, 1997

Zheng CR, Tu C, Chen HM (1998) Preliminary experiment on purification of eutrophic water with vetiver, Proceedings of International Vetiver Workshop, Fuzhou, China, Oct. 21-26, 1997

# 15

# The Role of Macrophytes in Nutrient Removal using Constructed Wetlands

Margaret Greenway

School of Environmental Engineering, Griffith University, Nathan, Queensland 4111, AUSTRALIA, Email: m.greenway@griffith.edu.au

## 1. Introduction

### 1.1 Overview

This chapter reviews the role of aquatic plants (macrophytes) in the removal of nutrients from wastewater using constructed wetlands, with particular emphasis on surface-flow wetlands in tropical-subtropical climates. Nutrients (nitrogen and phosphorus) are potential contaminants in many wastewater effluent streams in urban, industrial and rural areas. This review focuses on the ecological requirements of macrophytes, the suitability of species for nutrient bioaccumulation and biomass production, and the overall performance and limitations of macrophytes in nutrient removal from the constructed wetlands.

A recent review of the phytoremediation potential of wetland plants focussing on natural wetland ecosystems has been undertaken by Williams (2002) and is complementary to this review. Williams' review addresses the role of wetland plants in the phytoremediation of metals, volatile organic compounds, pesticides, herbicides, TNT (and other explosives) and hydrocarbons, but does not include nutrient removal.

What are wetlands? Natural wetlands are areas that are permanently or periodically inundated or saturated by surface or groundwater and support the growth of aquatic vegetation (Mitsch and Gosselink 2000). Natural wetland types include saltwater wetlands (e.g. mangroves, salt marshes) and freshwater wetlands (e.g. sedgelands, reed beds, swamp forests and shallow lagoons). Wetlands are at the interface between terrestrial and aquatic environments and are strategically placed in the catchment, where they can intercept runoff water from uplands and floodwater from lowlands. Because of their strategic transitional location, floodplain wetlands are highly fertile areas. Globally, these natural wetlands have now disappeared due to cultivation for crops (Gopal 1999; Mitsch and Gosselink 2000).

The use of specifically designed constructed wetlands for the treatment of wastewater (municipal, industrial, urban and agricultural) has been widely accepted over the past 20 years. However, the use of natural wetlands to assist in water purification has been in existence in many parts of the world for centuries. The functional processes were not understood until ecological research focused on the nutrient dynamics of wetland systems in the 1960s and 1970s. It is the interaction between abiotic and biotic components which are vital for water-quality improvement by either removing, recycling or storing contaminants (Reddy and D'Angelo 1997; Mitsch and Gosselink 2000; Wetzel 2001; Williams 2002). The plants and micro-organisms remove and recycle nutrients and metals either from the water or sediments. The sediments, biotic components and detritus (dead organic matter) are major storage components. For constructed wetlands to be effective in water pollution control, they must function as "pollutant" sinks for sediment, nutrients, metals, i.e. these pollutants must be removed from the wastewater and stored within the wetland either in the sediment or the plants.

## 1.2 Wetland Processes to Improve Water Quality

Various processes of wetlands are improving water quality are summarised in Table 1. Thus, the effectiveness of water-quality improvement is dependent upon an array of complex and interacting processes which can broadly be classified into three categories - physical, biological and chemical. Most processes are facilitated by the wetland vegetation.

**Table 1.** Role of wetlands in improving water quality

| Potential Pollutant | Role of the Wetland |
|---|---|
| Suspended solids including biodegradable particulates (BOD) | Sedimentation is facilitated by the vegetation. The vegetation reduces water velocity and turbulence causing settlement. Finer particles adhere to the biofilm surface of the vegetation. The root system binds and stabilises deposited particulates. The leaf litter and vegetation reduce resuspension. |
| Nutrients - nitrogen and phosphorus | Direct uptake by plants and micro-organisms. Inorganic nutrients converted to organic biomass. Microbial processes facilitate the removal and transformation of nutrients, especially nitrogen removal. |
| Metals | Direct uptake by plants and micro-organisms. Microbial bioremediation of metals. Metals immobilised by adsorption onto sediments or by precipitation. |
| Hydrocarbons | Microbial hydrocarbon degradation |
| Pathogens | Natural UV disinfection. Natural biocontrol by microbial predators in the wetland ecosystem. Adsorption to fine particles and sedimentation. Natural death and decay. |

### 1.2.1 Physical Processes

Emergent macrophyte vegetation decreases water velocity, enabling the sedimentation of particles. Both submerged and emergent macrophytes are particularly effective in removing finely graded particles which will adhere directly onto the plant surface. The vegetation also distributes the flow and reduces turbulence, thereby allowing settlement of particles. The root system binds and stabilises deposited particles.

### 1.2.2 Biological Processes

Plants and photosynthetic micro-organisms remove soluble inorganic nutrients (ammonium, nitrite, nitrate, phosphate) and heavy metals by direct uptake. Rooted macrophytes remove these nutrients from the sediment, whereas submerged and floating macrophytes and algae remove the nutrients directly from the water column. These inorganic nutrients are assimilated and converted into organic matter (biomass) and rendered relatively unavailable until death and decay.

Macrophytes and photosynthetic micro-organisms also improve overall water quality by producing oxygen during photosynthesis which diffuses into the water column. Emergent macrophytes transport oxygen down their stems into the roots where it diffuses into the sediment to produce an aerobic micro-environment around the root zone (rhizosphere).

The interaction between macrophytes and microbes is essential for nitrogen removal. Microbial processes of significance for the removal and transformation of nitrogen are ammonification, nitrification and denitrification. Ammonification is a decomposition process whereby dead organic matter (proteins) is converted to amino acids and then ammonia. Ammonification occurs under both aerobic and anaerobic conditions. Ammonium ions can either be assimilated by plants or nitrified under aerobic conditions by nitrifying bacteria to nitrites and nitrates. Sediments being waterlogged are often anaerobic, and therefore nitrification cannot proceed and ammonium ions dominate. However, in aerobic micro-environments around the rhizosphere of macrophytes, nitrification occurs. These nitrates can then be taken up directly by the roots. The dead organic matter of macrophytes provides a carbon source for the heterotrophic denitrifying bacteria.

### 1.2.3 Chemical Processes

Chemical processes facilitate the adsorption and desorption of phosphorus onto and from sediment particles. Diffusion of oxygen from the roots of emergent macrophytes maintains an oxidised sediment surface layer and micro-environment around the root zone. This modifies the sediment redox conditions facilitating aerobic microbial processes including nitrification.

### 1.3 Applications of Constructed Wetlands

Because of the ability of constructed wetlands to remove, recycle, transform and/or immobilise a wide range of potential contaminants, there are an ever expanding number of applications of constructed wetland technology. The most widespread use of constructed wetlands is in the treatment of domestic and municipal wastewater. Constructed wetlands can provide secondary treatment (after primary treatment of screening, sedimentation) and final polishing, i.e. advanced or tertiary treatment (after activated sludge process, trickling filters and/or oxidation ditches). In recent years, constructed wetlands to treat urban stormwater from housing estates and shopping centres have been incorporated into the urban landscape. Constructed wetlands are also being designed to hold and treat runoff from major roads and highways. Wetlands have also been constructed to intercept crop runoff in agricultural areas, particularly where there are sensitive downstream aquatic ecosystems. Dairy farms, cattle feed lots, piggeries and poultry farms generate concentrated animal waste which can be treated by constructed wetlands. Aquaculture farms are also pre-treating their water through wetlands prior to discharge into streams and rivers.

Industrial applications are also increasing, and many of these have recently been reported in the literature (IWA 2000). They include:

- mining (acid coal mine drainage with high concentrations of dissolved iron, manganese, aluminium and sulphate; metal-mine drainage from lead, zinc, silver, copper, nickel and uranium mines)
- food processing wastes (peeling, pre-cooking and processing fruit and vegetables; sugar production; poultry and meat processing)
- petrochemicals (polishing of secondarily treated refinery wastewater, treatment of washdown runoff)
- pulp and paper-mill wastewater
- treatment of landfill leachate and wastewater sludges

### 1.4 Wetland Plants

Vegetation is the most conspicuous feature of wetlands. Wetland plants are morphologically and physiologically adapted to seasonal and/or permanent water inundation (Mitsch and Gosselink 2000). Aquatic plants are usually herbaceous and are referred to as hydrophytes. In constructed wetland technology, these aquatic plants are termed "macrophytes" (IWA 2000; US EPA 1988, 2000). Woody shrubs and trees also dominate many natural forested wetlands, e.g. mangrove swamps, *Cypress* swamps, *Melaleuca* swamps, riparian wetlands (Mitsch and Gosselink 2000).

Macrophytes can be classified according to their morphological form or functional type (Fig. 1). There are two broad functional types: (i) rooted plants, and (ii) non-rooted plants. Rooted plants are anchored in the sediment and remove nutrients for the interstitial pore water. Non-rooted plants are not anchored, and either float on the surface or are suspended in the water column.

Rooted plants can be further classified as:

- Emergent macrophytes, i.e. roots/rhizomes in the sediment, and emergent stems and leaves which rise above the water (e.g. reeds, bulrush, sedges). Rooted emergent macrophytes are restricted to shallow water from a few centimetres to a maximum depth of 1 m.
- Floating-leaved macrophytes, i.e. roots/rhizomes in the sediment, stems submerged and leaves floating on the water surface (e.g. water lilies). Maximum depth 1.5 m.
- Submerged macrophytes, i.e. stems and leaves submerged (e.g. *Potamogeton*, *Triglochin*, *Vallisneria*). The depth distribution of submerged plants is restricted by light and oxygen availability.
- Creepers or vines, i.e. anchoring roots in shallow sediment, floating stems and leaves, with adventitious roots in the water (e.g. *Bacopa monniera*, *Ipomopa aquatica*, *Ludwigia peploides*, *Persicaria strigosum*).
- Trees and shrubs, i.e. woody plants that dominate seasonally inundated swamp forests, riparian zones and floodplains (e.g. *Melaleuca*, *Salix*, *Alnus*).

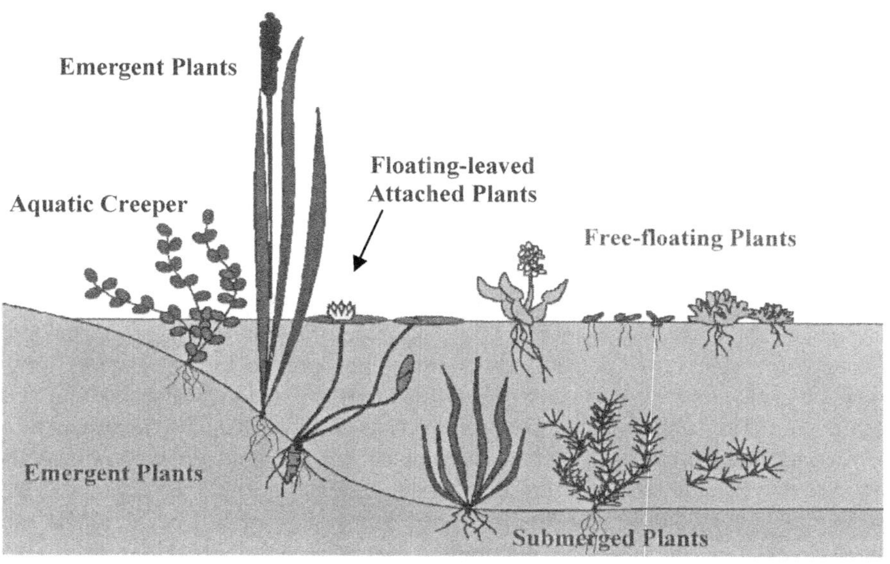


**Fig. 1.** Forms of aquatic plants found in constructed wetlands

Non-rooted plants can be further classified as:

- Floating plants, i.e. surface leaves (and stems) with roots which hang down into the water (e.g. water hyacinth (*Eichornia*), duckweed (*Lemna*), *Azolla*).
- Submerged plants, i.e. stems and leaves below the water surface; adventitious roots may be present but non-attached (*Ceratophyllum*, *Hydrilla*).
- Creepers or vines, i.e. floating stems and leaves with adventitious roots in the water (e.g. *Ipomea aquatica*, *Paspalum distichum*).

### 1.5 Types of Constructed Wetland Systems

There are basically two types of constructed wetlands - Free Water Surface Systems (FWS) and Sub-surface Flow Systems (SSF).

Freewater surface flow (FWS) wetlands resemble natural wetlands in appearance and are composed of shallow (20-50 cm) vegetated channels or basins and deeper (50 cm - 2 m) open-water ponds. Vegetated shallow areas are often referred to as marshes. Marshes are typically dominated by emergent macrophytes, i.e. plants with roots in the sediment and emergent stems and leaves (reeds, sedges, rushes). However, floating-leaved attached macrophytes, i.e. plants with roots in the sediment and floating leaves (water lilies), submerged macrophytes (pond weeds) and floating macrophytes (e.g. duckweed) are also found in these shallow wetlands. Deeper ponds support floating macrophytes or submerged macrophytes if there is sufficient light for growth. FWS wetlands are typically used to provide tertiary wastewater treatment after conventional secondary treatment involving trickling filters or oxidation ponds, which remove most of the organic pollutants. They are most suitable for mild temperate or tropical/subtropical climates where freezing of the water does not occur in winter and continuous aquatic plant growth can occur.

An FWS consists of channels or free-form shallow basins with a natural or constructed base of clay or impervious geotechnical material to prevent seepage, and a layer of suitable substrate to support rooted emergent macrophytes. Water depth can vary to suit the plant species used; lagoon configurations can also support floating aquatic plants. Substantial areas of land may be required to establish successful FWS.

Sub-surface flow (SSF) wetlands, also known as vegetated submerged bed systems (US EPA 2000) in the USA, and reed-bed or root-zone wastewater treatment systems (IWA 2000) in Europe, are gravel and/or soil/sand-filled trenches, channels or basins with no standing water, and support emergent vegetation. They are typically used in Europe to provide secondary treatment after screening and primary settlement. Because of the potential for clogging of the media, they are mostly used for small communities or single households. They are suitable for cold climates as microbial processes can still occur in the root zone in winter. The absence of standing water, however, precludes the use of many aquatic plant species; only emergent species can survive in the waterlogged gravel or soil media. There is generally a higher treatment performance efficiency per unit area of land. Therefore, less land is required for the construction of SSF as compared to an FWS wetland. However, there is a higher capital cost associated with media supply and maintenance, if clogging occurs.

An SSF consists of trenches with impermeable liners and a substrate of gravel and/or soil supporting emergent macrophytes. The systems can be designed to allow the wastewater to flow horizontally through the root zone, maximising filtration and sorption in the substrate, nutrient uptake by plants and micro-organisms and microbial degradation. Horizontal flow (HF) constructed wetlands are also termed reed bed treatment systems (RBTS) in Europe, because the reed *Phragmites australis* is commonly used. In North America, the term vegetated submerged bed (VSB) is used. Another type of SSF system is the vertical flow (VF) system. These layered gravel-sand reed beds are dosed intermittently with wastewater which is fed from the top, causing surface flooding. The wastewater drains vertically down, and the bed is then allowed to aerate before the next dosing.

FWS systems are more suitable in subtropical/tropical conditions where year-round plant growth occurs. A wide range of plant species can be used (Greenway 2003). SSF systems are most prevalent in Europe and temperate regions of North America. Although the plants die back in winter, microbial activity continues. The range of suitable plants species for SSF systems is limited *Phragmites australis, Phalaris arundinacea, Glyceria maxima, Typha* spp. and *Scripus* spp. have been used in Europe and North America (IWA 2000; Kvet et al. 1999). In sub-tropical Australia, *Baumea articulata, Carex fasicularis, Phylidrum languinosum* and *Schoenoplectus mucronata* are being trialled (Browning and Greenway 2003). Bolton used *Melaleuca* tree species (Greenway and Bolton 2002).

The choice of plant species depends upon the physical structure of the constructed wetland which is governed primarily by the type of wetland system (FWS or SSF) and the pollutant characteristics, i.e. chemical composition of the wastewater effluent. The type of wetland system is determined by the extent of treatment required, i.e. secondary or tertiary wastewater treatment, the mass loading, climatic conditions, area of land available and cost (Greenway 2004; IWA 2000).

## 1.6 Ecological Requirements of Macrophytes in Treatment Constructed Wetland Systems

Water, light, nutrients and oxygen are essential resources for the plant growth. Water and nutrients are the products associated with wastewater, in particular, sewage effluent, animal husbandry (piggeries, dairies), food processing, agricultural and urban stormwater runoff. A summary of the North American Treatment Database (IWA 2000) found average municipal wastewater effluent concentrations entering constructed wetlands were 3.8 mg/L $PO_4$-P, 5.49 mg/L

$NO_x$-N, and 4.97 mg/L $NH_4$-N. Thus constructed wetlands are ideal candidates for promoting plant growth. However, at high concentrations, some nutrients, especially ammonium and phosphate, may become toxic to plant growth. For example, concentrations of $NH_4$-N in animal wastewater are extremely high, with average concentrations of 105 mg/L for dairy, 74 mg/L for poultry, and 366 mg/L for piggery (CH2M HILL and Payne Engineering 1997). Not only are such high concentrations of ammonium toxic to most macrophytes, but rapid oxygen depletion occurs due to nitrification.

Water depth determines the different functional types of macrophytes found in constructed wetlands (Fig. 1). In natural wetlands, the distribution of the types and species of aquatic plants is usually governed by the water depth. Zonation is common with emergent, seasonally inundated species occurring at the landward interface, and submerged species or water lilies occurring in deeper permanent water. Free-floating species occur where there is open water, regardless of the water depth. Light is essential for photosynthesis and can limit the growth of submerged species where light penetration is reduced either through turbidity, shading, or very deep water. Oxygen is essential for aerobic respiration, and aquatic plants have morphological, anatomical and physiological adaptations for coping with the relatively low concentrations of dissolved oxygen in the water column and sediment. Emergent species can transport oxygen through special air spaces in their leaves and stems to the roots and rhizomes in the sediment. However, submerged species are unable to survive under anaerobic conditions. The distribution of species is also affected by substrate type and water quality (pH, salinity, toxic contaminants).

In FWS treatment, wetlands water depth does not generally fluctuate, and is maintained between 20-50 cm depth. Water quality is often high in TSS, BOD and nutrients, and the sediment can become very anaerobic. Thus, the physico-chemical conditions in treatment wetlands can be very different from natural wetlands. In treatment wetland systems, the plants need to be adapted to permanent waterlogging, and able to tolerate high nutrient concentrations in the water and sediment (Greenway 2003).

The layout or configuration of wetland zones is important for treating all forms of wastewater (IWA 2000; Greenway 2004). Deep ponds or lagoons are appropriate as retention basins for stormwater wetlands or for treating wastewater effluent using floating plants, such as duckweed or water hyacinth. Large-scale treatment systems using floating plants require regular harvesting. Harvesting not only removes bioaccumulated nutrients (and metals), but also provides a potential resource as fodder for cattle or other livestock. Treatment lagoons can also function as aquaculture ponds for fish.

Emergent macrophytes are an essential component of most constructed wetlands and play a major role in facilitating physical and biological processes in pollutant removal (Table 1). Emergent macrophytes, however, are restricted to shallower water, usually less than 50 cm deep, and not all species can tolerate permanent flooding.

Surface Flow Wetland Systems for the treatment of steady-flow wastewater streams exhibit a little variation in water levels and are usually designed for a depth of 30-50 cm of water. Thus, sedges and reeds tolerant of permanent inundation need to be planted. By contrast, huge fluctuations in water levels occur in stormwater wetlands, necessitating a range of shallow (< 10 cm) and deeper (50 cm) macrophyte zones. The shallower zones will completely dry out during low rainfall periods. Therefore, plant species, that can tolerate a wetting and drying cycle, should be selected for these areas. A diversity of vegetation zones can also enhance the overall wildlife value of the wetland as well as the landscape amenity.

## 2. Role of Macrophytes in Nutrient Removal

### 2.1 Overview

One of the primary factors that has attributed to the use of constructed wetland systems for municipal wastewater treatment is "recognition of the natural treatment functions of aquatic plant systems and wetlands, particularly as nutrient sinks and buffering zones" (US EPA Design Manual 1988). As outlined in Section "Wetland processes to improve water quality", nutrient removal, transformation, recycling and retention are largely biologically mediated. The macrophytes either directly or indirectly play an important role in nutrient removal and storage. The removal of soluble inorganic nitrogen and phosphorus via absorption from either the water column or the sediment, assimilation and storage in plant tissue, is a direct mechanism of nutrient sequestration. The provision of plant surfaces (leaves, stems and roots) for attached microbiota, epiphytic microflora and associated biofilm communities enables microbial assimilation, transformation and storage of nutrients. Although there is still debate about the relative importance of macrophytes versus microbes in nutrient removal (Brix 1997; IWA 2000; Tanner 2001; Wetzel 2001), plant biomass still accounts for substantial removal and storage of N and P (Rejmankova et al. 1990; IWA 2000; Greenway and Woolley 2001).

Since inorganic nitrogen and phosphorus are essential for the plant growth, it is possible to maximise the amount of nutrients removed from wastewater effluent by selecting macrophytes with a high capacity for inorganic nutrient absorption and conversion to organic plant biomass. They should have a long or continuous growing season and be highly productive and capable of accumulating large amounts of nutrients in plant biomass. Rooted plants remove nutrients directly from the sediments, whereas floating plants remove nutrients from the water column. Some emergent species, such as *Phragmites,* have adventitious "water roots", and the water snow flake *Nymphoides* produces roots from the floating leaf base, thereby enabling these species to remove nutrients from both sources. Many submerged species obtain nutrients directly from the water column via leaf absorption, particularly in species with poorly developed root systems, such as *Ceratophyllum*. Once nutrients have been absorbed, they can be translocated to other parts of the plant. Below-ground storage in rhizomes is common in emergent macrophytes.

Plant uptake is an important nutrient removal mechanism in wastewater treatment systems. Contact between the active zones of nutrient absorption and the wastewater or sediment must be maximised to optimise nutrient removal and incorporation into plant biomass.

## 2.2 Suitability of Macrophyte Species

In North America, 851 wetland plant species have been identified (Knight et al. 2001), of which 593 species have been recorded in constructed treatment wetlands. In subtropical-tropical Queensland, Australia, 150 wetland plant species have been identified as "potential aquatic plants for use in freewater surface flow constructed wetlands" (QDNR 2000), of which 72 have been found growing (planted or self-colonised) in treatment wetlands (Table 2) (Greenway 2003).

These two examples demonstrated the huge potential of using aquatic plant species which occur naturally in wetlands or waterways. However, as discussed earlier, for successful growth, the species selected must be able to tolerate permanent waterlogging, higher nutrient concentrations, lower dissolved oxygen due to high BOD (and COD) loads, higher turbidity due to high TSS loads, and potentially toxic contaminants depending on the source of the wastewater.

Greenway (2003) found that all species listed in Table 2 were able to grow successfully in secondary-treated sewage effluent with $PO_4$-P concentrations 2.5-8.7 mg/L, $NO_3$-N concentrations 9.7-15.8 mg/L, and $NH_4$-N concentrations 7.7-18.6 mg/L. The lowest species richness, however, occurred in a wetland receiving effluent with 22-30 mg/L $NH_4$-N. *Typha domingensis*, *Ludwigia peruviana*, the aquatic creepers *Ludwigia peploides*, *Paspalum distichum* and *Persicaria orientalis*, and duckweed were the only species to spread successfully. *Typha* was planted, but the other species were natural invaders and colonisers.

While Table 2 provides a list of species suitable for surface-flow wetlands in tropical-subtropical Australia, most of these genera and several species are cosmopolitan in distribution.

Many macrophyte species have been trialled successfully for use in SSF CW around the world (Tables 3 and 4). The most widespread and commonly used emergent species is the reed *Phragmites australis*. Species used in Europe include *Phalaris arundinaceae* (reed canary grass), *Glyceria maxima* (sweet manna grass) and *Typha* spp., and in the USA *Scirpus* spp. (IWA 2000). Commonly used species in Australia and New Zealand include *Phragmites*, *Schoenoplectus* and *Juncus* spp. (Browning and Greenway 2003).

**Table 2.** Macrophyte species occurring in constructed surface-flow tertiary-treatment wetlands in Queensland, Australia

| Family | Species and Genus |
|---|---|
| Alismataceae | **Sagittaria graminea* (E) |
| Apiaceae | *Hydrocotyle bonariensis* (FF) |
| Amaranthaceae | ***Alternanthera philoxeroides* (E or FF) |
| Araceae | *Pistia stratiotes (FF), *Colocasia esculenta (E)* |
| Asteraceae | *Eclipta prostrata (E or FF)* |
| Azollaceae (fern) | *Azolla sp (FF)* |
| Cannaceae | **Canna sp. (E)* |
| Ceratophyllaceae | *Ceratophyllum demersum (S)* |
| Convolvulaceae | *Ipomoea aquatica, Ipomoea diamantinensis (FF)* |
| Cyperaceae (all emergents) | *Baumea articulata, B. rubiginosa, Bolboschoenus fluviatilis, B. caldwelli, Cyperus alopercuroides, C. eragrostis, C. exaltatus, *C. papyrus, *Cyperus involucratus, Eleocharis acuta, E. dulcis, E. phillippinensis, E. sphacelata, Rhynchospora corymbosa, Scirpus* sp., *Scheonoplectus mucronatus, S. validus, Scleria poiformis, Schoenus apogon* |
| Gramineae (all emergents) | **Brachiara mutica, *Echinochloa crus-galli, *E. colona, E. polystachya, Hymenachne acutigluma, *H. amplexicaulis, Leersia hexandra, *Pennisetum alopercuroides, Phragmites australis, Paspalum distichum (FF)* |
| Hydrocharitaceae | *Vallisneria gigantea (S)* |
| Juncaginaceae | *Triglochin procera (S)* |
| Juncaceae | *Juncus planifolius, J. polyanthemus, J. prismatocarpus, J. kraussii, J. usitatus* (all E) |
| Lemnaceae | *Lemna* spp., *Spirodela* spp., *Wolffia* spp. *(all FF)* |
| Limnocharitaceae | **Hydrocleys nymphoides (FL)* |
| Marantaceae | **Thalia dealbata (E)* |
| Marsileaceae (fern) | *Marsilea mutica (FL), Marsilea drummondii (FL)* |
| Menyanthaceae | *Nymphoides indica ((FL)* |
| Myrtaceae | *Melaleuca quinquenervia (T)* |
| Nymphaeaceae | *Nymphaea capensis (FL), Nymphea gigantea (FL)* |
| Onagraceae | *Ludwigia peploides (FL or FF), *Ludwigia peruviana (E), L. octovalvis (E)* |
| Parkeriaceae (fern) | *Ceratopteris thalictroides (FF or E)* |
| Philydraceae | *Philydrum lanuginosum (E)* |
| Polygonaceae | *Persicaria attenuata, *P. orientale, P. strigosum (E or FF)* |
| Pontederiaceae | *Monochoria cyanea (FF)* |
| Potamogetonaceae | *Potamogeton crispus, P.pectinalis (S)* |
| Salvinaceae (fern) | ***Salvinia molesta (FF)* |
| Scrophulariaceae | *Bacopa monnieri (FF)* |
| Typhaceae | *Typha domingensis (E), Typha orientalis (E)* |

E = emergent, S = submerged, FL = floating leaved attached, FF = free-floating or aquatic creepers, T = tree. *Exotic (including introduced naturalised species), **Noxious weeds in Australia

**Table 3.** Macrophyte species used in subsurface-flow wetlands in temperate climates

| Family | Species and Genus |
|---|---|
| Alismataceae | *Sagittaria, Sagittifolia* |
| Cyperaceae | *Carex* spp., *Rhynchospora* spp., *Cyperus* spp., *Scirpus* spp., *S. acutus, S californicus , S. lacustris, Schoenoplectus validus, S. tabernaemontoni* |
| Gramineae | *Glyceria maxima, Panicum* spp., *Phalaris arundinacea, Phragmites australis, P. communis* |
| Iridaceae | *Acornus calamus, Iris pseudacorus, I. versicolor* |
| Juncacaeae | *Juncus* spp., *J. effusus* |
| Lythraceae | *Lythrum* spp., *L. salicaria* |
| Onagraceae | *Epilobium* spp. |
| Polygonaceae | *Polygonum amphibium, Rumex* spp., *R. crispus* |
| Sparganiaceae | *Sparganium erectum, S. americanum* |
| Typhaceae | *Typha* spp., *T. latifolia, T. angustifolia, T. domingensis* |
| Umbelliferae | *Oenanthe* spp. |

**Table 4.** Macrophyte species used in subsurface-flow (SSF) wetlands in subtropical/tropical climates (Americas, Southern Asia, Australasia, Africa)

| Family | Species and Genus |
|---|---|
| Cyperaceae | *Carex fasicularis, Cyperus articulatus, C. flabelliformis, C. immensus, C. papyrus, Schoenoplectus validus, Scirpus californicus, S. lacustris* |
| Graminaceae | *Miscanthidium violaceum, Paspalum penisetum, Phragmites* spp., *P. australis, P. karka, P. mauritianus, Vetiveria zizanioides, Zizaniopsis bonariensis* |
| Typhaceae | *Typha* spp., *T. dominguensis, T. latifolia, T. subulata* |

## 2.3 Nutrient Bioaccumulation

One of the many roles macrophytes play in CW for wastewater treatment includes their capacity for nutrient bioaccumulation, i.e. the direct uptake and storage of nutrients (Brix 1997). Desirable plant characteristics for macrophyte species to maximise nutrient uptake in a constructed wetland treating secondary effluent, include rapid growth, high plant-tissue content, and the ability to attain a high standing crop (Reddy and De Busk 1987; Greenway 2003). Additional to these characteristics is the ability to recover following cropping, and an attractive species is also desirable to increase wetland aesthetics.

The rate of nutrient uptake by macrophytes is limited by its growth rate and the concentration of nutrient within the plant tissues, with nutrient storage dependent on plant-tissue nutrient concentrations and potential for biomass accumulation (maximum standing crop) (Reddy and De Busk 1987). At low to medium nutrient concentrations, plant growth (biomass) is proportional to nutrient supply (Fig. 2). Increases in nutrients above this may result in the luxury uptake of nutrients by plants, but does not increase plant growth. Nutrient accumulation will eventually plateau. Beyond this point, increases in nutrient supply may cause nutrient toxicity.

A few plant species have been only grown successfully in constructed wetlands receiving high ammonium concentrations. Hunt and Poach (2001) stated that *Scirpus*, *Typha* and *Juncus* were the most commonly used plant genera in animal wastewater treatment. However, Kantawanichkul et al. (2003) reported retarded growth of *Scirpus grossus* in experimental tanks after 120 days.

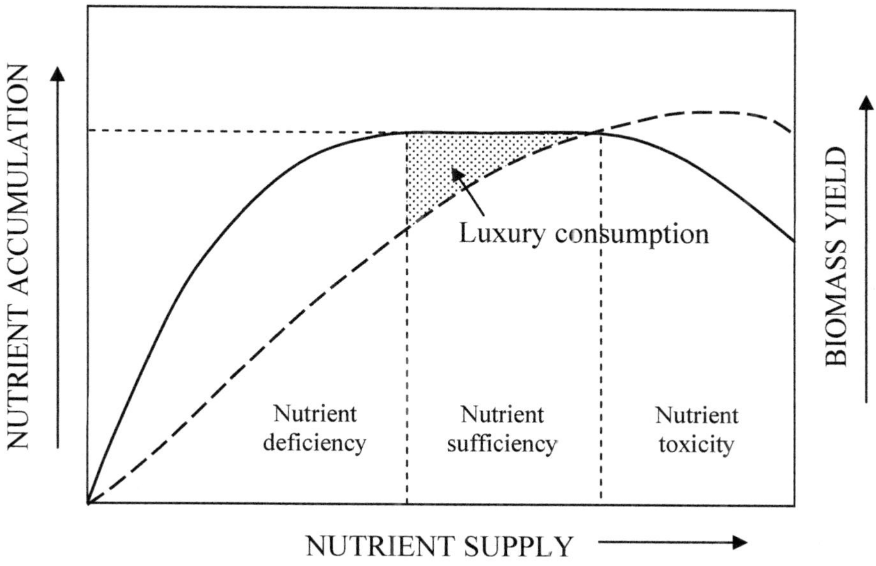


**Fig. 2.** Relationship between nutrient supply and nutrient accumulation (– – –) and biomass yield ( ____ ) (after Reddy and DeBusk 1987)

Nutrient content also changes with the plant and leaf age. Young plants and leaves often have the highest nutrient content, especially nitrogen. As the plant (or the leaf) reaches maturity, its nutrient content decreases. However, since plant biomass increases with maturity, total nutrient storage (bioaccumulation) also increases. Upon senescence, nutrients from the mature leaves are translocated to the growing shoots or storage organs. Thus, dead shoots have less nitrogen and phosphorus levels. Harvesting dead shoots will, therefore, not optimise nutrient removal.

A study by Greenway and Woolley (1999) of seven municipal wastewater treatment wetlands found no significant difference in the nutrient content of plants sampled from the inlet and outlet zones, suggesting nutrient sufficiency even in the final effluent.

In a comparative study of the nutrient content of plants of the same species collected from constructed wetlands and natural waterways, Greenway (1997) found that the plants growing in constructed treatment wetlands had a higher N and P tissue content (Table 5). For most pairs of species, the statistical difference was not significant for nitrogen content, but phosphorus content in the treatment plants was almost double for many species. The variability among control plants may have been due to higher nutrient concentrations even in the natural waterways, particularly nitrogen. These results, however, demonstrate the capacity for increased nutrient accumulation with increased nutrient supply (Fig. 2).

**Table 5.** Comparison of P and N content (mg/g dry wt) in leaf/stem tissue in selected macrophyte species in constructed (treatment) and natural (control) wetlands. (mean ± SD)

| Species | Type | Treatment | | Control | |
|---|---|---|---|---|---|
| | | P | N | P | N |
| *Phragmites australis* | E | 2.0 ± 0.6 | 20.4 ± 8.0 | 1.4 ± 0.6 | 12.0 ± 7.3 |
| *Typha domingensis* | E | 2.3 ± 1.0 | 15.8 ± 6.0 | 1.5 ± 0.9 | 9.2 ± 6.2 |
| *Baumea articulata* | E | 3.7 ± 1.9 | 13.1 ± 4.2 | 2.4 ± 0.7 | 12.6 ± 2.4 |
| *Schoenoplectus validus* | E | 2.6 ± 1.2 | 14.6 ± 5.0 | 0.8 ± 0.5 | 10.5 ± 5.0 |
| *Eleocharis sphacelata* | E | 2.7 ± 1.1 | 15.8 ± 4.0 | 1.3 ± 0.6 | 11.3 ± 5.6 |
| *Ludwigia peploides* | FF | 5.4 ± 1.8 | 36.9 ± 10.2 | 3.1 ± 1.0 | 27.0 ± 10.2 |
| *Persicaria orientalis* | E | 4.4 ± 1.3 | 33.2 ± 10.1 | 1.4 ± 0.6 | 13.5 ± 3.3 |
| *Nymphoides indica* | FL | 6.6 ± 2.0 | 25.8 ± 11 | 2.5 ± 0.6 | 22.1 ± 3.6 |
| *Nymphaea gigantea* | FL | 4.2 ± 2.0 | 28 ± 10 | 2.2 ± 0.5 | 22 ± 8 |
| *Paspalum distichum* | FF | 2.8 ± 1.2 | 12.5 ± 3 | 0.9 ± 0.4 | 11 ± 2.6 |
| *Marsilea mutica* | FL | 8.0 ± 1.6 | 28.8 ± 3 | 2.2 ± 0.5 | 14.7 ± 3.4 |

After: Greenway 1997

Greenway and Bolton (2002) and Bolton and Greenway (1997, 1999) also found that the leaves of *Melaleuca* trees growing in sewage effluent had a significantly higher P content than the leaves and trees growing in a natural *Melaleuca* wetland. However, in P-enriched effluent (12 mg P $L^{-1}$), decreased growth rates indicated P toxicity, with senescent leaves having the highest P content.

### *2.3.1 Nutrient Content of Plant Components*

A comparison of phosphorus and nitrogen in both root/rhizomes and leaf/stem tissue for a varie2ty of macrophytes has been shown in Table 6. Nitrogen content is highest in the leaves, and phosphorus in the root/rhizomes. From Tables 5 and 6, it can be seen that the highest nutrient content occurs in duckweed, followed by *Ludwigia peploides, Ceratopteris, Monocharia, Bacopa, Ipomoea, Ceratophyllum, Nymphaea* and *Nymphoides* - all these species remove nutrients from the water column. There was not a large variation in the mean values of the emergent macrophytes.

**Table 6.** Phosphorus and nitrogen contents (mg/g dry wt) (mean ± SD) in plant parts of native macrophytes in FWS wetlands in Queensland, Australia.

| Species | Root/Rhizome | | Leaf/Stem | |
|---|---|---|---|---|
| | P | N | P | N |
| **Emergent macrophytes** | | | | |
| *Baumea articulata* | 3.5 ± 1.0 | 20.0 ± 4.2 | 1.9 ± 0.5 | 16.3 ± 3.8 |
| *Bolboschoenus caldwellii* | 4.3 ± 1.5 | 13.5 ± 5.0 | 3.0 ± 1.4 | 14.3 ± 5.4 |
| *Cyperus eragrostis* | 3.7 ± 1.8 | 20.0 ± 9.6 | 3.7 ± 1.7 | 16.5 ± 4.2 |
| *Cyperus exaltatus* | 5.0 ± 4.0 | 15.0 ± 7.0 | 3.8 ± 1.8 | 16.7 ± 6.8 |
| *Eleocharis acuta* | 4.0 ± 2.7 | 14.0 ± 5.0 | 3.4 ± 1.5 | 18.8 ± 5.4 |
| *Eleocharis phillipensis* | 4.4 ± 2.0 | 14.0 ± 4.8 | 3.5 ± 1.2 | 17.2 ± 4.9 |
| *Eleocharis sphacelata* | 4.3 ± 2.5 | 13.5 ± 5.7 | 2.7 ± 1.0 | 15.8 ± 4.0 |
| *Rhynochosporus corymbosa* | 2.5 ± 1.9 | 13.9 ± 0.6 | 2.5 ± 0.6 | 16.9 ± 2.4 |
| *Scheonoplectus validus* | 4.0 ± 1.9 | 14.5 ± 7.0 | 3.9 ± 1.3 | 18.2 ± 4.1 |
| *Scleria poiformis* | 2.8 ± 0.8 | 12.3 ± 2.5 | 2.6 ± 1.0 | 16.0 ± 4.0 |
| *Phragmites australis* | 3.2 ± 1.4 | 17.3 ± 7.0 | 2.0 ± 0.6 | 20.4 ± 8.0 |
| *Typha* sp. | 4.0 ± 1.7 | 16.8 ± 10 | 2.3 ± 0.8 | 15.8 ± 6.0 |
| **Floating-leaved attached** | | | | |
| *Nymphaea sp.* | 7.1 ± 0.7 | 30.3 ± 1.7 | 4.0 ± 1.0 | 30 ± 8.6 |
| *Nymphoides indica* | 12.1 ± 3.7 | 19.8 ± 6.3 | 6.6 ± 2.0 | 25.8 ± 11 |
| **Free-floating macrophytes** | | | | |
| *Azolla* sp. | Whole plant | | 7.4 ± 1.0 | 40.0 ± 4.0 |
| Duckweed (*Spirodela* sp.) | Whole plant | | 12.4 ± 4.1 | 39.6 ± 10.3 |
| *Ceratopteris thalicoides* | Whole plant | | 8.3 ± 1.0 | 30.0 ± 9.0 |
| *Monocharia cyanea* | Whole plant | | 7.7 ± 3.1 | 20.9 ± 9.0 |
| **Submerged macrophytes** | | | | |
| *Ceratophyllum demersum* | Whole plant | | 15.4 ± 4.4 | 27.0 ± 6.3 |
| **Vines/creepers** | | | | |
| *Bacopa monnieri* | Whole plant | | 4.8 ± 1.1 | 30.0 ± 11.2 |
| *Ipomoea aquatica* | Whole plant | | 6.4 ± 1.2 | 30.7 ± 12.4 |
| *Ipomoea diamentinenis* | Whole plant | | 7.5 ± 2.1 | 37.3 ± 11.9 |
| *Ludwigia peploides* | Whole plant | | 5.3 ± 2.3 | 32.7 ± 12.3 |
| *Persicaria attenuatum* | Whole plant | | 4.2 ± 1.7 | 24.9 ± 11.4 |
| *Paspalum distichum* | Whole plant | | 4.1 ± 2.0 | 18.6 ± 11.0 |

## 2.4 Biomass Production

### 2.4.1 Plant Biomass

The rate of removal of nitrogen and phosphorus by plants and the incorporation of these nutrients into plant biomass are important for accessing the suitability of macrophyte species for phytoremediation. The turnover rates for plant biomass, individual plants, leaves/stems, and the nutrient storage capacity also need to be considered. The harvesting potential of plant biomass and the rate of regrowth following cropping are important for permanent removal of nutrients from the system.

Emergent macrophytes, in particular *Phragmites* and *Typha*, have very high plant biomass (IWA 2000). Vymazal et al. (1999) reviewed the literature on *Phragmites australis* and found that natural stands of *Phragmites*, growing in eutrophic waters, can achieve a total biomass of 12,700 g/m$^2$. They suggested that in constructed wetlands, high organic loads might stress the plants. However, the nutrient contents of plants from both natural stands and constructed wetlands were comparable.

Tables 7 and 8 provide information on whole plant biomass (shoots, rhizomes/roots) of selected macrophytes growing in a surface-flow constructed wetland in Cairns, tropical Australia (Greenway and Woolley 1999 2001). Biomass for floating macrophytes - duckweed (*Spirodela* and *Wolffia*) and *Azolla* was determined in sections of open water and where the plants grew among emergents. Biomass standing stock of duckweed in open water was 100% higher.

Biomass production values can be used as an indicator to estimate the nutrient uptake capacity of the plants. The uptake capacity of emergent macrophytes is in the range of 30 to 150 kg P/ha/y and 200 to 2500 kg N/ha/y (Brix 1997). Turnover rates for floating macrophytes, however, can be in the order of days or weeks. Under optimal conditions, some species of duckweed can double their biomass in 24 hours (Landolt 1996). Rejmankova et al. (1990) predicted optimum growth of 5.9 g/m$^2$/d by removing 25% of duckweed cover every four days and 2.1 g/m$^2$ by removing 75% of cover. Doubling times of 7-12 days have also been reported in temperate (10-15°C) wastewater ponds (Ozimek 1996). Without any harvesting, however, optimal growth is unlikely.

By comparing the mass removal of N and P from the effluent with the N and P content in plant biomass, it is possible to get some indication of how much N and P may have been removed directly by the plants themselves. In a three-year study, Greenway and Woolley (2001) found that the uptake capacity ranged from 134 to 162 kg P/ha/y and 380 to 474 kg N/ha/y. This represented between 67 and 80% of $PO_4$ removal, and 65-80% of soluble inorganic N removal which had been incorporated into plant biomass.

**Table 7.** Plant biomass (g dry wt/m$^2$), nutrient content of plant tissue (mg/g) and nutrient storage (g/m$^2$) in selected macrophytes (source: Greenway and Woolley 2001)

| Species | Biomass g/m$^2$ mean ± SD | P Content mg P/g | P Storage g P/m$^2$ | N Content mg N/g | N Storage g N/m$^2$ |
|---|---|---|---|---|---|
| *Typha orientalis* | | | | | |
| 1 shoot | 125 ± 75 | 3.85 | 0.48/shoot | 13.5 | 1.69/shoot |
| 14 shoots/m$^2$ | 1750 ± 750 | | 6.74 | | 23.63 |
| *Eleocharis* | | | | | |
| Dense | 1000 ± 250 | 4.2 | 4.20 | 15 | 15.0 |
| mid dense | 500 ± 300 | 4.2 | 2.10 | 15 | 7.5 |
| sparse | 300 ± 140 | 4.2 | 1.26 | 15 | 4.5 |
| *Schoenoplectus validus* | 800 ± 500 | 3.5 | 2.80 | 14.5 | 11.6 |
| *Schoenoplectus validus* (among Typha) | 360 ± 390 | 3.5 | 1.01 | 14.5 | 5.22 |
| *Marsilea* spp | 270 young | 9.5 | 2.57 | 27 | 7.29 |
| | 370 mature | 9.5 | 3.52 | 27 | 9.99 |
| | 470 old | 7.3 | 3.42 | 19 | 9.04 |
| *Nymphoides indica* | 83 ± 20 | 8.2 | 0.68 | 22 | 1.83 |
| *Paspalum distichum* | 860 ± 110 | 2.8 | 2.39 | 16 | 13.30 |
| *Alternathera philoxeroides* | 780 ± 170 | 3.2 | 2.50 | 16 | 12.48 |
| Duckweed (open water) | 40 ± 10 | 14.4 | 0.58 | 43 | 1.72 |
| *Azolla* and Duckweed (open water) | 33 ± 7 | 8.0 | 0.26 | 41 | 1.36 |
| Duckweed (among emergents) | 20 ± 4 | 14.4 | 0.29 | 43 | 0.86 |
| *Azolla* and Duckweed (among emergents) | 16 | 8 | 0.13 | 41 | 0.66 |
| *Ceratophyllum demersum* | 90 ± 30 | 18.95 | 1.71 | 31 | 2.74 |

Plant biomass and annual production rates are high in the tropical/subtropical regions (Greenway and Woolley 2001; Browning and Greenway 2003). In an experimental mesocosm band planted with *Phragmites*, *Schoenoplectus* and *Eleocharis*, mean shoot biomass after 12 months growth was 2200 g DW/m$^2$ (7 g P; 38 g N) (QDNR 2000). Regrowth from cropping yielded an annual production rate of 3500 g/m$^2$ and 10 g P; 58 g N. These growth rates are comparable to the Cairns study (Greenway and Woolley 2001), and further indicate the potential of these macrophyte species for nutrient

**Table 8.** Biomass production (growth rate g/m$^2$/y) after harvesting shoots in selected macrophytes from a tropical surface-flow wetland and a subtropical subsurface-flow wetland

| Species | Whole Plant Biomass (g/m$^2$/y) | | | Harvestable Shoot Biomass (g/m$^2$/y) | | |
|---|---|---|---|---|---|---|
| | g dry wt | g N | g P | g dry wt | g N | g P |
| **Surface flow** | | | | | | |
| *Typha orientalis* | 4000 | 53.7 | 15.3 | 2264 | 30.6 | 8.7 |
| *Eleocharis sphacelata* | 3210 | 46.8 | 13.1 | 918 | 13.8 | 3.9 |
| *Schoenoplectus validus* | 1000 | 14.5 | 3.5 | 581 | 8.4 | 2.0 |
| **Subsurface flow** | | | | | | |
| *Phragmites australis* | 6700 | 127.0 | 16.1 | 3564 | 71.0 | 6.7 |
| *Baumea articulata* | 3470 | 55.5 | 9.0 | 2512 | 41.0 | 4.8 |
| *Carex fasicularis* | 3920 | 60.0 | 9.1 | 2424 | 41.0 | 6.5 |
| *Schoenoplectus mucronatus* | 890 | 15.4 | 3.3 | 842 | 15.3 | 3.3 |
| *Philydrum lanuginosum* | 1000 | 17.1 | 2.5 | 947 | 16.9 | 2.2 |

(After: Greenway and Woolley 2001; Greenway 2002)

removal. In New Zealand, Tanner (2001) reported maximum nutrient accumulations of 8.8-13.4 g P/m$^2$ and 48-69 g N/m$^2$ in total biomass (shoots and below ground) in *Schoenoplectus*, but due to the higher nutrient loading rates compared to the Cairns studies, this only accounted for 2-8% total N removal from the system. Tanner concluded that "uptake and storage of N and P in live plant biomass can usually only account for a fraction of the improved performance of the planted systems". However, in assessing the importance of macrophytes in nutrient removal, it should be recognised that plants have a maximum removal and storage capacity (Fig. 2). Thus, if effluent loading rates are low, then the relative removal efficiency by macrophytes will be higher than constructed wetlands receiving high nutrient loading rates.

## 3. Conclusion

Vegetation is the dominant feature of constructed wetlands; the aquatic macrophytes and their associated microbial biofilms play several vital roles in removing, transforming and storing nutrients. The stems and leaves reduce water velocity and turbulence, causing filtration and settlement of particles (sediment, organic particulates), and provide an increased surface area for the attachment of epiphytic algae and micro-organisms. Oxygen produced in photosynthesis aerates the water. Inorganic bioavailable nutrients for plant and algal growth are removed either from the water column or the sediments. Oxygen transfer from aerial stems to the roots is released into the rhizosphere, facilitating the nitrification/denitrification process. Thus, macrophytes play a major role either directly or indirectly in the removal of nutrients from wastewater.

The performance efficiency of constructed wetlands depends on several variables - these include the quality and quantity of effluent to be treated; the extent of physical, biological and chemical processes functioning within the wetland; the contact time of wastewater with sites of biological and physical activity. Reactive biological surfaces include the plants and associated biofilms, the litter layer and/or sediment and associated microbial communities. The flows and storage volume determine the detention time (hydraulic retention time - HRT) and thus the opportunity for interactions between wastewater contaminants and the wetland ecosystem. In temperate climates, water temperature can influence biological processes. Nutrient removal can be optimised by using macrophytes with a high capacity for inorganic nitrogen and phosphorus absorption and conversion into plant biomass.

However, it should be recognised that all plant species have a maximum nutrient uptake and storage capacity in plant biomass. Species should, therefore, be selected with high nutrient removal capabilities, which means a mixture of macrophyte types (submerged, floating and emergent species) should be used. In tropical/subtropical climates with continuous growing seasons, a mixture of emergent macrophytes, submerged and floating species (duckweed) in surface-flow systems can contribute significantly to the removal of nitrogen and phosphorus from wastewater. However, to maximise removal efficiencies, effluent loading rates should not be too high. Harvesting of plant biomass may be suitable for floating species, such as duckweed, or water hyacinth, which can offer a resource benefit.

## References

Bolton KGE, Greenway M (1997) A feasibility study of *Melaleuca* trees for use in constructed wetlands in subtropical Australia. Wat Sci Tech 35(5):247-254

Bolton KGE, Greenway M (1999) Nutrient sinks in a constructed *Melaleuca* wetland receiving secondary effluent. Wat Sci Tech 40(3):341-347

Brix H (1997) Do macrophytes play a role in constructed treatment wetlands? Wat Sci Tech 35(5):11-17

Browning K, Greenway M (2003) Nutrient removal and plant growth in a subsurface flow constructed wetland in Brisbane, Australia. Wat Sci Tech 48(5):183-190

CH2M HILL, Payne Engineering (1997) Constructed wetlands for livestock wastewater management. Literature review, data base and research synthesis. Gulf of Mexico Program, Stennis Space Center, MI, USA

Gopal B (1999) Natural and constructed wetlands for wastewater treatment: potentials and problems. Wat Sci Tech 40(3):27-35

Greenway M (1997) Nutrient content of wetland plants in constructed wetlands receiving municipal effluent in tropical Australia. Wat Sci Tech 35(5):135-142

Greenway M, Woolley A (1999) Constructed wetlands in Queensland: performance efficiency and nutrient bioaccumulation. Ecol Eng 12:39-55

Greenway, Woolley A (2001) Changes in plant biomass and nutrient removal over three years in a constructed wetland in Cairns, Australia. Wat Sci Tech 44:303-310

Greenway M (2002) Seasonal Phragmites biomass and nutrient storage in a subtropical subsurface flow wetland receiving secondary treated effluent in Brisbane, Australia. 8th IWA International Conference on Wetland Systems for Water Pollution Control, Tanzania, September 2002

Greenway M, Bolton K (2002) Role of constructed *Melaleuca* wetlands in water pollution control in Australia. In: Treatment Wetlands for Water Quality Improvement. CH2M HILL, Canada Ltd

Greenway M (2003) Suitability of macrophytes for nutrient removal from surface flow constructed wetlands receiving secondary treated effluent in Queensland, Australia. Wat Sci Tech 48(2):211-218

Hunt PG, Poach ME (2001) State of the art for animal wastewater treatment in constructed wetlands. Wat Sci Tech 44(11-12):19-25

International Water Association (2000) Constructed Wetlands for Pollution Control: processes, performance, design and operation. IWA Specialist Group on Use of Macrophytes in Water Pollution Control. IWA Publishing

Kantawanichkul S, Somprasert S, Aekasin U, Shutes RBE (2003) Treatment of agricultural wastewater in two experimental combined constructed wetland systems in a tropical climate. Wat Sci Tech 48(5):199-206

Knight RL, Clarke RA, Bastian RK (2001) Surface flow (SF) treatment wetlands as a habitat for wildlife and humans. Wat Sci Tech 44(11-12):27-37

Kvet J, Dusek J, Husak S (1999) Vascular plants suitable for wastewater treatment in temperate zones. In: Vymazal J (ed) Nutrient Cycling and Retention in Natural and Constructed Wetlands. Backhuys Publishers, Leiden, Netherlands, pp 101-110

Landolt E (1996) Duckweeds (Lemnaceae): Morphological and ecological characteristics and their potential for recycling of nutrients. Environmental Research Forum 5-6:289-296

Mitsch WJ, Gosselink JG (2000) *Wetlands*. 3rd ed. John Wiley and Sons Inc

Ozimek (1996) Usefulness of *Lemna minor* in wastewater treatment in temperate climates - myth or fact? Env Res Forum 5-6:297-302

QDNR (2000) Queensland Department of Natural Resources. Guidelines for Using Freewater Surface Constructed Wetlands to Treat Municipal Sewage. QDNR, Brisbane, Australia

Reddy KR, De Busk WF (1987) Nutrient Storage Capabilities of Aquatic and Wetland Plants. In: Reddy, KR, Smith WH (eds) Aquatic Plants for Water Treatment and Resource Recovery. Magnolia Publishing, Orlando, Florida, pp 337-357

Reddy KR, D'Angelo EM (1997) Biogeochemical indicators to evaluate pollutant removal efficiency in constructed wetlands. Wat Sci Tech 35:1-10

Rejmankova E, Kvet J, Rejmanek M (1990) Maximising Duckweed (Lemnaceae) Production by Suitable Harvest Strategy. In: Whigham DF, Good RE, Kvet J (eds) Wetland Ecology and Management: Case Studies. Academic Publishers, Den Haag, The Netherlands, pp 39-43

Tanner CC (2001) Plants as ecosystem engineers in subsurface flow treatment wetlands. Wat Sci Tech 44(11-12):9-18

US EPA (1988) United States Environment Protection Agency Design Manual: Constructed Wetlands and Aquatic Plant Systems for Municipal Wastewater Treatment. EPA/625/1-88/022. Center for Environmental Research Information, Cincinnati, Ohio, USA

US EPA (2000) United States Environment Protection Agency Manual: Constructed Wetlands Treatment of Municipal Wastewaters. U.S. Environmental Protection Agency, Cincinnati, Ohio, USA, pp 152

Vymazal J, Dusek J, Kvet J (1999) Nutrient uptake and storage by plants in constructed wetlands with horizontal sub-surface flow: a comparative study. In: Vymazal J (ed) Nutrient Cycling and Retention in Natural and Constructed Wetlands. Backhuys Publishers, pp 85-100

Wetzel RC (2001) Fundamental processes within natural and constructed wetland ecosystems: short-term versus long-term objectives. Wat Sci Tech 44(11-12):1-8

Williams JB (2002) Phytoremediation in wetland ecosystems: progress, problems, and potential. Critical Reviews in Plant Sciences 21(6):607-635

# 16

# Nitrate Pollution and its Remediation

U.N. Dwivedi[1], Seema Mishra[2], Poorinima Singh[2] and R.D. Tripathi[2]

[1]Department of Biochemistry, University of Lucknow, Lucknw-226007, INDIA;
[2]National Botanical Research Institute, Rana Pratap Marg, Lucknow, INDIA

## 1. Introduction

Groundwater, due to its relative purity, enjoys a privileged place as a potable water source world wide. Among the selected chemical threats to groundwater in the world, nitrate ($NO_3^-$) is listed as second most common pollutant of groundwater next to pesticides (Spalding and Exner 1993; Bachmat 1994). Out of the total earth surface water almost 95% is in ocean, seas, ice caps, glaciers or buried deep under ground leaving only a small fraction i.e. 5% as fresh water, suitable for human consumption. Out of this small 5% of fresh water, approximately 68% is groundwater. It is thus very important to protect the groundwater resources from pollutants which threaten its quality. If not taken care of, it may pose serious problems for human and animal life and whole environment (Babiker et al. 2004).

Nitrate pollution usually originates from diffused sources, like intensive agriculture and unsewered sanitation or point sources, such as irrigation of land by sewage effluent (Bouchard et al. 1992; Eckhardt and Stackelberg 1995; McLay et al. 2001). Nitrate pollution of groundwater caused by agricultural activities has been encountered in almost all regions of the World (Dillon et al. 1991; Bernhard et al. 1992; Spalding and Exner 1993; Lagerstedt et al. 1994; Zhang et al. 1996; Levallois et al. 1998; Hudak 2000). It may also originate from industrial effluents, including paper and munitions maufacturing (Nitrate in News), septic tanks and human and animal wastes, due to biochemical activity of nitrifying bacteria. In groundwater recharge areas with large portions of agricultural land, the nitrate concentration of well water has shown rising trend in many countries within last two three decades. Nitrate leaching from agricultural land must be considered as a non-point source for nitrate pollution of the groundwater (Strebel et al. 1989). High levels of nitrate can leach through soils that have received heavy application of Manure. Water from farm ponds, road ditches or other surface depressions, which collect drainage from poultry houses, feedlots, heavily fertilized fields, septic tanks, or manure lagoons may contain high concentration of nitrate. Nitrate is one of the potential contaminants of groundwater, because it is soluble and moves readily with soil water (Salameh Al-Jamal et al. 1997). Hack-ten Broeke et al. (1996) defined the term 'nitrate leaching risk' as the number of days during the year with a $NO_3^-$ N concentration exceeding a predefined threshold value, for which the EC-directive for drinking water is used (i.e. 50 mg/l nitrate). Nitrate leaching potential is defined as the downward soil water flux from the root zone, possibly causing solute leaching (Hack-ten Broeke et al. 1993). In terrestrial ecosystems, nitrate is subjected to mass flow and leaching.

Groundwater with nitrate concentration exceeding the threshold of 3 mg/l Nitrate Nitrogen ($NO_3^-$ N) or 15 mg/l $NO_3^-$ is considered contaminated due to human activities (so called human affected value (Burkart and Kolpin 1993; EcKhardt and Stackelberg 1995; Agrawal et al. 1999). However, the maximum acceptable concentration of nitrate for potable water according to World Health Organisation (WHO) is 11.3 mg/l $NO_3^-$ N or 50 mg/l $NO_3^-$ (Strebel et al. 1989; Power and Schepers 1989). The current limits have been modified to 10 mg/l $NO_3^-$ N for drinking water which is equivalent to about 45 mg/l of nitrate (Agrawal et al. 1999). Hygienic and toxicological considerations are the decisive reasons for assessing the standard, particularly the risk of methemoglobinemia on infants and of carcinogenic effects of nitrosamino compounds possibly formed (Selenka 1985).

Recent studies in Australian arid zones have shown that bacteria associated with certain soil termites can cause considerable nitrate pollution of shallow groundwater under flash desert precipitation events (Barnes et al. 1992). No such information has yet been gathered from Indian arid zones. Also, denitrification of nitrate leads to production of nitrous oxide causing problem of global warming. In addition, the loss of nitrate from the field has to be considered as the loss of a resource of whose production is linked to the consumption of energy (ca. 47 MJ/ kg N fertilizer) and the emission of atmospherically active substances. On the average, 2500 g $CO_2$, 10 g $N_2O$ and 1 g $CH_4$ are emitted to produce 1 kg of N fertilizer (Kaltschmitt 1997).

## 2. Methods for Estimation of Nitrate Pollution

In order to find out the extent of nitrate pollution, it is essential to have methods for estimating nitrate contamination of the sites. A number of approaches have been used. Thus, traditionally $NO_3^-$ N leaching has been determined using lysimeteres where the drainage water is collected and $NO_3^-$ N content measured (Chapman et al. 1949; Owens 1960; Pratt and Chapman 1961), however, it is expensive method. Pratt et al. (1978) described a cheaper method where the ratio of chloride in the irrigation water, corrected for plant uptake, the chloride below the root zone is used to estimate leaching fraction (LF). The LF, seasonal evapotranspiration and $NO_3^-$ N concentration below the root zone are combined to estimate the $NO_3^-$ N leaching. Difficulties associated with solute leaching model are said to be calibration and their boundary conditions which could not be easily satisfied in complex land use systems and non-uniform strata (Addiscott and Wagenet 1985; McLay et al. 2001).

Another approach is looking for the correlation between the dominant land use in an area and the actual nitrate concentration measured in the underlying aquifers (Barringer et al. 1990; Burkart and Kolpin 1993; Eckhardt and Stackelberg 1995; Levallois et al. 1998; Ahn and Chon 1999; McLay et al. 2001). This approach is based on the assumption that land use influences the nitrogen flow in the surface soil and its consequent leaching out into the groundwater system. Among all these studies, agriculture stands as the most commonly correlated land use with nitrate contamination of groundwater. Severe nitrate contamination is found to be mainly associated with vegetable cultivation, orchards, and floriculture, due to the high rate of application of chemical and organic fertilizers (Salam Al-Jamal et al. 1997; McLay et al. 2001).

Geographical Information System (GIS) is recently being recognized as a powerful tool in environmental studies and modeling (Goodchild et al. 1996). However, they are also subjected to error and uncertainty introduced at almost every step of the spatial information generation and processing, from the data collection to the interpretation of the results (Aronoff 1993). Furthermore, the high value of GIS products in the evaluation, communication, and management of environmental problems is unambiguous. A study employed the GIS technology to investigate nitrate contamination of groundwater by geochemical fertilizers in the Kakamigahara Heights, Central Japan. Data was analyzed to study the extent and variation of nitrate contamination and to establish spatial relationship with responsible land use types. Ninety percent of the water samples showed nitrate concentration above the human affected value (3 mg/l $NO_3^-$), while more than 30% have exceeded the maximum acceptable level (44 mg/l $NO_3^-$) according to Japan regulation. The study indicated the association of pollution level specifically with vegetable fields, which were significantly higher than the under urban land or paddy fields (Babiker et al. 2004).

For isolated sample analysis for nitrate contamination in soil or water bodies, easy and quick methods have been developed. Many commercial nitrate test kits are available which use the heavy metal cadmium to reduce nitrate in the process of nitrate testing. As cadmium is a toxic chemical, nitrate testing done with cadmium puts the person running the test at risk of being exposed to a toxic chemical and the waste generated during the test puts the environment at risk of pollution with a toxic metal. While proper waste disposal can reduce the risk of environmental pollution, only an alternative method will eliminate the risk for the person doing the testing. In order to assess the nitrate pollution problem, Nitrate Test Kits (NTK), based on nitrate reductase, which are environment and user friendly have recently been developed by a company called Nitrate Elimination Company, Inc (NECi). Nitrate reductase used in the kit is very stable making enzyme-based nitrate testing easier than ever. (Campbell et al. 2002, 2004; Patton et al. 2004).

## 3. Sources of Nitrate Pollution

Agricultural activities are considered as a major source of nitrate pollution. With regard to agricultural activities, following three potential sources, causing nitrate pollution of groundwater, may be of high importance.

1. The intensification of crop production is connected with increasing level of nitrogen fertilization. The increasing nitrogen application means a higher soil nitrogen pool, more nitrogen in the nitrogen turnover and thus an increased risk of nitrogen loss.
2. Intensification of livestock production with the consequence of increasing livestock densities and an enormous production of liquid manure per acre cultivated land.
3. Conversion of large areas with permanent grassland to arable land (Strebel et al. 1989).

Nitrogen fertilizers are the N source most frequently cited as the cause for nitrate accumulation in groundwater, however some other sources collectively can contribute hundreds of kg of $NO_3$ N / ha each year. On the regional scale, rivers and lakes receive half of their total N load from agriculture e.g. in the European Union rivers receive 55% (Isermann and Isermann 1997) and in Germany 44% (Werner 1994) of total N from agriculture. Agricultural activities account for 64% of N input into the Lake of Constance (Switzerland) and to natural background concentrations for only 36% (Prasuhn et al. 1996).

Potential sources of nitrate pollution are discussed in details as below:

### 3.1 Fertilization

Compared to natural ecosystems, agro ecosystems are leaky systems with greater amounts of nutrients flowing in and out (Hendrix et al. 1992; Magdoff et al. 1997). In order to make complex nitrogen compounds the plant need a supply of simple nitrogen compounds. So, as agriculture has developed, man has applied more fertilizer to crops to enhance their growth and productivity. Worldwide fertilizer usage peaked in 1989 in terms of total million of tons at 146 after an almost continuous increase since 1950.The decline was reversed in 1996 and may have reached a peak again in 1998 at 137 million tons. While fertilizer used in the U.S. leveled off in 1980 and remained steady at about 20 million tons/year and declined dramatically in the former Soviet Union to about 5 million tons. However, there has been a steady increase in China and India. Data concluded that fertilizer use, on average, would remain constant on a per capita basis as the world population grows unless more efficient fertilizer usage is incorporated in the agricultural practices. (Nitrate in news). Since plants often cannot utilize all the nitrogen applied to the fields, some of it remains left in the soil and it can leach into groundwater. In addition, not all the applied nitrogen gets into the soil and some is washed off the fields in the form of run-off and it flows into the surface water such as streams and rivers. The run-off problem is often greatest when manure is used as a fertilizer such as in US many large commercial farms are used to produce pigs and chickens and companies provide the manure to farms (Nitrate in News).

In US, until after World War II, a diversified crop-livestock production system was commonly utilized in which several crops were rotated to provide feed and forage for several types of livestock. Within the last generation or two, however, these farms have become completely mechanized, eliminating the need for animals, fertilizers have largely replaced legumes as a source of nitrogen, and, often, monocultures have replaced diversified cropping systems (Power and Follett 1987). Especially in the semi-arid and arid regions, millions of hectares have been developed for irrigation. Livestock enterprises have often intensified and are limited to confined areas.

Nitrate commonly accumulates in soils because of fertilizer addition or when a crop demand is much less than the rate of $NO^{3-}$ N production (Jacinthe et al. 2000). Nitrate concentration in well water correlated positively with amount of nitrogenous fertilizers added per unit area per year (Singh and Sekhon 1976; Schepers et al. 1991). Of the various N sources, the farmer has most control over fertilizer N and animal wastes, so control of groundwater nitrates can be achieved most easily through judicious use of these two inputs.

Numerous field studies have shown that seldom is more than 50% of the N input into grain crops removed in the harvested crop (Bock 1984; Nelson 1985). This often leaves 100 kg N/ ha or more either stored in the soil or lost into the environment. Consequently, with annual environmental loadings of this magnitude, it is not surprising that a significant amount of this N eventually migrates to the groundwater. About 50% of the N input either remains in the soil - plant ecosystem or is lost to the environment. For example, Power (1981) estimated that total nitrogen input into US agriculture was approximately 21.1 MT (9.5 MT from fertilizer) and removal in harvested crops equaled 7.9 MT in the year 1977. Most crops remove less than 50% of nitrogen applied (Martin et al. 1970). Kimble et al. (1972) estimated that corn rarely removed more than 50% of nitrogen applied to the clay soil under irrigation and Bingham et al. (1971) found that less than 50% of nitrogen applied to orange trees planted in a sandy loam soil was taken up by the plants. In a study, an average 42% of nitrogen applied as fertilizer is leached from arable soils in England and Wales (Burns and Greenwood 1982) Because fertilizer N is the predominant N input, most of the adjustments in management practices needed to control environmental degradation would probably come from adjustments in fertilizer practices.

The study by the Central Groundwater Board (CGWB) (Mehta et al. 1990) attributed high nitrate levels in dug well waters of Ganjam district (Orissa) to high nitrogen fertilizer use. This was the first study to directly intricate agricultural diffused pollution. High nitrate level in groundwater of Udaipur district of Rajasthan increased five fold by the use of nitrogen fertilizers in the district during 79-89 (Gupta 1992). High nitrate concentrations in 41 samples from villages around Nagpur, metropolitan city and 49% samples from Gulbarga district, Karnataka might also reflect contributions of N leaching from agricultural sources (Bulusu and Pande 1990).

### 3.2 Geologic Origin

Nitrate comes in environment not only from anthropogenic activities but some geological formations also contribute. Studies conducted in early nineties in semiarid regions of North America suggested that it was not unusual for relatively large amount of plant available nitrogen to be present beneath root zones of native prairie vegetation. Concentrations of $NO_3^-$ N as great as 36 μg / g soil 150 cm beneath native range in eastern Montana, at a time when very little of that land was cultivated (Buckman 1910). These results suggest that, in regions where relatively unweathered sedimentary deposits exist beneath the root zone, there is potential for the presence of residual exchangeable ammonium, which is readily oxidized to $NO_3^-$ N when exposed to proper conditions. An additional source of sub-soil $NO_3^-$ N accumulations may result from sub-surface seepage through parched water tables. Water and nitrates could leach through fallow sandy soils until they reached a permeable aquifer. Nitrates would then flow essentially horizontally through the shallow aquifer and exit the soil by a hillside seep. A concentration range of 50-100 mg $NO_3^-$ N/L of seep water is very common.

### 3.3 Precipitation

An appreciable quantity of N is added to most soils annually through precipitation. This N is often in both nitrate and ammonium forms, both of which are commonly washed out of the atmosphere by precipitation. Much of the $NO_3^-$ N in the atmosphere originates from combustion, so values are often greatest downwind from power plants or major industrial areas. Major agricultural sources of atmospheric ammonium are ammonia volatilization from soils, fertilizers, animal wastes and vegetation. Demead et al. (1978) and others have shown that appreciable quantities of ammonia may escape through stomata of plant leaves in the transpiration stream. This process is particularly important during senescence of well-fertilized vegetation. Some of the ammonia escaping the soil and plant surfaces may be reabsorbed and utilized by other plant leaves, with the balance escaping to the atmosphere. Harper et al. (1983) showed that atmospheric ammonia concentrations above the plant canopy are often near ~10g $m^{-3}$, but that these values can temporarily increase to well over ~100g $m^{-3}$ after fertilization with urea.

Total quantity of N added to the soil through precipitation is highly variable and depends on surrounding agricultural and industrial activities. In temperate regions natural ecosystem, where precipitation is the major source of nitrogen, the nitrogen quantity ranged between 10-14 kg/ha/yr.

### 3.4 Waste Disposal

Disposal of wastes is recognized as a major concern not only for health maintenance, but also for environmental protection. Waste disposal from animal production units presents big problems which are common in many countries. Studies in North India showed that animal waste appeared to be the major contributor of nitrate pollution in village environment. The nitrate content of well water near village areas was significantly higher than in cultivated areas (Singh and Sekhon 1976). Little $NO_3^-$ N leached beneath well-stocked and well-managed cattle feedlots because of the hoof action of the animals, coupled with the salt in the diets, resulted in the dispersion of the surface soil, drastically reducing permeability. Presumably, such conditions would enhance denitrification of N in the animal wastes; however, abandoned or poorly managed feedlots were often aerobic, resulting in considerable nitrate production with high potential for leaching (Power and Schepers 1989). Intensively grazed grassland systems showed a high nitrate concentration compared to cut grassland because under grazing more than 75% of the nitrogen ingested by ruminants is excreted as urine and faeces (Strebel et al. 1989).

Many of the large dairy and poultry enterprises are concentrated to markets or otherwise well suited areas for such activity. Often there is very little other agricultural activity in these areas, so agricultural land on which to dispose of these animal wastes is limited. This results in overloading of available land with animal wastes, with considerable N ending up in groundwater or surface water. Such type of problem is most acute in the north-eastern and Great Lakes states of the U.S. because of the relatively high densities of dairy and poultries and the limited availability of agricultural land for waste disposal (Power and Papendick 1985). European countries are also suffering with such problems, like eastern part of the Netherlands and north-western Germany. Thus there are several reasons with nitrogen production by livestock of more than 250 kg nitrogen /ha/yr (Strebel et al. 1989).

### 3.5 Cultivation

Cultivation also contributed to groundwater nitrate pollution by leaching of nitrate beneath the root zones. Evidence of nitrate movement below the root zone for cultivated soils receiving essentially no manure or fertilizer N inputs has been presented by a number of investigators (Buckman 1910; Stewart et al. 1967; Boyce et al. 1976; Brown et al. 1982). This N may amount to several hundred kg / ha and can contribute significantly to groundwater contamination with nitrates. In European countries, conversion of permanent grasslands to arable land causes strongly enhanced leaching for a limited time period. Mean $NO_3^-$ N concentration of the annual groundwater recharge show rather high concentration for sandy soil with arable crops, intensively managed grazed grasslands and field cropping of vegetables. The $NO_3^-$ N concentration exceeded drinking water limit of 11.3 mg N/L by a factor of between 2 and more than 4 (Landreau and Roux 1984; Overgaard 1984; Rohmann and Southeiner 1985; De Smedt and Loy 1985; Foster et al. 1986). Linn and Doran (1984) showed that rates of mineralization and nitrification of organic sources of N in the soil increase as water-filled pore space increases to near 60 % of total pore volume (approximate water content at field capacity). At higher water-filled pore space values, mineralization and other aerobic processes decline sharply, and anaerobic processes, such as denitrification, begin. Doran (1987) found that, compared with native sod, water-filled pore space in ploughed soil often favoured rapid mineralization and nitrification for several days or weeks after ploughing. This resulted in a rapid accumulation of $NO_3^-$ N in the surface of ploughing soil, which could have leached below the root zone with sufficient precipitation.

Crop residues produced each year contain 3-4 million metric tons of N, most of which is recycled annually (Power and Papendick 1985). Types of crop residue (legume versus non-legume) and crop residue management system used to determine to a large extent the fate of this N. Residues from a legume, such as soybean, decompose relatively rapidly, and much of the N in legume residues is mineralized and utilized by the next crop grown (Power et al. 1986). Residues from non-legumes, such as cornstalks and wheat straw, decompose much slower and often initially result in immobilization of inorganic N in the microbial biomass associated with the decomposition process. The subsequent mineralization of this N is a relatively slow process. Consequently, seldom do appreciable quantities of soil nitrates accumulate in the soil after addition of non-legume residues. Method of cultivation can also have a major effect on cycling of N and the accumulation of nitrates. Disturbing the soil with tillage (ploughing, disking) increases aeration and mixes crop residues with readily available carbon sources intimately with soil organisms. With access to ample supplies of both oxygen and energy from the carbon source, microbiological activity is usually greatly enhanced after tillage until the soil becomes too dry (Doran and Power 1983).

### 3.6 Irrigation

A special mention is made of irrigated agriculture because nitrate contamination of groundwater is especially prevalent in irrigated areas. For sustained irrigation, some leaching must occur periodically to remove soluble salts brought in with the irrigation water. Unlike rain-fed agriculture, a significant quantity of salt is introduced with all irrigation waters, and these must be flushed out of the root zone every year or two. If the leaching occurs at a time when appreciable nitrates are present in the root zone, these nitrates are then leached into the vadose zone and, eventually, into the water table (Power and Schepers 1989). In irrigated regions of the Great Plains in US, much of the leaching occurs during the winter and spring months, when actively growing crops are absent (Schepers et al. 1985; Hergert 1986). Ideally, for both reduced cost of operation and maintenance of groundwater quality, a farmer would like to use management practices that minimize the amount of residual nitrate in the soil during this non-crop period.

## 4. Landscape Physiology Affecting Nitrate Flux

Haag and Kaupenjohann (2001) reviewed extensively how landscape components either facilitate or impede N translocation from the field to the stream (headwater). They have categorized landscape in two components, ecotones/retention compartments and conduits/corridors. Retention compartments, like the capillary fringe/ saturated zone and riparian vegetation eliminate N through denitrification, whereas conduits, such as macropores, preferential interflow-paths, drainage, tiles and streams, rapidly relocate nitrate to headwaters. Thus retention compartments play an important role for denitrification process and eliminate nitrate pollution. However they have also emphasized on adverse effects arising from denitrification.

Leached nitrate passes a number of compartments and landscape elements prior to discharge to the aquatic system. Having left the root zone, nitrate passes the vadose zone (sub-soil) and a capillary fringe, eventually reaching an aquifer. Often distinct aquifer storeys co-exist, in particular an unconfined shallow aquifer may be underlained by (semi-) confined, deeper aquifers. Lateral transport of nitrate takes place in interlow, drainage tiles and aquifers. Retention of nitrate is either due to plant uptake or to denitrification. While the first represents temporary storage in the system, the latter leads to the elimination of N from the system.

Ecosystem theory conceives various landscape compartments within a nested hierarchy constituting holons as basic units (Ahl and Allen 1996). Holons are delimited by boundaries acting as differentially permeable membranes (Wiens et al. 1985). Boundaries increase landscape resistance and they are important control points for material flux. Corridors are conduits which connect holons on a large scale (Allen and Hoekstra 1992) and are expressed as preferential flow-paths. Different landscape compartments provide resistance or free flow of nitrates and play an important role in nitrate cycling. Retention time of leachate in soil and underlying substrates varies from days (karst) to decades (fine-textured, thick substrates without fissures), thus N passage to aquifers may be retarded considerably (Hölting et al. 1995). Aquifers are of three types (Davis and DeWiest 1991; Hölting 1980): unconsolidated, porous aquifers (gravel, sand), consolidated aquifers (cracks in solid rock) and karst aquifers (fractures). Retention takes place in transition zones while fissures and fractures serve as conduits. Groundwater transport is usually slow and can retard discharge of nitrate to streams for years to decades. Riparian zones improve water quality due to sedimentation, plant uptake, retention in soil and microbial processes (Correll 1997). A major factor, however, for the realization of retention potentials and the effectiveness of buffer zones, is hydrological setting (Addiscott 1997; Correll 1997; Haycock et al. 1997). It determines the residence time. Riparian forests of different hydrological positions thus vary in nutrient retention (Risser 1990). Zone of contact of groundwater and surface water are called as hyporheic zone where connections are bidirectional (Bencala 1993). Flow through conduits, i.e. preferential flow paths, is generally quite fast reducing time of contact with soil surface, minimizing retention and conveying nitrate rapidly into aquatic systems, and thus contributing to water pollution (Kohl et al. 1971; Mosley 1982; Bouma 1992; Bach et al. 1997). When considered globally, nitrate from contaminated local streams is conveyed to large rivers and ultimately to sea, which are ultimate sink. Average discharge to North Sea is 1450 kg N/km$^2$/yr (Howarth et al. 1996).

## 5. Role of Nitrifying and Denitrifying Microbes in Nitrate Pollution

Urea ($NH_2CONH_2$) is the most common nitrogenous fertilizer applied to agricultural fields in India and is also a major intermediate product of protein metabolism. Urobacterial and other microbes hydrolyze urea by the following reaction:

$$\underset{\text{Urea}}{CO(NH_2)_2} + \underset{\text{Water}}{2\,H_2O} \longrightarrow \underset{\text{Ammonium carbonate}}{(NH_4)_2CO_3}$$

This reaction can take place both in aerobic and anaerobic conditions. Ammonium carbonate being a salt of the weak acid and weak base, easily hydrolyzes as follows:

$$\underset{\text{Ammonium Carbonate}}{(NH_4)_2CO_3} + \underset{\text{Water}}{HOH} \longrightarrow \underset{\text{Ammonium hydroxide}}{NH_4OH} + \underset{\text{Ammonium bicarbonate}}{NH_4HCO_3}$$

Dissociation of the ammonium hydroxide is expressed by the equilibrium:

$$\underset{\text{Ammonium hydroxide}}{NH_4OH} \rightleftharpoons \underset{\text{Ammonia}}{NH_3} + \underset{\text{Water}}{H_2O}$$

The ammonia so derived is bacterially oxidized in two distinct phases under the action of microorganisms *Nitrosomonas* and *Nitrobacter*. The first step is activated by *Nitrosomonas*:

$$\underset{\text{Ammonia}}{2\,NH_3} + \underset{\text{Oxygen}}{3\,O_2} \longrightarrow \underset{\text{Nitrous acid}}{2\,HNO_2} + \underset{\text{Water}}{2\,H_2O} + 158\ \text{Kcal}$$

The second step takes place in presence of *Nitrobacter*:

$$\underset{\text{Nitrous Acid}}{2\,HNO_2} + \underset{\text{Oxygen}}{O_2} \longrightarrow \underset{\text{Nitric acid}}{2\,HNO_3} + 43\ \text{Kcal}$$

Both the above reactions are exothermic, very slow and mildly temperature dependent. Conversion of 10 mg of ammonium nitrogen into nitrite may take 15 days and 10 mg of nitrite may take 40 days to change into nitrate. The variation of the temperature range 9-26°C does not result in a change in speed of reaction but it is slowed down as the temperature drops below 9°C. At 6°C the reaction is substantially impeded and at 0°C nitrification practically stops (Voznaya 1981). In the tropical to subtropical Indian climate, nitrification proceeds at relatively faster rate compared to those in the temperate climate of western countries

The reverse process, i.e. reduction of nitrate to nitrite and finally to free nitrogen gas, is induced by facultative anaerobes and denitrifying bacteria. Denitrification occurs in presence of nitrogen free organic compounds like carbohydrates, cellulose, salts of volatile fatty acids etc. which are oxidized by the oxygen liberated from nitrates providing energy for the reaction. Denitrifying bacteria can convert the nitrate back to nitrite and nitrogen by anaerobic reduction, but in the absence of such a process, nitrate in filtering deep into aquifers may remain as such for a long time. Denitrifying bacterial genera include *Achromabacter, Alacligenes, Bacillus, Chromobacter, Cornebacterium, Halobacterium, Moraxella, Paracoccus, Propionibacterium, Pseudomonas, Spirillium, Thiobacillus, Xanthonas.*

This process of denitrification takes place in the absence of free oxygen and in presence of organic matter can be expressed by the following generalized equation:

$$\underset{\text{Pot. Nitrate}}{4\,KNO_3} + \underset{\text{Carbon}}{5C} \longrightarrow \underset{\text{Pot. Carbonate}}{2K_2CO_3} + \underset{\text{Carbon dioxide}}{3\,CO_2} + \underset{\text{Nitrogen}}{2N_2}$$

The process of denitrification can also take place in absence of organic matter when sulphur is present:

| $6\,KNO_3 + 5S + 2CaCO_3 \longrightarrow$ | $3K_2SO_4 + 2CaSO_4 + 2CO_2 + 3N_2$ |
|---|---|
| Pot. Sulphur Calcium carbonate | Pot. sulphate Calcium sulphate Carbon dioxide Nitrate |

The above reaction is catalysed by *Thiobacterium denitrificans.*

Organic carbon is the key limiting factor for denitrification in sub-soils, so that movement of carbon from the soil surface is necessary to support denitrificatin (Rice and Rogers 1993). Anaerobic conditions are another precondition. Depending on soil type and agricultural land use denitrification losses ranged from 1 to 223 N kg/ha/yr in a number of field experiments (Wendland 1992), however, it may be insignificant under certain conditions (Rice and Rogers 1993). Denitrification can be substantial or very little depending on the type of aquifers (Rice and Rogers 1993; Mariotti 1994). Autotrophic denitrification, requiring an inorganic source for oxidation e.g. pyrite, is uncommon in groundwater (Hiscock et al. 1991). Denitrification is considered an important mechanism attenuating nitrate concentration in shallow unconfined aquifers (Lowrance and Pionke 1989; Montgomery et al. 1997). Denitrification is very common in wetland ecotones, the riparian zones (Gilbert et al. 1990). In riparian zones of the river Garonne in France, denitrification was so intensive that approximately 30 m of groundwater flow under a woodlot were enough to remove the entire nitrate (Pinay et al. 1990).

Denitrification though helps in nitrate removal from environment, there are associated risk and problems with it. Production of nitrous oxide due to denitrification leads to problems on a global scale as nitrous oxide is a very efficient greenhouse gas (Houghton 1994) and plays a role in stratospheric ozone depletion (Crutzen 1970). Depending on fertilizer type, 0.07-2.7% may evolve as $N_2O$ (Eichner 1990). Shallow aquifers are supposed to be more likely sources of $N_2O$ than confined aquifers (Rice and Rogers 1993) and aquifers on whole could account for 5-10% of total global nitrous oxide source.

## 6. Nitrate Assimilation by Plants

Acquisition and assimilation of nitrogen is a fundamental process, crucial for the growth and development of the plant. An adequate amount of nitrogen is needed for the synthesis of amino acids, nucleic acids and other cellular constituents necessary for the plants. This nitrogen is available in the soil as ammonium ($NH_4^+$), nitrate ion ($NO_3^-$) or as reduced nitrogen derived from the degradation of dead plants and animals. However, as $NH_4^+$ is rapidly nitrified in the soil, $NO_3^-$ is the dominant form of nitrogen available to the plants (Schmidt 1982). Active transport of $NO_3^-$ across the plasma membrane and into the cytoplasm is the first step in the acquisition of $NO_3^-$ (Krapp et al. 1998). $NO_3^-$ thus taken up is reduced either in the roots or in the shoots. Reduction of $NO_3^-$ usually takes place more efficiently in the leaves, due to the availability of the photosynthates namely carbon, reductant and energy (Oaks 1992). The nitrate after reduction to $NH_4^+$ is subsequently incorporated into glutamine and glutamate, which serve to translocate organic nitrogen to various parts of the plant (Lam et al. 1996). Thus, the key enzymes of nitrogen assimilation, catalyzing assimilation of nitrate to amino acids glutamate and glutamine are nitrate reductase (NR), nitrite reductase (NiR), glutamine synthetase (GS) and glutamate oxoglutarate amino transferase (glutamate synthase; GOGAT) and glutamate dehydrogenase (GDH) (Fig. 1).

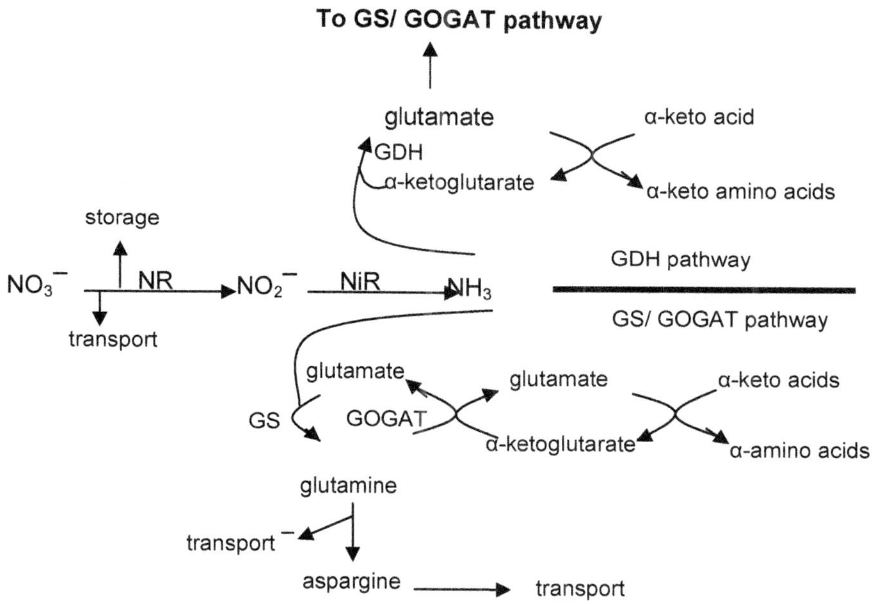


**Fig. 1.** Nitrate assimilation pathway in plants

A detailed description of nitrate assimilation pathway in plants is presented here. The first committed step of this pathway is catalyzed by NR, a cytosolic enzyme that brings about two electron reduction of $NO_3^-$ to $NO_2^-$ using NAD(P)H as the electron donor. The NADH specific (E.C. 1.6.6.1) NR is present in most of the higher plants and algae, while the NAD(P)H bispecific NR (E.C. 1.6.6.2) is present in a few higher plants (Warner and Kleinhofs 1992; Campbell 1999). The NADPH specific NR (E.C. 1.6.6.3) is found only in fungi. NADH NR is a homodimeric enzyme with each subunit of 100 kD. Each subunit has a FAD, Cyt c $b_{557}$ and a molybdenum cofactor as the prosthetic group. The enzyme possesses two active sites, one at which NADPH donates the electron to FAD, initiating the transport of electron via heme Fe and Mo/ Mo pterin and the second active site, where $NO_3^-$ is reduced to $NO_2^-$ (Campbell 1999). Since the $NO_2^-$, thus formed, is toxic due to its mutagenic property owing to its ability to diazotize the amino groups, elaborate mechanisms exist to regulate the activity of NR. The amount of NR present in the plants is 0.05 % of the total extractable protein present in the plant, with the major portion being confined to the leaves. The enzyme is also characterized by a short half-life of only few hours (Campbell 1996). Both inducible and constitutive NR are present in higher plants. Inducible NR is regulated by several environmental factors with light and $NO_3^-$ being of prime importance (Lillo 1994).

The regulation of NR involves, both transcriptional and post transcriptional mechanisms which regulate the amount of NR transcript, while the post translational mechanism regulates NR activity in immediate response to its environment (Su et al. 1996). The post-translational mechanism involves the inactivation regulation. Such inactivation regulation is by the phosphorylation of NR, binding of $Mg^{2+}$ or another divalent cation and an inhibitor protein (Kaiser and Huber 1994; Glaab and Kaiser 1995; MacKintosh et al. 1995). The binding of the inhibitory protein causes a conformational change that impedes the electron flow from heme Fe to Mo pterin cofactor domain (Huber et al. 1996).

The subsequent conversion of $NO_2^-$ to $NH_4^+$ is a six electron reduction process, catalyzed by the plastidic enzyme NiR, which uses ferredoxin or NADH as the electron donor. In most plants, NiR is a monomeric polypeptide of about 60-63 kD and contains a siroheme prosthetic group (Warner and Kleinhofs 1992). Since the product $NH_4^+$ is phytotoxic, due to its ability to uncouple respiration even at low concentration, it must be assimilated rapidly into non-toxic organic compounds. Apart from the $NH_4^+$ generated via reduction of $NO_3^-$, $NH_4^+$ can also be generated by a variety of metabolic pathways such as photorespiration, phenyl propanoid metabolism, utilization of nitrogen transport compounds, amino acid catabolism and symbiotic nitrogen fixation (Cren and Hirel 1999). Both exogenous and endogenous $NH_4^+$ thus obtained are further assimilated by enzymes glutamine synthetase and glutamate synthase, acting in concert (Lea and Miflin 1974; Miflin and Lea 1982).

The enzyme GS catalyzes the ATP-dependent conversion of glutamate to glutamine, using $NH_4^+$ as the substrate. GS either with Ferredoxin dependent GOGAT (E.C.1.4.7.1) or NAD(P)H dependent GOGAT (E.C.1.4.1.14) catalyzes the reductive transfer of amido group of glutamine to the α-keto position of 2-oxoglutarate, yielding two molecules of glutamate. One of the two molecules of glutamate thus formed, can function as the $NH_4^+$ acceptor during another GS/GOGAT cycle, while the second molecule can act as substrate for transaminating enzymes. Thus, a flow of nitrogen from $NH_4^+$ into glutamine and glutamate is maintained by the concerted action of these two enzymes namely GS and GOGAT. This reaction is also considered a major route for incorporation of inorganic $NH_4^+$ into organic molecule. These are subsequently utilized for several aminotransferase reactions (Lea et al. 1990).

Two forms of GS are known in higher plants namely GS1, the cytosolic form which is localized in the cytoplasm of the leaf and non-photosynthetic tissue and GS2, the chloroplastic form, present in the chloroplast of the photosynthetic tissue (Cren and Hirel 1999). In both root and shoot of nitrate assimilator C3 plants, majority of the cytosolic GS is confined to the roots. On the other hand, in C4 plants a large proportion of cytosolic GS is present in shoots also. Both the cytosolic and plastidic isoforms of GS are octameric polypeptide of about 39 to 45 kD and 43 to 45 kD respectively. The GS2 subunits have a 50 amino acid signal peptide at the N-terminal end, which target the protein to the chloroplastic compartment (Cren and Hirel 1999). It is generally believed that the GS activity is regulated at the level of transcription (Roche et al. 1993; Sukanya et al. 1994; Temple et al. 1995). However, recent literature shows that GS activity can be regulated at the level of protein turnover. The oxidative modification is the first step in regulation process. The oxidized form of the enzyme is more susceptible to degradation than non-oxidized GS. Histidine and cysteine protect the GS against the metal catalyzed oxidation indicating oxidative modification of the GS active site and the cysteine and histidine being the site of modification. Similarly ATP and ATP/ glutamate in particular provide greatest protection (Ortega et al. 1999).

The higher plants also possess two antigenically distinct forms of GOGAT that use ferredoxin and NADH as the electron carrier respectively. The NADH-GOGAT (E.C. 1.4.1.14) is localized in the plastids of the non-photosynthetic tissue such as roots, while the Fd-GOGAT (E.C. 1.4.7.1) is localized in the leaf chloroplast, where light leads to an increase in the Fd-GOGAT protein and activity. Since GOGAT functions in conjunction with GS during the assimilation of $NH_4^+$ derived from $NO_3^-$ and catabolic reactions, the expression of genes for the cytosolic GS1 and NADH-GOGAT appear to be coordinately regulated. Fd-GOGAT is proposed to be essential for assimilating photorespiratory ammonia as the mutants in Fd-GOGAT die in conditions promoting photorespiration (Lam et al. 1996). The Fd-GOGAT is a Fe-S flavoprotein with subunit of 130-180 kD and functions as a monomer. While NADH-GOGAT also a Fe-S flavoprotein, exists as a monomer with a mass of 225-230 kD (Temple et al. 1998). The Fd-GOGAT in spinach is recognized as a new member of the thioredoxin dependent enzymes in higher plants, thus showing light dependent regulation (Lichter and Häberlein 1998). In higher plants, cells possess up to six different thioredoxins, disseminated between the cytosol, mitochondria and chloroplast (Follmann and Häberlein 1996). All thioredoxins have a single active site in common, which is formed by a tetrapeptide and two adjacent cysteine residue (-C-G-P-C-). Reversible redox changes between the reduced dithiol form and the oxidized disulfide form enable the thioredoxins to transfer or perceive the reducing equivalents. The thioredoxins in the chloroplast, are reduced by the electron supplied by PS I via ferredoxin and ferredoxin-thioredoxin reductase. Finally the thioredoxin transfers the reducing equivalents to regulate the disulfide bridge of the target enzyme. In the chloroplast metabolism, thioredoxins are responsible for light dependent regulation of the enzymes.

Two major forms of GDH have been reported namely, NADH dependent in the mitochondria and NADPH dependent in the chloroplast. Though a high level of GDH exists in plants, its high $K_m$ for ammonia contradicts its active role in assimilation of $NH_4^+$ (Lam et al. 1996). Glutamate and glutamine thus formed serve as an important nitrogen donor in many cellular reactions including biosynthesis of aspartate and asparagine. While aspartate, an integral part of the malate-aspartate shuttle, transfers reducing equivalents from the mitochondria and chloroplast into the cytoplasm, asparagine serves as an important compound for the transport and storage of nitrogen, due to its high stability and high N: C ratio (Lam et al. 1996).

Thus there are two pathways of ammonia assimilation in plants namely GS/GOGAT and GDH pathway. The GS/GOGAT pathway is regarded as the primary pathway for the assimilation of ammonia (Lea and Miflin 1974; Miflin and Lea 1982), while the GDH pathway is operative under conditions of stress, when excess ammonia is produced by catabolic reactions (Cammaerts and Jacobs 1985; Shargool and Jain 1989). The nitrogen assimilation pathway relies on an elaborate regulatory network that responds to a wide range of environmental and internal signals, including nitrate, light, carbon dioxide level, nitrogen and carbon metabolites and phytohormones. Thus, a complex network coordinates $NO_3^-$ assimilation with other metabolic processes (Hoff et al. 1994; Huber et al. 1994; Crawford 1995).

## 7. Biological Toxicity Due to Nitrate Pollution

Nitrate concentration in drinking water at 10-45 mg/L or more is considered to be carcinogenic and the causative factor for Blue babies. The functioning of central nervous system (CNS) and cardiovascular system (CVS) may also be adversely affected by nitrate rich water. Nitrous compounds formed from nitrate and nitrite are potentially carcinogenic (Clifford and Liu 1993b). Nitrate also affects the health of the pregnant women, the elderly people with compromised immune systems and anyone already in poor health.

Many drinking water wells contain $NO_3^-$ N at levels above the accepted concentration of 10 mg/L (Clifford and Liu 1993b). Water containing high concentration of nitrate is unfit for human consumption, and if discharged to fresh water or marine habitat, can contribute to algal blooms and eutrophication (Thorburn et al. 2003). The dead zone in the Gulf of Mexico near the US coast was caused by excess nitrate, which resulted in the growth of algae. When algae die, water is depleted of oxygen as the algae rot and are consumed by bacteria. This drives the fishes out of this region and kills the animals, which cannot escape the oxygen deficient water.

The world's seagrass is shrinking because of urban and industrial pollution, rising the specter of under water prairies turning into marine desert of shifting sand. Australia's estimated 20,000 sq miles of seagrass contain half of the world's estimated 70 species more than anywhere in the world. Scientists have estimated that in the past 15 years, 173 sq miles of seagrass has been lost. CSIRO reports that Australia's urban sprawl was responsible for major seagrass loss. It resulted in disturbed marine ecology.

Nitrate is a potential human health threat especially to infants, causing a condition known as methemoglobenamia, also called "blue baby syndrome. Nitrate is converted in the gut to nitrite, which then combines with hemoglobin to form methemoglobin, thus decreasing the ability of the blood to carry oxygen .In addition to human, it also causes methemoglobinemia in cattles when fed or grazed forager containing high level of nitrate. Nitrate is converted to ammonia in the rumen of cattles by microbes.

Nitrate toxicity is a function of amount and rate at which nitrate is consumed. Under certain conditions, the rate at which nitrite is converted to ammonia becomes limiting and nitrite accumulates and absorbed from rumen, causing methemoglobin (Smith and Gutherie). Chronic consumption of high level of nitrate may also cause other health problem in human such as gastric cancer, bladder cancer, tetragenic effects and non Hodgkin's lymphoma. Low-level exposure to nitrate for long time period could cause certain type of cancer. Nitrate is produced naturally within the body; environmental sources include food (including many vegetables), contaminated drinking water, cigarette smoking and certain medications. Upto 20% of ingested nitrate is transformed in body to nitrite, which can then undergo transformation in the stomach, colon and bladder to form nitroso compounds. These compounds are known to cause cancer in a variety of organs in more than 40 animal species including high primates (Nitrate in News).

Thus, the scope of N impacts ranges from adverse effects on water quality, over acidification and eutrophication of aquatic ecosystems to loss of biological diversity, and to impacts on atmosphere and climate, e.g. nitrous oxide as greenhouse gas as well as health hazards (Lehn et al. 1995; Vitousek et al. 1997).

## 8. Problem Areas for Nitrate Pollution

The groundwater is becoming more polluted with nitrate in the Mid Atlantic US, Northern China plains, Western Europe and Northern India. Terrestrial waters include polar icecaps water frozen at high altitudes, water stored in lakes and rivers and sub-surface rock formations (groundwater). Nitrate level in relatively pollution free areas of continent, such as high altitude lakes and rivers and snow clad mountains helps in understanding the anthropogenic nature of this pollutant. In Central Himalayan snow and ice, the nitrate level is about 0.5 mg/L. The average nitrate level in world rivers is about 1 mg/L which is close to its content in Himalayan rivers e.g. the Bhagirati and the Alaknanda and in the Ganga at Rishikesh, the slight increase in nitrate in river waters compared to that of frozen waters may be due to contributions from biochemically derived soil nitrate (Table 1). In Dal lake of Kashmir (Table 1), nitrate was below detectable range (i.e. $< 0.01$ mg/L) (Handa et al. 1991). No nitrate was detected in rain water lake Dal. It thus emerges that in a relatively pure system, surface water contains less than 1 mg/L nitrate and its higher concentrations in groundwater may therefore reflect anthropogenic or geologic concentrations. In oceans, which are the ultimate sink for terrestrial waters, the average nitrate level is 0.67 mg/L (Mason and Moore 1985). This is slightly lower than in average river water indicating possible biochemical removal by marine plankton. In USA, nitrate in terrestrial water ranges from 1-20 mg/L, the higher content being in groundwater.

Some cases of high nitrate in groundwater in India have been reported by Handa (1975), Kumar (1983) and Kakar (1985). Handa (1975) observed that nitrate in deep waters was only 1-2 mg/L whereas in shallow waters upto 100 mg/L in humid areas and upto 100 mg/L arid and semi-arid regions. One of the reasons for high nitrate content of groundwater is use of unbalanced and excessive fertilizer. Thus, the present consumption of nitrogenous fertilizers in India is 62% of total (N+P+K) fertilizers. In Punjab and Haryana, average annual consumption of fertilizers has attained a level of 162.33 kg/ha and 91.06 kg/ha, respectively in 1991, much higher than other states.

**Table 1.** Nitrate in terrestrial waters in relatively pollution free areas

| Locality | Nitrate (mg/Ll) |
|---|---|
| Central Himalayan snow* | 0.496 |
| Cenral Himalayan ice* | 0.436 |
| Himalayan rivers* | |
| Bhagirathi | 0.310-0.992 |
| Alaknanda | 0.992 |
| Ganga at Rishikesh | 0.806 |
| Lake Dal water** | 1.07 |
| World average river water*** | 1.00 |
| World average ocean water*** | 0.67 |

*Agrawal et al. 1999; ** Handa et al. 1991; *** Mason and Moore 1985

Study undertaken by NEERI reported nitrate in groundwaters from selected districts in 17 states of India including Rajasthan and Karnataka but excluding North Eastern states, Goa and kerala (Bulusu and Pande 1990) found high nitrate content ($> 45$ mg/L) in most of the samples. Mathur and Ranganathan (1990) found high nitrate content (45 to $> 600$ mg/L) in shallow and deep tubewells due to seepage from industrial effluents and urban sewage around Jodhpur city in Rajasthan and Lucknow city in Uttar Pradesh (Singh et al. 1991). In Banglore district, Karnataka, $NO_3^-$ N in water was found to be 50-200 mg/L from nitrogen bearing effluents (Tamta et al. 1992). Khanna et al. (1994) studied 23 samples from shallow dug wells located in Sirsa, Hisar, MahendraGarh, Bhivani and Rohtak district and it was observed that 13 of samples contained more than 45 mg/L nitrate, in 9 samples, nitrate was above 100 mg/L, while it was above 300 mg/L in 6 samples and the maximum being 1030 mg/L in one sample.

In some 58 samples of water collected from topsoil, gravel and laterite horizons (2-14 m depth) in Jhansi district of Central India, high nitrate levels (50 - 108 mg/L) were found in 7 samples (Chandu et al. 1995). Bhinde (1990) has studied the nitrate pollution potential of Indian urban solid waste capable of affecting water table aquifers by leaching.

Elevated levels of nitrates have been found in wells in most countries, giving rise to health concerns. Uncontaminated groundwater usually contains less than 3 mg $NO_3^-$ N / L and highly contaminated water contains over 10 mg $NO_3$ N / L (Madison and Brunett 1985; Agrawal et al. 1999). In US, 1,24,000 well water samples were analyzed in which 80.4% found $NO_3^-$ N concentration upto 3 mg/L, 13.2% had 3-10 mg/L and 6.4% more than 10 mg/L (Madison and Brunett 1985).

Average nitrate leaching from terrestrial ecosystems in Central Europe is 15 kg/ha/yr: N leaching is 15.9 kg/ha/yr in Germany (Werner 1994), 15.0 kg/ha/yr in the watershed Lake of Constance, the second largest European lake (Prasunhn et al. 1996), and 14.7 kg/ha/yr in the canton Bern in Switzerland (Prasuhn and Braun 1995).

In general, areas with the greatest problems with nitrates in groundwater were either those in heavily populated or those in relatively large areas under irrigation. Because agriculture is implicated in the nitrate pollution problem, farmers and rural communities are most threatened populations. In U.S. problem is concentrated in the mid -west and far-west with large areas of Iowa, Illinois, Kansas, Michigan, Wisconsin, Washington, and California being heavily effected (Nitrate in News; Power and Schepers.1989).

Nitrate pollution of groundwater in coastal North-eastern Australia is of particular concern because of its proximity to environmentally sensitive areas e.g. The Great Barrier Reef. 1454 wells in North-eastern Australia have been examined to determine nitrate contamination. The likely sources of nitrate were investigated by comparing $\delta^{15}N$ values of groundwater to those of possible industrial or organic N contaminants. In Berdekin, Mackay, and Bundaberg areas 11% of wells had elevated nitrate concentrations i.e.$\geq$ 20mg/L in which approximately 50% come directly from fertilizer and only eight was likely to have come from organic sources, such as sewage, septic or feedlots overflows (Thorburn et al. 2003).

Various aquifers in The Netherlands, West Germany, and Chalk aquifer in France and Eastern-Central England Showed increasing nitrate concentration with a maximum of 1.3 mg/L/yr increment; some aquifers containing more than 13 mg $NO_3^-$ N/L (Strebel et al. 1989 and references therein). Nitrogen content below the root zone tends to increase with sand content. Nitrogen concentration below the root zone of silt loam was less than 9 mg/L (Saxton et al.1977) compared to 20-175 mg/L for sand loam soil (Smika et al. 1977; Lembke and Thorone 1980; Lund 1982; Ritter et al. 1990). Study on onion (Shallow rooted), chile and alfalfa (deep rooted), field with irrigation efficiency range from 77-80% for onion, 70-76% for chile and 97% for alfalfa, showed nitrogen loading bellow root zone for chile field varied from 290 kg/ha/year (for sand loam soil) to 64 kg/ha (for clay). Nitrogen loading below the root zone of onions varied from 199 kg/ha/year (loamy sand field) to 161 kg/ha/year (clay) and that of alfalfa was only 42 kg/ha/yearin sandy loam (Salameh Al-Jamal et al.1997) The low amount of nitrogen measured below the alfalfa root zone also because alfalfa uses $NO_3^-$ nitrogen that has been leached from previous crop (Stewart et al.1968).

## 9. Management Options for Nitrate

For nitrates to leach through the root zone, vadose zone, and into the water table, two conditions are required: (1) an accumulation of nitrate N in the root zone; (2) excess water available for leaching at the same time. Trends in agricultural practices that contribute to these two conditions are discussed in this section.

### 9.1 Nitrate Accumulation Management

As mentioned previously, the primary N inputs into a production system controlled by the farmers are fertilization and manuring practices. To a some extent, cropping practices (especially those involving a legume crop) could also affect nitrate accumulation. For example, there have been reports of nitrate accumulations and leaching after ploughing in alfalfa fields (Power and Schepers 1989; Salameh Al-Jamal et al. 1997). Fertilizer N has largely replaced legume N as the primary source of N for World agriculture. With these larger fertilizer N inputs, there is no surprise that frequency of reports of nitrates in groundwater is increasing. Most fertilizer N is added in relatively large quantities between harvest of the previous crop and 1 month after planting the next crop. Thus a large pool of $NO_3^-$ N frequently exists in the soil during the non-crop period when much of the leaching occurs. Nelson (1985) outlined several agronomic practices that can be used to reduce the probability of the existence of large nitrate pools in the soil at times when water percolation is most likely to occur. These include reduced fertilizer N rates, split applications, side dressing, or fertigation (applying fertilizers with irrigation water) during the active growing season, and the use of nitrification inhibitors. Another possibility might be the use of legume cover crops to reduce fertilizer N inputs and control rate of soil nitrate accumulation (Power 1987).

Commercialization of eeding operations increased number of livestock production, and consequently huge amount of animal waste. Proper handling of manure is very important for controlling nitrogen pollution.

## 9.2 Water Management

The second factor necessary for nitrate leaching - percolating water - is controlled by water management. Except for sandy soils, most nitrate leaching from cropped fields occurs during the non-crop period. The relatively high water requirements of an actively growing crop, especially after canopy development, usually preclude significant water movement below the root zone during the growing season. Possible exceptions include not only sandy soils (Mielke et al. 1979; Schepers and Mielke 1983), but also surface-irrigated soils that are poorly managed.

Irrigation also offers another unique problem relating to groundwater quality. All irrigation water contains some salts, which remain in the soil after the water added is lost by evapotranspiration. Consequently, continued irrigation followed by evapotranspiration of water will eventually result in the accumulation of an unacceptable quantity of salts in the soil. These excess salts are best disposed of by flushing beneath the root zone with excess water. Thus, periodic leaching is an essential part of a well-managed irrigation system. If nitrates are among these salts, they too will move below the root zone. Consequently, in irrigated agriculture, it is especially important to reduce soil $NO_3^-$ N concentrations during the period leaching occurs i.e. non-crop season.

Nitrate leaches mainly through irrigation water, thus managing irrigation time, frequency, and quantity may help in controlling nitrate leaching. There are many methods for irrigation scheduling, ranging from planning based on soil water measurements (Phene et al. 1981; Merriam 1996), crop stress indicators (English et al. 1981), or crop calendars (Hill and Allen 1995) to computer-based systems combined with measurements (Wesseling and Van den Broek 1988; Malano and Wood 1995; Chang et al. 1996; D'Urso et al. 1999). It is generally expected that irrigation will result in a higher download soil water flux and as a consequence, nitrate loss to groundwater (Nguyen et al. 1996; Schneekloth et al. 1996; Chang 1997), but it may sometime reduce nitrate leaching as a result of improved water and nitrogen (N) uptake by the crop (Dijkstra and Hack-ten Broeke 1995).

In irrigated areas that have high groundwater nitrate levels, nitrates in the groundwater can be a significant source of N for the crop (Schepers et al. 1985). For example, applying 30 cm of irrigation water containing 20 mg / L $NO_3$ N results in the addition of 60 kg N / ha to the crop. This quantity of available N applied in the irrigation water needs to be included in any calculation of fertilizer requirements.

A second approach would be to reduce soil water and $NO_3^-$ N contents during likely periods for leaching. In many regions, double cropping or winter cover crops are appropriate for this purpose.

Several courses of action are available for short-term solutions to problems of groundwater nitrates, which include soil testing, fertilizer recommendations, and irrigation scheduling, by which groundwater nitrate accumulation under irrigated agriculture can generally be controlled. Soil tests for residual nitrates to at least 1-m depth were made on all participant fields, and fertilizer N was applied accordingly, based on field calibrated algorithms developed by the University of Nebraska-Lincoln (1984). Likewise, water meters were installed on irrigation systems, and irrigation water was applied in accordance with climatologically based irrigation scheduling models adapted by the University of Nebraska-Lincoln (1984). Results of this demonstration project, conducted on an area of over 16000 ha of predominantly irrigated corn, showed that corn yields could be maintained while reducing fertilizer N inputs by an average of 88 kg N / ha and reducing irrigation water inputs by 47 mm annually. During this period, $NO_3^-$ N concentration was measured in 589 irrigation wells within the demonstration area, and results indicated that the previous 20-year trend for increased groundwater $NO_3^-$ N concentrations was stopped (Power and Schepers 1989). Best management practices (BMPs) decrease nitrate application amounts by applying fertilizer in a schedule matching crop nitrogen uptake and minimising the leaching fraction through proper irrigation scheduling (Salameh Al-Jamal et al. 1997).

In order to reduce the leaching of nitrate, it is very important to minimise the residual nitrate content in the root zone at harvest time or in other words make the uptake of the plant available nitrogen in the root zone by the crop as effective as possible and to preserve the available nitrogen during the main leaching period in the form of biologically fixed plant nitrogen within the nitrogen cycle.

Under semi-arid or arid conditions, agricultural needs intensified irrigation and fertilization practices which introduces a long-term risk of groundwater pollution by unused fertilizers, e.g. nitrogen, salts and pesticides, herbicides, leached from irrigated fields. There is a need to optimize the use of water and fertilizers applied to any field to match the crop requirement to be grown on that field and to suit to the conditions prevailing in any particular area (Hadas et al. 1999). In general, from agricultural benefits and pollution point of view, conditions should be optimized for each and every area separately. Hadas et al. (1999) conducted a study in Israel to determine the amounts of water and salts leached below several agricultural areas and to try to relate them to the yields obtained. The results showed that intensified agriculture leads to increased hazards to surface and groundwater pollution and this can be diminished provided balanced irrigation-fertilization programs are developed for different crops.

Simulation models have been applied to study the effect of certain agricultural measures on emissions on field scale (Line et al. 1993; Dijkstra and Hack-ten

Broeke 1995; Rode et al. 1995). However, the simulation of N dynamics and the assessment of output potentials neither address the path nor the fate of nitrate emissions. Attempts are made to adapt life cycle assessment procedures to agricultural production systems (Vito 1998). Life cycle approaches assess the impact of agricultural production systems on the environment in terms of effect potentials; they disregard the spatial dimension and setting. Most models however take no account of the spatial setting into which agricultural sites are embedded. Budget models should thus also encompass to reduce site specific risk, agricultural activity risk, headwater contamination risk and regional and global scale risk (Haag and Kaupenjohann 2001).

Planning for managing the land use requires at least the evaluation of crop productivity and the environmental consequences. Crop-simulation models are one means to do this. In Hungary, a long term crop rotation experiment with various N fertilizer applications was conducted from 1958-1988 which provided an excellent data set to test the capability of crop simulation models for examining biomass production, yield and nitrate leaching. The study gave good accepted results and showed that when 150 kg/ha/yr or less nitrogen fertilizer was added, leaching gets reduced to level as no fertilizer. However, with a 250 kg/ha/yr application, there was about 100 kg/ha/yr leached and yields were not improved over the 150 kg/ha/yr treatment (Kovács and Németh 1995).

In India, Singh et al. (2002) conducted a study at a coastal site near Machilipatnam, Andhra Pradesh to estimate the nitrogen loss through drainage effluent in sub-surface drained farmer's fields. The nitrogen loss in three forms, namely, $NH_4$-N, $NO_2$-N, and $NO_3$-N was studied from 15, 35, 55 m drain spacing areas. 15 m drain space area was already reclaimed decreasing its salinity to low level but not the 35, and 55 m drain space areas. Studies showed that predominantly, $NO_3^-$ N leached (82%) in 15 m drain spacing areas whereas the ammonium form contributed (93% and 82%) leaching in 35 m and 55 m drain spacing areas, respectively.

Allaire-Leung et al. (2001) has demonstrated a contradictory result in a study where they studied the nitrate leaching in fields of carrot crops in California, USA. As these fields were subjected to non-uniformity of irrigation, it was expected that an uniform irrigation practice would improve nitrate loss. However, study showed that nitrate leaching correlated with $NO_3^-$ N content of soil, but not correlated to irrigation depth, irrigation uniformity and deep percolation, thus demonstrating that irrigation non-uniformity has less effect on nitrate leaching.

## 9.3 Remediation of Existing Nitrate Pollution

### *9.3.1 Physical Remediation Methods*

There is a need to develop approaches for removing nitrate from groundwater because of its adverse effect in aquatic environment (Vitousek et al. 1997).

Application of treatment technology to restore contaminated groundwater is becoming increasingly important particularly where alternative water species are not available (Bouwer 1989). Different approaches have been proposed and used to remove nitrate from groundwater. These include membrane separation, ion-exchange, biological denitrificatin etc. Abiotic remediation methods are also being used for remediation of nitrate from soil by using electrokinetics coupled with zero valent iron (Fe) treatment wall. In electrokinetics, iron wall process relies on ability to remove water and dissolved contaminations (i.e. Trichloroethene, $NO_3^-$ N) through low permeability soils by electrosmosis and electromigration and nitrogen allocation where they can be treated within the iron treatment wall. In this method, 54 to 87% $NO_3^-$ N got transformed to ammonia and nitrogen gases (Chew and Zhang 1998).

In catalytic reduction process, nitrate can be removed selectively from groundwater in contrast to ion-exchange or membrane separation process. In a study, among three catalysts, paladium (Pd), platinum (Pt) and rhodium (Rh), rhodium was found most effective in removing nitrate. Rhodium catalyzesd at -400 mv and 6 hrs reaction time to decrease nitrate concentration from 40 to 11.9 mg/L (Reddy and Lin 2000).

Thus, the best option to maintain high $NO_3^-$ N removal rates and to reduce the proportion of $N_2O$ in the emitted gases is to maintain the high water table for a prolonged period in the most biologically active portion of the soil profile (Jacinthe et al. 2000).

### *9.3.2 Bioremediation*

In recent years, there has been growing interest on bacteria which can remove nitrate from hypersaline wastes. A denitrifying moderately halalkalophylic bacterium, *Halomonas campisalis* was isolated and characterized (Mormile et al. 1999). Liu and Clifford (1996) demonstrated a hybrid biological denitrification ion exchange process. Clifford and Liu (1993a) isolated a salt tolerant denitrifying organism from sewage sludge. Recently nitrate reducing and sulfide oxidising bacteria were characterized from oil field brines (Gevertz et al. 2000). The transformation of uranyl nitrate and other compounds in high ionic strength brines by a halomomas sp. (WIPPIA) under denitrifying conditions has also been demonstrated (Francis et al. 2000). Perchlorate reducing bacteria such as *Perclace* can also reduce nitrate (Herman and Frankenberger Jr 1999). The nitrate and perchorate ions can be co-removed by a co-culture of *Perclace* and a salt tolerant bacterial isolate, *Citrobacter* sp. which significantly reduced both perchlorate and nitrate to 34.9 and 15.6%, respectively in one week (Okeke et al. 2002). In a study, *Shewanella oneidensis* strain MR-1, a metabollically versatile bacterium that can use a diversity of compounds to obtain energy needed for growth and survival, have been shown to clean nitrate polluted water bodies (Shewanella Federation).

Crop rotation may prove to be a good management option to reduce the $NO_3^-$ N below the root zone e.g. deep rooted alfa-alfa in rotation with shallow rooted crops would appear to be a good management option to reduce the $NO_3^-$ N that passes below the root zone (Salameh Al-Jamal et al. 1997).

Porous treatment walls are increasingly used for remediation of contaminated groundwater. These walls were constructed below the water table and perpendicular to the groundwater flow. Successful nitrate removal from groundwater has been demonstrated in porous walls amended with sawdust (to promote anaerobic environment and provide energy source for denitrifying bacteria). Constructed denitrification walls offer a good approach for nitrate removal from shallow groundwater (Robertsen and Cherry 1995; Schipper and Vojvodić-Vuković 1998).

Water table management has been proposed as a way of removing excess nitrate from soil and protecting sub-surface water from $NO_3^-$ N pollution. Water table management stimulates denitrifying bacteria thus removing the accumulated $NO_3^-$ N by converting it to $N_2O$ (a green house gas) and $N_2$. It involves creating saturated conditions in upper portion of the profile by raising the water table as a result oxygen is rapidly depleted in the soil pores thus creating conditions favorable for denitrification. Denitrification is the biological process whereby $NO_3^-$ N is used as an alternative electron acceptor by soil microbes and is converted into nitrous oxide and nitrogen gas. Smith (1980) indicated that the ratio of $N_2O$ : $N_2$ emitted during denitrification depends upon the balance between the rate of nitrous oxide diffusion from the site where it is produced and the rate of $N_2O$ reduction. When $N_2O$ diffusion was restricted, this gas was converted to $N_2$ (Letey et al. 1980). If denitrification occurs deeper in the soil profile, the mole fraction of $N_2O$ would be smaller than if the process took place near the soil surface.

Phytoremediation is the use of plants to remove pollutants and other toxic materials from both soil and water. Each remediation situation is unique and when treated successfully with the correct quantity and species of plants, restoration can reach up to a 100% effectiveness. In the arena of water pollution, wetland treatment systems have been a popular choice for a variety of pollutants. These systems have been implemented all over the world. They can offset the cost of chemical treatments and are an alternative to regions too remote, too small, or too economically disadvantaged to support standard waste water treatment plants. Depending on the plants used, these systems can remove bacteria, improve dissolved oxygen content, and reduce the level of nitrates along with other human-generated pollutants.

Duckweeds can be used for wastewater treatment. These plants absorb pollution causing nutrients from waste. Duckweeds refer two species of free floating; stem less aquatic plants appreciated for their use in waste treatment, animal feed and pharmaceuticals. These can be used to cleanse wastewater-reducing nitrogen and phosphorus in human waste, can be daily harvested, and dried out. The resulting olive green material would be sold as feed for livestock, and for fishes. Duckweed contains some kind of protein levels as the soybean has (Nitrate in News). Three aquatic plant species namely, *Cladophora sp* (a filamentous algae), *Scirpus pungens* (a member of bulrush family) and *Elodea canadenesis* (a water weed) have been shown to effectively remove nitrates from nutrient enriched water bodies (Seewane). Plants, such as bulrush, water lilies, arrowhead, cattail, sweet flag, water hyacinths, bamboo and poplar have been shown to clean the water polluted with nitrates and make it safe for wild life and human alike (USEPA 1996; Schnoor et al. 1995).

A riparian zone located below and adjacent to a field-sized watershed planted with soyabeans eliminated up to 93% of groundwater nitrate (Line et al. 1993). In a large number of studies, riparian nitrate removal exceeded 93% (Hill 1996) and removals of 90% seem to be common. Sustainability of riparian buffers may be affected, however, by declining availability of organic carbon for denitrification and decreasing uptake by old vegetation (Haycock et al. 1993).

A novel biotechnological method for removing nitrate from contaminated water has been developed. In this method, nitrate is reduced to nitrogen gas with no residuals left to contaminate the water. Thus the method can be applied in a on-line fashion to purify drinking water. Nitrate containing groundwater is passed through a column where three enzymes namely, nitrate reductase, nitrite reductase and nitrous oxide reductase are coimmobilized along with electron-carrying dye which are energized by electrical current, which provide the reducing power to drive the conversion of nitrate to nitrogen gas and water. Prototype columns for field tests are made to process 500 liters of water per min. Nitrate can be completely removed with a single pass and no contaminating residues are left in water (Mobetec GmbH, German Patent Application 1990).

The biotechnological method of nitrate removal, as outlined above, is also relevant to the cleaning up of huge nitrate wastes accumulated in defense laboratories devoted to explosive manufacturing. The majority of explosives, such as TNT (trinitrotoluene), RDX (royal demolition explosive) and HMX (high melting explosive, 1,3,5,7-tetranitro-1,3,5,7-tetrazacyclooctane) are organic nitro-compounds. These compounds are used in various military applications and to implode fissionable material in nuclear devices. Contaminated wastewaters generated at the explosives handling facilities are usually released into the environment leading to the pollution problem. The HMX-wastewater contains high levels of HMX; up to 350 mg/L alongwith high levels of inorganic nitrates (as high as 100,000 mg/L). Therefore, prior to biological treatment of the HMX-wastewater, the nitrates need to be reduced to concentrations tolerable to the microorganisms or plants used for the treatment.

## 10. Conclusion

Humans have had a major impact on the earth's water reservoirs: rivers, lakes, oceans as well as groundwater. Nitrate is listed as second most common pollutant of groundwater next to pesticides. Whether it is by deforestation of riparian zones, inundating agricultural fields with fertilizer, faulty septic systems or poorly designed waste water overflow systems, the detrimental effects of human activities have started to become apparent. With the growing awareness of the increasing nitrate problem and its impact on ecosystems as well as human health, the question remains: what alternatives do we have? Are our only choices to reduce the human population, dig millions upon millions of miles of tunnels underneath towns and cities, prohibit the use of fertilizers, or fund tertiary waste water treatment? Some of these suggestions are more far-fetched than others. In the past fifty years or so, strides have been made using processes which incorporate physico-chemical or biological means to help restore an area or remove this pollutant from soil and water. The fact is that most of the above actions are either extremely expensive or completely unethical, a much less expensive, and more environment friendly alternative could be phytoremediation. Taking into consideration the world's growing population and the adverse effect humans have had on the nitrate concentrations of water bodies, more measures both effective and eco-friendly are needed to remedy the menace of growing $NO_3^-$ pollution in groundwater.

## References

Addiscott TM (1997) A critical review of the value of buffer zone environment as a pollution control tool. In: Naycock NE, Burt TP, Goulding KW, Pinay G (Eds.), Buffer zones: their processes and potential in water protection. Quest Environment, Hertfordshire, pp. 236-243

Addiscott TM, Wagenet RJ (1985) Concepts of solute leaching in soils: a review of modeling approaches. J Soil Sci 36:411-424

Agrawal GD, Lunkad SK, Malkhed T (1999) Diffuse agricultural nitrate pollution of groundwaters in India. Wat Sci Tech 39:67-75

Ahl V, Allen TFH (1996) Hierarchy Theory: A Vision, Vocabulary, and pistemology. Columbia University Press, New York

Ahn H-I, Chon H-T (1999) Assessment of groundwater contamination using geographic information systems. Environ Chem Health 21:273-289

Allaire-Leung SE, Wu L, Mitchell JP, Sanden BL (2001) Nitrate leaching and soil nitrate content as affected by irrigation uniformity in a carrot field. Agric Water Manage 48:37-50

Allen TFH, Hoekstra TW (1992) Toward a Unified Ecology. Columbia University Press, New York

Aronoff S (1993)Geographical information systems: a management perspective. Ottawa, Ontario, Canada: WDL Publications; p. 294

Babiker IS, Mohamed MAA, Terao H, Kato K, Ohta K (2004) Assessment of groundwater contamination by nitrate leaching from intensive vegetable cultivation using geographical information system. Environ Intl 29:1009-1017

Bach M, Fabis J, Frede H-G (1997) Filterwirkung von Uferstreifen für Stoffeinträge in Gewässer in unterschiedlichen Landschaftsräumen, Vol. 28. DVWK, Bonn

Bachmat Y (1994) Groundwater as part of the water system. In: Groundwater Contamination and Control. Ed. U Zoller, Marcel Dekker, Inc, New York

Barnes CJ, Jacobson G, Smith GD (1992) The origin of high-nitrate waters in the Australian arid zone. J Hydrol 137:181-197

Barringer T, Dumn D, Battaglin W, Vowinkel E (1990) Problems and methods involved in relating land use to groundwater quality. Water Res Bull 26:1-9

Bencala KE (1993) A perspective on stream–catchment connections. J N Am Benthol Soc 12:44–47

Bernhard C, Carbiener R, Cloots AR (1992) Nitrate pollution of groundwater in the Astian Plain (France)- a multidisciplinary study of an agricultural area: the Central Ried of the Ill River. Envrion Geol Water Sci 20:125-137

Bhinde AD (1990) Groundwater pollution due to the solid awstes. BHU-JAL News, Quarterly Journal of Central Groundwater Board 5(2):13-15

Bingham FT, Davis S, Shade E (1971) Water relations, salts balance, and nitrate leaching losses of a 960-acre citrus watershed. Soil Sci 112:410-418

Bock BR (1984) Efficient use of nitrogen in cropping systems. In: R.D. Hauck (Editor), Nitrogen in Crop Production. Am Soc Agron, Madison, WI, pp. 273-294

Bouchard DC, Williams MK, Surampalli RY (1992) Nitrate combination of groundwater sources and potential health effects. J Am Med Assoc 7:85-90

Bouma J (1992) Influence of soil macroporosity on environmental quality. Adv Agron 47:1-37

Bouwer H (1989) Agricultural contamination: problems and solutions. Water Environ Technol 10:292-297

Boyce JS, Muir J, Edwards AP, Seim EC, Olson RA (1976) Geologic nitrate in pleistocein loess in Nebraska. J Environ Qual 5:93-96

Brown PL, Halverson AD, Siddoway FH, Mayland HF, Miller MR (1982) Saline-seep diagnosis, control, and reclamation. USDA Conservation Research Report No. 30, pp. 22

Buckman HO (1910) Moisture and nitrate relations in dry land agriculture. Proc Am Soc Agron 2:121-138

Bulusu KR, Pande SP (1990) Nitrates- A serious threat to groundwater pollution. BHU-JAL News, Quarterly Journal of Central Groundwater Board 5(2):39

Burkart MR, Kolpin DW (1993) Hydrologic and land use factors associated with herbicides and nitrates in near-surface acquifers. J Environ Qual 22:646-656

Burns IG, Greenwood DJ (1982) Estimation of the year-to-year variations in nitrate leaching in different soils and regions of England and Wales. Agric Environ 7:35-45

Cammaerts D, Jacobs M (1985) A study of the role of glutamate dehydrogenase in the nitrogen metabolism of *Arabidopsis thaliana*. Planta 163:517-526

Campbell WH (1996) Nitrate reductase biochemistry comes of age. Plant Physiol 111:355-364

Campbell WH (1999) Nitrate reductase structure, function, and regulation: bridging the gap between biochemistry and physiology. Annu RevPlant Physiol Plant Mol Biol 50:277-303

Campbell WH, Kinnunen-Skidmore T, Brodeur-Campbell MJ, Campbell ER (2004) New and improved nitrate reductase for enzymatic nitrate analysis. American Laboratory 22(10): 12. Patton, Charles J., Anne E. Fischer

Campbell WH, Campbell Ellen R (2002) Corn leaf nitrate reductase: A nontoxic alternative to cadmium for photometric nitrate determinations in water samples by air-segmented continuous-flow analysis. Environ Sci Tech 36:729-35

Chandu SN, Subbarao NV, Ravi Prakash S (1995) Suitability of groundwater for domestic and irrigational purposes in some part of Jhansi in UP. BHU-JAL News, Quarterly Journal of Central Groundwater Board 10(1):12-18

Chang Y (1997) A multiple objective analysis in On-Farm irrigation scheduling. Ph.D. thesis. Katholieke Universiteit Lewven, Leuven

Chang Y, Vanclooster M, Hubrechts L, Feyen J (1996) Multicriteria decision analysis in irrigation scheduling. In: Camp CR, Sadler EJ, Yoder RE (Eds.), Evapotranspiration and Irrigation Scheduling. Proceedings of the International Conference, 3-6 November, San Antonio Convention Centre, San Antonio, Texas, pp. 1128-1133

Chapman HD, Liebig GF, Rayner DD, (1949) A lysimeter investigation of nitrogen gains and losses under various systems of cover cropping and fertilization and a discussion of error sources. Hilgardia 19:57-128

Chew CF, Zhang TC (1998) *In-situ* remediation of nitrate contaminated groundwater by electrokinetics/iron wall process. Wat Sci Tech 38:135-142

Clifford D, Liu X (1993a) Biological denitrification of spent regenerant brine using a sequencing batch reactor. Water Res 27:1477– 1484

Clifford D, Liu X (1993b) Ion exchange for nitrate removal. Amer Water Works Assoc J 85:135-143

Correll DL (1997) Buffer zones and water quality protection: general principles. In:, Buffer Zones: Their Processes and Potential in Water Protection (Naycock NE, Burt TP, Goulding KW, Pinay G Eds). Quest Environment, Hertfordshire, pp. 7-20

Crawford NM (1995) Nitrate: nutrient and signal for plant growth. Plant Cell 7:859-868

Cren M, Hirel B (1999) Glutamine synthesis in higher plants: regulation of gene and protein expression from the organ to the cell. Plant Cell Physiol 40:1187-1193

Crutzen PJ (1970) The influence of nitrogen oxides on the atmospheric ozone content. Quart J R Meteorol Soc 96:320-325

D'Urso G, Menenti M, Santini A (1999) Regional application of one-dimentional water flow models for irrigation management. Agric Water Manage 40:291-302

Davis SN, De Wiest RJM (1991) Hydrogeology. Krieger, Malabar, FL

De Smedt P, Loy W (1985) Les nitrates dans l'eau souterrains en Belgique. Memoires 18.Congr. Int. Assoc. Hydrogeol. (Hydrogeology in the service of man), Cambridge, Part 3, pp. 178-187

Demead OT, Nielsen R, Thurtell GW (1978) Ammonia exchange over a corn crop. Soil Sci Soc Am J 42:840-842

Dijkstra JP, Hack-ten Broeke MJD (1995) Simulation of different management options within integrated arable farming affecting nitrate leaching. In: Scenario Studies for the Rural Environment (Schoute JFTh, Finke PA, Veeneklaas FR, Wolfert HP Eds). Kluwer, Dordrecht, pp. 329-333

Dillon PJ, Ragusa SR, Richardson SB (1991) Biochemistry of a plume of nitrate contaminated groundwater. In: Bogardi I, Kuzelka RD. editors. Nitrate contamination: exposure, consequence and control. NATO ASI serial G: ecological sciences, vol. 309, Berlin, Springer, pp. 173-180

Doran JW, Power JF (1983) The effects of tillage on the nitrogen cycle in corn and wheat production. In: Nutrient Cycling in Agricultural Ecosystems (Lowrance R, Todd R, Asmussen L, Leonard R Eds). Univ. GA Coll. Agric. Spec. Pub. No. 23, Athens, GA, pp. 441-455

Doran JW (1987) Microbial biomass and mineralizable nitrogen distributions in no-tillage and plowed soils. Biol Fertil Soils 5:68-75

Eckhardt DAV, Stackelberg PE (1995) Relation of ground-water quality to land use on Long Island, New York. Groundwater 33:1019-1033

Eichner MJ (1990) Nitrous oxide emissions from fertilized soils: summary of available data. J Environ Qual 19:272-280

English M, Glenn M, VanSickle J (1981) Scheduling for optimum water use. In: Irrigation Scheduling for Water and Energy Conservation in the 1980's. Proceedings of the Irrigation Scheduling Conference. December 14-15, Chicago, ASAE Publication 23-81, pp. 61-72

Follmann H, Häberlein I (1996) Thioredoxin: Universal, yet specific thiol- disulfide redox cofactors. Biofactors 5:147-156

Foster SSD, Bridge LR, Geake AK, Lawrence AR, Parker JM (1986) The groundwater nitrate problem. Hydrogeol Rep 86/2, Br Geol Surv, Wallingford, pp 95

Francis AJ, Dodg CJ, Gillow JB, Papenguth HW (2000) Biotransformation of uranium compounds in high ionic strength brine by a halophilic bacterium under dentrifying conditions. Environ Sci Technol 34:2311-2317

Gevertz D, Telang AJ, Voordouw G, Jenneman GE (2000) Isolation and characterization of strains CVO and FWKOB, two novel nitrate-reducing, sulfide-oxidizing bacteria isolated from oil field brine. Appl Environ Microbiol 66:2491-2501

Gilbert J, Dole-Olivier M, Marmonier P, Vervier P (1990) Surface-groundwater ecotones. In: Ecology and Management of Aquatic-terrestrial Ecotones (Naiman RJ, Décamps H Eds). UNESCO, Paris, pp. 7-21

Glaab J, Kaiser WM (1995) Inactivation of nitrate reductase involves NR-Protein phosphorylation and subsequent binding of an inhibitor protein. Planta 195:514-518

Goodchild MF, Steyaert LT, Parks BO, Johnston C, Maidment D, Crane M, Glendinning S (1996) GIS and environmental modeling: progress and research issues. Fort Collins, CO: GIS'World Books

Gupta SC (1992) Udaipur Kshetra ke bhujal mein nitrate ka badhata star. BHU-JAL News, Quarterly Journal of Central Groundwater Board 7(1):17-19

Haag D, Kaupenjohann M (2001) Landscape fate of nitrate fluxes and emissions in Central Europe: A critical review of concepts, data, and models for transport and retention. Agric Ecosys Environ 86:1-20

Hack-ten Broeke MJD, Van Lanen HAJ, Bouma J (1993) The leaching potential as a land quality of two Dutch soils under current and potential management conditions. Geoderma 60:73-88

Hack-ten Broeke MJD, De Groot WJM, Dijkstra JP (1996) Impact of excreted nitrogen by grazing cattle on nitrate leaching. Soil Use Manage 12:190-198

Hadas A, Hadas A, Sagiv B, Haruvy N (1999) Agricultural practices, soil fertility management modes and resultant nitrogen leaching rates under semi-arid conditions. Agric Water Manage 42:81-95

Handa BK (1975) High nitrate and potassium ion concentration as indicators for groundwater pollution in India. Abstract, International Symposium on Geochemistry of Natural waters. Canada Centre for Inland Waters, Burlington, Ontario, Canada

Handa BK, Kumar A, Bhardwaj RK (1991) Studies on Dal lake, Srinagar, J and K: eutrophication status. BHU-JAL News, Quarterly Journal of Central Groundwater Board 6(4):26-35.

Harper LA, Catchpole VR, Davis R, Weier KL (1983) Ammonia volatilization: Soil, plant, and microclimate effects on diurnal and seasonal fluctuations. Agron J 75:212-218

Haycock NE, Pinay G, Burt TP, Goulding KWT (1997) Buffer zones: current concerns and future directions. In: Buffer Zones: Their Processes and Potential in Water Protection (Naycock NE, Burt TP, Goulding KW, Pinay G Eds). Quest Environment, Hertfordshire, pp. 236-243

Haycock NE, Pinay G, Walker C (1993) Nitrogen retention in river corridors: European perspective. Ambio 22:340-346

Hendrix PF, Coleman DA, Crossley DA (1992) Using knowledge of soil nutrient cycling to design sustainable agriculture. J Sustainable Agric 2:63-81

Hergert GW (1986) Nitrate leaching through sandy soils as affected by sprinkler irrigation method. J Environ Qual 15:272-278

Herman DC, Frankenberger Jr WT (1999) Bacterial reduction of perchlorate and nitrate in water. J Environ Qual 28:1018-1024

Hill RW, Allen RG (1995) Developing simple irrigation scheduling calendars. In: Irrigation Scheduling from Theory to Practice. ICID/FAO workshop, Rome

Hill AR (1996) Nitrate removal in stream riparian zones. J Environ Qual 25:743-755

Hiscock KM, Lloyd JW, Lerener DN (1991) Review of natural and artificial denitrification of groundwater. Water Res 25:1099-1111

Hoff T, Truon HN, Caboche M (1994) The use of mutants and transgenic plants to study nitrate assimilation. Plant Cell Environ 17: 489-505

Hölting B (1980) Hydrogeologie. Enke, Stuttgart

Hölting B, Haertlé T, Hohberger K, Nachtigall K, Villinger E, Weinzierl W, Wrobel J (1995) Konzept zur Ermittlung der schutzfunktion der Groundwasseruberdechung. Geol. jb. C 63, 5-24

Houghton JT (1994) Climate change 1994: radiative forcing of climate change and an evaluation of the IPCC IS92 emission scenarios. Reports of Working Groups I and III of the International Panel on Climate Change, Cambridge University Press, Cambridge, pp 339.

Howarth RW, Billen G, Swaney D, Townsend A, Jaworski L, Lajtha K, Downing JA, Elmgren R, Caraco N, Jordan T, Berendse F, Freney J, Kudeyarov V, Murdoch P, Zhao-Liang Z (1996) Regional nitrogen budgets and riverine N and P fluxes for the drainages to the North Atlantic Ocean: natural and human influences. Biogeochemistry 35:75-139

Huber JL, Redinbaugh MG, Huber SC, Campbell WH (1994) Regulation of maize leaf nitrate reductase activity involves both gene expression and protein phosphorylation. Plant Physiol 106:1667-1674

Huber SC, Buchmann M, Huber JL (1996) Post-translational regulation of nitrate reductase activity: a role for $Ca^{++}$ and 14-3-3 proteins. Trends Plant Sci 1:432-438

Hudak PF (2000) Regional trends in nitrate content of Texas groundwater. J Hydrol 228:37-47

Isermann K, Isermann R (1997) Globale, territoriale und betriebliche Nährstoffbilanzierung. In: Umweltbundesamt (Ed.), Stoffbilanzierung in der Landwirtschaft, Vol. 20. Umweltbundesamt, Wien, pp. 241-313

Jacinthe P-A, Dick WA, Brown LC (2000) Bioremediation of nitrate-contaminated shallow soils and waters via water table management techniques: evolution and release of nitrous oxide. Soil Biol Biochem 32:371-382

Kaiser WM, Huber SC (1994) Posttranslational regulation of nitrate reductase in higher plants. Plant Physiol 106:817-821

Kakar YP (1985) Nitrate pollution of groundwater in Haryana. Proceedings of the seminar on Water quality and its management, Central Board of Irrigation and Power pp. 59-68

Kaltschmitt M (1997) Nachwachsende Energieträger: Grundlagen, Verfahren, ökologische Bilanzierung. Vieweg, Braunschweig, Wiesbaden

Khanna SP, Kumar A, Kumar S (1994) Boron in groundwater of Haryana state. BHU-JAL News, Quaterly Journal of Central Groundwater Board 9(3/4):15-19

Kimble JM, Bartlett RJ, McIntosh JL, Varancy KE (1972) Fate of nitrate from manure and inorganic nitrogen in a clay soil cropped to continuous corn. Journal of Environmental Quality 1(4):413-415

Kohl DH, Shearer GB, Commoner B (1971) Fertilizer nitrogen: contribution to nitrate in surface water in a cornbelt watershed. Science 174:1331-1334

Kovács GJ, Németh T (1995) Testing simulation models for the assessment of crop production and nitrate leaching in Hungary. Agric. Systems 49:385-397

Krapp A, Fraisier V, Scheible W-R, Quesada A, Gojan A, Stitt M, Caboche M, Daniel-Vedele F (1998) Expression studies of Nrt 2:1 Np, a putative high affinity nitrate transporter: evidence for its role in nitrate uptake. Plant J 14:723-731

Kumar A (1983) Pollution of groundwater by nitrate in Uttar Pradesh. Proceedings of CGWB seminar on assessment, development and management of groundwater vol. 11

Lagerstedt E, Jacks G, Sefe F (1994) Nitrate groundwater and N circulation in eastern Botswana. Environ Geol 23:60-64

Lam H-M, Coschigano KT, Oliveira IC, Melo-Oliveira R, Coruzzi GM (1996) The molecular - genetics of nitrogen assimilation into amino acids in higher plants. Annual Rev Plants Physiol Plant Mol Biol 47:569-593

Landreau A, Roux JC (1984) Les nitrates dans les eaux souterraines. Rap. 84 SGN 361 ENV, Bur. Rech. Geol. Min., Orleans, pp. 51

Lea PJ, Miflin BJ (1974) Alternative route for nitrogen assimilation in higher plants. Nature 251:614-616

Lea PJ, Robinson SA, Stewart GR (1990) The enzymology and metabolism of glutamine, glutamate and asparagine. In: Biochemistry of plants: Intermediary nitrogen metabolism Vol.16 (Miflin BJ, Lea PJ Eds), Academic Press, pp 121-159

Lehn H, Flaig H, Mohr H (1995) Vom Mangel zum Überfluß: Störungen im Stickstoffkreislauf. Gaia 4:13-25

Lembke WD, Thorone MD (1980) Nitrate leaching and irrigated corn production with ot-ganic and inorganic fertilizers on sandy soils. Transactions of the ASAE 23:1153-l156

Letey J, Valoras N, Hadas A, Focht DD (1980) Effect of airfield porosity, nitrate concentration, and time on the ratio $N_2O/N_2$ during denitrification. J Environ Qual 9:227-231

Levallois P, Thiriault M, Rouffignat S, Tessier R, Landry P, Dyatte M, Girard S, Gingras D, Ganvin Chaisson C (1998) Goundwater contamination by nitrate associated with intensive potato culture in Québec. Sci Total Environ 217:91-101

Lichter A, Häberlein I (1998) A light- dependent redox signal participates in the regulation of ammonia fixation in chloroplasts of higher plants - ferredoxin: glutamate synthase is thioredoxin - dependent enzyme. J Plant Physiol. 153:83-90

Lillo C (1994) Light regulation of nitrate reductase in green leaves of higher plants Physiol Plant 90:616-620

Line DE, Arnold JA, Osmond S, Coffey D, Gale JA, Spooner A, Jennings GD (1993) Nonpoint sources. Water Environ Res 64:558-571

Linn DM, Doran JW (1984) Effect of water-filled pore space on $C0_2$ and $N_20$ production in tilled and nontilled soils. Soil Sci Soc Am J 48:1267-1272

Liu CX, Clifford DA (1996) Ion exchange with denitrified brine reuse. J Amer Water Works Assoc 88:88-99

Lowrance RR, Pionke HB (1989) Transformations and movement of nitrate in aquifer systems. In: Nitrogen Management and Groundwater Protection, Vol. 21. (Follett RF Ed.) Elsevier, Amsterdam, pp. 373-392

Lund LJ (1982) Variation in nitrate and chloride concentration below selected agricultulal fields. Soil Science Society of America Jounal 46:1062-1066

Mackintosh C, Douglas P, Lillo C (1995) Identification of protein that inhibits the phosphorylated form of nitrate reductase from spinach (*Spinacia oleracea*) leaves. Plant Physiol 107:451-457

Madison RJ, Brunett JO (1985) Overview of the occurrences of nitrates in groundwater of the United States. US Geological Survey Water Supply Paper 2275, pp. 93-105

Magdoff F, Lanyon L, Liebhardt B (1997) Nutrient cycling, transformations, and flows: implications for a more sustainable agriculture. Adv Agron 60:1-73

Malano HM, Wood ML (1995) Surface irrigation management in real time in south eastern Australia: Irrigation Scheduling. In: Irrigation Scheduling from Theory to Practice. ICID/FAO workshop, Rome

Mariotti A (1994) Dénitrification in situ dans les eaux souterraines, processus naturels ou provoqués: une revue. Hydrogéologie 3:43-68

Martin WP, Fenster WE, Hanson ID (1970) Fertilizer management for pollution control. In: Agricultural Practices and Water Quality (Willrich TL, Smith GE Eds), Iowa State University Press, Ames, pp. 142-158

Mason B, Moore CB (1985) Principles of geochemistry. Wiley Eastern Ltd., pp. 350

Mathur AK, Ranganathan S (1990) Jodhpur, Salawas. Rajasthan Kshetra mein bhumi jal, prashudhan samasya-sthiti adhyayan. BHU-JAL News, Quarterly Journal of Central Groundwater Board 5(2):16-20

McLay CDA, Dragten R, Sparling G, Selvarajah N (2001) Predicting groundwater nitrate concentrations in a region of mixed agricultural land use: a comparison of three approaches. Enivorn Pollut 115:191-204

Mehta BC, Singh RV, Srivastava SK, Das S (1990) Impact of fertilizer use on grondwater quality in parts of Ganjam district, Orissa BHU-JAL News, Quarterly Journal of Central Groundwater Board 5(2):44-48

Merriam JL (1996) Simple irrigation scheduling technique using a soil probe. In: Evapotranspiration and Irrigation Scheduling. Proceedings of the International Conference (Camp CR, Sadler EJ, Yoder RE Eds.), 3-6 November, San Antonio Convention Centre, San Antonio, Texas, pp. 93-96

Mielke LN, Schepers JS, Richards KA (1979) Nitrogen leaching under center-pivot irrigation on sandy loam soil. Proc Tech Conf, The Irrig Assoc, San Francisco, CA, 18-21 Feb., pp 85-94

Miflin, B.J and Lea,P.J .(1982). Ammonia assimilation and amino acid metabolism, In: Boulter D, Partheir, B (ed) Encyclopedia of plants physiology Vol. 14 A, Springer-Verlag, Berlin, pp 5-64

Montgomery E, Coyne MS, Thomas GW (1997) Denitrification can cause variable $NO_3$ concentration in shallow groundwater. Soil Sci 162:148-156

Mormile MR, Romine MF, Garcia MT, Ventosa A, Bailey TJ, Peyton BM (1999) *Halomonas campisalis* sp nov, a denitrifying, moderately alkaliphilic bacterium. Syst Appl Microbiol 22:551-558

Mosley MP (1982) Subsurface flow velocities through selected forest soil, South Island, New Zealand. J Hydrol 55:65-92.

Nelson D (1985) Minimizing nitrogen losses in non-irrigated eastern areas. In: Plant Nutrient Use and the Environment Symposium Proceedings. The Fertilizer Institute, Washington, DC, pp 173-209

Nguyen HV, Nieber JL, Misra D (1996) Modeling BMP impacts on groundwater quality. In: Camp CR, Sadler EJ, Yoder RE (eds), Evapotranspiration and Irrigation Scheduling. Proceedings of the International Conference, 3-6 November, San Antonio Convention Centre, San Antonio, Texas, pp 762-768

Nitrate in News. http://www.nitrate.com.nitrate1.html

Oaks A (1992) A re-evaluation of nitrogen assimilation in roots. Bioscience 42:103-111

Okeke BC, Giblin T, Frankenberger Jr WT (2002) Reduction of perchlorate and nitrate by salt tolerant bacteria. Environ Pollut 118:357-363

Ortega JL, Roche D, Sengupta-Gopalan C (1999) Oxidative turnover of soybean root glutamine synthetase. *in vitro* and *in vivo* studies. Plant Physiol 119:1483-1495

Our garden gang, www.ourgardengang.com/whsuckitup

Overgaard K (1984) Trends in nitrate pollution of groundwater in Denmark. Nordic Hydrol 15:177-184

Owens LD (1960) Nitrogen movement and transformation in soils as evaluated by a lysimeter study utilizing isotopic nitrogen. Soil Science Society of America Proceedings 24:372-376

Patton CJ, Kryskalla J, Campbell ER, Campbell WH (2004) Replacing Toxic Cadmium with Environmentally Benign Nitrate Reductase in Automated Continuous Flow and Batch Determinations of Nitrate in Environmental Water Samples: An Overview. PittCon 2004, Chicago, IL, March 9, 2004. Abstract: 200

Phene CJ, Fouss JL, Howell TA, Patton SH, Fisher MW, Bratcher JO, Rose JL (1981) Scheduling and monitoring irrigation with the new matric potential sensor. In: Irrigation Scheduling for Water and Energy Conservation in the 1980's. Proceedings of the Irrigation Scheduling Conference. December 14-15, Chicago, ASAE Publication 23-81, pp 91-105

Pinay G, Décamps H, Chauvet E, Fustec E (1990) Functions of ecotones in fluvial systems. In: Naiman RJ, Décamps H (eds), Ecology and Management of Aquatic–terrestrial Ecotones. UNESCO, Paris, pp 7-21

Power JF, Schepers JS (1989) Nitrate contamination of groundwater in North America. Agric Ecosys Environ 26:165-87

Power JF (ed) (1987) The Role of Legumes in Conservation Tillage Systems. Soil Conserv Soc Am, Ankeny, IA, pp 153

Power JF, Follett RF (1987) Monoculture Sci Am 256:79-86

Power JF, Papendick RI (1985) Organic sources of nutrients. In: Engelstad OP (ed), Fertilizer Technology and Use, 3rd edn. Soil Sci Soc Am, Madison, WI, Chap 14, pp 503-520

Power JF (1981) Nitrogen in the cultivated ecosystem. In: Clark FE, Rosswall T (eds), Terrestrial Nitrogen Cycles-Processes, Ecosystem Strategies and Management

Impacts. Ecol Bull No 33, Swedish Natural Science Research Council, Stockholm, Sweden, pp 529-546

Power JF, Wilhelm WW, Doran JW (1986) Crop residue effects on soil environment and dryland maize and soybean production. Soil Tillage Res 8:101-111

Prasuhn V, Braun M (1995) Regional differenzierte Abschätzung diffuser Phosphor- und Stickstoffeinträge in die Gewässer des Kantons Bern (Schweiz). Z Kulturtechnik Landentwickl 36:309-314

Prasuhn V, Spiess E, Braun M (1996) Methoden zur Abschätzung der Phosphor- und Stickstoffeinträge aus diffusen Quellen in den Bodensee

Pratt PF, Chapman HD (1961) Gains and losses of mineral elements in an irrigated soil during a 20.year lysimeter investigation. Hilgardia 30:445-467

Pratt PF, Lund LJ, Rible JM (1978) An approach to measurin g leaching of nitrate from freely drained irrigated fields. In: Nielsen DR, MacDonald JG (eds), Nitrogen in the Environment, vol 1. Nitrogen Behavior in Field Soil. Academic Press, pp 223-265

Reddy KJ, Lin J (2000) Nitrate removal from groundwater using catalytic reduction. Wat Res 34:995-1001

Rice CW, Rogers K (1993) Denitrification in subsurface environments: potential source for atmospheric nitrous oxide. In: ASA (ed), Agricultural Ecosystem Effects on Trace Gases and Global Climate Change, Vol. 55. ASA, Madison, pp 121-132

Risser PG (1990) The ecological importance of land–water ecotones. In: Naiman RJ, Décamps H (eds.), Ecology and Management of Aquatic-terrestrial Ecotones. UNESCO, Paris, pp 7-21

Ritter WF, Chirnside AE, Scarborough RW (1990) Soil nitrate profiles under irrigation on coastal plain soil. Journal of the Irrigation and Drainage Division of the ASCE 1 16(6):738-751

Robertson WD, Cherry JA (1995) *In situ* denitrification of septic-system nitrate using reactive porous media barriers: field trials. Groundwater 33:99-111

Roche D, Temple SJ, Sengupta-Gopalan C (1993) Two classes of differentially regulated glutamine synthetase genes are expressed in the soybean nodule: a nodule specific class and constitutively expressed class. Plant Mol Biol 22: 971-983

Rode M, Grunwald S, Frede H-G (1995) Methodik zur GIS-gestützten Berechnung von Nährstoffeinträgen in Fließgewässer durch Oberflächenabfluß mit dem Modell AGNPS. Z. Kulturtechnik Landentwickl 36:63-68

Rohmann U, Sontheimer H (1985) Nitrat im Grundwasser: Ursachen, Bedeutung und LSsungswege. DVGW-Forschungsstelle/Engler-Bunte-Inst., Univ. Karlsruhe, pp 468

Salameh Al-Jamal M, Sammis TW, Jones T (1997) Nitrogen and chloride concentration in the deep soil cores related to fertilization. Agric Water Manage 34:1-16

Saxton KE, Schuman GE, Burwell RE (1977) Modelling nitrate movement and dissipation in fertilized soils. Soil Sci Soc Am J 41:265-271

Schepers JS, Moravek MG, Alberts EE, Frank KD (1991) Maize production impacts on groundwater quallity. J Environ Qual 20:12-16

Schepers JS, Mielke LN (1983) Nitrogen fertilization, mineralization, and leaching under irrigation in the Midwest. In: Lowrance R, Todd R, Asmussen L, Leonard R (eds), Nutrient Cycling in Agricultural Ecosystems. Univ GA Coll Agric Spec Pub No 23, Athens, GA, pp 325-334

Schepers JS, Frank KD, Watts DG (1985) Influence of irrigation and nitrogen fertilization on groundwater quality. In: Dunin FX, Matthess G, Gras RA (eds),

Relation of Groundwater Quantity and Quality. IAHS Press, Inst. Hydrology, Wallingford, Oxfordshire, IAHS Pub. No. 146, pp 21-32

Schipper LA, Vojvodić-Vuković M (2000) Nitrate removal from groundwater and denitrification rates in a porous treatment wall amended with sawdust. Ecol Eng 14:269-278

Schmidt EL (1982) Nitrification in soil. In: Stevenson FJ (ed) Nitrogen in Agricultural soils. American Society of Agronomy, Madison, pp 253-288

Schneekloth JP, Klocke NL, Clark RT, Norton NA (1996) Irrigation management strategies to reduce leaching. In: Camp CR, Sadler EJ, Yoder RE (eds), Evapotranspiration and Irrigation Scheduling. Proceedings of the International Conference, 3-6 November, San Antonio Convention Centre, San Antonio, Texas, pp 775-780

Schnoor JL, Licht LA, McCutcheon SC, Wolfe NL, Carreira LH (1995) Phytoremediation of Organic and Nutrient Contaminants. *Environ Sci Technol* 29:318-323

Selenka F (1985) Nitrate in drinking water: The basis for a regulatory limit. In: Winteringham FPW (ed), Environment and Chemicals in Agriculture. Elsevier Appl Sci Publ, Amsterdam, pp 87-104

Shargool PD, Jain JC (1989) Purification and immunological properties of an NAD(H) dependent glutamate dehydrogenase from soybean cells (*Glycine max* L.). Plant Sci 60:173-179

Shewanella Federation www.shewanella.org/why_shewanella.sjsp

Singh B, Sekhon GS (1976) Nitrate pollution of groundwater from nitrogen fertilizers and animal wastes in the Punjab, India. Agric Environ 3:57-67

Singh BK, Pal OP, Pandey DS (1991) Groundwater pollution: a case study around Northeastern Railway City station, Lucknow, Uttar Pradesh. BHU-JAL News, Quaterly Journal of Central Groundwater Board 6(2):46-49

Singh M, Bhattacharya AK, Nair TVR, Singh AK (2002) Nitrogen loss through subsurface drainage effluent in coastal rice field from India. Agric Water Manage 52:249-260

Smika DE, Heimian DF, Duhe HR, Batchalolen AR (1977) Nitrate-N percolation through irrigated sandy soil as affected by water management. Agron J 69(41):623-626

Smith JW, Gutherie LD. Nitrate toxicity and prussic acid poisoning in dairy cattle. http://www.ces.uga.edu/pubcd/b1155-w.num

Smith KA (1980) A model of the extent of anaerobic zones in agricated soils, and its potential application to estimates of denitrification. J Soil Sci 31:263-277

Spalding RF, Exner ME (1993) Occurrence of nitrate in groundwater- a review. J Environ Qual 22:392-402

Stewart BA, Viets FG Jr, Hutchinson GL, Kemper WD, Clark FE, Fairbourn ML, Strauch F (1967) Distribution of nitrates and other water pollutants under fields and corrals in the middle of the Platte Valley of Colorado. USDA-ARS Research Report No. 41-134

Stewart BA, Vrets FG, Hutchinson GL (1968) Agriculture's effect on nitrate pollution for groundwater. J Soil Water Conserv 23:13-15

Strebel O, Duynisveld WHM, Böttcher J (1989) Nitrate pollution of groundwater in Western Europe. Agric Ecosys Environ 26:189-214

Su W, Huber SC, Crawford NM (1996) Identification *in vitro* of a post –translation regulatory site in the hinge1 region of *Arabidopsis* nitrate reductase. Plant Cell 8:519-527

Sukanya R, Li M-G, Snustad DP (1994) Root and shoot specific responces of individual glutamine synthetase gene of maizeto nitrate and ammonium Plant Mol Biol 26:1935-1946

Tamta SR, Kapoor SL, Govardhan T (1992) Quality assessment of groundwater in Banglore district of Karnataka. BHU-JAL News, Quaterly Journal of Central Groundwater Board 7(2/3):5-8

Temple SJ, Heard J, Ganter G, Dunn K, Sengupta-Gopalan C (1995) Characterization of a nodule- enhanced glutamine synthetase from alfaalfa : nucleotide sequence, *in situ* localization and transcript analysis. Mol Plant Microbe Interact 8:218-227

Temple SJ, Vance CP, Gantt JS (1998) Glutamate synthase and nitrogen assimilation. Trends Plant Sci 3:51-56

Thorburn PJ, Biggs JS, Weier KL, Keating BA (2003) Nitrate in groundwaters of intensive agricultural areas in coastal Northeastern Australia. Agric Ecosys Environ 94:49-58

United States Environmental Protection Agency (1996) A Citizen's Guide to Phytoremediation. http://clu-in.com/citguige/phyto.htm

University of Nebraska Institute of Agriculture and Natural Resources (1984) Nitrogen and irrigation management - Hall County Water Quality Special Project. Final Report. University of Nebranska-Lincoln Cooperative Extension Service, South Central Station, Clay Centre, NE, pp 22

Vito M (1998) Proceedings of the International Conference on Life Cycle Assessment in Agriculture, Agro-Industry and Forestry, December 3-4, 1998, Brussels, Belgium

Vitousek PM, Aber J, Howarth RW, Likens EG, Matson P, Schindler DW, Schlesinger WH, Tilman D (1997) Human alteration of the global nitrogen cycle: causes and consequences. Ecol Appl 7:737-750

Voznaya NF (1981) Role of microorganisms in matter cycle. In: Chemistry of Water and Microbiology, Mir Publishers, Moscow pp 221-273

Warner RL, Kleinhofs A (1992) Genitics and molecular biology of nitrate metabolism in higher plants. Physiol Plant 85:245-252

Wendland F (1992) Die Nitratbelastung in den Grundwasserlandschaften der alten Bundesländer (BRD), Vol. 8. Forschungszentrum Jülich, Jülich

Werner W (1994) Stickstoff-und Phosphateintrag in Fließgewässer Deutschlands unter besonderer Berücksichtigung des Eintragsgeschehens im Lockergesteinsbereich der ehemaligen DDR Agrarspectrum, Vol. 22. DLG-Verlag, Frankfurt, pp 243

Wesseling JG, Van den Broek BJ (1988) Prediction of irrigation scheduling with the numerical model SWATRE. Agric Water Manage 14:299-306

Wiens JA, Crawford CS, Gosz JR (1985) Boundary dynamics: a conceptual framework for studying landscape ecosystems. Oikos 45:412-427

Zhang WL, Tian ZX, Li XQ (1996) Nitrate pollution of groundwater in northern China. Agric Ecosyst Environ 59:223-231

# 17

# Bioremediation of Petroleum Sludge using Bacterial Consortium with Biosurfactant

K.S.M. Rahman[1], T.J. Rahman[2], I.M. Banat[2], R. Lord[1] and G. Street[1]

[1]Clean Environment Management Centre, School of Science and Technology, University of Teesside, Middlesbrough – TS13BA, Tees Valley, UK, Email: P.Rahman@tees.ac.uk; [2]Biotechnology Research Group, School of Biomedical Sciences, University of Ulster, Coleraine – BT52 1SA, Northern Ireland, UK

## 1. Introduction

Petroleum hydrocarbon continues to be used as the principle source of energy and hence an important global environmental pollutant. Apart from accidental contamination of the ecosystem, the vast amounts of oil sludge, generated in refineries from water oil separation systems and accumulation of waste oily materials in crude oil storage tank bottoms, pose great problems because of the expensive disposal methods (Ferrari et al. 1996; Vasudevan and Rajaram 2001). Despite decades of research, successful bioremediation of petroleum hydrocarbon contaminated soil remains a challenge. Petroleum is a complex mixture of non-aqueous and hydrophobic components like n-alkane, aromatics, resins and asphaltenes. Bioavailability might be the limiting factor in the biodegradation of such compounds.

Biosurfactants are amphiphilic compounds that reduce surface and interfacial tensions by accumulating at the interface of immiscible fluids or of a fluid and a solid and increase the surface areas of insoluble compounds leading to increased mobility, bioavailability and subsequent biodegradation. They are produced by many bacterial strains that can degrade or transform the components of petroleum products. They are non-toxic, non-hazardous, biodegradable and environmentally friendly compounds (Banat et al. 2000), which may be produced cost effectively under *ex-situ* conditions, while *in-situ* production may be stimulated at the site of contamination and can be recovered and recycled (Moran et al. 2000). There have been recent successful reports on using them in enhanced oil recovery and in the release of bitumen from tar sands (Mulligan et al. 2001). Hence, reclamation of petroleum hydrocarbon polluted sites can be carried out by the bioremediation, which is an enhanced natural process of biodegradation, using biosurfactant producing and oil degrading bacterial cultures. Bioremediation technologies generally aim at providing favourable conditions of aeration, temperature and nutrients to enhance biological hydrocarbon breakdown (Rahman et al. 2002a,b). In the present study, we investigated the effect of rhamnolipid biosurfactant (RL) produced by a *Pseudomanas aeruginosa* strain and addition of nutrients, such as nitrogen, phosphorus and potassium (NPK) and a bacterial consortium (BC) to augment natural fertility of the polluted site on the bioremediation of crude oil tank bottom sludge (TBS).

## 2. Methods

### 2.1 Soil and Microbial Cultures Preparation

Seashore sand samples from the Portrush coastal area of Northern Ireland and garden soil from the University of Ulster campus were collected. Both were sieved using a 1mm sieve and used at 1:1 ratio for the preparation of a composite soil sample. Part of the soil was sterilized in a hot air oven at 180°C for 2 h and a part kept as normal condition (non-sterile). The sterility of the soil was confirmed by pour plate technique on plate count agar (Merck, UK). An oil degrading bacterial consortium containing five bacterial strains (*Micrococcus* sp. GS2-22 ($21.7\pm1.4 \times 10^5$ CFU/ml), *Bacillus* sp. DS6-86 ($30.3\pm0.9 \times 10^5$ CFU/ml), *Corynebacterium* sp. GS5-66 ($27.4\pm4.7 \times 10^5$ CFU/ml, *Flavobacterium* sp. DS5-73 ($18.9\pm3.6 \times 10^5$ CFU/ml), *Pseudomonas* sp. DS10-129 ($32.6\pm0.8 \times 10^5$ CFU/ml) previously isolated on hydrocarbon containing medium were inoculated in 200 ml of nutrient broth and kept in a shaker for 24 h at room temperature. The strain name designated with GS was isolated from gasoline station and DS from diesel station soils, followed by its strain number, were depicted in our strains (Rahman et al. 2002a). Members of the bacterial consortium were selected depending on their efficiency of crude oil degradation (Rahman et al. 2002b). For the preparation of amendments, the rhamnolipid, produced by a *Pseudomonas aeruginosa* strain available at University of Ulster, was used.

### 2.2 Preparation of Amendments

To both sterile (sterilized in an oven at 180°C for 3 h) and non-sterile soil samples, 10% and 20% of tank bottom sludge (TBS) with 87.4% oil and grease at pH 6.7 was added and mixed thoroughly. To find out the role of indigenous microbial populations present in soil and tank bottom sludge, controls were set up with sterile and non-sterile soil with no amendments. Other amendments containing bacterial consortium, NPK solution and rhamnolipid were set up to test the effects of these additives on biodegradation (Table 1).

**Table 1.** Preparation of the different treatments of sterile and non-sterile soil samples

| Amendments | NS or SS (g) | TBS (%) | RL (mg) | NPK (mg) | BC (ml) | Moisture content (%) |
|---|---|---|---|---|---|---|
| NS +TBS | 100 | 10 or 20 | | | | 1.2 |
| NS +TBS +RL | 100 | 10 or 20 | 4 | | | 1.2 |
| NS +TBS+NPK | 100 | 10 or 20 | | 0.1 | | 1.2 |
| NS +TBS+BC | 100 | 10 or 20 | | | 1 | 1.2 |
| NS +TBS+RL+NPK+BC | 100 | 10 or 20 | 4 | 0.1 | 1 | 1.2 |
| SS+TBS | 100 | 10 or 20 | | | | 1.2 |
| SS+TBS +RL | 100 | 10 or 20 | 4 | | | 1.2 |
| SS+TBS+NPK | 100 | 10 or 20 | | 0.1 | | 1.2 |
| SS+TBS+BC | 100 | 10 or 20 | | | 1 | 1.2 |
| SS+TBS+RL+NPK+BC | 100 | 10 or 20 | 4 | 0.1 | 1 | 1.2 |

NS - Non-sterile soil; SS - Sterile soil; TBS - Tank Bottom Sludge; BC - Bacterial Consortium; RL - Rhamnolipid; NPK - Nitrogen, Phosphorus and Potassium solution

The treatments were set up in sets of screw cap glass universal bottles as microcosms containing 10 g of soil samples and moisture content was adjusted at 12%. All microcosm tubes were incubated at 30°C. Triplicate sets of experimental samples were analysed at 0, 28, 56 and 84 days to enumerate total heterotrophic bacterial counts and to estimate protein content, percentage of n-alkane degradation, pH and surface tension (ST).

### 2.3 Enumeration of Bacterial Population

Total heterotrophic bacteria were enumerated by using a pour plate technique on plate count agar (Merck, UK) after 24 h incubation at 30°C, which also allowed growth of all members of the added bacterial consortium. Identity of the individual bacterial isolate was confirmed by biochemical test as described in our earlier report (Rahman et al. 2002a).

### 2.4 Total Protein Estimation

For the estimation of total protein, 1 ml supernatant without any soil particle was taken from soil: water mixture (1:10 ratio). It was centrifuged at 13000 rpm for 10 min and to the pellet obtained was added 1 ml of 3 N NaOH solution and boiled for 3 min. After cooling at room temperature, 1 ml of 1 M $H_3PO_4$ solution was added. A 50 μL aliquot was taken and mixed with 950 μL Coomassie protein assay reagent (Pierce, Rockford, USA) and incubated at 30°C for 10 min and the optical density was measured at 595 nm using UV-visible spectrophotometer (Shimadzu model no. UV – 2101PC, Shimadzu Europe Ltd., UK). The total protein was estimated using a standard curve prepared with albumin (Bradford 1976).

### 2.5 Characterization of Rhomnolipid using Mass Spectrometry

Rhamnolipid fraction from culture free supernatant was extracted by adding equal volume of Chloroform: Methanol (2:1) solvent mixture and mixed thoroughly. Then the organic layer was separated using separatory funnel, air dried and dissolved in methanol. Mass spectrometry characterization and detection of the rhamnolipid fractions under investigation were performed using an LCQ™ quadrupole ion-trap mass spectrometer (Finnigan MAT, San Jose, California, USA) with electrospray ionization (ESI). Standard solutions and samples under investigation were infused into the mass spectrometer at a flow rate of 10 μl/min. In the ESI, source nitrogen sheath and auxiliary gas flows were maintained at 50 and 5, respectively and referred to arbitrary values set by the software. The heated capillary temperature was 250°C and the spray voltage set to 5 kV. Negative ion mode was used throughout and scans initiated over the 50-2000 m/z range.

### 2.6 Surface Tension Analysis and Measurement of pH

The surface tension of the soil extract (soil: water 1:10) was measured using a digital tensiometer (Kruss digital tensiometer model no. K9) equipped with a 6 cm De Nuoy platinum ring. To increase the accuracy, average of triplicates was used for the study. The pH of the soil extract (soil:water 1:10) was estimated using pH meter (Microcomputer pH meter model no. 6171, Jenco Instruments Inc., SanDiago, USA).

### 2.7 Hydrocarbon Estimation

The hexane soluble n-alkanes (nC8-nC40) in the soil samples were determined using gas chromatography (Perkin-Elmer GC model no. 8310). Soil and hexane (1:100 ratio) were mixed for 5 minutes in a vortex mixture and soil free hexane extract was separated using membrane filter and then used for GC analysis. A 30 m fused silica capillary column (Restek Corporation, USA) and GC with flame ionisation detector were used for analysis. The injection temperature was 250°C; detector temperature 250°C; column temperature was programmed as 50°C/4 min, then increased at the rate of 10°C/min to 330°C and maintained at 330°C for 20 min. Total recoverable petroleum hydrocarbon standard with purity of 99.9999% obtained from Restek Corporation, USA, was used to identify the n-alkanes. Degradation was estimated as the difference between the initial and final concentrations of the n-alkane fractions.

### 2.8 Statistical Analysis

The experiment was set up as a factorial design consisting of two concentrations they were 10% and 20% sludge contaminated soil x 10 treatments; 1) NS+TBS,

2) NS+TBS+RL, 3) NS+TBS+NPK, 4) NS+TBS+BC, 5) NS+TBS+RL+NPK+BC, 6) SS+TBS, 7) SS+TBS+RL, 8) SS+TBS+NPK, 9) SS+TBS+BC, 10) SS+TBS+RL+NPK+BC x four time periods (0, 28, 56 & 84 days) x three replicates per treatment. Statistical analysis was carried out using Analysis of Variance (ANOVA). Mean of the various treatments were tested for level of significance at 1% and 5% probability by Duncan's multiple range test (DMRT) (Gomez and Gomez 1984).

## 3. Results and Discussion

### 3.1 Effect of Bacterial growth on Biodegradation

Sandy soil was used along with garden soil to increase the porosity and thus aeration for enhanced bioremediation. An initial bacterial population of about $2.1 \pm 0.7 \times 10^3$ CFU/g was observed in the non-sterile soil amended with 10% of tank bottom sludge. Low bacterial numbers may be because of the use of sandy soil with low nutrients and microflora. An increase in bacterial population was encountered in all amended soil samples particularly with rhamnolipid solution (Table 2). This may be due to the biosurfactant induced desorption of hydrocarbons from soil to the aqueous phase of soil slurries leading to increased microbial mineralization, either by increasing hydrocarbon solubility or by increasing the contact surface with hydrophobic compounds (Moran et al. 2000; Rahman et al. 2002d). Two orders of magnitude increase in the bacterial population were observed in soil samples amended with 10% petroleum TBS after 56 days of incubation. The available nutrients were rapidly assimilated by soil microbes, thus depleting the nutrient reserves. In fact the objective of augmenting NPK solution to the soil samples was to restore the availability of essential nutrients. Several researchers have also described an increase in microbial activity and rate of biodegradation following addition of inorganic nutrients (Radwan et al. 2000; Del 'Arco and de Franca 2001; Vasudevan and Rajaram 2001).

### 3.2 Change in Protein Concentration during Degradation

The protein estimation by Bradford's method was effective in monitoring the microbial population in the hydrocarbon contaminated soil sample. In the non-sterile control, the initial concentration of protein observed was $1.25 \pm 0.16$ mg/g of soil, whereas in sterile soil it was $0.001 \pm 0.0$ mg/g. This reduction may be due to the proteins destroyed in the soil during sterilization. The various amendments and mixed consortium caused proliferation of bacteria up to 56 days of incubation and resulted in an increased protein content in these treatments up to a value of 6.24 mg/g in soil samples amended with 10% TBS (Table 3).

**Table 2.** Bacterial growth during degradation of n-alkane in oil sludge treated with different amendments

| S. No. | Amendments/ Days | Bacteria (CFU/g) | | | | | | | |
|---|---|---|---|---|---|---|---|---|---|
| | | 10% sludge | | | | 20% sludge | | | |
| | | 0 | 28 | 56 | 84 | 0 | 28 | 56 | 84 |
| 1 | NS+TBS | 2.1±0.7[B] x $10^3$e[A] | 6.1±0.3 x $10^3$e | 7.2±0.2 x $10^3$e | 2.4±0.4 x $10^3$e | 2.7±0.3 x $10^3$e | 4.1±0.2 x $10^3$e | 7.3±0.6 x $10^3$e | 6.7±0.6 x $10^3$e |
| 2 | NS+TBS+RL | 7.9±0.9 x $10^3$c | 8.1±0.5 x $10^3$d | 89.0±2.3 x $10^3$d | 59.0±1.2 x $10^3$d | 92.0±4.9 x $10^3$c | 31.0±1.8 x $10^3$d | 56.0±4.1 x $10^3$d | 39.0±0.1 x $10^3$d |
| 3 | NS+TBS+NPK | 2.8±0.4 x $10^3$d | 39.0±1.1 x $10^3$c | 660.0±15 x $10^3$c | 440.0±16 x $10^3$c | 6.4±2.3 x $10^3$d | 43.0±2.6 x $10^3$c | 91.0±6.3 x $10^3$c | 63.0±2.5 x $10^3$c |
| 4 | NS+TBS+BC | 240.0±11 x $10^3$b | 1.8±0.2 x $10^7$b | 4.3±0.1 x $10^8$a | 3.8±0.5 x $10^8$b | 220.0±16 x $10^3$b | 3.8±0.1 x $10^6$b | 5.6±0.2 x $10^7$b | 2.8±0.3 x $10^7$b |
| 5 | NS+TBS+RL+NPK+BC | 810.0±17 x $10^3$a | 6.8±0.4 x $10^8$a | 3.8±0.3 x $10^8$b | 4.1±0.5 x $10^{10}$a | 500.0±37 x $10^3$a | 1.7±0.1 x $10^7$a | 2.6±0.2 x $10^8$a | 2.1±0.1 x $10^8$a |
| 6 | SS+TBS | 0.12±0.01 x $10^3$e | 0.80±0.07 x $10^3$c | 0.97±0.8 x $10^3$e | 0.27±0.04 x $10^3$e | 0.14±0.02 x $10^3$e | 0.37±0.02 x $10^3$d | 0.68±0.04 x $10^3$d | 0.51±0.04 x $10^3$c |
| 7 | SS+TBS+RL | 0.18±0.01 x $10^3$c | 0.28±0.01 x $10^3$e | 2.50±0.3 x $10^3$d | 1.10±0.04 x $10^3$d | 0.19±0.01 x $10^3$d | 0.27±0.01 x $10^3$e | 0.99±0.01 x $10^3$c | 0.42±0.03 x $10^3$d |
| 8 | SS+TBS+NPK | 0.16±0.02 x $10^3$d | 0.56±0.04 x $10^3$d | 6.4±0.5 x $10^3$c | 5.2±0.6 x $10^3$c | 0.22±0.02 x $10^3$c | 0.84±0.08 x $10^3$c | 0.32±0.02 x $10^3$e | 0.12±0.01 x $10^3$e |
| 9 | SS+TBS+BC | 210.0±1.3 x $10^3$b | 640.0±49 x $10^3$b | 290.0±19 x $10^3$b | 170.0±14 x $10^3$b | 18.0±0.1 x $10^3$b | 6.7±0.04 x $10^6$b | 9.1±0.9 x $10^6$b | 8.9±0.7 x $10^6$b |
| 10 | SS+TBS+RL+NPK+BC | 370.0±55 x $10^3$a | 9.1±0.7 x $10^6$a | 3.0±0.1 x $10^7$a | 2.7±0.1 x $10^7$a | 270.0±16 x $10^3$a | 4.6±0.02 x $10^7$a | 3.9±0.2 x $10^8$a | 1.9±0.01 x $10^8$a |

NS - Non-sterile soil; SS - Sterile soil; TBS - Tank Bottom Sludge; BC - Bacterial Consortium; NPK - Nitrogen, Phosphorus and Potassium solution; RL - Rhamnolipid

[A]a, b, c, d, e: Arithmetic means within row with the same letter are not significantly different from each other at 5% probability level by DMRT; [B]Standard error

**Table 3.** Protein concentration during degradation of n-alkane in oil sludge treated with different amendments for a period of up to 84 days

| S. No. | Amendments/ Days | Protein (mg/g) | | | | | | | |
|---|---|---|---|---|---|---|---|---|---|
| | | 10% sludge | | | | 20% sludge | | | |
| | | 0 | 28 | 56 | 84 | 0 | 28 | 56 | 84 |
| 1 | NS+TBS | 1.2e[A] ±0.16[B] | 1.72d ±0.15 | 2.19d ±0.13 | 2.23d ±0.29 | 0.08d ±0.00 | 1.12e ±0.09 | 1.97e ±0.11 | 2.10e ±0.17 |
| 2 | NS+TBS+RL | 1.74c ±0.11 | 2.07c ±0.08 | 2.56c ±0.24 | 2.58c ±0.17 | 1.20c ±0.02 | 1.88c ±0.06 | 2.12d ±0.17 | 2.32d ±0.21 |
| 3 | NS+TBS+NPK | 1.29d ±0.07 | 1.58e ±0.04 | 1.58e ±0.08 | 2.25d ±0.09 | 0.08d ±0.01 | 1.24d ±0.10 | 2.30c ±0.20 | 2.40c ±0.28 |
| 4 | NS+TBS+BC | 2.15b ±0.19 | 3.99b ±0.24 | 4.24b ±0.21 | 4.83b ±0.16 | 1.70b ±0.11 | 3.10b ±0.17 | 3.70b ±0.24 | 3.98b ±0.11 |
| 5 | NS+TBS+RL+NPK+ BC | 2.41a ± 0.21 | 4.93a ±0.21 | 6.24a ±0.16 | 6.00a ±0.37 | 2.01a ±0.15 | 3.50a ±0.29 | 4.12a ±0.55 | 4.51a ±0.24 |
| 6 | SS+TBS | 0.01d ±0.00 | 0.05d ±0.01 | 0.07c ±0.00 | 0.08c ±0.00 | 0.02c ±0.00 | 0.06c ±0.00 | 0.09c ±0.01 | 0.09c ±0.01 |
| 7 | SS+TBS+RL | 0.01d ±0.00 | 0.05d ±0.00 | 0.07c ±0.00 | 0.09c ±0.01 | 0.02c ±0.00 | 0.06c ±0.00 | 0.07c ±0.00 | 0.08c ±0.00 |
| 8 | SS+TBS+NPK | 0.02c ±0.00 | 0.06c ± 0.00 | 0.07c ±0.00 | 0.07c ±0.00 | 0.03c ±0.00 | 0.05c ±0.00 | 0.06c ±0.00 | 0.07c ±0.00 |
| 9 | SS+TBS+BC | 1.87b ±0.06 | 3.20b ±0.24 | 3.50b ±0.27 | 3.59b ±0.27 | 1.70b ±0.08 | 2.70b ±0.15 | 3.05b ±0.09 | 3.21b ±0.24 |
| 10 | SS+TBS+RL+NPK+BC | 2.73a ±0.18 | 3.98a ±0.18 | 4.12a ±0.39 | 4.37a ±0.46 | 2.91a ±0.24 | 3.52a ±0.30 | 3.98a ±0.27 | 4.10a ±0.35 |

NS – Non sterile soil; SS – Sterile soil; TBS - Tank bottom sludge; BC – Bacterial consortium; NPK – Nitrogen, Phosphorus, Potassium solution; RL – Rhamnolipid biosurfactant solution

[A]a, b, c, d, e: Arithmetic means within row with the same letter are not significantly different from each other at 5% probability level by DMRT; [B]Standard error

### 3.3 Biodegradation versus Surface Tension

The indigenous microbial community of non-sterile and sterile soil caused a slight decrease in the surface tension, indicating that those microorganisms could not produce sufficient biosurfactant activities. Surface tension of the soil extract was 69.7±0.4 - 71.1±0.6 mN/m (milli-Newton/meter), which was reduced to 52.3±2.2 and 48.1±1.8 mN/m in NS+TBS+RL and SS+TBS+RL amended with 10% TBS respectively. A reduction in surface tension occurred because of the presence of rhamnolipid (RL) in NS+TBS+RL and SS+TBS+RL with 10% TBS amendment (Table 4). Furthermore, in soil samples augmented with a bacterial consortium and amended with rhamnolipid and NPK, a significant reduction in surface tension was noted after 56 days of incubation. A possible reason for this may be the rhamnolipid-mediated desorption of petroleum hydrocarbons, which increased their solubility and hence the biological activity of indigenous microflora or added hydrocarbon degrading bacterial consortium. In a study by Oberbremer and Muller-Hurtig (1989), a positive correlation was obtained between reduction in the surface tension of the fluid phase in a stirred soil bioreactor and the onset of biodegradation of hydrophobic petroleum hydrocarbons. It has also been reported that a rhamnolipid biosurfactant can mediate reduction in the surface tension (Banat et al. 2000; Noordman et al. 2000).

### 3.4 Effect of Degradation on pH

A range of pH 7.2 ± 0.3 to 7.2 ± 0.4 was estimated in the sterile and non-sterile soil samples. Alternatively, in soil samples amended with mixed consortium, rhamnolipid or NPK, an increase in pH was observed after 56 days of incubation suggesting the release of by-products during hydrocarbon degradation (Table 5).

### 3.5 Biodegradation of n-alkanes

Gas chromatographic analyses revealed all hexane soluble n-alkanes in the range of nC8–nC40, which were relatively abundant in tank bottom crude oil sludge. The degradation of the above was discussed in four different ranges, such as nC8–nC11, nC12–nC21, nC22–nC31 and nC32–nC40. The nC8–nC11 range consisted of volatile hydrocarbons. A percentage of hydrocarbon degradation of approximately 100% (nC8–nC11), 83-98% (nC12-nC21), 80-85% (nC22-nC31) and 57-73% (nC32-nC40) was noted in non-sterile soil samples with 10% TBS amended with RL+NPK+BC (Fig. 1). Among the different treatments, in NS+TBS+RL+NPK+BC amended with 10% TBS, all the hydrocarbons in the range of nC8- nC11 were degraded, whereas in SS+TBS+RL+NPK+BC with 10% TBS, NS+TBS+RL+NPK+BC and SS+TBS+RL+NPK+BC with 20% TBS, only 81-87%, 64-83% and 55-61% degradation was observed, respectively (Figs. 4-6).

**Table 4.** Surface tension of samples during degradation of n-alkane in oil sludge treated with different amendments for a period of up to 84 days

| S. No. | Amendments/ Days | Surface tension (mN/m) | | | | | | | |
|---|---|---|---|---|---|---|---|---|---|
| | | 10% sludge | | | | 20% sludge | | | |
| | | 0 | 28 | 56 | 84 | 0 | 28 | 56 | 84 |
| 1 | NS+TBS | 69.7c[A] ±0.4[B] | 70.3a ±0.9 | 65.5b ±2.7 | 67.7b ±0.9 | 70.1b ±0.5 | 67.1b ±0.4 | 63.1c ±1.9 | 70.5a ±0.4 |
| 2 | NS+TBS+RL | 52.3d ±2.2 | 69.8b ±0.4 | 69.7a ±3.1 | 65.1c ±1.1 | 57.1c ±2.1 | 69.1a ±0.2 | 66.8a ±0.3 | 69.9b ±1.0 |
| 3 | NS+TBS+NPK | 71.5a ±0.4 | 66.7d ±1.4 | 62.9d ±1.2 | 62.9d ±0.4 | 70.2b ±0.1 | 61.8e ±1.1 | 59.8e ±0.5 | 67.4e ±1.4 |
| 4 | NS+TBS+BC | 70.5b ±0.5 | 68.8c ±1.4 | 63.3c ±2.1 | 69.7a ±0.3 | 70.5a ±0.4 | 65.1c ±2.3 | 63.3b ±0.7 | 69.5c ±0.4 |
| 5 | NS+TBS+RL+NPK+BC | 32.1e ± 1.6 | 62.7e ±2.9 | 57.2e ±3.0 | 61.5e ±1.1 | 41.2d ±2.1 | 63.1d ±2.4 | 61.1d ±1.2 | 68.1d ±2.3 |
| 6 | SS+TBS | 70.1b ±1.5 | 70.6a ±0.2 | 69.4a ±0.6 | 69.2a ±0.9 | 71.1b ±0.6 | 69.2a ±1.3 | 68.9a ±2.0 | 67.5b ±0.7 |
| 7 | SS+TBS+RL | 48.1d ±1.8 | 61.1c ±3.1 | 62.9b ±2.4 | 57.4e ±2.3 | 67.1d ±1.2 | 64.5e ±3.4 | 64.7d ±3.4 | 65.5d ±1.5 |
| 8 | SS+TBS+NPK | 69.4c ±0.1 | 69.9b ±1.2 | 61.7c ±1.5 | 67.9b ±1.7 | 70.1c ±0.2 | 67.8b ±2.9 | 66.9b ±1.6 | 66.9c ±3.4 |
| 9 | SS+TBS+BC | 71.7a ±0.4 | 70.4a ±0.6 | 62.9b ±3.1 | 64.1c ±2.0 | 71.5a ±0.5 | 64.9d ±3.1 | 66.5c ±3.3 | 67.6a ±2.9 |
| 10 | SS+TBS+RL+NPK+BC | 40.1e ±2.6 | 59.3d ±1.7 | 61.9c ±0.4 | 62.4d ±1.6 | 47.2e ±2.1 | 65.5c ±4.0 | 61.3e ±0.9 | 58.9e ±3.7 |

NS – Non sterile soil; SS – Sterile soil; TBS - Tank bottom sludge; BC – Bacterial consortium; NPK – Nitrogen, Phosphorus, Potassium solution; RL – Rhamnolipid biosurfactant solution

[A]a, b, c, d, e: Arithmetic means within row with the same letter are not significantly different from each other at 5% probability level by DMRT; [B]Standard error.

**Table 5.** pH of the soil during degradation of n-alkane in oil sludge treated with different amendments for a period of up to 84 days

| S. No. | Amendments/ Days | pH | | | | | | | |
|---|---|---|---|---|---|---|---|---|---|
| | | 10% sludge | | | | 20% sludge | | | |
| | | 0 | 28 | 56 | 84 | 0 | 28 | 56 | 84 |
| 1 | NS+TBS | 7.2a[A]±0.4[B] | 7.1c ±0.4 | 6.9d ±0.6 | 6.9c ±0.4 | 7.2a ±0.1 | 7.1c ±0.5 | 6.7c ±0.2 | 6.9c ±0.4 |
| 2 | NS+TBS+RL | 6.9c ±0.2 | 7.0d ±0.1 | 7.0c ±0.2 | 7.0b ±0.3 | 6.9c ±0.5 | 7.0d ±0.1 | 7.1a ±0.4 | 6.9c ±0.6 |
| 3 | NS+TBS+NPK | 7.1b ±0.3 | 7.6a ±0.3 | 7.2b ±0.4 | 7.0b ±0.1 | 7.1b ±0.3 | 7.6a ±0.2 | 7.2a ±0.5 | 7.2a ±0.5 |
| 4 | NS+TBS+BC | 7.2a ±0.1 | 7.1c ±0.2 | 7.0c ±0.3 | 7.0b ±0.5 | 7.2a ±0.3 | 7.1c ±0.4 | 68b ±0.3 | 6.9c ±0.3 |
| 5 | NS+TBS+RL+NPK+BC | 6.9c ± 0.3 | 7.3b ±0.4 | 7.3a ±0.7 | 7.5a ±0.3 | 6.9c ±0.1 | 7.3b ±0.6 | 7.1a ±0.7 | 7.1b ±0.4 |
| 6 | SS+TBS | 7.2a ±0.3 | 7.1c ±0.4 | 6.9 ±0.6 | 7.0 ±0.4 | 7.2 ±0.5 | 7.0 ±0.5 | 6.8 ±0.4 | 7.0 ±0.7 |
| 7 | SS+TBS+RL | 6.8c ±0.2 | 7.2b ±0.3 | 7.1 ±0.3 | 6.9 ±0.5 | 6.7 ±0.6 | 7.1 ±0.6 | 7.2 ±0.5 | 7.1 ±0.6 |
| 8 | SS+TBS+NPK | 6.9b ±0.5 | 7.4a ± 0.4 | 7.2 ±0.2 | 7.3 ±0.1 | 6.9 ±0.4 | 7.3 ±0.3 | 7.8 ±0.3 | 7.1 ±0.4 |
| 9 | SS+TBS+BC | 6.9b ±0.1 | 7.2b ±0.5 | 7.0 ±0.4 | 6.9 ±0.3 | 6.9 ±0.3 | 7.2 ±0.4 | 7.0 ±0.2 | 7.0 ±0.3 |
| 10 | SS+TBS+RL+NPK+BC | 6.9b ±0.6 | 7.4a ±0.6 | 7.4 ±0.5 | 7.3 ±0.4 | 6.9 ±0.4 | 7.3 ±0.1 | 7.5 ±0.4 | 7.2 ±0.2 |

NS – Non sterile soil; SS – Sterile soil; TBS - Tank bottom sludge; BC – Bacterial consortium; NPK – Nitrogen, Phosphorus, Potassium solution; RL – Rhamnolipid biosurfactant solution

Aa, b, c, d, e: Arithmetic means within row with the same letter are not significantly different from each other at 5% probability level by DMRT; B Standard error.

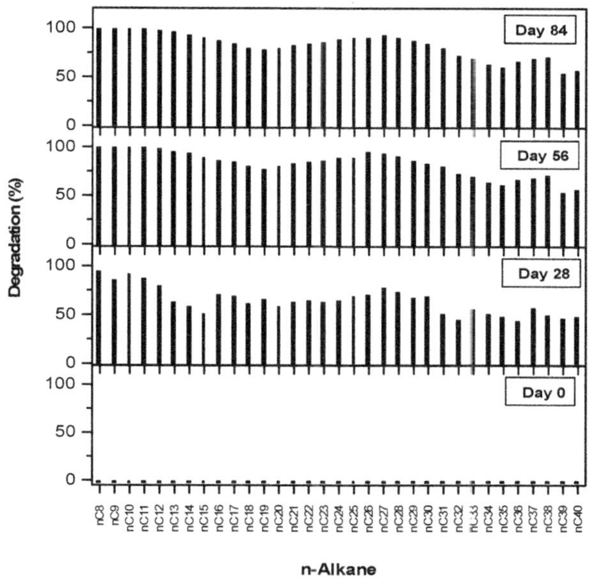


**Fig. 1.** n-Alkane degradation in non-sterile soil with 10% of tank bottom sludge and BC+NPK+RL at various time intervals

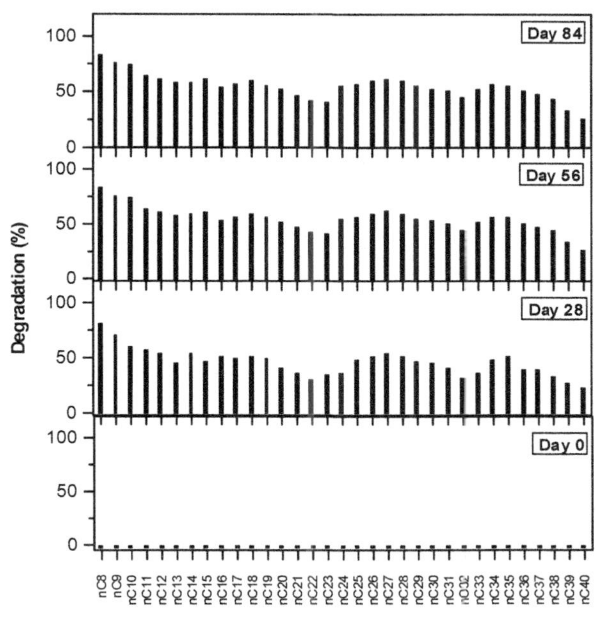


**Fig. 2.** n-Alkane degradation in non-sterile soil with 20% of tank bottom sludge and BC+NPK+RL at various time intervals

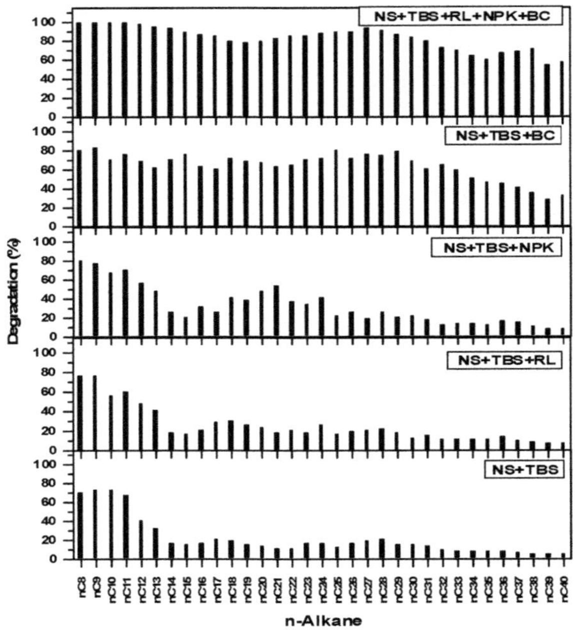


**Fig. 3.** n-Alkane degradation in non-sterile soil with 10% of tank bottom sludge and BC+NPK+RL on 56th day of treatment

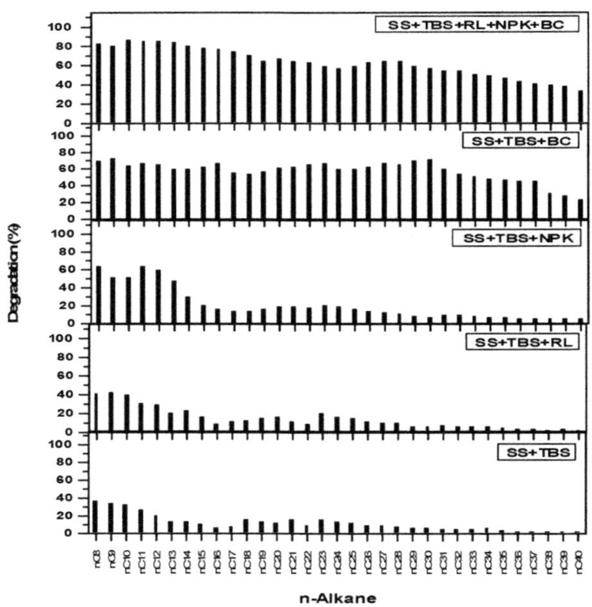


**Fig. 4.** n-Alkane degradation in sterile soil with 10% of tank bottom sludge and BC+NPK+RL on 56th day of treatment

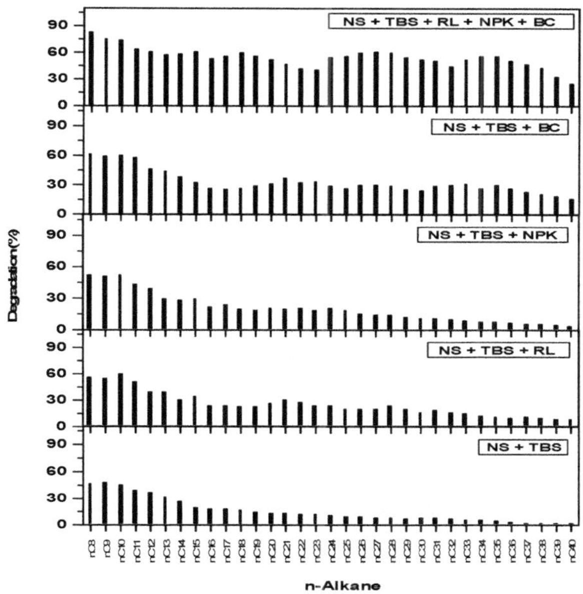


**Fig. 5.** n-Alkane degradation in non-sterile soil with 20% of tank bottom sludge and BC+NPK+RL on 56th day of treatment

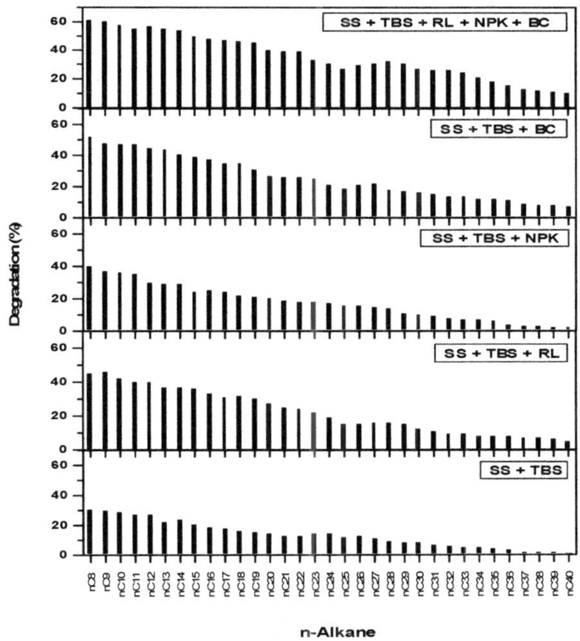


**Fig. 6.** n-Alkane degradation in sterile soil with 20% of tank bottom sludge and BC+NPK+RL on 56th day of treatment

The decreasing utilization trend after 56 days of incubation observed with soil samples amended with 10% TBS was not only due to the substrate depletion but also to the fact that the remaining hydrocarbons were relatively more resistant to biodegradation. The rate of petroleum biodegradation and quantity of hydrocarbon degraded depend on environmental conditions, chemical structure of the pollutant compounds, type and amount of oil present at the contaminated site (Del 'Arco and de Franca 2001). At 20% TBS concentration, the decrease in microbial degradation activity may be due to the toxicity caused by higher hydrocarbon contamination (Fig. 2).

The bacterial consortium enhanced the degradation of all the fractions of hydrocarbons from nC8-nC40 to various degrees in sterile and non-sterile samples supplemented with 10% and 20% TBS. This observation is in general agreement with the earlier report regarding the use of bioaugmentation (Mulligan et al. 2001). When compared to all the sets, different treatments of non-sterile soil (NS+TBS, NS+TBS+RL, NS+TBS+NPK, NS+TBS+BC and NS+TBS+RL+NPK+BC) amended with 10% TBS exhibited a higher percentage of hydrocarbon degradation (Fig. 3). The degree of degradation observed with SS+TBS was lower than that in the NS+TBS. These results indicated the ubiquitous distribution of diversified hydrocarbon structures, originating in particular from plants in the environment and consequently the presence of specific bacterial hydrocarbon degraders. Furthermore, the TBS amended soil samples treated with rhamnolipid or NPK lost substantially fewer hydrocarbons in the range of nC12–nC40 than those treated with bacterial consortium. In our study, no lag period was observed preceeding petroleum hydrocarbon mineralisation in sterile soil samples amended with TBS, suggesting the presence of an active hydrocarbon degrading population in the TBS. Addition of NPK solution alone had only a minor effect on hydrocarbon degradation compared to other soil amendments which may be due to a slight increase in biological activity of the microflora present in soil and sludge. The addition of rhamnolipid however, significantly enhanced the rate of biodegradation of hydrocarbon fractions by the bacterial consortium and the NPK solution in all the treatments.

When hydrocarbons are present in non-inhibitory concentration (available or desorbed form) in the soil, it may affect the rate of biodegradation by enhancing the biodegradation activity of the indigenous microbial population. Adding surfactants to soil contaminated with hydrophobic contaminants may increase the bioavailability of these compounds to hydrocarbon degrading microorganisms (Banat et al. 1991; Banat 1995). Complete degradation of nC8-nC11 and 73-98% of nC12 - nC40 was observed with the mixed bacterial consortium amended with rhamnolipid and NPK solution in 10% TBS amended soil samples at 56 days of incubation (Figs. 3 and 5), which was higher than all the earlier reports.

Dave et al. (1994) achieved 70% bioremediation of a slop oil contaminated soil using oil degrading cultures. One of the main reasons for the prolonged persistence of hydrophobic hydrocarbons in the contaminated environments is their strong adsorption even on coarse-grained and organic free soils by microporosity, which makes them less available for hydrocarbon degrading microorganisms and remain even after bioremediation. Hence, for efficient and complete biodegradation, solubilization of these hydrocarbons with biosurfactants prior to bioaugmentation is advantageous. Moreover, use of biosurfactant producing hydrocarbon degrading microorganisms for bioaugmentation to enhance hydrocarbon degradation offers an advantage of a continuous supply of a non-toxic and biodegradable surfactant at a low cost (Moran et al. 2000; Rahman et al. 2002c). The biosurfactant used in this study is a dirhamnolipid type of surfactant. Mass spectrometry using electrospray ionization is an efficient method to characterize rhamnolipid biosurfactant and since *Pseudomonas* sp. DS10-129 had highest production, we analysed its fermentation broth (Rahman et al. 2002d). Daziel et al. (1999) reported about different rhamnolipid species produced by *Pseudomonas* sp. 57RP with mannitol and naphthalene as carbon source. We detected a presence of mono and dirhamnolipids the Rha-C10-C10 and the Rha-Rha-C10-C-10 (MW=504 and 650) (Fig. 7).

However, the potential benefits of *in situ* application of surfactants must be weighed against the possibility of increased ground water contamination due to surfactant-mediated enhanced mobility of contaminants. Hence, repeated use of smaller dose schedule should be investigated as means to control contaminant mobility together with careful monitoring of the rate and extent of hydrocarbon degradation.

All the results were statistically analyzed using ANOVA and DMRT procedures to determine significant parameters. The results presented in Table 6

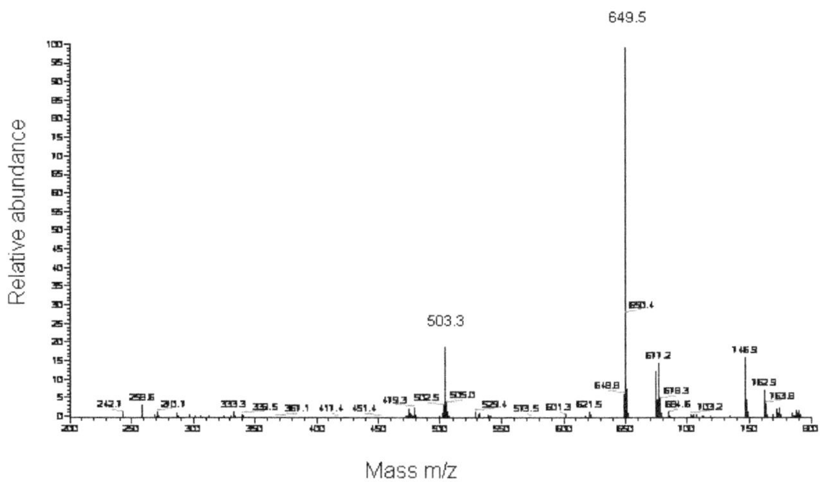


**Fig. 7.** Mass spectrum of rhamnolipids produced by *Pseudomonas aeruginosa* DS10-129 using soybean oil as substrate

**Table 6.** Significance level for the different parameters tested within our treatments computed by DMRT

| Parameter | Bacteria (x$10^3$ CFU/g) | | | Protein (mg/g) | | | Surface tension (mN/m) | | | pH | | |
|---|---|---|---|---|---|---|---|---|---|---|---|---|
| Factorial Effect | SE | CD | SL | SE | CD | SL | SE | CD | SL | SE | CD | SL |
| Concentration (C) | 9.24 | 18.48 | ** | 0.02 | 0.03 | ** | 0.17 | 0.29 | ** | 0.14 | 0.24 | ** |
| Amendment (A) | 23.60 | 47.2 | ** | 0.09 | 0.16 | ** | 0.43 | 0.74 | ** | 0.20 | 0.46 | ** |
| Days (D) | 36.10 | 72.2 | ** | 0.17 | 0.31 | ** | 1.54 | 2.93 | * | 0.39 | 0.61 | ** |
| C x A | 54.30 | 108.6 | ** | 0.27 | 0.53 | ** | 1.90 | 3.48 | * | 0.43 | 0.83 | * |
| C x D | 61.20 | 122.4 | ** | 0.34 | 0.65 | ** | 2.36 | 4.31 | ns | 0.35 | 0.67 | ns |
| A x D | 86.40 | 172.8 | ** | 0.39 | 0.74 | ** | 2.68 | 5.16 | ns | 0.67 | 1.24 | ns |
| C x A x D | 100.0 | 197.5 | ** | 0.44 | 0.85 | ** | 3.91 | 7.57 | ns | 0.62 | 1.29 | ns |

SE - Standard Error; CD -Cumulative Difference; SL - Significant level; ns - not significant at 1% or 5% probability levels.
* Significant at 5% probability level (within column); ** Significant at 1% probability level (within column)

revealed that all the above parameters were highly influenced by single factors (concentration (C), amendments (A), number of days (D) treated); two factor combinations (C x A, C x D and A x D) and three factor combinations (C x A x D) at a 1% probability level. However, the number of days treated (D), and the two factor combination C x A for surface tension and pH were significant at 5% probability level. Moreover, the two factor combinations C x D and A x D and the three factor combination C x A x D were not significant at 1% or 5% probability levels for surface tension and pH.

## 4. Conclusion

Several strategies have been attempted for bioremediation of hydrocarbon-polluted sites. Bioaugmentation with designed bacterial consortium, followed by the addition of rhamnolipid biosurfactant and NPK solution to soils contaminated with up to 10% tank bottom sludge, enhanced the rate of biodegradation over a period of 56 days. Pre-treatment of hydrocarbon contaminated soil with biosurfactants enhanced bioavailability of the hydrocarbons to microbial population. Furthermore, supplementation with inorganic nutrients like NPK solution enhanced the secondary successions of crude petroleum utilizers. For bioremediation, a single inoculation with the biosufactant-producing hydrocarbon degrading bacterial consortium at the beginning of the process would reduce the cost of inoculum preparation considerably. Hence we suggest a combined treatment as a possible bioremediation technology for the reclamation of oil sludge polluted soils.

Acknowledgements. We wish to thank Northern Ireland Environment and Heritage Service and European Regional Development Fund for financial supports.

## References

Banat IM, Samarah N, Murad M, Horne R, Banerjee S (1991) Biosurfactant production and use in oil tank clean-up. World J Microbiol Biotechnol 7:80-84

Banat IM (1995) Biosurfactants production and possible uses in microbial enhanced oil recovery and oil pollution remediation: a review. Bioresource Technol 51:1-12

Banat IM, Makkar RS, Cameotra SS (2000) Potential commercial applications of microbial surfactants. Appl Microbiol Biotechnol 53:495-508

Bradford MM (1976) A rapid and sensitive method for the quantification of microgram quantities of protien utilizing the principle of protein-dye binding. Anal Biochem 72:248- 254

Dave H, Ramakrishna C, Bhatt BD, Desai JD (1994) Biodegradation of slope oil from petroleum Industry and bioreclamation of slop oil contaminated soil. World J Microbiol Biotechnol 10:653-656

Del 'Arco JP, de Franca FP (2001) Influence of oil contamination levels on hydrocarbon biodegradation in sandy sediments. Environ Pollut 110:515-519

Ferrari MD, Neirotti E, Albornoz C, Mostazo MR, Cozzo M (1996) Biotreatment of hydrocarbons from petroleum tank bottom sludges in soil slurries. Biotechnol Lett 18:1241-1246

Gomez KA, Gomez AA (1984) Statistical procedure for agricultural research. John Wiley and Sons, New York, pp 680

Moran AC, Olivera N, Commendatore M, Esteves JL, Sineriz F (2000) Enhancement of hydrocarbon waste biodegardation by addition of a biosurfactant from *Bacillus subtilis* O9. Biodegradation 11:65-71

Mulligan CN, Yong RN, Gibbs BF (2001) Surfactant enhanced remediation of contaminated soil: a review. Eng Geol 60:371-380

Noordman WH, Brusseau ML, Janssen DB (2000) Adsorption of a multicomponent rhamnolipid surfactant to soil. Environ Sci Technol 34:832-838

Oberbremer A, Muller-Hurtig R (1989) Aerobic stepwise hydrocarbon degradation and formation of biosurfactants by an original soil population in a stirred reactor. Appl Microbiol Biotechnol 31:582-586

Radwan SS, Al-Mailem D, El-Nemr I, Salamah S (2000) Enhanced remediation of hydrocarbon contaminated desert soil fertilized with organic carbons. Int Biodeterior Biodegrad 46:129-132

Rahman KSM, Rahman TJ, Lakshmanaperumalsamy P, Banat IM (2002a) Occurrence of crude oil degrading bacteria in gasoline and diesel station soils. J Basic Microbiol 42:286-293

Rahman KSM, Banat IM, Rahman TJ, Thayumanavan T, Lakshmanaperumalsamy P (2002b) Bioremediation of gasoline contaminated soil by bacterial consortium amended with poultry litter, coir-pith and rhamnolipid biosurfactant. Bioresource Technol 81:25-32

Rahman KSM, Rahman TJ, Lakshmanaperumalsamy P, Banat IM (2002c) Towards efficient crude oil degradation by mixed bacterial consortium. Bioresource Technol 85:257-261

Rahman KSM, Rahman TJ, McClean S, Banat IM (2002d) Rhamnolipid biosurfactants production by strains of *Pseudomonas aeruginosa* using low cost raw materials. Biotechnol Prog 18:1277-1281

Vasudevan N, Rajaram P (2001) Bioremediation of oil sludge- contaminated soil. Environ Int 26:409-411

18

# Diversity, Biodegradation and Bioremediation of Polycyclic Aromatic Hydrocarbons

Sumeet Labana, Manisha Kapur, Deepak K. Malik, Dhan Prakash and R.K. Jain

Institute of Microbial Technology, Sector 39-A, Chandigarh 160 036, INDIA, Email: rkj@imtech.res.in

## 1. Introduction

Polycyclic aromatic hydrocarbons (PAHs) consist of a class of chemicals with two or more fused benzene rings in linear, angular or cluster arrangements. Among the most abundant environmental pollutants, the aromatic compounds are of major concern because of their persistence and toxicity. PAHs are ubiquitous in nature found throughout the environment in air, water and soil. They are produced during fossil fuel combustion, waste incineration, or as by-products of industrial processes, such as coal gasification and petroleum refining, and often released in large quantities into the environment (Finlayson-Pitts and Pitts 1997). There are more than 100 different PAHs which occur as complex mixtures, not as single compound.

One of the main reasons for the prolonged persistence of hydrophobic hydrocarbons in the environments is their low water solubility which increases their sorption to soil particles and limits their availability to biodegrading microorganisms (Cerniglia 1993). The decontamination of PAH-polluted sites is mandatory because many PAH compounds are known or suspected to be toxic, mutagenic or carcinogenic (Patnaik 1992). Therefore, PAHs are considered to be environmental pollutants that can have a detrimental effect on the flora and fauna of affected habitats, resulting in the uptake and accumulation of toxic chemicals in food chains which cause serious health problems and/or genetic defects in humans. The high molecular weight (HMW) PAHs (four or more fused rings) are of particular environmental concern, because of their potential mutagenicity and carcinogenicity (Goldman et al. 2001). On the basis of their abundance and toxicity, 16 PAH compounds have been identified by the U.S. Environmental Protection Agency (EPA) as priority pollutants (Keith and Telliard 1979) of which chemical structures are shown in Figure 1.

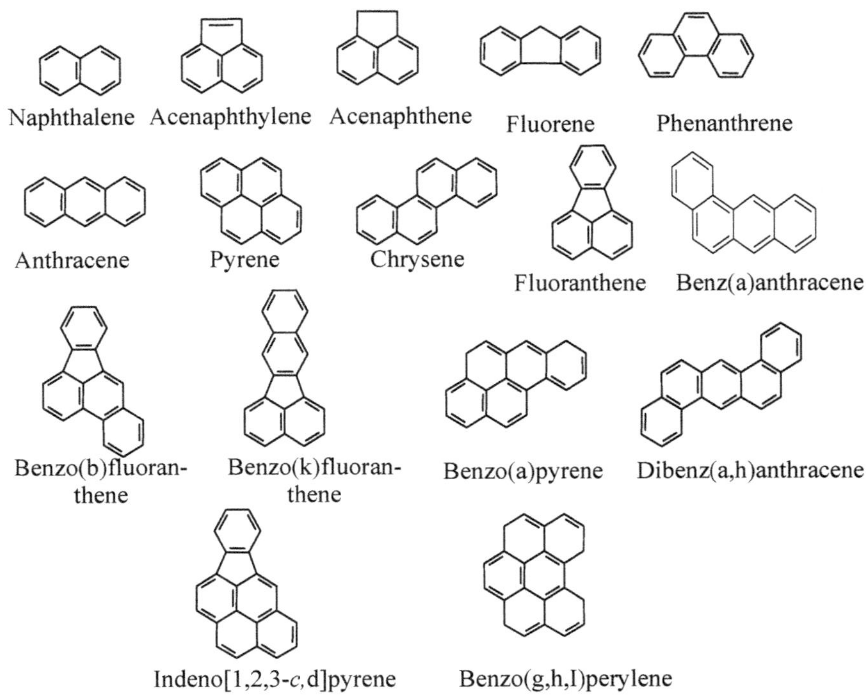


**Fig. 1.** Structures of the 16 polycyclic aromatic hydrocarbons listed by the U.S. Environmental Protection Agency as priority pollutants.

## 2. Natural Sources of PAHs in the Environment

PAHs are predominantly distributed in nature as components of surface waxes of leaves, plant oils, cuticles of insects and lipids of microoganisms (Millero and Sohn 1991). Petroleum and coal provide the largest source of mononuclear and polynuclear compounds. Studies in terrestrial and marine environments show that PAHs can also occur from geochemical origin. They are formed whenever organic substances are exposed to high temperatures. The aromatic rings so formed are more stable than their precursors, usually alkylated benzene rings.

The arrangement of the aromatic rings affects their environmental stability and hence their natural distribution (Blumer 1976). For example, linearly arranged benzene rings, such as in anthracene and tetracene, are the least stable and generally do not survive in nature unless sequestered into certain organic or inorganic matrices. The most stable arrangement is that of the annular types seen in phenanthrene, chrysene and picene. Such PAHs abound when organics have been exposed to elevated temperatures (Blumer 1976).

## 3. Anthropogenic Sources of PAHs in the Environment

Historically, PAHs have been released into the environment from three sources: biosynthetic (biogenic), geochemical and anthropogenic (National Research Council 1983). Anthropogenic sources are of two types: one is the result of accidental spillage and intentional dumping of such materials as creosote, coal tar and petroleum products, while the other type is derieved from the incomplete combustion of organic matter, such as wood burning, municipal incineration, automobile emissions and industrial discharges. Later sources of PAHs are the current focus of many environmental clean up programs and consequently from the basis for the development of effective bioremediation technologies. Atmospheric PAH depositions are usually from very dispersed sources, but they cover significant area of land surface. PAH concentrations from these sources are typically quite low in soil and they are absorbed strongly to soil particles.

Another type of broad, non-point source introduction of PAHs into soils is through land treatment procedures. For example, in US and Europe, sewage slude is applied to agricultural land as fertilizer and this has been shown to contain significant concentrations of PAHs (Wild et al. 1990a,b).

All the available basic information on PAHs has led to the development of bioremediation as a cost-effective approach for cleaning up contaminated soils, waters and sediments. Considerable efforts are currently underway to develop the necessary field application techniques that make this a profitable endeavour. This chapter provides an overview of the current activities associated with PAH biodegradation and bioremediation, including the latest on the genetic diversity of PAH degrading bacteria and how this technology is being successfully used in the field application.

## 4. Biodegradation of PAHs

Microbial transformation is a major environmental process affecting the fate of PAHs in both terrestrial and aquatic ecosystems. The microbial degradation of PAHs, having two or three rings, is well documented, but in the last decade, a number of bacteria, that metabolize larger PAH molecules, have also been isolated. Biological technologies are now being explored for their potential in the remediation of contaminated sites. However, their successful application demands a broader understanding of the biochemical pathways by which PAHs are degraded, both individually and in mixtures. An extensive literature describing the degradation of individual PAHs by microorganisms which are able to utilize them as sole sources of carbon and energy, does exist. These studies have yielded fundamental information about the biodegradability of individual compounds (Gibson and Subramanian 1984; Cerniglia 1992). The rates of biodegradation of PAHs are highly variable and are dependent not only on PAH structure, but also on the physico-chemical parameters of the site as well as the number and types of microorganisms present. PAHs sorb to organic matter in soils and sediments, and the rate of their desorption strongly influences the rate at which microorganisms can degrade the pollutants (Shuttleworth and Cerniglia 1995). Much of the research is focused on techniques to enhance the bioavailability and consequently the degradation rates of PAHs at polluted sites. Degradation products of PAHs are, however, not necessarily less toxic than the parent compounds. Therefore, toxicity assays need to be incorporated into the procedures used to monitor the effectiveness of PAH bioremediation (Shuttleworth and Cerniglia 1995). Aerobic bacteria have been extensively studied for use in remediation processes and both enzymologic and genetic studies are being carried out for the purpose of effective biodegradation. PAHs are degraded by microorganisms either in metabolism or co-metabolism (Habe and Omori 2003). Co-metabolism is very important for degradation of mixtures of PAHs and high molecular weight PAHs. In contrast, several two-, three- and four-ring PAHs have been known to be growth substrates for bacteria.

A few microorganisms have been shown to utilize four ring PAHs for their growth in the absence of co-factors or surfactants (Weisenfels et al. 1990; Walter et al. 1991; Boldrin et al. 1993; Thibault et al. 1996). However, a *Mycobacterium* sp., isolated from PAH contaminated freshwater sediments, was found to be capable of mineralizing phenanthrene, pyrene and fluoranthene without co-factors, out of which phenanthrene and pyrene were used as the sole sources of carbon and energy. No DNA hybridization was detected with the *nahAc* gene probe, indicating that enzymes involved in PAH metabolism were not related to the well characterized naphthalene dioxygenase gene (Churchill et al. 1999).

The catabolism of PAHs, possessing three or less fused aromatic rings, has been well studied, while the metabolism of higher PAHs containing four or more rings has not been investigated extensively. The processes involving biodegradation are proportional to the ring size of PAH molecules. The lower molecular weight PAHs are degraded more rapidly than the higher weight PAHs. Till the late 1980s, there were no reports of axenic microbial cultures utilizing PAHs containing four or more fused rings as the sole source of carbon and energy. Since then, a number of pure cultures have been reported which are capable of degrading higher PAHs, such as fluoranthene, pyrene, chrysene and benz[a]anthracene. The biochemical pathways involved in the catabolism of these PAHs have been well identified. However, microorganisms capable of degrading PAHs containing five benzene rings have been difficult to obtain. The very low solubility of more complex PAHs strongly reduces their bioavailability, due to which they do not serve as amenable substrates for microbial metabolism.

### 4.1 Degradation of PAHs by Bacteria

Bacterial degradation of PAHs involves an initial oxidation step in which both atoms of the oxygen molecule are incorporated into the aromatic ring to form cis-dihydrodiol. This initial hydroxylation step of unsubstituted PAHs is catalyzed by a dioxygenase (Fig. 2). Since PAHs, such as phenanthrene, pyrene, benzo[a]pyrene and benz[a]anthracene, are complex fused ring structures, bacteria metabolize PAHs at multiple sites to form isomeric cis-dihydrodiols (Mueller et al. 1996). Monooxygenases have also been shown to be involved in oxidation to form trans-dihydrodiols (Heitkamp et al. 1988b; Kelley et al. 1991). The cis-dihydrodiols undergo rearomatization by dehydrogenases to form dihydroxylated intermediates (Patel and Gibson 1974). Further, catabolism involves ring cleavage by dioxygenases to form aliphatic intermediates. Cleavage of these *ortho*-dihydroxylated intermediates occurs either between the two hydroxyl groups (intradiol or *ortho*-fission) or adjacent to one of the hydroxyl groups (extradiol or *meta* fission) (Mueller et al. 1996). There are different enzymes for different ring fission substrates, each forming a different aliphatic product. The aromatic ring dioxygenases are multi-component enzymes which consist of a reductase, a ferredoxin and a third component consisting of two proteins, large and small iron sulfur protein subunits (Ensley and Gibson 1983; Suen and Gibson 1993; Suen et al. 1996).

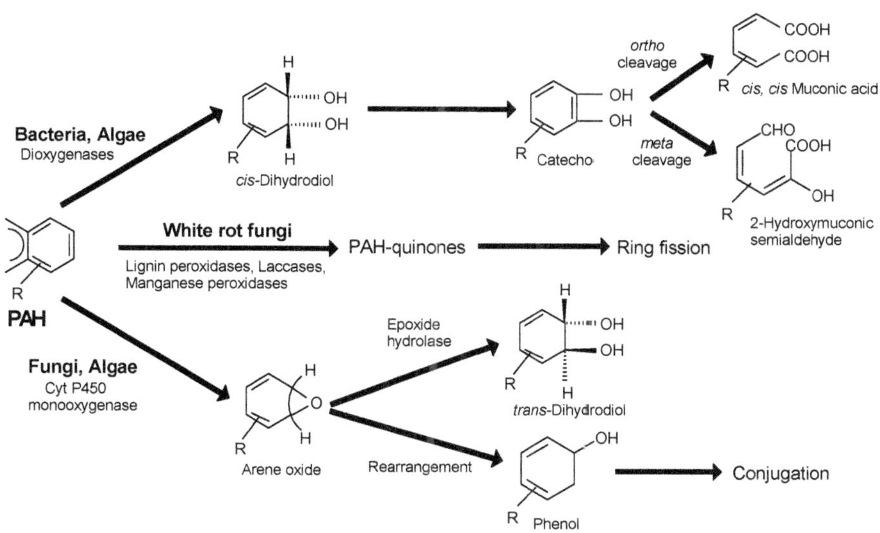


**Fig. 2.** Major pathways involved in the metabolism of polycyclic aromatic hydrocarbons by bacteria, fungi and algae (adapted from Mueller et al. 1996).

Bacterial genera, capable of degrading PAHs commonly, include species of *Pseudomonas, Alcaligenes, Rhodococcus, Sphingomonas* and *Mycobacterium*. This is a relatively small range of genera considering the prevalence of PAHs in the environment. Most of these bacteria have been enriched based on their ability to grow on low molecular PAHs naphthalene, phenanthrene, fluorene, anthracene and acenaphthene. But in the last few years, several bacteria have been isolated that are able to grow on four ring PAHs, particularly fluoranthene and pyrene.

It is interesting to note that *Mycobacteria* have been repeatedly isolated as the bacteria capable of degrading PAHs containng four or more fused aromatic rings. This is probably due to the hydrophobic cell surface which allows their adhesion to hydrophobic PAHs, thus facilitating mass transfer of the substrates inside the cells (Harayama 1997). Although *Mycobacteria* are usually slow growing, their growth on PAHs is faster than that of other microorganisms. For example, the growth rate of *Mycobacterium* sp. BBI on pyrene was 0.056/h, whereas that of *Rhodococcus* sp. UW1 was 0.023/h (Heitkamp et al. 1988b). *Mycobacterium* sp. strain CH1, isolated from PAH contaminated freshwater sediments, is capable of mineralizing three and four ring PAHs including phenanthrene, pyrene and fluoranthene, and can utilize phenanthrene and pyrene as the sole carbon and energy sources (Churchill et al. 1999). *Mycobacterium* sp. strain PYRI is a versatile organism which has been shown to mineralize anthracene, fluoranthene, pyrene, 1-nitropyrene, phenanthrene and benzo[a] pyrene (Heitkamp et al. 1988a; Heitkamp and Cerniglia 1988; Kelley et al. 1991; Rafii et al. 1992; Kelley and Cerniglia 1995; Wang et al. 1995). It is known to have an inducible system for PAH degradation (Heitkamp et al. 1988a). Two newly isolated strains *Mycobacterium austroafricanum* GTI-23 and *M. vanbaalenii* have also been shown to mineralize a range of PAHs including fluorene and benzo[a]pyrene, both in liquid and in soil environments (Bogan et al. 2003; Moody et al. 2004).

The microbial degradation of a few representative low molecular weight and high molecular weight PAHs is discussed below.

### 4.1.1 Naphthalene

Naphthalene is the simplest homologue in the polycyclic series and has been extensively studied as a model compound for understanding the microbial metabolic pathway of more complex PAHs. Naphthalene degradation was first studied as early as 1943 by Strawinski and Stone who isolated salicylic acid from culture filtrates of naphthalene grown *Pseudomonas aeruginosa*. *Pseudomonas* bacteria, capable of degrading naphthalene along with other PAHs, can be readily isolated from soils contaminated with PAHs (Heitkamp et al. 1988a). All *Pseudomonas* strains tested have been found to degrade naphthalene to salicylate via similar biochemical pathways (Yen and Serdar 1988; Sayler et al. 1990). The initial reaction in the bacterial oxidation of naphthalene involves the formation of a dihydrodiol intermediate. A number of microorganisms, such as *Pseudomonas putida*, *P. fluorescens*, *P. cepacia*, *P. testosteroni*, *Alcaligenes denitrificans*, *Mycobacterium* sp., *Rhodococcus* sp., *Corynebacterium* sp., *Moraxella* sp., *Bacillus* sp. etc., which can degrade naphthalene, have been isolated (Gibson and Subramanian 1984; Smith 1990; Rosenberg and Ron 1996). The enzymatic mechanism and genetic plasmid-encoded naphthalene degradation pathway regulation of pseudomonads have been well characterized (Zylstra and Gibson 1991).

Virtually all that is known of the biochemistry and genetics of bacterial naphthalene metabolism was gained from the analysis of pseudomonads, such as *P. putida* and its plasmid NAH7. Studies on naphthalene metabolism by pseudomonads have been instrumental in developing an understanding of aromatic hydrocarbon metabolism. The initial reaction is catalyzed by naphthalene dioxygenase, a multicomponent enzyme system, to form 1,2-dihydroxynaphthalene (Patel and Gibson 1974; Jerina et al. 1976) which undergoes oxidation to form salicylaldehyde and pyruvate (Barnsley 1976). Salicylaldehyde further undergoes oxidation catalyzed by an $NAD^+$ dehydrogenase to form salicylate. Salicylate is, in most cases, oxidized by a hydroxylase to form catechol which is then cleaved either via *ortho*- or *meta*-cleavage pathways. In the *ortho*-pathway, catechol is converted to *cis*, *cis*-muconate by the enzymatic action of catechol-1,2-dioxygenase, followed by hydrolysis to form β ketoadipate. In the *meta*-pathway, catechol is converted to acetaldehyde and pyruvate by the enzyme catechol-2,3-dioxygenase (Gibson and Subramanian 1984). Salicylate can also be oxidized through gentisic acid (Yen and Serdar 1988). The naphthalene utilizing phenotype of *P. putida* strain G7 is specified by the plasmid NAH7 which contains the genes for 11 enzymes involved in naphthalene degradation (Dunn and Gunsalus 1973). The genes are organized into two operons in which the first cluster (*nah* operon) includes genes *nahABCDEF* encoding the conversion of naphthalene to salicylate and pyruvate, and the second cluster (*sal* operon) includes *nahGHIJK* encoding genes for oxidation of salicylate to catechol and for *meta* cleavage pathway. The genes for the individual components of naphthalene dioxygenase in *P. putida* PpG7 have been designated *nahAa*, *nahAb* and *nahAcAd*. The *nahAcAd* gene has been cloned and enzyme activity expressed in *E. coli* (Yen and Serdar 1988). Salicylate has been reported to be an inducer of all the enzymes of naphthalene oxidation pathway (Barnsley 1975).

A strain of *Pseudomonas putida* capable of mineralizing naphthalene ($Nap^+$) via salicylate ($Sal^+$) was isolated in our laboratory, and all regulatory and structural genes for the whole pathway were found to be encoded on a 25 kb EcoRI fragment of an approximately 83 kb plasmid present in this strain (Samanta et al. 1998).

Recently, a *Rhodococcus opacus* strain M213, capable of growing on naphthalene as a sole carbon source, was described, in which salicylate does not appear to be an intermediate suggesting a different degradation pathway (Uz et al. 2000). Further, at least a part of the naphthalene catabolic pathway was encoded by a very large linear plasmid unlike the circular plasmids typical of naphthalene metabolizing pseudomonads.

### *4.1.2 Phenanthrene*

Phenanthrene, a three ringed PAH, is an ideal model system to study various aspects of microbial metabolism and physiology. In general, the initial reaction of phenanthrene degradation involves the action of a dioxygenase, followed by oxidation to form 3,4-dihydroxyphenanthrene, which subsequently undergoes *meta* cleavage and is converted to 1-hydroxy-2-naphthoic acid. This is the common upper route of phenanthrene degradation pathway. 1-hydroxy-2-naphthoic acid can be further degraded via two routes. In one route, it undergoes ring cleavage to form *o*-phthalic acid protocatechuic acid, which is finally cleaved to form pyruvic acid and eventually enters the TCA cycle (Kiyohara et al. 1976; Ghosh and Mishra 1983; Houghton and Shanley 1994). In the other route, 1-hydroxy-2-naphthoic acid undergoes oxidative decarboxylation to form 1,2-dihydroxynapthalene, which is then subjected to *meta* cleavage to form salicylic acid (Evans et al. 1965; Gibson and Subramanian 1984). Salicylic acid is further degraded via the formation of either catechol or gentisic acid. Both catechol and gentisic acid undergo ring fission to form TCA cycle intermediates (Houghton and Shanley 1994).

*Mycobacterium* species metabolize phenanthrene at different sites of the molecule, presumably via both the dioxygenase and monooxygenase attacks on the aromatics nucleus. Recent studies have revealed alternative pathways in *Mycobacterium* sp. strain PYR-1 in addition to the previously known routes, which were suggested to be due to the presence of different dioxygenases or to a relaxed specificity of the same dioxygenase for the initial attack (Moody et al. 2001). In this strain, phenanthrene was metabolized with initial attack in the K region to form the *cis* and *trans*-9,10 dihydrodiols which were further metabolized to form 2,2′-diphenic acid. Dioxygenase attack also occurred at the C-3 and C-4 positions of phenanthrene to form a *cis*-3,4- dihydrodiol that was dehydrogenated to form 3,4-dihydroxyphenanthrene and eventually 1-hydroxy-2-naphthoic acid. The formation of *trans*-9,10-dihydrodiol was suggestive of a monooxygenase attack on the phenanthrene nucleus to form phenanthrene 9,10-epoxide, followed by the action of epoxide hydrolase to form the trans dihydrodiol. In another *Mycobacterium* sp. strain KR-2, phenanthrene degradation pathway was similar to that of *Mycobacterium* sp. strain PYR-1 except that phenanthrene trans-9, 10, dihydrodiol was not detected in the former (Rehmann et al. 1996).

Another novel route of phenanthrene metabolism was proposed recently by Prabhu and Phale (2003) wherein *Pseudomonas* sp. strain PP2 initiated phenanthrene degradation by double hydroxylation, resulting in the formation of 3,4-dihydroxyphenanthrene. The diol was oxidized via successive formation of 1-hydroxy-2-napthoic acid, α-naphthol, 1,2 dihydroxy naphthalene, salicylate and catechol to eventually form 2-hydroxymuconicsemialdehyde.

In our laboratory, four PAH-degrading bacteria, namely *Arthrobacter sulphureus* RKJ4, *Acidovorax delafieldii* P-1, *Brevibacterium* sp. HL4 and *Pseudomonas* sp. DLC-P11 were found to use phenanthrene as the sole source of carbon and energy (Samanta et al. 1999). Analysis of degradation pathway revealed that strain P4-1 degraded phenanthrene via *o*-phthalic acid whereas strain RKJ4 degraded it via *o*-phthalic acid and protocatechuic acid, both of which are the conventional lower pathway intermediates (Samanta et al. 1999). On the other hand, strains HL4 and DLC-P11 were found to degrade phenanthrene via novel pathways. In case of HL4, degradation proceeded via formation of 1-hydroxy-2-naphthoic acid, 1-napthol and salicylic acid, whereas DLC-P11 degraded phenanthrene via 1-hydroxy-2-naphthoic acid, 1-naphthol and *o*-phthalic acid (Samanta et al. 1999).

### 4.1.3 Anthracene

A number of bacterial species with the ability to utilize anthracene, another tricyclic PAH, as the sole carbon and energy source, have been isolated. *Pseudomonas* sp. and *Sphingomonas yanoikuyae* B1 initially oxidize anthracene in the 1,2 position to form (+)- (1*R*,2*S*)-*cis*-1,2-dihydroxy-1,2-dihydroanthracene which is subsequently converted into 1,2-dihydroxyanthracene that undergoes *meta* ring cleavage. The cleavage product is further degraded to form 2-hydroxy-3-naphthoic acid, salicylate and catechol in a manner similar to the naphthalene degradation pathway (Fernley et al. 1964; Evans et al. 1965; Akhtar et al. 1975; Jerina et al. 1976).

Two recent papers have proposed new pathways for the degradation of anthracene. Moody et al. (2001) reported oxidation of anthracene by *Mycobacterium* sp. strain PYR-1 to anthracene cis-1,2 dihydrodiol in a reaction similar to those previously reported in *Pseudomonas* and *Sphingomonas* sp. Anthracene cis-dihydrodiol was then dehydrogenated to form 1, 2-dihydroxyanthracene which was further metabolized to form either 1-methoxy-2-hydroxyanthracene, a novel metabolite in anthracene degradation; or 3-(2-carboxyvinyl) naphthalene-2-carboxylic acid, another novel *ortho* ring fission product; or 6,7-benzocoumarin which was further degraded. An alternate route of enzymatic attack by strain PYR-1 was at the C-9 and C-10 positions of anthracene to form anthracene-9,10-dihydrodiol which was further metabolized to form the dead end product 9,10-anthraquinone. Another novel anthracene degradation pathway was proposed recently by Herwijnen et al. (2003) in *Mycobacterium* sp. strain LB501T which utilizes anthracene as sole carbon and energy source. Mutants (generated by UV light), which were impaired in anthracene utilization, were studied along with the wild type strains to determine the pathway. They observed that in addition to the known degradation pathway of anthracene via formation of 3-hydroxy-2-naphthoic acid to eventually form salicylate and catechol, there exists a novel anthracene catabolic pathway which proceeds through the formation of *o*-phthalic acid and protocatechuic acid. The authors proposed that a cleavage reaction similar to that of 1-hydroxy-2 naphthoic acid in the *Aeromonas* degradation pathway of phenanthrene occurred to perform the cleavage of 3-hydroxy-2-naphthoic acid. Aldolase reactions of the cleavage product and subsequently a dehydrogenase reaction would give rise to *o*-phthalic acid. Known metabolic pathways of o-phthalic acid through protocatechuic acid then enter the central metabolism by *ortho* or *meta* cleavage of protocatechuic acid.

#### 4.1.4 Pyrene

Pyrene has often been used as a model compound of higher molecular weight PAH degradation. Heitkamp et al. (1988a) described for the first time a bacterial isolate that mineralized pyrene and now many pyrene-degrading bacteria have been reported.

Although both *Mycobacterium* sp. strain PYR-1 (Heitkamp et al. 1988b) and *Rhodococcus* sp. strain UW1 (Walter et al. 1991) can degrade pyrene via initial dioxygenation at the 1,2-positoin, one primary pathway is the major catabolic pathway of pyrene that produces both *cis*- and *trans*-4,5-pyrenedihydrodiols (Heitkamp et al. 1988b). Rearomatization of the dihydrodiols and subsequent *ortho*-cleavage leads to the formation of 4,5-phenanthrene dicarboxylic acid which is further metabolized to 4-phenanthroic acid. The subsequent intermediate cis-3,4-phenenthrenedihydrodiol-4-carboxylic acid is formed by a second dioxygenase reaction. Rearomatization of the metabolite yields 3, 4, dihydroxyphenanthrene, which is also an intermediate in bacterial phenanthrene degradation and further metabolism proceeds via catabolic pathways similar to those of phenanthrene. Recently, the genes encoding a novel polycyclic ring dioxygenase were cloned and sequenced from M*ycobacterium* sp. strain PYR-1 (Khan et al. 2001).

#### 4.1.5 Benzo[a]pyrene

Currently, there is limited information on bacterial degradation of PAHs with five or more rings in both environmental samples and pure or mixed cultures. Benzo[a]pyrene has been widely studied due to its high toxicity. This compound is highly recalcitrant and turnover times of greater than 3.3 years in oil contaminated freshwater sediments and more than 60 years in uncontaminated sediments have been reported (Herbes and Schwall 1978). However, in recent years, extensive cometabolic mineralization of $^{14}C$ benzo[a]pyrene has been reported to occur in soil (Kanaly and Bartha 1999). Bacteria, capable of utilizing benzo[a]pyrene as the sole source of carbon and energy, have not been isolated till date. Co-metabolism is thus an important feature in the bacterial degradation of benzo[a]pyrene. The metabolites of benzo[a]pyrene degradation identified in *Mycobacterium* sp. strain RJGH-135 included cis-7,8-benzo[a]pyrenedihydrodiol; 4,5-chrysene dicarboxylic acid; cis-4-(8-hydroxypyren-7-yl)-2-oxobut-3-enoic acid {or cis-4-(7-hydroxypyren-8-yl)-2-oxobut-3-enoic acid} and 7,8-dihydropyrene-7-carboxylic acid (or 7,8-dihydropyrene-8-carboxylic acid). The authors were not able to distinguish between the *meta* fission products through the 7, 8 bond and the 9,10 bond, thus the possibility of two products for two of the metabolites (Schneider et al. 1996). A very recent study on metabolism of benzo[a]pyrene in the bacterium *Mycobacterium vanbaalenii* PYR-1 showed that this organism initially oxidized benzo[a]pyrene with dioxygenases and monooxygenases at C-4,5, C-9,10, and C-11,12. The major intermediates of benzo[a]pyrene metabolism, that had accumulated in the culture media after 96 h of incubation, were cis-4,5-dihydro-4,5-dihydroxybenzo[a]pyrene (benzo[a]pyrene cis-4,5-dihydrodiol), cis-11,12-dihydro-11,12-dihydroxybenzo[a]pyrene (benzo[a]pyrene cis-11,12-dihydrodiol), trans-11,12-dihydro-11,12-dihydroxybenzo[a]pyrene (benzo[a]pyrene trans-11,12-dihydrodiol), 10-oxabenzo[def]chrysen-9-one, and hydroxymethoxy and dimethoxy derivatives of benzo[a]pyrene. The ortho-ring fission products 4-formylchrysene-5-carboxylic acid and 4,5-chrysene-dicarboxylic acid and a monocarboxylated chrysene product were formed when replacement culture experiments were conducted with benzo[a]pyrene cis-4,5-dihydrodiol.

Reports of microbial growth on chrysene, benz[a]anthracene and other PAHs containing more than 5 rings are scarce. Chrysene, a typical four ring compound has been reported to be used as a sole source of carbon and energy by *Pseudomonas fluorescens* (Caldini et al. 1995). The biodegradation rate followed first order kinetics. However, no studies to identify its degradation pathway have been carried out. This strain could also grow on and degrade other four ring PAHs benz[a]anthracene and benzo[b]naphthothiophene.

## 4.2 Degradation of PAHs by Fungi

In contrast to bacteria, fungi generally do not utilize PAHs as their sole carbon and energy source, but transform them co-metabolically to detoxified metabolites (Sutherland 1992). A diverse group of fungi, both ligninolytic and non-ligninolytic, are able to degrade PAHs. Non-ligninolytic fungi metabolize PAHs in pathways that are similar to those used by mammalian enzyme systems (Cerniglia et al. 1992; Sutherland 1992, Holland et al. 1986). There are two main enzyme groups involved in fungal degradation of PAHs. These are the cytochrome P-450 monooxygenases and lignin peroxidases. The cytochrome P-450 monooxygenases are complex multicomponent systems like the bacterial aromatic ring dioxygenases. They are usually membrane bound and have broad substrate specificities. One atom of molecular oxygen is incorporated into the PAH by the monooxygenase to form an arene oxide, while the other atom is reduced to water (Cerniglia 1984). The arene oxide formed undergoes spontaneous isomerization to form a phenol which can be conjugated with glucuronic acid, glucose, sulfate or glutathione. The enzymatic hydration of the arene oxide leads to the formation of a *trans*-dihydrodiol catalyzed by epoxide hydrolase (Fig. 2). In addition to lignin peroxidases, other extracellular enzymes produced by white rot fungi, such as laccases and manganese peroxidases, are also involved in PAH degradation. The lignin peroxidases oxidize PAHs with ionization potentials of less than about 7.6 eV (Hammel et al. 1986). They initiate a free radical attack on PAHs by a single electron transfer to form an aryl cation radical which undergoes further oxidation to form a quinone. The best studied white rot fungus, *Phanerochaete chrysosporiuim*, produces multiple lignin peroxidases and manganese peroxidases. Purified lignin peroxidase from *P. chrysosporium* has been shown to oxidize benzo[a]pyrene, benz[a]anthracene, pyrene, anthracene and perylene (Haemmerli et al. 1986; Hammel et al. 1986; Sanglard et al. 1986). Other white rot fungi, such as *Trametes versicolor, Bjerkandera* sp. and *Pleurotus ostreatus*, have been shown to mineralize PAHs to $CO_2$ more rapidly than *P. chryosporium* (Field et al. 1992).

*Cunninghamella elegans*, a non ligninolytic fungus, metabolizes a wide range of PAHs containing two to five aromatic rings as well as several nitro-PAHs (Cerniglia et al. 1992; Sutherland 1992; Pothuluri et al. 1992a,b). Like other fungi, *C. elegans* does not utilize PAHs as the sole source of carbon and energy, but biotransforms or co-metabolizes them to products that are generally less mutagenic or toxic than the parent compounds (Cerniglia et al. 1985a,b). Although this fungus metabolizes PAHs in a manner similar to mammalian systems, there are differences in the regio-and stereo-specificities of the fungal and mammalian enzymes (Cerniglia et al. 1983, 1990; Sutherland et al. 1993). Ectomycorrhizal fungi have also been reported to degrade some PAHs, particularly benzo[a]pyrene (Braun-Lüllemann et al. 1999). Other non-ligninolytic fungi, such as *Penicillium janthinellium* and *Syncephalastrum* sp., can also transform a variety of PAHs including pyrene, chrysene and benzo[a]pyrene to polar metabolites (Pothuluri et al. 1994; Launen et al. 1995; Kiehlmann et al. 1996).

Boonchan et al. (2000) have reported degradation and mineralization of high molecular weight PAHs by fungal-bacterial co-cultures. A co-culture containing *P. janthinellum* VUO10,201 and bacterial consortium VUN10,009 was able to mineralize and grow on benzo[a]pyrene as a sole carbon and energy source. Higher rates of benzo[a]pyrene mineralization and degradation were achieved, when *P. janthinellum* VUO10,201 was cocultured with *S. maltophila* VUN10,010. No significant microbial growth or benzo[a]pyrene mineralization was observed with axenic cultures, suggesting co-operative catabolism between the fungi and bacteria for degradation of PAHs. In another study, the rate of benzo[a]pyrene mineralization by a pure culture of white rot fungus *Bjerkanera* sp. B0555 was enhanced after it was inoculated with a PAH adapted sediment sludge containing indigenous bacteria (Kotterman et al. 1998). In fact, some other reports have also suggested that PAH degradation in nature is a consequence of sequential breakdown by fungi and bacteria with the initial oxidation step being carried out by the fungi (Wiesche et al. 1996; Meulenberg et al. 1997; Sack et al. 1997).

### 4.3 Degradation of PAHs by Algae

Both prokaryotic and eukaryotic algae could be important in the degradation of PAHs as they are widely distributed in aquatic environments which may be a major sink for degradation and/or transformation of PAHs. Cyanobacteria (blue green algae) and eukaryotic algae oxidize PAHs under photoautotrophic conditions to form hydroxylated intermediates (Fig. 2) (Mueller et al. 1996). Naphthalene and phenanthrene are oxidized by cyanobacteria to metabolites which are similar to those formed by mammals and fungi (Narro et al. 1992a,b). On the other hand, the green alga *Selenastrum capricornutum* under photoautotrophic conditions, oxidizes benzo[a]pyrene to isomeric *cis*-dihydrodiols which is suggestive of dioxygenase catalyzed reactions similar to those found in heterotrophic prokaryotes, rather than monooxygenase catalyzed reactions occurring in fungi and mammals (Warshawsky et al. 1990). Naphthalene was shown to be oxidized by the marine cyanobacterium *Oscillatoria* sp. strain JCM via an arene oxide intermediate that isomerized with a concomitant non-enzymatic rearrangement shift (Narro et al. 1992a). Another marine cyanobacterium, *Agmenellum quadruplicatum* PR6 metabolized phenanthrene to trans-9,10-dihydroxy-9,10-dihydrophenanthrene and 1-methoxyphenanthrene (Narro et al. 1992b). However, there is virtually no information about the enzymes involved in degradation of PAHs by cyanobacteria.

## 5. Bioremediation Studies

The environmental fate of PAHs includes volatilization as well as biotic and abiotic transformations. Volatilization is important only for two-ring PAHs, such as naphthalene, whereas it is the biotic mechanism which is responsible for removal of PAHs with three or more rings. Bioremediation is the process whereby biodegradative abilities of microorganisms are harnessed or exploited to remove or detoxify environmental pollutants. Different bioremediation technologies used for cleaning up of contaminated soils, sediments and groundwater are shown in Figure 3. These techniques have been discussed in details (Mueller et al. 1996). The rate and extent of biodegradation of PAHs in soils and sediments is affected by multiple factors (Table 1). The major factor limiting the bioremediation of soils and

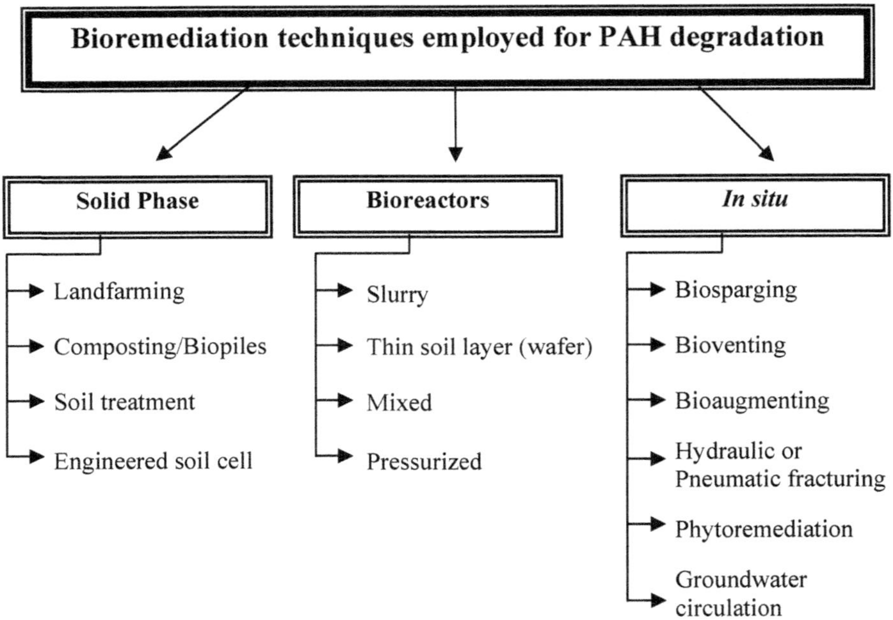


**Fig. 3.** Different technologies used for bioremediation of sites contaminated with polycyclic aromatic hydrocarbons

**Table 1.** Factors affecting bioremediation of PAH-contaminated sites

| Physico-chemical factors | Biological factors | Environmental factors |
|---|---|---|
| • Physical/chemical properties of PAHs (number of rings, log Kow)<br>• Organic content of soil<br>• Structure/particle size of soil<br>• Presence of contaminants | • Characteristics of the microbial population (diversity, genetic/ catabolic potential) | • Temperature<br>• Moisture<br>• pH<br>• Sorption<br>• Degree of contamination |

sediments contaminated with PAHs is the poor availability of these hydrophobic contaminants to microorganisms (Mihelcic et al. 1993; Hughes et al. 1997). Bioavailability i.e. the ability of a compound to be freely transported across the cell membrane for intracellular or available for extracellular metabolism, may be the most important factor in determining the feasibility of bioremediation of PAHs. In most cases, mass transfer limitations prevent the full exploitation of the microbial degradative potential (Bosma et al. 1997). Limited bioavailability is due to low water solubility and consequently the tendency to partition onto soil mineral surfaces and to sorb strongly to the soil or sediment matrix (Harms and Bosma 1997). Several mechanisms work together to influence bioavailability, and different mechanisms predominate in any given situation (Pignatello and Xing 1996), however, they are still not fully understood. It is usually assumed that the water-dissolved fraction of chemicals is the only one available to microorganisms (Thomas et al. 1986; Stucki and Alexander 1987). Therefore, degradation rates are dependent on the mass transfer rates of PAHs from solid or soil bound phase to the aqueous phase (Volkering et al. 1992) and desorption of PAHs from soil is considered as the controlling factor in their biodegradation. The principal approach for increasing the mass transfer to the aqueous phase is based on enhancing the solubilization or dissolubilization rates. This can be achieved by increasing the total surface area between the substrate and the aqueous phase and is usually carried out by the addition of surface active agents i.e. surfactants (Edwards et al. 1991). A variety of synthetic surfactants, both ionic and non-ionic have been shown to increase the bioavailability of PAHs as well as other hydrophobic contaminants (Aronstein and Alexander 1993; Zhang and Miller 1995) and have contributed to our understanding of the mechanisms by which surfactants increase solubility (Volkering et al. 1995; Willumsen et al. 1998). However, some synthetic surfactants can actually inhibit PAH biodegradation via toxic interactions, stimulation of surfactant degraders, or sequestration of PAHs into surfactant micelles. Microbially produced surfactants represent a promising alternative to chemical surfactants (Desai and Banat 1997). The isolation of microroganisms producing biosurfactants, when grown on PAHs, has been reported (Déziel et al. 1996; Prabhu and Phale 2003). Biosurfactants have been shown to have many of the positive effects of synthetic surfactants, but without the drawbacks. They are biodegradable and non-toxic, and many biosurfactants do not produce true micelles, thus facilitating direct transfer of the surfactant-associated PAHs to bacteria. Several biosurfactants have been used to enhance biodegradation of many hydrophobic contaminants, for example, in an oil contaminated beach (Harvey et al. 1990), soils (Van Dyke et al. 1993; Bai et al. 1997) and soil slurried in bioreactors (Oberbremer et al. 1990).

Newer alternative methods to improve bioavailability and biodegradation of hydrophobic compounds include the two liquid phase (TLP) bioreactors which are used for the bioconversion of hydrophobic/toxic substrates into products of commercial interest (Van Sonsbeek et al. 1993; Nikolova and Ward 1993). In this system, a water immiscible liquid is added to enhance its efficiency by increasing the substrate bioavailability or by decreasing substrate toxicity (Déziel et al. 1999). This liquid acts as a non-biodegradable and biocompatible liquid phase reservoir in which hydrophobic organic substrates are dissolved and then released to the microorganisms. Villemur et al. (2000) used silicone oil as a non-biodegradable and biocompatible solvent to increase the bioavailability of PAHs by promoting PAH desorption from soil and their subsequent transfer to microorganisms, thus allowing the development of a high molecular weight PAH-degrading consortium. They studied the biodegradation of high molecular weight PAHs (pyrene, chrysene and benzo[a]pyrene) in soil by combining the soil slurry concept with a TLP biosystem, which resulted in initial biodegradation rates that were two to five times faster than in a classical slurry bioreactor. Addition of low molecular weight PAHs naphthalene and phenanthrene stimulated the biodegradation of pyrene, chrysene, benzo[a]pyrene and perylene by the consortium in the TLP bioreactor (Marcoux et al. 2000). The 16S rRNA analysis of the consortium identified four of the isolates as *Mycobacterium gilvum* B1, *Bacillus pumilus* B44, *Microbacterium esteraromaticum* B21 and *Porphyrobacter* sp. B51 (Gauthier et al. 2003). Characterization of bacteria in the consortium indicated that it was composed of microorganisms with different abilities to grow at the interface or in the aqueous phase according to the culture conditions. In another study, fast and complete microbial degradation of phenanthrene and pyrene was achieved in a TLP partitioning bioreactor in which silicone oil was used as the organic phase (Guieysse et al. 2001). The use of a bioavailable solvent bis(ethyl-hexyl) sebacate in TLP for the degradation of mixtures of PAHs by *Mycobacterium* sp. PYR1 led to significant degradation of pyrene, phenanthrene, naphthalene and anthracene (Macleod and Daugulis 2003). 1g of pyrene was completely degraded within 4 days at the rate of 138 mg/l/day which is the highest pyrene degradation rate reported in literature so far. Thus biphasic reactors represent a very attractive treatment option for remediation of PAH mixtures resulting from contaminated soil extraction.

## 6. Diversity of PAHs Degrading Bacteria

As a group, the bacteria are well known for their metabolic diversity. One consequence of this diversity is the fact that many biohazardous or persistent anthropogenic chemical compounds are degraded by microbial activities. Although individual species of bacteria and bacterial consortia have been shown to metabolize PAHs in laboratory culture, but evaluating their potential in a community of microorganisms at sites is more difficult. Biodegrading organisms may or may not be the predominant species which directly affects the ability to identify and quantify their presence. In addition, the physio-chemical properties of the immediate environment can have a major influence on microbial physiology as well as contaminant bioavailability.

Rapid analysis of diversity in complex microbial communities has remained elusive, but is an important goal in microbial ecology. Community diversity can be examined at several levels. The simplest analysis uses DNA profiles to identify differences in the composition of communities. More refined approaches describe differences not only in community composition, but also in community organization by measuring the number (*richness*) and relative abundance (*structure or evenness*) of species or phylotypes. The richness and evenness of biological communities reflect selective pressures that shape diversity within communities. Measuring these parameters is most useful while assessing treatment effects (eg. physical disturbance, pollution, nutrient addition, predation, climate change etc.) on community diversity. Shifts in microbial community composition can be induced by changes in environmental factors, such as temperature, pH, moisture content, nutrient levels etc.

To fully identify the nature of a contaminant's impact on an extant microbiota, a polyphasic approach that combines phenotypic and genotypic measurements is necessary. Changes in microbial community structure and reduction in bacterial diversity in response to environmental stress and contamination have been well documented. Using phospholipid fatty acid analysis, MacNaughton et al. (1999) demonstrated a community shift in a crude-oil-contaminated coastal site. Shi et al. (1999) reported differences in the microbial community structures of uncontaminated and fuel-contaminated sand aquifers. Bååth et al. (1998) also demonstrated by phospholipid fatty analysis that the species composition changed in soils amended with high levels of metal-rich sludge. Torsvik et al. (1996) compared the total bacterial diversities in agricultural and forest soils and found that diversity in the agricultural soil was 2 to 5 times lower than that in the forest soil. A reduction in soil microbial diversity was also observed by Øvreås et al. (1998), when they incubated agricultural soil with a mixture of methane and air.

It has been proposed that polluted soils typically show low biodiversity due to selective pressures presented by high levels of chemical contamination (Atlas et al. 1991). Reports indicate that environmental stresses including contamination, not only reduce the biodiversity of the original community, but may also selectively enrich specific microorganisms that are more adapted to the new environment. Naphthalene, which is composed of two fused aromatic rings, has long been used in enrichment cultures to isolate PAHs-catabolizing bacteria from soil and fresh water. Naphthalene-degrading bacteria commonly isolated from terrestrial environments include *Pseudomonas* and *Burkholderia* strains. In addition, naphthalene-degrading *Pseudomonas*, *Comamonas*, *Acinetobacter* and *Sphingomonas* strains have been isolated from soil enrichment cultures by using other PAHs (Mueller et al. 1997; Zylstra et al. 1997). Since naphthalene-degrading *Pseudomonas* strains are abundant in PAH-contaminated terrestrial and fresh water sites (Sanseverino, et al. 1993), studies of these bacteria in the laboratory may yield information relevant to the dominant PAH degradation events at these sites. In contrast, although naphthalene degrading *Pseudomonas* (Garcia-Valdez' et al. 1988) and *Sphingomonas* (Zylstra et al. 1997) strains have been enriched marine sediments, it is not clear why PAHs-catabolizing *Pseudomonas*, *Burkholderia*,

*Comanomonas, Sphingomonas* strains are abundant in the marine environment. In fact, a few studies that have focused on isolating numerically important PAH-degrading bacteria from marine sites, both polluted and non-polluted. These bacteria have been identified as members of different genera including *Cycloclasticus*, *Vibrio* and *Pseudoalteromonas* (Geiselbrecht et al. 1996, 1998; Hedlund et al. 1996). All of these bacteria are obligately marine and thus it seems possible that a significant portion of the PAHs degradation, that occurs in marine environments, is degraded by obligately marine microorganisms. In another study by Hedlund et al. (1999), a new marine bacterium *Neptunomonas naphthovorans* gen. nov., sp. nov. was isolated that was capable of utilizing naphthalene as a sole carbon and energy source. Shi et al. (1999) observed a proliferation of minor phylotypes within the fuel-contaminated aquifer upon toluene exposure. A study by Langworthy et al. (1998) on a freshwater sedimentary microbial community demonstrated higher frequencies of PAHs-degradative genes at contaminated sites. Studies on pristine soils and soils with a known history of PAHs contamination revealed that pristine soils did not yield PAHs degraders, whereas contaminated soils harbored closely related PAHs-degrading bacteria (Mueller et al. 1994). Using denaturing gradient gel electrophoresis, Rooney-Varga et al. (1999) also noted a selective enrichment of microorganisms in a petroleum contaminated aquifer. Furthermore, Ferris et al. (1997) reported that the disturbance of a hot spring cyanobacterial mat community led to the colonization by previously absent cyanobacterial populations in the disturbed areas.

## 7. Diversity of PAHs Metabolic Genes

In recent decades, a large number of xenobiotics have been released into the environment. While many of these chemicals are rapidly degraded by microorganisms in the environment, some resist attack and remain recalcitrant. Given time, however, most microorganisms, in particular bacteria, are able to adapt to using these compounds as carbon and energy sources. The biochemical versatility is largely due to the plasticity of the microbial genomes. By modifying the existing genes, a novel metabolic capacity can be developed that allows xenobiotics to be metabolized. This requires the alteration and exchange of genetic information and recombination processes, such as gene conversion; duplication and transposition that play crucial roles in the reassortment of discrete genetic modules and their expression (van der Meer et al. 1992).

Recently, several genes encoding PAH-catabolic enzymes have been characterized (Habe and Omori 2003). Analysis of the PAHs catabolic genes in different species of bacteria can give useful information about the evolution of enzyme structure-function relationships and the evolution and diversity of catabolic pathway genes via horizontal transfer, transposition events, DNA rearrangement, gene fusion, point mutation and so on. In applied studies, genetic information is useful for monitoring of bacterial populations that degrade PAHs in the contaminated soils.

Different repertoire of catabolic/metabolic genes have been detected in various species of Gram negative as well as Gram positive bacteria. The details of the genes isolated and identified so far are given below:

## 7.1 PAHs Catabolic Genes of Gram-negative Bacteria

### *7.1.1 nah-like Genes of Pseudomonas sp.*

The aerobic catabolism of low-molecular-weight PAHs by bacteria has been extensively studied. Napthalene is one of the aromatic hydrocarbons commonly found in the environment and often selected as a model compound for the study of PAHs degradation because of its highly aqueous solubility. Since the first report of a biochemical pathway for naphthalene oxidation by *Pseudomonas* species in 1964 (Davies and Evans 1964), extensive studies have rigorously defined the metabolic pathway, genes and enzymes involved (Cerniglia 1984; Gibson and Subramanian 1984; Eaton and Chapman 1992).

The metabolism of naphthalene has been well studied genetically in a *Pseudomonas putida* strain G7 and a transmissible plasmid coding for naphthalene catabolism NAH7 has been isolated (Dunn and Gunsalus 1973). The catabolic genes are organized in three operons on the 83 kb plasmid, NAH7, one encoding the upper pathway enzymes involved in the conversion of naphthalene to salicylate, the second encoding the lower-pathway enzymes involved in the conversion of salicylate to a TCA cycle intermediate via *meta*-ring cleavage and the third encoding a regulatory protein Nah R (Yen and Gunsalus 1982, 1985; Grund and Gunsalus 1983). Both the upper and lower operons are regulated by a *trans*-acting positive control regulator encoded by the *nahR* gene, which is between the two operons. Nah R is needed for the high level expression of the *nah* genes and their induction by salicylate (Schell 1985, 1986; Schell and Wender 1986).

Some of the genes for metabolism of naphthalene are also found on transposons. Transposons are discrete DNA segments that are able to move in the absence of genetic homology from one genetic location (donor site) to another (target site) (Berg and Howe 1989). This process requires a transposase that is encoded by the genetic element itself. The transposase interacts with the ends of the transposon in a site-specific manner, cuts the DNA at both ends of the elements and proceeds with the strand-transfer reaction (Hallet and Sherratt 1997). The *nah* gene cluster for naphthalene catabolism was found to be a part of a 38 kb class II transposon, Tn*4655*, on the 83-kb plasmid NAH7 in *P. putida* PpG7 (Yen and Serdar 1988; Tsuda and Lino 1990). Tn*4655* contains the *tnpR* gene and *res* region, but is defective in the cointegration step requiring a complementing transposase from other Tn*1722*-type transposons in order to transpose. The resolution function of Tn*4655* is unique, as it cannot complement other class II transposons nor be replaced by their resolvases. Analysis and experimental observations suggest the Tn*4655* TnpR protein to be a site-specific integrase able to catalyze both integration and resolution reactions (Berg and Howe 1989; Abremski and Hoess 1992).

Nucleotide sequences of genes encoding the naphthalene upper-catabolic enzymes from several *Pseudomonas* strains have been reported. These include the *ndo* (naphthalene dioxygenation) genes from *P. putida* strain NCIB 9816 (Kurkela et al. 1988), *nah* (naphthalene degradation) genes from *P. putida* strain G7 and NCIB 9816-4 (Simon et al. 1993; Eaton 1994), *dox* (dibenzothiophene oxidation) genes from *Pseudomonas* sp. strain C18 (Denome et al. 1993), *pah* (polycyclic aromatic hydrocarbon (phenanthrene) degradation) genes from *P. putida* strain OUS82 and *P. aeruginosa* strain PaK1 (Takizawa et al. 1994, 1999), *nah* (naphthalene degradation) genes from *P. putida* strain BS202 and from *P. stutzeri* strain AN10 (Bosch et al. 1999a). Among these genes, the upper pathway gene sequences were completely designated for strains OUS82, PaK1 and AN10, but only partial sequences were analyzed for other strains. The gene organization and sequence similarity (about 90%) among the upper catabolic pathway genes of these strains were similar to those of the *nah* genes from the NAH 7 plasmid of strain G7. These genes are usually called 'classical *nah*-like genes'.

For the lower pathway, the complete gene sequence was determined in only strain AN10 (Bosch et al. 1999a; 2000) and partial sequences were analyzed in strains G7 (You et al. 1991; Grimm and Harwood 1999) and NCIB 9816 (Platt et al. 1995). In the lower pathway, various genes are present, such as those encoding for salicylate hydroxylase (*nahG*), chloroplast ferredorin-like protein (*nahT*), catechol 2,3-dioxygenase (*nahH*), hydroxymuconic semialdehyde dehydrogenase (*nahI*), hydroxymuconic semialdehyde hydrolase (*nahN*), 2-oxopent-4-enoate hydratase (*nahL*), acetaldehyde dehydrogenase (*nahO*), 2-oxo-4-hydroxypentanoate aldolase (*nahM*), 4-oxalocrotonate decarboxylase (*nahK*) and 4-oxalocrotonate isomerase (*nahJ*), present in this order. Also, another salicylate hydroxylase gene (*nahW*) was found to be present in strain AN10 (Bosch et al. 1999b).

### *7.1.2. nag Genes of Ralstonia sp. Strain U2*

The naphthalene utilizing bacterium *Ralstonia* sp. strain U2 was isolated from oil-contaminated soil in Venezuela (Fuenmayor et al. 1998). The naphthalene dioxygenase genes (*nag* gene) were cloned and characterized. They were present in the order: genes encoding ferredoxin reductase (*nagAa*), ferredoxin (*nagAb*), the α subunit of ISP (*nag Ac*), the β subunit of ISP (*nagAd*), *cis*-dihydrodiol dehydrogenase (*nagB*) and aldehyde dehydrogenase (*nagF*) and two ORFs (*nagG* and *nagH*), that were very similar to *nahAc2* and *nahAd2* of strain GZ42, were inserted between *nagAa* an *nagAb*. The *nagG* product was identical to the α subunit of other aromatic ring dioxygenases, but the *nagH* product had limited similarity to the β subunit of other aromatic-ring dioxygenases. Recently, Zhou et al. (2001) reported the whole gene organization of the *nag* operon. NagG and NagH were structural subunits of salicylate 5-hydroxylase linked to electron transport proteins consisting of *NagAb* and *NagAa* (Zhou et al. 2002). The genes for the conversion of naphthalene to gentisate (*nagAaGHAbAcAdBFCQED*) in strain U2 were similar to and in the same order as the genes in the classical *nah*-like operon of *Pseudomonas* strains, with the exception of the *nagGH* insertion. A further difference between the *nag* and *nah* (NAH 7 plasmid) operons is the location of the regulatory gene (*nagR*) and the putative chemotaxis gene (*nagY*). In strain U2, both *nagY* and *nagR* genes were upstream from *nagAa*, but in the *nah* operon, *nahR* and *nahY* genes were downstream from the upper-catabolic pathway (Grimm and Harwood 1999).

### 7.1.3 phd Genes of Comamonas Testosteroni Strains GZ39

*Comamonas testosteroni* strains GZ38A, GZ39 and GZ42 were isolated and were found all capable of utilizing phenanthrene as sole carbon source. Cloning of the genes responsible for the initial conversion of naphthalene and phenanthrene (*phd* genes) from strain GZ39 revealed that these strains did not contain any genes very similar to the classical *nah*-like genes from *P. putida* strain NCIB 9816-4 (Goyal and Zylstra 1996). Therefore, the genes for phenanthrene degradation in strain GZ38A are similar, but not identical to those from strain GZ39, but strain GZ42 did not have any *phd* genes similar to those from strain GZ39. The three *C. testosteroni* strains thus possess at least two new classes of genes involved in PAHs degradation (Goyal and Zylstra 1996; Zylstra et al. 1997). The order of genes present in strain GZ39 is as follows: genes coding for ferredoxin (*phdAb*), ferredoxin reductase (*phdAa*), cis-dihydrodiol dehydrogenase (*phdB*), the α subunit of ISP (*phdAc*), the β subunit of ISP (*phdAd*), isomerase (*phdD*), an unknown ORF, glutathione-S-transferase and hydratase-aldolase (*phdE*). Comparison of the *phd* genes with known genes indicated that the *PhdAc* sequence fells into the family of naphthalene dioxygenases (although very distantly related), but that *PhdAd* and *PhdAb* sequences have little similarity to isofunctional proteins of other aromatic-ring dioxygenases (Zylstra et al. 1997).

### 7.1.4 phn Genes of Burkholderia sp. Strain RP007

*Burkholderia* sp. strain RP007 was isolated from a PAH-contaminated site in New Zealand on the basis of its ability to degrade phenanthrene as a sole carbon and energy source. This strain was also found to utilize low-molecular weight PAHs like naphthalene and anthracene as sole carbon sources. Napthalene and phenanthrene are degraded through a common upper pathway via salicylate and 1-hydroxy-2-napthoic acid, respectively. The *phn* locus was cloned and the *phn* genes were found to be different in sequence similarity and gene organization from the previously characterized PAHs-catabolic genes (Laurie and Lloyd-Jones 1999a, 1999b). The different genes were found to be present in the following order: genes for regulatory protein (*phnR*), regulatory protein (*phnS*), aldehyde dehydrogenase (*phnF*), hydratase-aldolase (*phnE*), extradiol dioxygenase (*phnC*), isomerase (*phnD*), ISP α subunit of initial dioxygenase (*phnAc*), ISP β subunit of initial dioxygenase (*phnAd*), dihydrodiol dehydrogenase (*phnB*). The *phn* gene locus lacks the ferredoxin and reductase components. The *phnB* gene encoding for *cis*-diol dehydrogenase was found to be more closely related to the corresponding genes from biphenyl catabolic pathways than to those found in the classical *nahB*-like genes. Also, the *phnC* gene encoding the PAHs extradiol dioxygenase, had a phylogeny not seen before among extradiol dioxygenases from any PAH or biphenyl catabolic pathways. Besides this, two catechol 2,3-dioxygenase genes, which are predicted to be involved in lower pathways for aromatic degradation, have been also cloned and characterized (Laurie and Lloyd-Jones 1999b).

### 7.1.5 PAHs Catabolic Genes of Sphingomonas and its Related Species

The members of the genes *Sphingomonas* and related species are able to utilize a wide variety of aromatic compounds, including PAHs as carbon and energy sources. For example, *Novosphingobium aromaticivorans* (formerly *Sphingomonas aromaticivorans*) strain F199 can grow on toluene, all isomers of xylene, *p*-cresol, biphenyl, napththalene, dibenzothiophene, fluorene, salicylate and benzoate (Fredrickson et al. 1991, 1995). Similarly, *S. yanoikuyae* strain B1 can grow on 1,2,4-trimethylbenzene, toluene, *p*-ethyltoluene, *m*- and *p*-xylene, biphenyl, naphthalene and anthracene (Gibson et al. 1973; Zylstra and Kim 1997) and *S. paucimobilis* strain EPA505 utilizes fluoranthene, naphthalene and phenanthrene as sole carbon and energy sources (Mueller et al. 1990).

Recently, the complete sequence of a 184-kb catabolic plasmid pNL1 from strain F199 was identified (Romine et al. 1999). At least 13 gene clusters were predicted to encode enzymes associated with the degradation of aromatic compounds that were completely arranged on the plasmid pNL1. Seven sets of oxygenase components seemed to interact with the only set of ferredoxin and reductase components in pNL1. Several parts of the DNA sequence in pNL1 regions encoding aromatic catabolic genes were similar to those in *S. yanoikuyae* strain B1 (Zylstra and Kim 1997; Kim and Zylstra 1995, 1999), *S. paucimobilis* strain EPA505 (Story et al. 2000), *S. paucimobilis* strain Q1 (Taira et al. 1988), *Sphingomonas* sp. strain HV3 (Yrjala et al. 1997), *Sphingomonas* sp. strain DJ77 (Kim et al. 1997a,b; Shin et al. 1997), *S. paucimobilis* strain TNE12 (Shuttleworth et al. 2000), These results suggest that the unusual arrangement of various genes from different catabolic pathways may be typical of *Sphingomonas* species.

### 7.2 PAHs Catabolic Genes of Gram-positive Bacteria

#### *7.2.1 nar Genes of Rhodococcus sp. Strain NCIMB 12038*

The genus *Rhodococcus* is a diverse group of Gram-positive soil bacteria that degrade many xenobiotic compounds. Although *Rhodococcus* species utilize naphthalene as their sole carbon and energy source (Uz et al. 2000), but PAH catabolic genes have not been characterized until recently.

The nucleotide sequence analysis of *narAa* and *narAb* genes encoding the α and β subunits of ISP from *Rhodococcus* sp. strain NCIMB12038 revealed that the genes are 31-39% identical to the α and β subunits of a number of aromatic-ring dioxygenases (Larkin et al. 1999) Another gene *narB*, encoding *cis*-naphthalene dihydrodiol dehydrogenase was found to have 39% amino acid identity with *NahB* from *P. putida* strain G7 (Kulakov et al. 2000).

#### *7.2.2 phd Genes of Nocardioides sp. Strain KP7*

The *phd* genes of *Nocardioides* sp. strain KP7 are the most studied PAHs-catabolic genes in Gram positive bacteria and belong to a new class of PAHs-catabolic genes because of differences in gene organization and sequence similarity. Strain KP7 was isolated on the basis of its ability to grow on phenanthrene at 40°C from marine samples and was found to degrade phenanthrene via the phthalate pathway (Iwabuchi et al. 1998). The *phdIJK* gene cluster is responsible for the transformation of 1-hydroxy-2-naphthoate to phthalate (Iwabuchi and Harayama 1998). Also *phdA* and *phdB* genes, encoding for the α and β subunits of ISP of phenanthrene dioxygenase, had less than 60% sequence identity to the α- and β- subunits of other aromatic–ring dioxygenases *phdC* and *phdD* genes encoding for ferredoxin and ferredoxin reductase components and were found downstream of *phdB* gene. All three components *PhdABCD* are necessary for the efficient dioxygenase activity that converts phenantherene to its *cis*-diol compound (Saito et al. 2000). Further, PhdE, PhdF, PhdG and PhdH had much similarity to dihydrodiol dehydrogenase, extradiol dioxygenase, hydratase–aldolase and aldehyde dehydrogenase, respectively.

## 8. Conclusion

Degradation of organic pollutants by microorganisms has been studied for many decades. Over the past few years, an extensive database has been developed on the environmental biodegradation of PAHs by a wide variety of bacteria, fungi and algae. This has resulted in a remarkable understanding of the biochemical pathways and molecular genetics involved in the catabolism of a relatively small number of intensively studied pollutants by a relatively small group of microorganisms. Bioremediation, which exploits the catabolic versatility of microorganisms to accelerate the degradation of environmental pollutants, is an important industry in alleviating environmental contamination. Bioremediation can be viewed as an extension of the metabolism that occurs within the microorganisms in which they utilize the organic pollutant as a sole source of carbon and energy for their growth.

It seems to be a general rule in soil that the concentration of PAHs with two or three rings decreases faster than the concentration of PAHs with four to six rings. The low molecular weight, more water-soluble organics, such as naphthalene, are more readily biodegradable than the high molecular weight PAHs, such as benzo[a]pyrene. It does not seem likely at this time that indigenous microflora will facilitate the bioremediation of PAHs-contaminated soils at a rate that would be acceptable to public and to regulatory agencies. Therefore, innovative strategies for the selection of communities of microorganisms with the capacity to degrade highly condensed PAHs are necessary. For successful bioremediation, a deeper understanding is needed at the molecular level, as to how bacterial biodegradation proceeds in PAHs. As a result of extensive studies, the number of known PAHs-catabolic genes for degrading PAHs composed of two or three rings has been increasing. It is important not only to understand the function of these genes, but also to construct gene probes for monitoring the degraders in a contaminated environment. In addition, the efficiency of PAHs degradation could be increased by molecular cloning of the genes for PAHs degradation into high copy number plasmids in bacteria capable of expressing the cloned genes. Considering that PAHs exist in the environment as complex mixtures, further genetic studies on degradation of PAHs other than naphthalene and phenanthrene are needed.

The diversity of PAHs degrading bacteria in the environment is only now being explored with the advent of advanced molecular biology tools. There may still be many more unidentified bacterial genera including unculturable ones in the environment. Exploring our enormous diversity and microbial wealth could provide us with quick and easy solutions to solve the problems of contamination. For effective bioremediation, a critical interplay between the biotic and abiotic factors is necessary. Therefore, with a careful balance between basic and applied research, environmental biotechnology will rapidly emerge as a routine remedial tool for the restoration of PAH-contaminated environments.

Acknowledgements. The authors are grateful to Mr. Gunjan Pandey, Ms. Debarati Paul, Ms. Harsimran K. Kehal and Mr. Nikhil Kirtipal for their assistance.

## References

Abremski KE, Hoess RH (1992) Evidence for a second conserved argnine residue in the integrase family of recombination proteins. Protein Eng 5:87-91

Akhtar NM, Boyd DR, Thompson MJ, Koreeda M, Gibson DT, Mahadevan V, Jerina DM (1975) Absolute stereochemistry of the dihydroxyanthracene-cis and trans-1,2-diols produced from anthracene by mammals and bacteria. J Chem Soc Perkin Trans 1:2506-2511

Aronstein BN, Alexander M (1993) Effect of a non-ionic surfactant added to the soil surface on the biodegradation of aromatic hydrocarbons within the soil. Appl Microbiol Biotechnol 39:386-390

Atlas RM, Horowitz A, Krichevsky M, Bej AK (1991) Response of microbial populations to environmental disturbance. Microb Ecol 22:249-256

Bååth E, Díaz-Raviña M, Frostegård Å, Campbell CD (1998) Effect of metal- rich sludge amendments on the soil microbial community. Appl Environ Microbiol 64:283-245

Bai G, Brusseau ML, Miller RM (1997) Biosurfactant-enhanced removal of residual hydrocarbon from soil. J Contam Hydrol 25:157-170

Barnsely EA (1976) Naphthalene metabolism by pseudomonads: the oxidation of 1,2-dihydroxynaphthalene to 2-hydroxychromene-2-carboxylic acid and formation of 2-hydroxylbenzalpyruvate. Biochem Biophys Res Commun 72:1116-1121

Barnsley EA (1975) The induction of the enzymes of naphthalene metabolism in pseudomonads by salicylate and 2-aminobenzoate. J Gen Microbiol 88:193-196

Berg DE, Howe MM (1989) Mobile DNA. American Society for Microbiology, Washington D.C.

Blumer M (1976) Polycyclic aromatic hydrocarbons in nature. Sci Am 234:35-45

Bogan BW, Lahner LM, Sullivan WR, Paterek JR (2003) Degradation of straight-chain aliphatic and high-molecular-weight polycyclic aromatic hydrocarbons by a strain of *Mycobacterium austroafricanum*. J Appl Microbiol 94:230-239

Boldrin B, Tiehm A, Fritzsche C (1993) Degradation of phenanthrene, fluorene, fluoranthene and pyrene by a *Mycobacterium* species. Appl Environ Microbiol 59:1927-1930

Boonchan S, Britz ML, Stanley GA (2000) Degradation and mineralization of high-molecular-weight polycyclic aromatic hydrocarbons by defined fungal-bacterial cocultures. Appl Environ Microbiol 66:1007-1019

Bosch R, Garcia-Valdés E, Moore ERB (1999a) Genetic characterization and evolutionary implications of a chromosomally encoded naphthalene-degradation upper pathway from *Pseudomonas stutzeri* AN10. Gene 236:149-157

Bosch R, Garcia-Valdés E, Moore ERB (2000) Complete nucleotide sequence and evolutionary significance of a chromosomally encoded naphthalene-degradation lower pathway from *Pseudomonas stutzeri* AN10. Gene 245:65-74

Bosch R, Moore ERB, Gracia-Valdés E, Pieper DH (1999b) Nah W, a novel, inducible salicylate hydroxylase involved in mineralization of naphthalene by *Pseudomonas stutzeri* AN10. J Bacteriol 181:2315-2322

Bosma TNP, Middeldorp PJM, Schraa G, Zehnder AJB (1997) Mass transfer limitation of biotransformation: quantifying bioavailability. Environ Sci Technol 31:248-52

Braun-Lüllemann A, Hüttermann A, Majcherczyk A (1999) Screening of ectomycorrhizal fungi for degradation of polycyclic aromatic hydrocarbons. Appl Microbiol Biotechnol 53:127-132

Caldini G, Cenci G, Manenti R, Morozzi G (1995) The ability of an environmental isolate of *Pseudomonas fluorescens* to utilize chrysene and other four-ring polynuclear aromatic hydrocarbons. Appl Microbiol Biotechnol 44:225-229

Cerniglia CE (1984) Microbial metabolism of polycyclic aromatic hydrocarbons. Adv Appl Microbiol 30:31-71

Cerniglia CE (1992) Biodegradation of polycyclic aromatic hydrocarbons. Biodegradation 43:156-164

Cerniglia CE (1993) Biodegradation of polycyclic aromatic hudrocarbons. Curr Opin Biotechnol 4:331-338

Cerniglia CE, Althaus JR, Evans FE, Freeman JP, Mitchum RK, Yang SK (1983) Stereochemistry and evidence for an arene-oxide-NIH shift pathway in the fungal metabolism of naphthalene. Chem Biol Interact 44:119-132

Cerniglia CE, Campbell WL, Fu PP, Freeman JP, Evans FE (1990) Stereoselective fungal metabolism of methylated anthracenes. Appl Environ Microbiol 56:661-668

Cerniglia CE, Freeman JP, White GL, Heflich RH, Miller DW (1985b) Fungal metabolism and detoxification of the nitropolycyclic aromatic hydrocarbon, 1-nitropyrene. Appl Environ Microbiol 50:649-652

Cerniglia CE, Sutherland JB, Crow SA (1992) Fungal metabolism of aromatic hydrocarbons. In: Winkelmann G (ed) Microbial Degradation of Natural Products, VCH Press, Weinheim, pp 193-217

Cerniglia CE, White GL, Heflich RH (1985a) Fungal metabolism and detoxification of polycyclic aromatic hydrocarbons. Arch Microbiol 143:105-110

Churchill SA, Harper JP, Churchill PF (1999) Isolation and characterization of a *Mycobacterium* species capable of degrading three- and four-ring aromatic and aliphatic hydrocarbons. Appl Environ Microbiol 65:549-552

Davies JI, Evans WC (1964) Oxidative metabolism of naphthalene by soil pseudomonads: the ring-fission mechanism. Biochem J 91:251-261

Denome SA, Stanley DC, Olson ES, Young KD (1993) Metabolism of dibenzothiophene and naphthalene in *Pseudomonas* strains: complete DNA sequence of an upper naphthalene catabolic pathway. J Bacteriol 175:6890-6901

Desai JD, Banat I (1997) Microbial production of surfactants and their commercial potential. Microbiol Mol Biol Rev 61:47-64

Déziel E, Comeau Y, Villemur R (1999) Two-liquid-phase bioreactors for enhanced degradation of hydrophobic/toxic compounds. Biodegradation 10:219-233

Déziel E, Paquette G, Villemur R, Lépine F, Bisaillon J-G (1996) Biosurfactant production by a soil *Pseudomonas* strain growing on polycyclic aromatic hydrocarbons. Appl Environ Microbiol 62:1908-1912

Dunn, N.W., Gunsalus, I.C. (1973) Transmissible plasmid coding early enzymes of naphthalene oxidation in *Pseudomonas putida*. J Bacteriol 114:974-979.

Eaton RW (1994) Organization and evolution of naphthalene catabolic pathways: sequence of the DNA encoding 2-hydroxychromene-2-carboxylate isomerase and *trans-o*-hydroxy benzylidenepyruvate hydratase–addolase from the NAH7 plasmid. J Bacteriol 176:7757-7762

Eaton RW, Chapman PJ (1992) Bacterial metabolism of naphthalene: construction and use of recombinant bacteria to study ring cleavage of 1,2-dihydroxynaphthalene and subsequent reactions. J Bacteriol 174:7542-7554

Edwards DA, Luthy RG, Liu Z (1994) Solubilization of polycyclic aromatic hydrocarbons in micellar nonionic surfactant solutions. Environ Sci Technol 25:127-133

Ensley BD, Gibson DT (1983) Naphthalene dioxygenase: purification and properties of a terminal oxygenase component. J Bacteriol 155:505-511

Evans WC, Fernley HN, Griffiths E (1965) Oxidative metabolism of phenanthrene and anthracene by soil pseudomonads. Biochem J 95:819-831.

Fernley HN, Griffiths E, Evans WC (1964) Oxidative metabolism of phenanthrene and anthracene by soil bacteria: the initial ring fission step. Biochem J 91:15-16

Ferris JM, Nold SC, Rersbech NP, Ward DM (1997) Population structure and physiological changes within a hot spring microbial mat community following disturbance. Appl Environ Microbiol 63:1367-1374

Field JA, de Jong E, Feijoo Costa G, de Bont JAM (1992) Biodegradation of polycyclic aromatic hydrocarbons by new isolates of white rot fungi. Appl Environ Microbiol 58:2219-2226

Finlayson-Pitts BJ, Pitts JN, Jr. (1997) Tropospheric air pollution: ozone, airborne toxics, polycyclic aromatic hydrocarbons and particles. Science 276:1045-1052

Fredrickson JK, Balkwill DL, Drake GR, Romine MF, Ringelberg DB, White DC (1995) Aromatic-degrading *Sphingomonas* isolates from the deep subsurfaces. Appl Environ Microbiol 61:1917-1922

Fredrickson JKJ, Brockman FJ, Workman DJ, Li SW, Stevens TO (1991) Isolation and characterization of a subsurface bacterium capable of growth on toluene, naphthalene and other aromatic compounds. Appl Environ Microbiol 57:796-803

Fuenmayor SL, Wild M, Boyes AL, Williams P (1998) A gene cluster encoding steps in conversion of naphthalene to gentisate in *Pseudomonas* sp. strain U2. J Bacteriol 180:2522-2530

Garcia-Valdéz E, Cozar E, Rotger R, Lalucat J, Ursing J (1988) New naphthalene degrading marine *Pseudomonas* strains. Appl Environ Microbiol 54:2478-2485

Gauthier E, Déziel E, Villemur R, Juteau P, Lépine F, Beaudet R (2003) Initial characterization of new bacteria degrading high-molecular weight polycyclic aromatic hydrocarbons isolated from a 2-year enrichment in a two-liquid-phase culture system. J Appl Microbiol 94:301-311

Geiselbrecht AG, Hedlund BP, Tichi MA, Stanley JT (1998) Isolation of marine polycyclic aromatic hydrocarbon-degrading *Cycloclasticus* strains from the Gulf of Mexico and comparison of their PAH degradation ability with that of Puget Sound strains. Appl Environ Microbiol 64:4703-4710

Geiselbrecht AG, Herwig RP, Deming JW, Stanley JT (1996) Enumeration and phylogenetic analysis of polycyclic aromatic hydrocarbon-degrading marine bacteria from Puget Sound sediments. Appl Environ Microbiol 62:3344-3349

Ghosh DK, Mishra AK (1983) Oxidation of phenanthrene by strain of *Micrococcus*: evidence of protocatechuate pathway. Curr Microbiol 9:219-224

Gibson DT, Subramanian V (1984) Microbial degradation of aromatic hydrocarbon. In: Gibson DT (ed) Microbial Degradation of Organic Compounds, Marcel Dekker, New York, pp 181-252

Gibson DT, Roberts RL, Wells MC, Kobal VM (1973) Oxidation of biphenyl by a *Beijerinckia* species. Biochem Biophys Res Commun 50:211-219

Goldman R, Enewold L, Pellizzari E, Beach JB, Bowman ED, Krishnan SS, Shields PG (2001) Smoking increases carcinogenic polycyclic aromatic hydrocarbons in human lung tissue. Cancer Res 61:6367-6371

Goyal AK, Zylstra GJ (1996) Molecular cloning of novel genes for polycyclic aromatic hydrocarbon degradation from *Comamonas testosteroni* GZ39. Appl Environ Microbiol 62:230-236

Grimm AC, Harwood CS (1999) NahY, a catabolic plasmid-encoded receptor required for chemotaxis of *Pseudomonas putida* to the aromatic hydrocarbon naphthalene. J Bacteriol 181:3310-3316

Grund AD, Gunsalus IC (1983) Cloning of genes for naphthalene metabolism in *Pseudomonas putida.* J Bacteriol 156:89-94

Guieysse B, Cirne MDTG, Mattiasson B (2001) Microbial degradation of phenanthrene and pyrene in a two-liquid phase-partitioning bioreactor. Appl Microbiol Biotechnol 56:796-802

Habe H, Omori T (2003) Genetics of polycyclic aromatic hydrocarbon metabolism in diverse aerobic bacteria. Biosci Biotechnol Biochem 67:225-243

Haemmerli SD, Leisola MSA, Sanglard D, Fiechter A (1986) Oxidation of benzo[a]pyrene by extracellular ligninase of *Phanerochaete chrysosporium*: veratryl alcohol and stability of ligninase. J Biol Chem 261:6900-6903

Hallet B, Sherratt DJ (1997) Transposition and site-specific recombination: adapting DNA cut-and-paste mechanisms to a variety of genetic rearrangements. FEMS Microbiol Rev 21:157-178

Hammel, K.E., Kalyanaraman, B., Kirk, T.K. (1986) Oxidation of polycyclic aromatic hydrocarbons and dibenzo[p]dioxins by *Phanerochaete chrysosporium* ligninase. J Biol Chem 261:16948-16952.

Harayama S (1997) Polycyclic aromatic hydrocarbon bioremediation design. Curr Opin Biotechnol 8:268-273

Harms H, Bosma TNP (1997) Mass transfer limitation of microbial growth and pollutant degradation. J Ind Microbiol Biotechnol 18:97-105

Harvey S, Elashvili I, Valdes JJ, Kamely D, Chakrabarty AM (1990) Enhanced removal of Exxon Valdez spilled oil from Alaskan gravel by a microbial surfactant. Bio/Technology 8:228-230

Hedlund BP, Geiselbrecht AD, Stanley JT (1996) Dioxygenase and phylogenetic diversity among marine PAH-degrading bacteria, Abstract No. Q339. In: Abstracts of the 96$^{th}$ General Meeting of the American Society for Microbiology 1996. American Society for Microbiology, Washington, D.C

Hedlund BP, Geiselbrecht AD, Bair TJ, Stanley JT (1999) Polycyclic aromatic hydrocarbon degradation by a new marine bacterium, *Neptumonas naphthovorans* gen, nov., sp. nov. Appl Environ Microbiol 65:251-259

Heitkamp MA, Cerniglia CE (1988) Mineralization of polycyclic aromatic hydrocarbons by a bacterium isolated from sediment below an oil field. Appl Environ Microbiol 54:1612-1614

Heitkamp MA, Franklin W, Cerniglia CE (1988a) Microbial metabolism of polycyclic aromatic hydrocarbons: isolation and characterization of a pyrene-degrading bacterium. Appl Environ Microbiol 54:2549-2555

Heitkamp MA, Freeman JP, Miller DW, Cerniglia CE (1988b) Pyrene degradation by a *Mycobacterium* sp.: identification of ring oxidation and ring fission products. Appl Environ Microbiol 54:2556-2565

Herbes SE, Schwall LR (1978) Microbial transformation of polycyclic aromatic hydrocarbons in pristine and petroleum-contaminated sediments. Appl Environ Microbiol 35:306-316

Herwijnen RV, Springael D, Slot P, Govers HAJ, Parsons JR (2003) Degradation of anthracene by *Mycobacterium* sp. strain LB 5-1T proceeds via a novel pathway, through *o*-phthalic acid. Appl Environ Microbiol 69:186-190

Holland HL, Khan SH, Richards D, Riemland E (1986) Biotransformation of polycyclic aromatic compounds by fungi. Xenobiotica 16:733-741

Houghton JE, Shanley MS (1994) Catabolic potential of pseudomonads: a regulatory prespective. In: Chaudhry RG (ed) Biological Degradation and Bioremediation of Toxic Chemicals, Chapman and Hall, London, pp 11-32

Hughes JB, Beckles DM, Chandra SD, Ward CH (1997) Utilization of bioremediation processes for the treatment of PAH contaminated sediments. J Ind Microbiol Biotechnol 18:152-160

Iwabuchi T, Harayama S (1998) Biochemical and molecular characterization of 1-hydroxy-2-naphthoate dioxygenase from *Nocardioides* sp. KP7. J Mol Biol 273:8332-8336

Iwabuchi T, Inomata-Yamoguchi Y, Katsuta A, Harayama S (1998) Isolation and characterization of marine *Nocardioides* capable of growing and degrading phenanthrene at 42°C. J Mar Biotechnol 6:86-90

Jerina DM, Selander H, Yagi H, Wells MC, Davey JF, Mahadevan V, Gibson DT (1976) Dihydrodiols from anthracene and phenanthrene. J Am Chem Soc 98:598-5996

Kanaly RA, Bartha R (1999) Cometabolic mineralization of benzo[a]pyrene caused by hydrocarbon addition to soil Environ. Toxicol Chem 18:2186-2190

Keith LH, Telliard WA (1979) Priority pollutants. I. A perspective view. Environ Sci Technol 13:416-423

Kelley I, Cerniglia CE (1995) Degradation of a mixture of high molecular weight polycyclic aromatic hydrocarbons by a *Mycobacterium* strain PYR-1. J Soil Contam 4:44-91

Kelley I, Freeman JP, Cerniglia CE (1991) Identification of metabolites from the degradation of naphthalene by a *Mycobacterium* sp. Biodegradation 1:283-290

Khan AA, Wang R-F, Cao W-W, Doerge DR, Wennerstrom D, Cerniglia CE (2001) Molecular cloning, nucleotide sequences and expression of genes encoding a polycyclic aromatic ring dioxygenase from *Mycobacterium* sp. strain PYR-1. Appl Environ Microbiol 67:3577-3585

Kiehlmann E, Pinto L, Moore M (1996) The transformation of chrysene to trans-1,2-dihydroxy-1,2-dihydrochrysene by filamentous fungi. Can J Microbiol 42:604-608

Kim E, Zylstra GJ (1999) Functional analysis of genes involved in biphenyl, naphthalene, phenanthrene, and *m*-xylene degradation by *Sphingomonas yanoikuyae* B1. J Ind Microbiol Biotechnol 23:294-302

Kim E, Zylstra GJ (1995) Molecular biochemical characterization of two *meta*-cleavage dioxygenases involved in biphenyl and *m*-xylene degradation by *Beijerinkia* sp. strain B1. J Bacteriol 177:3095-3103

Kim S, Kweon OK, Kim Y, Kim CK, Lee KS, Kim YC (1997a) Localization and sequence analysis of the *phnH* gene encoding 2-hydroxypent-2,4-dienoate hydratase in *Pseudomonas* sp. strain DJ77. Biochem Biophys Res Commun 238:56-60

Kim S, Shin HJ, Kim Y, Kim SJ, Kim YC (1997b) Nucleotide sequence of the *Pseudomonas* sp. DJ77 *phnG* gene encoding 2-hydroxymuconic semialdehyde dehydrogenase. Biochem Biophys Res Commun 240:41-45

Kiyohara H, Magao K, Momi R (1976) Degradation of phenanthrene through *o*-phthalate by an *Aeromonas* sp. Agri Biol Chem 40:1075-1082

Kotterman MJJ, Vis EH, Field JA (1998) Successive mineralization and detoxification of benzo[a]pyrene by the white rot fungus *Bjerkandera* sp. strain BOS555 and indigenous microflora. Appl Environ Microbiol 64:2853-2858

Kulakov EA, Allen CCR, Lipscomb DA, Larkin MJ (2000) Cloning and characterization of a novel *cis*-naphthalene dihydrodiol dehydrogenase gene (*narB*) from *Rhodococcus* sp. NCMB12038. FEMS Microbiol Lett 182:327-331

Kurkela S, Lehvaslaiho H, Palva ET, Teeri TH (1988) Cloning, nucleotide sequence and characterization of genes encoding naphthalene dioxygenase of *Pseudomonas putida* strain NCIB 9816. Gene 73:355-362

Langworthy DE, Stapleton RD, Sayler GS, Findlay RH (1998) Genotypic and phenotypic responses of a riverine microbial community to polycyclic aromatic hydrocarbon contamination. Appl Environ Microbiol 64:3422-3428

Larkin MJ, Allen CCR, Kulakov LA, Lipscomb DA (1999) Purification and characterization of a novel naphthalene dioxygenase from *Rhodococcus* sp. strain NCIMB 12038. J Bacteriol 181:6200-6204

Launen L, Pinto L, Weibe C, Kiehlmann E, Moore M (1995) The oxidation of pyrene and benzo[a]pyrene by nonbasidiomycete soil fungi. Can J Microbiol 41:477-488

Laurie AD, Lloyd-Jones G (1999a) The *phn* genes of *Burkholderia* sp. strain RP007 constitute a divergent gene cluster for polycyclic aromatic hydrocarbon catabolism J Bacteriol 181:531-540

Laurie AD, Lloyd-Jones G (1999b) Conserved and hybrid *meta*-cleavage operons from *Burkholderia* RP007. Biochem Biophys Res Commun 262:308-314

Macleod CT, Daugulis AJ (2003) Biodegradation of polycyclic aromatic hydrocarbons in a two-phase partitioning bioreactor in the presence of a bioavailable solvent. Appl Microbiol Biotechnol 62:291-296

MacNaughton SJ, Stephen JR, Venosa AD, Davis DA, Chang Y-J, White DC (1999) Microbial population changes during bioremediation of an experimental oil spill. Appl Environ Microbiol 65:3566-3574

Marcoux J, Déziel E, Villemur R, Lépine F, Bisaillon J-G, Beaudet R (2000) Optimization of high-molecular-weight polycyclic aromatic hydrocarbons' degradation in a two-liquid-phase bioreactor. J Appl Microbiol 88:655-662

Meulenberg R, Rijnaarts HHM, Doddema HJ, Field JA (1997) Partially oxidized polycyclic aromatic hydrocarbons show an increased bioavailability and biodegradability. FEMS Microbiol Lett 152:45-49

Mihelcic JR, Lueking DR, Mitzell RJ, Stapleton JM (1993) Bioavailability of sorbed- and separate-phase chemicals. Biodegradation 4:141-153

Millero FJ, Sohn ML (1991) *Chemical Oceanography*. CRC Press, Boca Raton, FL, pp 531

Moody JD, Freeman JP, Doerge DR, Cerniglika CE (2001) Degradation of phenanthrene and anthracene by cell suspensions of *Mycobacterium* sp. strain PYR-1. Appl Environ Microbiol 67:1476-1483

Moody JD, Freeman JP, Fu PP, Cerniglia CE (2004) Degradation of benzo[a]pyrene by *Mycobacterium vanbaalenii* PYR-1. Appl Environ Microbiol 70:340-345

Mueller JG, Cerniglia CE, Pritchard PH (1996) Bioremediation of environments contaminated by polycyclic aromatic hydrocarbons. In: Crawford RL, Crawford DL (eds) Bioremediation: Principles and Applications, Cambridge University Press, U.K., pp 1215-1294

Mueller JG, Chapman PJ, Blattmann BO, Pritchard PH (1990) Isolation and characterization of a fluoranthene utilizing strain of *Pseudomonas paucimobilis*. Appl Environ Microbiol 56:1079-1086

Mueller JG, Devereux R, Santavy DL, Lantz SE, Willis SG, Pritchard PH (1997) Phylogenetic and physiological comparisons of PAH-degrading bacteria from geographically diverse soils. Antonie van Leeuwenhoek 71:329-343

Mueller JG, Lantz SE, Devereux R, Berg JD, Pritchard PH (1994) Studies on the microbial ecology of polycyclic aromatic hydrocarbon biodegradation. In: Hinchee

RE, Seprini L, Ong SK (eds) Bioremediation of Chlorinated and PAH compounds, Lewis publishers, Boca Raton, Florida, pp 218-230

Narro ML, Cerniglia CE, Van Baalen C, Gibson DT (1992b) Metabolism of phenanthrene by the marine cyanobacterium *Agmenellum quadruplicatum,* strain PR-6. Appl Environ Microbiol 58:1351-1359

Narro ML, Cerniglia CE, Van Baalen C, Gibson DT (1992a) Evidence of NIH shift in naphthalene oxidation by the marine cyanobacterium, *Oscillatoria* species strain JCM. Appl Environ Microbiol 58:1360-1363

National Research Council. (1983) *Polycyclic aromatic hydrocarbons: Evaluation of Sources and Effects*. National Academy Press, Washington, D.C.

Nikolova P, Ward OP (1993) Whole cell biocatalysis in nonconventional media. J Ind Microbiol 12:76-86

Oberbremer A, Müller-Hurtig R, Wagner F (1990) Effect of the addition of microbial surfactants on hydrocarbon degradation in a soil population in a stirred reactor. Appl Microbiol Biotechnol 32:485-489

Øvreås Jensen LS, Daae FL Torsvik V (1998) Microbial community changes in perturbed agricultural soil investigated by molecular and physiological approaches. Appl Environ Microbiol 64:2739-2742

Patel TR, Gibson DT (1974) Purification and properties of (+)-cis-napthalene dihydrodiol dehydrogenase of *Pseudomonas putida*. J Bacteriol 19:879-888

Patnaik P (1992) Hydrocarbon, aromatic. In: A Comprehensive Guide to the Hazardous Properties of Chemical Substances, Van Nostrand Reinhold, New York, pp425-445

Pignatello JJ, Xing B (1996) Mechanism of slow sorption of organic chemicals to natural particles. Environ Sci Technol 25:372-379

Platt A, Shingler V, Taylor SC, Williams PA (1995) The 4-hydroxy-2-oxovalerate aldolase and acetaldehyde dehydrogenase (acylating) encoded by the *nahM* and *nahO* genes of the naphthalene catabolic plasmid pWW60-22 provide further evidence of conservation of meta-cleavage pathway gene sequences. Microbiology 141:2223-2233

Pothuluri JV, Freeman JP, Evans FE, Cerniglia CE (1992a) Fungal metabolism of acenaphthene by *Cunninghamella elogans.* Appl Environ Microbiol 58:3654-3659

Pothuluri JV, Heflich RH, Fu PP, Cerniglia CE (1992b) Fungal metabolism and detoxification of fluoranthene. Appl Environ Microbiol 58:937-941

Pothuluri JV, Selby A, Evans FE, Freeman JP, Cerniglia CE (1994) Transformation of chrysene and other polycyclic aromatic hydrocarbon mixtures by the fungus *Cunnighamella elegans*. Can J Bot 73:1025-1033

Prabhu Y, Phale PS (2003) Biodegradation of phenanthrene by *Pseudomonas* sp. strain PP2: novel metabolic pathway, role of biosurfactant and cell surface hydrophobicity in hydrocarbon assimilation. Appl Microbiol Biotechnol 61:342-351

Rafii F, Butler WR, Cerniglia CE (1992) Differentiation of a rapidly growing, scotochromogenic, polycyclic aromatic hydrocarbon-metabolizing strain of *Mycobacterium* sp. from other known *Mycobacterium* species. Arch Microbiol 157:512-520

Rehmann K, Steinberg CEW, Kettrup AA (1996) Branched metabolic pathway for phenanthrene degradation in a pyrene-degrading bacterium. Polycycl Aromat Comp 11:125-130

Romine MF, Stillwell LC, Wong K-K, Thurston SJ, Sisk EC, Sensen C, Gaasterland T, Fredrickson JK, Saffer JD (1999) Complete sequence of a 184-kilobase catabolic plasmid from *Sphingomonas aromaticivorans* F199. J Bacteriol 181:1585-1602

Rooney-Varga JN, Anderson RT, Fraga JL, Ringelberg D, Lovley DR (1999) Microbial community associated with anaerobic benzene degradation in petroleum-contaminated aquifer. Appl Environ Microbiol 65:3056-3063

Rosenberg E, Ron EZ (1996) Bioremediation of petroleum contamination. In: Crawford RL, Crawford DL (eds) Bioremediation: Principles and Applications, Cambridge University Press, UK, pp 100-124

Sack U, Heinze J, Deck J, Cerniglia CE, Martens R, Zadrazil F, Fritsche W (1997) Comparison of phenanthrene and pyrene degradation by different wood-decaying fungi. Appl Environ Microbiol 63:3919-3925

Saito A, Iwabuchi T, Harayama S (2000) A novel phenanthrene dioxygenase from *Nocardioides* sp. KP7: expression in *Escherichia coli.* J Bacteriol 182:2134-2141

Samanta SK, Chakraborti AK, Jain RK (1999) Degradation of phenanthrene by different bacteria:evidence for novel transformation sequences involving the formation of 1-naphthol. Appl Microbiol Biotechnol 53:98-107

Samanta SK, Rani M, Jain RK (1998) Segregational and structural instability of a recombinant plasmid carrying genes for naphthalene degrading pathway. Lett Appl Microbiol 26:265-269

Sanglard, D., Leisola, M.S.A., Fiechter, A. (1986) Role of extracellular ligninase in biodegradation of benzo[a]pyrene by *Phanerochaete chrysosporium.* Enzyme Microbiol Technol 8:209-212

Sanseverino J, Werner C, Fleming J, Applegate B, King JMH, Sayler GS (1993) Molecular diagnostics of polycyclic aromatic hydrocarbon degradation in manufactured gas plant soils. Biodegradation 4:303-321

Sayler GS, Hooper SW, Layton AC, Kind JMH (1990) Catabolic plasmids of environmental and ecological significance. Microb Ecol 20:1-20

Schell MA (1985) Transcriptional control of the *nah* and *sal* hydrocarbon degradation operons by the *nahR* gene product. Gene 36:301-309

Schell MA (1986) Homology between nucleotide sequences of promoter regions of *nah* and *sal* operons of the NAH 7 plasmid of *Pseudomonas putida.* Proc Natl Acad Sci USA 83:369-373

Schell MA, Wender PE (1986) Identification of the *nahR* gene product and nucleotide sequences required for its activation of the *sal* operon. J Bacteriol 166:9-14

Schneider J, Grosser R, Jayasimhulu K, Xue W, Warshawsky D (1996) Degradation of pyrene, benz[a]anthracene and benzo[a]pyrene by *Mycobacterium* sp. strain RJGII-135, isolated from a former coal gasification site. Appl Environ Microbiol 62:13-19

Shi Y, Zwolinski MD, Schreiber ME, Bahr JM, Sewell GW, Hickey WJ (1999) Molecular analysis of microbial community structures in pristine and contaminated aquifers: field and laboratory microcosm experiments. Appl Environ Microbiol 65:1118-1123

Shin HJ, Kim SJ, Kim YC (1997) Sequence analysis of the *phnD* gene encoding 2-hydroxymuconic semialdehyde hydrolase in *Pseudomonas* sp. strain DJ77. Biochem Biophys Res Commun 232:288-291

Shuttleworth KL, Cerniglia CE (1995) Environmental aspects of polycyclic aromatic hydrocarbon biodegradation. Appl Biochem Biotechnol 54:291-302

Shuttleworth KL, Sung J, Kim E, Cerniglia CE (2000) Physiological and genetic comparison of two aromatic hydrocarbon-degrading *Sphingomonas* strains. Mol Cell 10:199-205

Simon MJ, Osslund TD, Saunders R, Ensley BD, Suggs S, Harcourt A, Suen W, Cruden DL, Gibson DT, Zylstra GJ (1993) Sequences of genes encoding naphthalene dioxygenase in *Pseudomonas putida* strains G7 and NCIB 9816-4. Gene 127:31-37

Smith MR (1990) The biodegradation of aromatic hydrocarbons by bacteria. Biodegradation 1:784-791

Story SP, Parker SH, Kline JD, Tzeng TJ, Mueller JG, Kline EL (2000) Identification of four structural genes and two putative promoters necessary for utilization of naphthalene, phenanthrene, and fluoranthene by *Sphingomonas paucimobilis* var EPA 505. Gene 260:155-169

Strawinski RJ, Stone RW (1943) The utilization of hydrocarbons by bacteria. J Bacteriol 40:461-463

Stucki G, Alexander M (1987) Role of dissolution rate and solubility in biodegradation of aromatic compounds. Appl Environ Microbiol 153:292-297

Suen WC, Gibson DT (1993) Isolation and preliminary characterization of the subunits of the terminal component of naphthalene dioxygenase from *Pseudomonas putida* NCIB9816-4. J Bacteriol 175:5877-5881

Suen WC, Haigler BE, Spain JC (1996) 2,4-Dinitrotoluene dioxygenase from *Burkholderia* sp. strain DNT: similarity to naphthalene dioxygenase. J Bacteriol 178:4926-4934

Sutherland JB (1992) Detoxification of polycyclic aromatic hydrocarbons by fungi. J Ind Microbiol 9:53-62

Sutherland JB, Fu PP, Yang SK, Von Tungein LS, Casillus RP, Crow SA, Cerniglia CE (1993) Enantiomeric composition of the trans-dihydrodiols produced from phenanthrene by fungi. Appl Environ Microbiol 59:2145-2149

Taira K, Hayase N, Arimura N, Yamashita S, Miyazaki T, Furukawa K (1988) Cloning and nucleotide sequence of the 2, 3-dihydroxydioxygenase gene from the PCB-degrading strain of *Pseudomonas paucimobilis* Q1. Biochemistry 27:3990-3996

Takizawa N, Iida T, Sawada T, Yamauchi K, Wang Y-W, Fukuda M, Kiyohara H (1999) Nucleotide sequences and characterization of genes encoding naphthalene upper pathway of *Pseudomonas aeruginosa* PaK1 and *Pseudomonas putida* OUS82. J Biosci Bioeng 87:723-731

Takizawa N, Kaida N, Torigoe S, Moritani T, Sawada T, Satoh S, Kiyohara H (1994) Identification and characterization of genes encoding polycyclic aromatic hydrocarbon dioxygenase and polycyclic aromatic hydrocarbon dihydrodiol dehydrogenase in *Pseudomonas putida* OUS82. J Bacteriol 176:2444-2449

Thibault SL, Anderson M, Frankenberger WT Jr (1996) Influence of surfactants on pyrene desorption and degradation in soils. Appl Environ Microbiol 62:283-287

Thomas JM, Yordy JR, Amador JA, Alexander M (1986) Rates of dissolution and biodegradation of water-insoluble organic compounds. Appl Environ Microbiol 52:290-296

Torsvik V, Sørheim R, Goksøyr J (1996) Total bacterial diversity in soil and sediment communities-a review. J Ind Microbiol 17:170-178

Tsuda M, Lino T (1990) Naphthalene degrading genes on plasmid NAH 7 are on a defective transposon. Mol Gen Genet 223:33-39

Uz I, Duan YP, Ogram A (2000) Characterization of the naphthalene degrading bacterium *Rhodococcus opacus* M213. FEMS Microbiol Lett 185:231-238

Van der Meer JR, de Vos WM, Harayama S, Zehnder AJB (1992) Molecular mechanisms of genetic adaptation to xenobiotic compounds. Microbiol Rev 56:677-694

Van Dyke MI, Couture P, Brauer M, Lee H, Trevors TJ (1993) *Pseudomonas aeruginosa* UG2 rhamnolipid biosurfactants: structural characterization and their use in removing hydrophobic compounds from soil. Can J Microbiol 39:1071-1078

Van Sonsbeek HM, Beeftink HH, Tramper J (1993) Two-liquid phase bioreactors. Enzyme Microb Technol 15:722-729

Villemur R, Déziel E, Benachenhou A, Marcoux J, Gauthier E, Lépine F, Beaudet R, Comeau Y (2000) Two-liquid-phase slurry bioreactors to enhance the degradation of high-molecular-weight polycyclic aromatic hydrocarbons in soil. Biotechnol Prog 16:966-972

Volkering F, Breure AM, van Andel JG, Rulkens WH (1995) Influence of nonionic surfactants on bioavailability and biodegradation of polycyclic aromatic hydrocarbons. Appl Environ Microbiol 61:1699-1705

Volkering F, Breure AM, Sterkenburg A, van Andel JG (1992) Microbial degradation of polycyclic aromatic hydrocarbons: effect of substrate availability on bacterial growth kinetics. Appl Microbiol Biotechnol 36:548-552

Walter U, Beyer M, Klein J, Rehm HJ (1991) Degradation of pyrene by *Rhodococcus* sp. UW1. Appl Microbiol Biotechnol 34:671-676

Wang RF, Cao W-W, Cerniglia CE (1995) Phylogenetic analysis of polycyclic aromatic hydrocarbon degrading mycobacteria by 16SrRNA sequencing. FEMS Microbiol Lett 130:75-80

Warshawsky D, Keenan TH, Reilman R, Cody TE, Radike MJ (1990) Conjugation of benzo[a]pyrene metabolites by freshwater green alga *Selenastrum capricornutum*. Chem Biol Interact 74:93-105

Weisenfels WD, Beyer M, Klein J (1990) Degradation of phenanthrene, fluorene and fluoranthene by pure bacterial cultures. Appl Microbiol Biotechnol 32:479-484

Wiesche C in der Martens R, Zadrazil F (1996) Two step degradation of pyrene by white-rot fungi and soil microorganisms. Appl Microbiol Biotechnol 46:653-659

Wild SR, Jones KC, Waterhouse KS, McGrath SP (1990a) Organic contaminants in an agricultural soil with a history of sewage sludge amendments: Polynuclear aromatic hydrocarbons. Environ Sci Tech 24:1706-1711

Wild SR, McGrath SP, Jones KC (1990b) The polynuclear aromatic hydrocarbon (PAH) content of archives sewage sludge. Chemosphere 20:703-716

Willumsen PA, Karlson U, Pritchard PH (1998) Response of fluoranthene-degrading bacteria to surfactant. Appl Microbiol Biotechnol 50:475-482

Yen K-M, Gunsalus IC (1982) Plasmid gene organization: naphthalene/salicylate oxidation. Proc Natl Acad Sci USA 79:874-878

Yen K-M, Gunsalus IC (1985) Regulation of naphthalene catabolic genes of plasmid NAH 7. J Bacteriol 162:1008-1013

Yen KM, Serdar CM (1988) Genetics of naphthalene catabolism in pseudomonads. CRC Crit Rev Microbiol 15:247-268

You I-S, Ghosal D, Gunsalus IC (1991) Nucleotide sequence analysis of the *Pseudomonas putida* PpG7 salicylate hydroxylase gene (*nahG)* and its 3'-flanking region. Biochemistry 30:1635-1641

Yrjala K, Paulin L, Romantschuk M (1997) Novel organization of catechol *meta*-pathway genes in *Sphingomonas* sp. HV3 pSKY4 plasmid. FEMS Microbiol Lett 154:403-408

Zhang YM, Miller RM (1992) Enhanced octadecane dispersion and biodegradation by a *Pseudomonas* rhamnolipids surfactant (biosurfactant). Appl Environ Microbiol 58:3276-3282

Zhou N-Y, Al-Dulayymi J, Baird MS, Williams PA (2002) Salicylate 5-hydroxylase from *Ralstonia* sp. strain U2: a monoxygenase with close relationship to and shared electron transport proteins with naphthalene dioxygenase. J Bacteriol 184:1547-1555

Zhou N-Y, Fuenmayor SL, Williams PA (2001) *nag* genes of *Ralstonia* (formerly *Pseudomonas)* sp. strain U2 encoding enzymes for gentisate catabolism. J Bacteriol 183:700-708

Zylstra GJ, Gibson DT (1991) Aromatic hydrocarbon degradation: a molecular approach. Genet Eng (NY) 13:183-203

Zylstra GJ, Kim E (1997) Aromatic hydrocarbon degradation by *Sphingomonas yanoikuyae* B1. J Ind Microbiol Biotechnol 19:408-414

Zylstra GJ, Kim E, Goyal AK (1997) Comparative molecular analysis of genes for polycyclic aromatic hydrocarbon degradation. Genet Eng 19:257-269

# 19

# Environmental Applications of Fungal and Plant Systems: Decolourisation of Textile Wastewater and Related Dyestuffs

Albino A. Dias, Ana Sampaio and Rui M. Bezerra

CETAV-Department of Biological and Environmental Engineering, University of Trás-os-Montes e Alto Douro, Apartado 1013, 5001-911 Vila Real, PORTUGAL

## 1. Introduction

Only a few groups of microorganisms are capable of bringing about the biodegradation of recalcitrant organic polluting matter, lignin and other aromatic compounds being a case in point. Due to its intrinsic properties, such as insolubility and chemical complexity, lignin protects structural plant cell wall carbohydrates (cellulose and hemicellulose) from microbial attack and enzymatic hydrolysis (Reid 1995; Pérez et al. 2002). From the chemical point of view, lignin is an amorphous heteropolymer derived from three phenylpropanoid monomers: *p*-coumaryl alcohol and its two methoxy-substituted derivatives, *p*-coniferyl alcohol and *p*-sinapyl alcohol. These basic units are randomly joined together by different types of carbon-carbon and ether linkages (Adler 1977; Douglas 1996). Despite natural high structural variability, lignin is essentially an aromatic nonphenolic substrate, since a typical lignin only contains 10-20% of phenolic hydroxyl radicals (Tuor et al. 1995; Youn et al. 1995).

Fungi, in general, but more specifically the wood-decaying basidiomycetes of white-rot type, are the most efficient microorganisms that perform the depolymerization and even complete lignin mineralization to carbon dioxide and water (Reid 1995; Leonowicz et al. 1999; ten Have and Teunissen 2001). The high performance exhibited by white-rot fungi (WRF) could be mainly ascribed to the production of a powerful extracellular ligninolytic system. Three main types of different oxidoreductase activities can be found in this enzymatic system: polyphenoloxidases, peroxidases and auxiliary $H_2O_2$-generating oxidases (Breen and Singleton 1999; Leonowicz et al. 1999). More specifically, the key ligninolytic enzymes synthesised by WRF are laccase (polyphenoloxidase; EC 1.10.3.2), manganese-dependent peroxidase (MnP; EC 1.11.1.13) and lignin peroxidase (LiP; EC 1.11.1.14). However, several ligninolytic fungal species only produce two of them, with the most usual combination of activities being laccase and MnP (Tuor et al. 1995; Tekere et al. 2001). Since the lignin-degrading system of basidiomycetous WRF has typically a broad substrate specificity, both whole cultures and their ligninolytic enzymes were found to be useful for the bioremediation of a wide number of environmental pollutants, ranging from natural compounds to (perhaps a bit surprisingly) xenobiotic ones, including textile dyes (Field et al. 1993; Reddy 1995; Fu and Viraraghavan 2001; Jarosz-Wilkolazka et al. 2002; Wesenberg et al. 2003).

## 2. Environmental Fate of Textile Dyeing and Treatment Difficulties

Several industrial activities, such as textile dyeing, olive oil extraction and the manufacture of pulp and paper are characterised by intensive water consumption rates. Concomitantly, they release huge amounts of more or less coloured effluents into the environment (Galeno and Agosin 1990; Wesenberg et al. 2002; Font et al. 2003; Dias et al. 2004). As far as synthetic dye release is concerned, textile dyeing facilities and the manufacture of dyestuffs are two major polluting sources. In addition, traditional textile dyeing processes generate a large amount of coloured effluents, because about 100 litres of water are required to process 1 Kg of dyed fabrics (Abadulla et al. 2000). Moreover, up to 15% of applied dyestuffs are lost to the effluents due to dyeing process inefficiencies (Jarosz-Wilkolazka et al. 2002). Colour itself could be very pernicious to the receiving water courses not only for aesthetic reasons and toxicity towards many aquatic organisms, but also because coloured compounds reduce water transparency, which, in turn, affects photosynthetic activity, thus causing severe damage to the ecosystems. As a consequence, this parameter has also a limit of discharge.

Industrial textile dyes have been designed and synthesised to be highly resistant to washing, chemical agents, including solvents, and environmental factors, such as the action of sunlight, water and microbial attack. On the other hand, heavy metal (e.g. copper, cobalt, chromium) complexed dyes are of public health concern (Nyanhongo et al. 2002; Blanquez et al. 2004). There are currently more than 10,000 different textile dyes commercially available in the world market (Campos et al. 2001; Keharia and Madamwar 2002), which can be classified according to the application processes (e.g. direct, reactive, vat, disperse) and chemical class. The chemical structures of selected textile dyes illustrating the following chromophoric groups - azo, indigoid, anthraquinone, triphenylmethane and phthalocyanine - are presented in Figure 1. Among them, azo, indigoid and anthraquinone are the major chromophores used in the textile industries (Zollinger 1991), azo dyes being the largest class with an estimated share of about 70% (Soares et al. 2002).

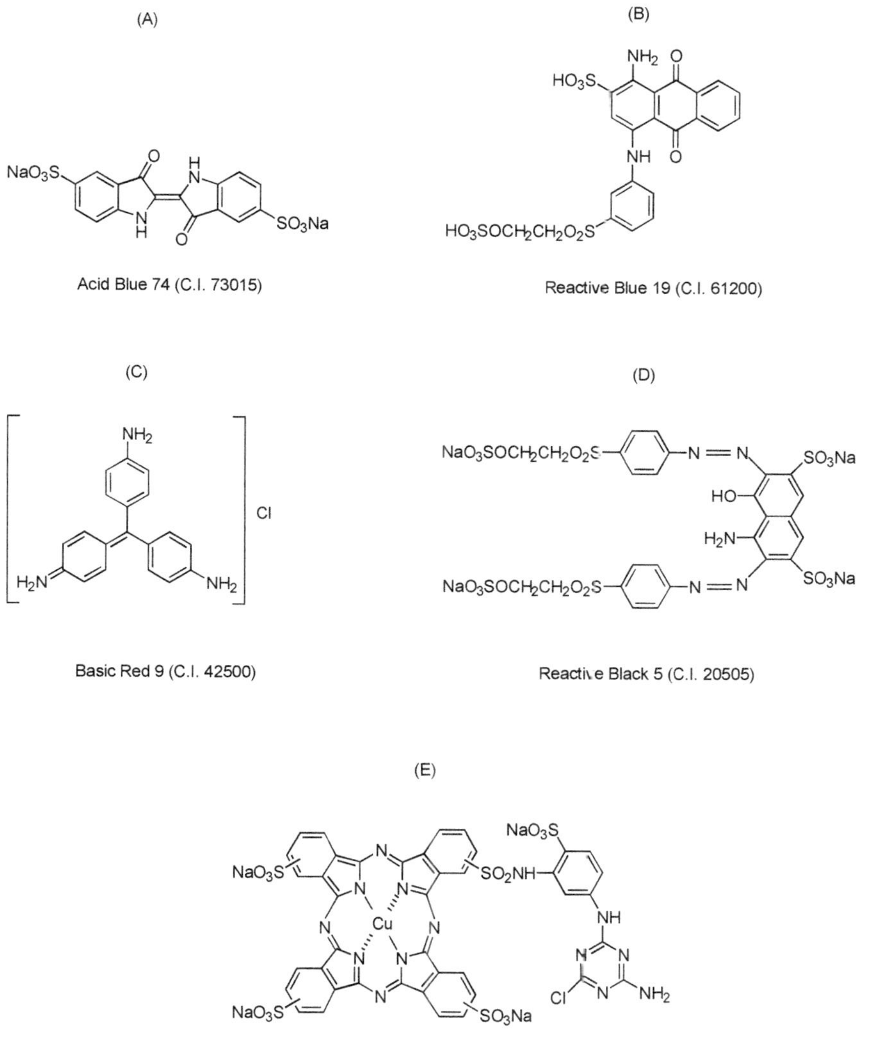


**Fig. 1.** Selected textile dyestuffs and their chromoforic classes: indigoid (A), anthraquinone (B), triphenylmethane (C), azo (D), phthalocyanine (E)

A large portion of dyes, that is lost during the dyeing process, could remain more or less intact, given the fact that both traditional physico-chemical and biological wastewater treatments are unable to perform an acceptable degradation and decolourization of the majority of the available dyes (Shaul et al. 1991; Gill et al. 2002). For example, Weber and Stickney (1993) have reported that the half-life of reactive blue 19 is 46 years at 25°C and pH 7.0. In addition, reactive dyes typically have poor fixation to fabrics, and dye concentrations up to 1,500 mg/L could be found in the liquor that is discharged into the sewers (Pierce 1994). Moreover, about 90% of reactive dyes persist after being subjected to activated sludge treatment. Thus, this textile dye class is, by far, one of the most recalcitrant to the depurating action of conventional wastewater treatments.

Azo dyes, besides being the most widely used class, are also frequent chromoforic moieties of reactive textile dyes. These xenobiotic compounds are characterized by the presence of one, two, three or more azo bonds (-N=N-) and aromatic rings, respectively monoazo, disazo, trisazo and polyazo. The main reason, which makes them generally recognised as recalcitrant compounds, can be mainly attributed to their complex aromatic structures joined by azo bonds and synthetic origin. On the other hand, textile dyeing activities use a wide range of chemical dyes in short time periods and hence their effluents are extremely variable in composition (Correia et al. 1995), which means that textile wastewater requires an unspecific treatment process.

## 3. Overview of Biological Treatments

The biosorption of dyes can be achieved by means of the biopolymer chitin. Azo dyes, such as orange G and orange IV, were successfully adsorbed by shrimp chitin (Longhinotti et al. 1998). These authors have reported that the adsorption capacity is highly dependent on pH and temperature values. Living yeast biomass can also be used for the biosorption of dyes, particularly at low pH values. For example, a *Candida tropicalis* strain isolated from sewage was successfully used for textile dyes (remazol blue, reactive black and reactive red) removal, being the maximum specific bio-accumulation capacity range from 79 mg/g for reactive red to 112 mg/g for remazol blue (Dönmez 2002). Dry biomass of aquatic plants could be used for the removal of dyes and/or heavy metals from textile effluents. Water hyacinth (*Eichornia crassipes*) dried roots removed methylene blue and basic blue dyes efficiently (Low et al. 1995). Also, the dried giant duckweed *Spirodela polyrrhiza* can be used to remove methylene blue at a broad range pH values (3-11) (Waranusantigul et al. 2003). This process is practical and has many economic advantages, in contrast with other adsorbent processes, like activated carbon, a very expensive method. In addition to dye removal ability, heavy metal ions are also adsorbed by dry biomass (Schneider and Rubio 1999).

Microbial treatments can be performed either in the presence (aerobic) or absence (anaerobic) of oxygen. Both processes have lower operational costs, when compared to chemical and/or physical treatments, which tend to concentrate pollutants rather than degrade them. More importantly, they pose less health and safety risks. However, the selection of the most adapted microorganisms to a particular operation is an arduous task.

Combined with physico-chemical treatments, activated sludge is the process most widely used by the textile industry. In this process, the effluent is mechanically agitated in the presence of air and microbial biomass. In spite of removing up to 80% of dye content, most of it (40 – 80%) is only adsorbed or absorbed into the biomass (Shaul et al. 1991; Pagga and Taeger 1994), producing sludge with high dye concentration, which prevents its further utilisation. Besides displaying high levels of sludge production, it is also very sensitive to effluent composition, particularly as far as the content of toxic substances is concerned. Moreover, activated sludge treatment is almost ineffective with reactive textile dyeing effluents.

Anaerobic treatment with the production of methane, carbon dioxide and water, requires less energy effort and produces low sludge quantities. It has been shown that reductive decolourisation of azo dyes could be achieved by the action of bacterial strains under anaerobic conditions (Stolz 2001). However, the production of potential carcinogenic aromatic amines, which resist further degradation, has been reported (Sweeney et al. 1994). Furthermore, re-colourization of anaerobic-treated effluents may take place upon exposure to air (Knapp and Newby 1995). These findings triggered the screening of alternative biological systems as well as its performance evaluation. In recent years, a growing number of research papers have been putting in evidence for the feasibility of dye decolourization by fungi and their oxidative enzymatic systems. Another promising process for the treatment or final polishment of textile effluents is the phytoremediation technology with constructed wetlands.

## 4. Extracellular Oxidoreductases Useful in Pollution Abatement

### 4.1 Laccase (E.C. 1.10.3.2)

Laccase is a metallo-enzyme included in the heterogeneous phenoloxidases group. This group shares the ability to catalyse the oxidation of aromatic substrates coupled to the reduction of molecular oxygen to water. Three distinct enzymatic activities have been characterised (Fig. 2): cresolase or monophenoloxidase (E.C. 1.14.18.1); catechol oxidase or *ortho*-diphenol: oxygen oxidoreductase (E.C. 1.10.3.1); and laccase or *para*-diphenol: oxygen oxidoreductase (E.C. 1.10.3.2).

However, it should be noted that several phenoloxidases exhibit oxidative activities against an overlapping range of substrates. Furthermore, enzymes displaying both cresolase and catecholase activities have different trivial names according to its biological origin. While tyrosinases are enzymes isolated from animal, fungal or bacterial sources and usually display both activities, catechol oxidase or polyphenoloxidase are enzymes identical to tyrosinases, but isolated from higher plants and usually have no cresolase activity (Das et al. 1997).

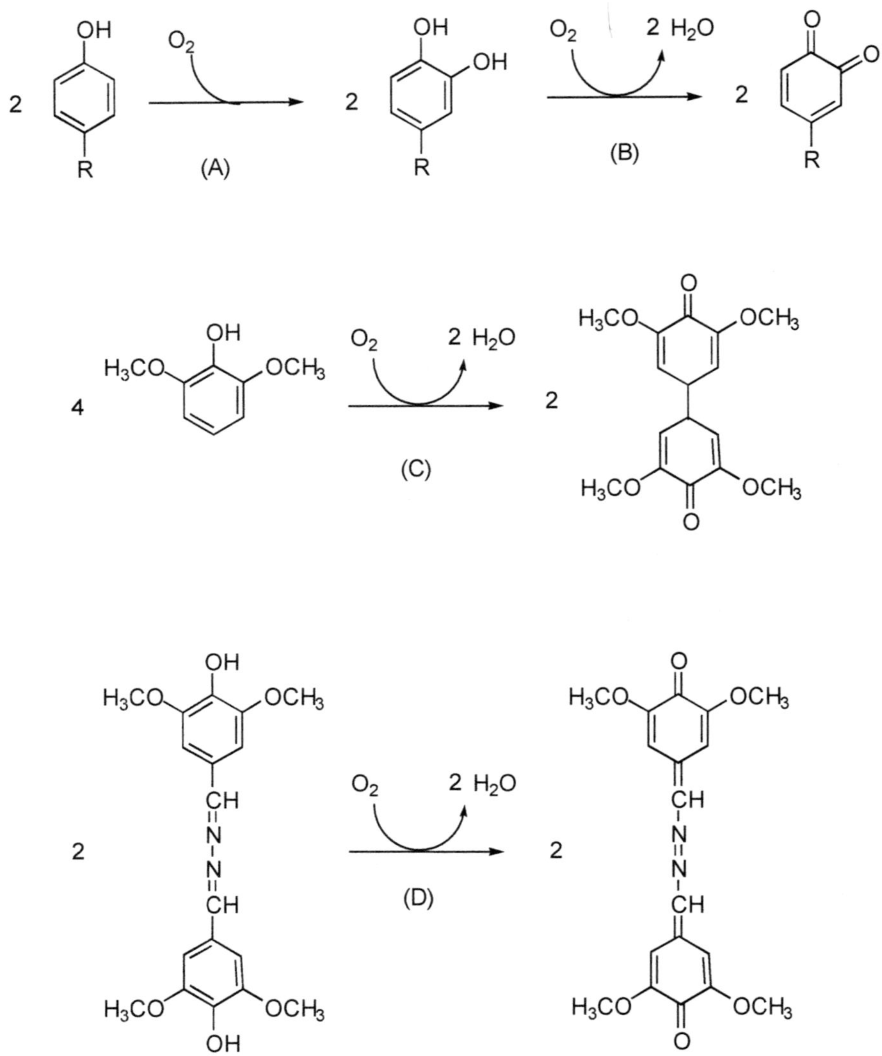


**Fig. 2.** Reactions catalysed by phenoloxidases: A – hydroxylation of monophenols to *o*-diphenols (cresolase activity), B – oxidation of *o*-diphenols to *o*-quinones (catechol oxidase activity), C – oxidation of 2,6-dimethoxyphenol to tetramethoxy-diphenyl-quinone (laccase activity), D – oxidation of syringaldazine to tetramethoxy-azo-bis-methylen-quinone (laccase activity)

As pointed out by Mayer and Staples (2002), clear and undoubted identification of laccase activity requires careful experimentation. As far as substrate specificity is concerned, laccase preferentially oxidises *p*-diphenols and the oxidation of methoxy-activated phenols (e.g. guaiacol, 2,6-

dimethoxyphenol, syringaldazine), is a typical reaction not observed in the other phenoloxidases. Moreover, it is generally accepted that syringaldazine and ABTS (2,2'-azinobis(3-ethylbenzthiazoline-6-sulfonate)) are specific substrates for laccases (Leonowicz and Grzywnowicz 1981; Niku-Paavola et al. 1990; Givaudan et al. 1993). However, they should be used with caution, especially in assays with crude or non-purified enzyme extracts, because in the presence of hydrogen peroxide, both syringaldazine and ABTS are also peroxidase (E.C. 1.11.1.7) substrates.

Laccase is one of the oldest known enzymes. It was discovered more than one century ago by Yoshida (1883) in the latex of the Japanese lacquer tree *Rhus vernicifera*. Laccase was also found in fungi thirteen years later by two investigators, Bertrand and Laborde, separately (Thurston 1994). This enzyme was named *laccase* by Bertrand (1894), who first recognised that it was a metal-containing oxidase and suggested the concept of metallic co-factor (Beinert 2002). However, only in the next century, did Keilin and Mann (1939) prove that laccase is a copper-dependent enzyme. Besides higher plants (e.g. *Rhus vernicifera*, *Acer pseudoplatanus*), laccase has already been described in basidiomycetous WRF (e.g. *Trametes versicolor*, *Pycnoporus cinnabarinus*), non-ligninolytic fungi (e.g. *Aspergillus nidulans*, *Neurospora crassa*), ectomycorrhizal fungi (e.g. *Lactarius* spp., *Russula* spp.), litter decaying fungi (e.g. *Agaricus* spp., *Lepista* spp., *Marasmius* spp.), actinomycetes (e.g. *Streptomyces griseus*, *S. lavendulae*), non-actinomycetous bacteria (e.g. *Azospirillum lipoferum*) and in the cyanobacterium *Anabaena azollae* (Sterjiades et al. 1992; Givaudan et al. 1993; Faure et al. 1995; Malliga et al. 1996; Eggert et al. 1997; Gramss, et al. 1998; Gianfreda et al. 1999; Diamantidis et al. 2000; Suzuki et al. 2003; Moldes et al. 2004).

A new age in laccase research has been emerging and several practical applications have gained a new impulse after the work of Bourbonnais and Paice (1990). These authors have demonstrated the feasibility of the laccase-mediator system (oxidation of recalcitrant non-substrates in the presence of an oxidizable primary substrate). As a consequence, the diversity of potential laccase substrates greatly expanded and, for the first time, the prospects of laccase use in both bioremediation of environmental pollutants (e.g. textile dyes) and biotechnological processes (e.g. biopulping and pulp bleaching) have become realistic. Besides, acknowledging the valuable work carried out by several research teams around the world, five events deserve special mention, namely: (*1*) the production of recombinant laccase (Saloheimo and Niku-Paavola 1991); (*2*) the mechanism of laccase-catalysed oxidation of phenolic azo dyes (Chivukula and Renganathan 1995); (*3*) the discovery of a fungal metabolite (3-hydroxyanthranilate) which is a natural laccase mediator (Eggert et al. 1996); (*4*) the three-dimensional structure of an inactive basidiomycetous laccase form from *Coprinus cinereus* (Ducros et al. 1998); (*5*) the three-dimensional structures of two fully active basidiomycetous laccases from *Trametes versicolor* (Bertrand et al. 2002; Piontek et al. 2002) and an ascomycetous laccase from *Melanocarpus albomyces* (Hakulinen et al. 2002). Taken together, these studies have provided valuable insights into the reaction mechanism of laccase-catalysed oxidation of aromatic substrates and its potentialities.

A typical laccase is a monomeric globular glycoprotein with three (A, B, C) cupredoxin-like domains (Bertrand et al. 2002; Hakulinen et al. 2002; Piontek et al. 2002), that belongs to a small enzyme family denominated multicopper polyphenoloxidases. Generally, it contains four copper ions that have been classified into the following three different spectroscopic sites (T1, T2, T3): one type 1 (T1) or blue copper, one type 2 (T2) or normal copper and two type 3 or binuclear copper site (Lee et al. 2002). The T1 copper belongs to domain C and is tightly coordinated to a cysteine residue (Cys S-Cu), which gives rise to a strong absorption band around 600 nm. The T2 copper, the most labile laccase copper, exhibits normal behaviour and could be detected by electron paramagnetic resonance (EPR). The type 3 site contains a pair of antiferromagnetically coupled coppers which make them EPR-undetectable; they exhibit an absorption band around 330 nm (usually a shoulder in the near-UV). Furthermore, the three copper ions from T2 and T3 sites actually form a trinuclear cluster arranged into a nearly isosceles triangle, which is located at the interface between domains A and C (Bertrand et al. 2002; Hakulinen et al. 2002; Piontek et al. 2002).

Laccase catalyses substrate oxidation by one-electron abstraction to form free radicals which could be re-oxidised again by laccase or undergo further abiotic radical coupling reactions. The catalytic cycle of laccase involves a stepwise abstraction of electrons from reducing substrates and concomitant oxygen reduction. Because four electrons are needed for the reduction of molecular oxygen, laccase could be viewed as a biological battery accumulating reducing power. Appropriated reducing substrates are oxidised at the mononuclear T1 copper site due to its high redox potential, which, for fungal laccases, ranges between 500 mV and 800 mV (Xu et al. 1998; Kumar et al. 2003). The electrons are further transferred to the trinuclear cluster (T2/T3), where the reduction of molecular oxygen to water takes place. The detailed mechanism of oxygen reduction is not fully understood, but probably occurs in two steps of two-electrons without release of intermediate reactive oxygen species (Palmer et al. 2001; Lee et al. 2002).

Chivukula and Renganathan (1995) have carried out detailed studies of the laccase-catalysed biodegradation of phenolic monoazo dyes. They found that the chemical nature of substituents on the phenolic ring profoundly affected dye biodegradability. Dyes containing chloro or nitro substituents were not oxidised, while those containing methyl or methoxy substituents (i.e., electron-donating groups) served as substrates for two one-electron oxidation steps catalysed by laccase from *Pyricularia oryzae*. According to the proposed mechanism (Fig. 3), a carbonium ion is formed and, after nucleophilic water attack to the carbon, in which the positive charge is localised, two products are obtained:

benzoquinone and sulfophenyldiazene. The latter is oxidised in the presence of oxygen, and an unstable phenyldiazene radical intermediate is formed which loses molecular nitrogen to ultimately produce a sulfophenylhydroperoxide. A similar reaction mechanism for the laccase-catalysed biodegradation of phenolic disazo dyes bearing an electron-donating carboxylic substituent has been proposed by Soares *et al.* (2002). Also, purified fungal laccases from *Trametes hirsuta* and *Sclerotium rolfsii* were able to catalyse the biodegradation of the textile dye indigo (Campos et al. 2001). According to the proposed reaction mechanism, indigo can be oxidised originating isatin (indole derived), which, after water attack decomposes into anthranilic acid (intermediate of the shikimic acid pathway).

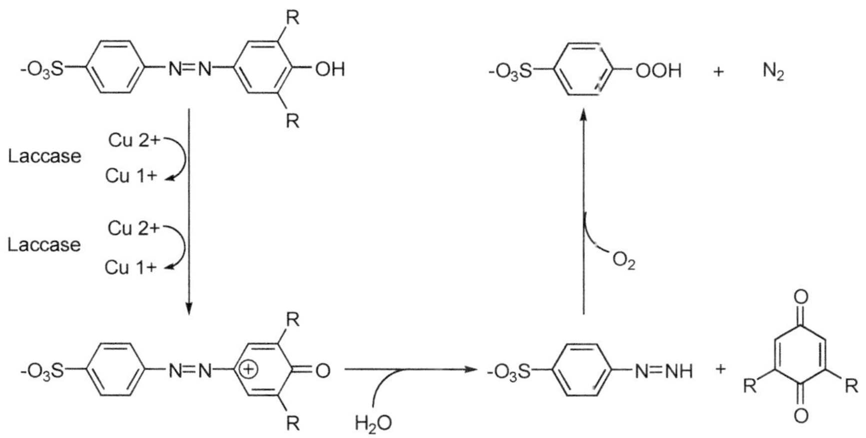


**Fig. 3.** Simplified reaction mechanism for the biodegradation of phenolic monoazo dyes by laccases (adapted from Chivukula and Renganathan 1995)

### 4.2 Lignin Peroxidase (E.C. 1.11.1.14) and Manganese Peroxidase (E.C. 1.11.1.13)

LiP and MnP are ligninolytic enzymes that contain heme (protoporphyrin IX) as a prosthetic group. Like other peroxidases, they exhibit a typical catalytic cycle (Fig. 4). Firstly, hydrogen peroxide promotes a two-electron oxidation of native enzyme giving the so-called Compound I and water. Compound I catalyses one-electron oxidation of a substrate molecule rendering a free radical and Compound II. Finally, another substrate molecule reacts with Compound II (one-electron oxidation) to give a free radical, while the enzyme is reduced back to the native state with concomitant water production.

Peroxidases can be irreversibly inactivated by hydrogen peroxide, because Compound II decays slowly. When this intermediate reacts with hydrogen peroxide, a highly reactive Compound III is formed which ultimately conducts to the oxidative destruction of the tetrapyrrol prosthetic group.

Native Peroxidase + $H_2O_2$ ⟶ Compound I + $H_2O$

Compound I + AH ⟶ Compound II + $A^{\bullet}$

Compound II + AH ⟶ Native Peroxidase + $A^{\bullet}$ + $H_2O$

**Fig. 4.** Simplified catalytic cycle for heme peroxidases [AH: reducing substrate; $A^{\bullet}$: free radical (oxidised substrate)]

LiP was first discovered in carbon and nitrogen-limited batch cultures of the WRF *Phanerochaete chrysosporium*. In the same year, three independent research groups reported their purification and biochemical characterisation (Glenn et al. 1983; Shimada and Higuchi 1983; Tien and Kirk 1983). Among several LiP-catalysed reactions, the breakdown of aromatic structures deserves a special mention. Also, their high redox potential and very low pH optimum are two unique characteristics (ten Have and Teunissen 2001). As far as the biodegradation of textile dyes is concerned, it has been shown that LiP purified from *P. chrysosporium* cultures oxidises some azo dyes by following a mechanism similar to that observed with laccase (Chivukula et al. 1995).

MnP was also first discovered in batch cultures of *P. chrysosporium* (Kuwahara et al. 1984). MnP shows a distinct characteristic, since it is dependent on $Mn^{2+}$ for completing their catalytic cycle. In fact, it has been proved that reduction of Compounds I and II is a $Mn^{2+}$-dependent reaction (Glenn et al. 1986). Concomitantly, Mn ions are oxidised to $Mn^{3+}$, which after chelation with some organic acids (e.g. malonate, oxalate) can then oxidise aromatic compounds through free radicals formation (Wariishi et al. 1992).

### 4.3 Extracellular $H_2O_2$-generating Enzymes

Aryl alcohol oxidase (E.C. 1.1.3.7) and glyoxal oxidase (E.C. 1.2.3.5) are two extracellular FAD-dependent enzymes. Because hydrogen peroxide is a product of its catalytic mechanism, they promote the activity of ligninolytic peroxidases. Aryl alcohol oxidase was discovered by Farmer *et al.* (1960) in the growth medium of a WRF. This enzyme performs the oxidation of various aromatic alcohols into the corresponding aldehydes, and simultaneously the reduction of molecular oxygen gives hydrogen peroxide. Glyoxal oxidase, first discovered by Kersten and Kirk (1987), preferentially oxidises glioxylate into oxalate using molecular oxygen, which is reduced to hydrogen peroxide.

## 5. Textile Dyes Decolourisation by Fungi and their Enzymes

The first report demonstrating the feasibility of fungal decolourisation of azo dyes was provided by Cripps *et al.* (1990), who founded that *P. chrysosporium* was able to decolourise the azo dyes acid orange 6, orange II and Congo red. Posterior intensive researches have found more potent fungal strains, such as *Trametes versicolor*, *Dichomitus squalens*, *Phlebia fascicularia*, *Irpex flavus* and *Polyporus sanguineus*, among others (Rodríguez et al. 1999; Pointing et al. 2000; Gill et al. 2002; Chander et al. 2004).

Despite wastewater treatment, carried out with enzymes, can be more advantageous than with whole fungal cultures (Karam and Nicell 1997), scientific articles reporting dye decolourisation by purified enzymes are scarce. However, their feasibility either by soluble or immobilised enzymes has been demonstrated (Rodríguez et al. 1999; Abadulla et al. 2000; Nagai et al. 2002; Dias et al. 2003). Also, there is a derth of information on the use of bioreactors employing fungal cultures for wastewater treatment. This situation probably reflects the existing wastewater treatment technology, which is almost adapted to non-filamentous and unicellular organisms like bacteria. Thus, recent evidences of textile dyes decolourisation, carried out by yeasts, are very encouraging findings, since in general they are non-filamentous and unicellular fungi (Yang et al. 2003; Ramalho et al. 2004).

## 6. New Tendencies in Textile Wastewater Treatments

### 6.1 Constructed Wetlands

Constructed wetlands can be successfully used in the treatment of several types of effluents: municipal, agricultural, industrial, and hospital wastewater. Pollutants can be eliminated by phytoremediation either by degradation or transformation by plant enzymes, or by microorganisms associated to root plants. It is well known that rhizosphere (i.e. the surrounding root system zone, rich in microorganisms) has high microbial activity due to the presence of several substrates exuded by the plant roots (Paul and Clark 1989). Enzymes secreted by plant roots or microbial community to the rhizosphere comprise esterases and different oxidoreductases (phenoloxidases and peroxidases).

Plant peroxidases are exuded to rhizosphere by some members of Fabacea, Graminae and Solanacea (Gramss et al. 2000). White radish (*Raphanus sativus*) and horseradish (*Armoracia rusticana*) contain peroxidase, whereas stonewort (*Nitella* spp.) and parrotfeather (*Myriophyllum aquaticum*) contain laccase. In spite of the fact that such enzymes can be synthesised by some plants, the most important effects of plants growing in constructed wetlands are physical, since they can provide surface area for microbial growth. Additionally, most plants in a solid matrix, can establish associations with bacteria and/or fungi. In fact, plants play a minor role, when compared to the impact caused by microbial community (Langergrabe 2000).

Microbial oxidoreductases are predominantly synthesised by saprophytic organisms, such as wood-degrading and terricolous fungi, and by actinomycetes. It has been also recognised that arbuscular mycorrhizal fungi (AMF), particularly ectomycorrhizal (ECM) and ericoid mycorrhizal fungi, have the ability to degrade lignin and phenolic compounds. However, their ligninolytic activities are limited compared to white rot fungi (Bending and Read 1997). In 17 isolates of ECM fungi (e.g. *Amanita* spp., *Laccaria* spp., *Lactarius* spp, *Suillus* spp., *Xerocomus* spp.) tested by these authors, none had polyphenoloxidase activity and 8 strains exhibited peroxidase activity. However, at the present time, polyphenoloxidase activities, such as laccase, catechol oxidase and tyrosinase, are known to be synthesised by some ECM and ericoid fungi (Burke and Cairney 2002). Axenic cultures of two ECM fungi *Suillus granulatus* and *Paxillus involutus*, grown in liquid media, exhibited different extracellular enzymatic activities (Gunther et al. 1998). In *S. granulatus*, tyrosinase, laccase and peroxidase activities were detected, whereas *Paxillus involutus* predominantly produced laccase. Both fungi, when in symbiosis with *Pinus sylvestris*, increased the level of peroxidase, and *S. granulatus* only had polyphenoloxidase activities (laccase and tyrosinase) like the mycorrhizal symbiont. Several studies have pointed out the potential of mycorrhizal fungi in bioremediation (Gunther et al. 1998; del Val et al. 2000), despite their role in degradation of recalcitrant compounds remains less understood.

Mycorrhizal fungi were found both in submerged (e.g. *Elatine hexandra*, *Littorella uniflora*, *Ranunculus flammula*) and in emergent (e.g. *Polygonum amphibium*, *Bernula erecta*, *Myosotis palustra*, *Veronica anagallis-aquatica*) species (Beck-Nielsen and Madsen 2001), suggesting that ECM fungi may be important in wetland ecosystems. More recently, Bauer *et al.* (2003) found mycorrhizes in *Carex* and *Scirpus* species, previously thought to be non-mycorrhizal plants. Also, these authors suggest that mycorrhizae colonisation is particularly widespread in the members of the Poaceae family.

## 6.2 Constraints on using Plants for Bioremediation

The limiting factors of phytoremediation treatment are related to the obvious fact that plants are living organisms. Therefore, this process is restricted to sites, where the pollutants concentration are below the toxic level to the plants used. Additionally, pollutants must be bio-available and accessible either to the plant root system, or to the rhizosphere microorganisms. The fact, that phytoremediation is a multidisciplinary technique, makes it difficult to demonstrate its feasibility. The role of plant enzymes, plant physiology, microbial enzymes, metabolites degradation pathways, microbial ecology (free or associated with plants), macrophyte and constructed wetlands system selection, has yet to be better understood for the success of phytoremediation technology.

### 6.3 Case Studies

Despite the fact that the working mechanism of constructed wetlands is still not clear, this process has been widely used in wastewater treatment with a great range of parameters and compositions. However, full-scale applications and use of wetlands for removal of recalcitrant or xenobiotic compounds from industrial effluents are scarce. One of the few examples is the vertical flow reed bed installed at the chemical industry Quimigal S.A. in Portugal. This wetland has a total planted area of 10,000 $m^2$, which efficiently removes nitroaromatic compounds, such as aniline, nitrophenols and nitrobenzene (Dias 2000).

Information on the use of constructed wetlands for textile wastewater treatment is also very limited. One of the first report trials was performed in an horizontal bed reed (*Phragmites* spp.) of 150 $m^2$, in Australia (Davies and Cottingham 1994). In USA (Georgia state), Coats American is currently using constructed wetlands as the final step in their wastewater treatment operations. Although both input and output colour levels remains identical, dyes were removed by 50% (Baughman and Perkins 2002; Perkins and Baughman 2002). These results surprised the authors, who concluded that most of the colour in the wastewater appeared to come from photochemical oxidation of naphthalene sulfonate formaldehyde, used as dye dispersing agent and diluent in dye products. Dispersing agents, resistant to biodegradation, are transformed by sunlight from almost colourless to bright yellow. The most important conclusions in this wetland treatment were: the remission of a large portion of residual dye content, an appreciable decrease in the total chemical oxygen demand (COD), and an apparent lower chronic toxicity from the textile effluent.

In addition to colour removal, macrophytes can also clean up the heavy metals which are present in several textile dyestuffs. In fact, aquatic plants, such as water hyacinth (*Eichhornia crassipes*), pennywort (*Hydrocotyle umbellata*) and duckweed (*Lemma minor*) can accumulate metals and metal-loaded plants are harvested and disposed of when saturated (Prasad and Freitas 2003).

## 7. Conclusion

Wastewater treatment plants, such as activated sludge and methanogenic reactors, are not the natural habitat of WRF, since these organisms prefer solid substrates and well-aerated environments. The fact, that constructed wetlands (e.g. sub-surface flow systems with rooted emergent macrophytes), are transitional environments, i.e. are intermediate between terrestrial and aquatic ecosystems, can be an advantage in the treatment of polluted effluents. The wetlands system treats wastewater by physical, chemical and biotic processes, in a close association of appropriated plants, microorganisms, macro-organisms and substrates. Macrophytes enhance physical filtration, prevent clogging in vertical flow systems, mediate oxygen transfer to the rhizosphere and favour microorganism colonisation (Brix et al. 1996; Brix 1997). In sub-surface systems, there is an oxygen gradient, with high partial pressures near the plant roots, to be replaced progressively by anaerobic and anoxic environments. The mixture of aerobic, anoxic and anaerobic zones stimulates different microbial communities that can degrade complex organic substances (such as azo dyes) almost to mineralisation. The extent of dyes biodegradation must be evaluated, since the formation of intermediate compounds can enhance toxicity (Sweeney et al. 1994). The use of constructed wetlands is a low cost technique, with low maintenance needs (Schwitzguébel et al. 2002; Susarla et al. 2002). It is able to tolerate high fluctuations in flow, temperature (Winthrop et al. 2002) and the composition and/or concentration of pollutants in wastewater. Finally, it is likely to find widespread acceptance with the public for its obvious technological and aesthetic qualities.

Acknowledgement. The authors are grateful to Dr. Sofia Sampaio for English revision of the draft manuscript.

## References

Abadulla E, Tzanov T, Costa S, Robra KH, Cavaco-Paulo A, Gubitz GM (2000) Decolorization and detoxification of textile dyes with a laccase from *Trametes hirsuta*. Appl Environ Microbiol 66:3357-3362

Adler E (1977). Lignin chemistry - past, present and future. Wood Sci Technol 11:169-218

Bauer CR, Kellog CH, Bridgham SD, Lamberti GA (2003) Mycorrhizal colonisation across hydrologic gradients in restored and reference freshwater wetlands. Wetlands 23:961-968

Baughman GL, Perkins WS (2002) Treatment of textile effluents in constructed wetlands: a case study. AATCC Review 2(3):24-25

Beck-Nielsen D, Madsen TV (2001) Occurrence of vesicular-arbustal mycorrhiza in aquatic macrophytes from lakes and streams. Aquatic Bot 71:141-148

Beinert H (2002) Bioinorganic chemistry: a new field or discipline? words, meanings, and reality. J Biol Chem 277:37967-37972

Bending GD, Read DJ (1997) Effects of the soluble polyphenol tannic acid on the activities of ericoid and ectomycorrhizal fungi. Soil Biol Biochem 28:1595-1602

Bertrand MG (1894) Sur le latex de l`arbe laque. CR Acad Sci (Paris) 118:1215-1218

Bertrand T, Jolivalt C, Briozzo P, Caminade E, Joly N, Madzak C, Mougin C (2002) Crystal structure of a four-copper laccase complexed with an arylamine: insights into substrate recognition and correlation with kinetics. Biochemistry 41:7325-7333

Blanquez P, Casas N, Font X, Gabarrell X, Sarra M, Caminal G, Vicent T (2004) Mechanism of textile metal dye biotransformation by *Trametes versicolor*. Water Res 38:2166-2172

Bourbonnais R, Paice MG (1990) Oxidation of non-phenolic substrates: an expanded role for laccase in lignin biodegradation. FEBS Lett 267:99-102

Breen A, Singleton FL (1999) Fungi in lignocellulose breakdown and biopulping. Curr Opin Biotechnol 10:252-258

Brix H (1997) Do macrophytes play a role in constructed treatment wetlands? Water Sci Technol 35:11-17

Brix H, Sorrell BK, Schierup HH (1996) Gas fluxes achived by *in situ* convective flow in *Phragmites australis*. Aquatic Bot 54:151-163

Burke RM, Cairney JWG (2002) Laccases and other polyphenol oxidases in ecto- and ericoid mycorrhizal fungi. Mycorrhiza 12:105-116

Campos R, Kandelbauer A, Robra KH, Cavaco-Paulo A, Gubitz GM (2001) Indigo degradation with purified laccases from *Trametes hirsuta* and *Sclerotium rolfsii*. J Biotechnol 89:131-139

Chander M, Arora DS, Bath HK (2004) Biodecolourisation of some industrial dyes by white-rot fungi. J Ind Microbiol Biotechnol 31:94-97

Chivukula M, Renganathan V (1995) Phenolic azo dye oxidation by laccase from *Pyricularia oryzae*. Appl Environ Microbiol 61:4374-4377

Chivukula M, Spadaro JT, Renganathan V (1995) Lignin peroxidase-catalyzed oxidation of sulfonated azo dyes generates novel sulfophenyl hydroperoxides. Biochemistry 34:7765-7772

Correia VM, Stephenson T, Judd SJ (1995) Characteristics of textile wastewaters - a review. Environ Technol 15:917-929

Cripps C, Bumpus JA, Aust SD (1990) Biodegradation of azo and heterocyclic dyes by *Phanerochaete chrysosporium*. Appl Environ Microbiol 56:1114-1118

Das JR, Bhat SG, Gowda LR (1997) Purification and characterization of a polyphenol oxidase from the kew cultivar of Indian pineapple fruit. J Agric Food Chem 45:2031-2035

Davies TH, Cottingham PD (1994) The use of constructed wetlands for treating industrial effluent (textile dyes). Water Sci Technol 29:227-232

del Val C, Guralchuk Z, Campos E, Azcón-Aguillar C, Barea JM (2000) Biodiversity of arbuscular mycorrhizal fungi in heavy metal contamined soils and its implications for bioremediation. In: Intercost Workshop on Bioremediation. Sorrento, Italy, pp 45-47

Diamantidis G, Effosse A, Potier P, Bally R (2000) Purification and characterization of the first bacterial laccase in the rhizospheric bacterium *Azospirillum lipoferum*. Soil Biol Biochem 32:919-927

Dias AA, Bezerra RM, Pereira AN (2004) Activity and elution profile of laccase during biological decolorization and dephenolization of olive mill wastewater. Bioresource Technol 92:7-13

Dias AA, Bezerra RM, Lemos PM, Pereira AN (2003) *In vivo* and laccase-catalysed decolourization of xenobiotic azo dyes by a basidiomycetous fungus: characterization of its ligninolytic system. World J Microbiol Biotechnol 19:969 - 975

Dias SM (2000) Nitroaromatic compounds removal in a vertical flow reed bed case study: industrial wastewater treatment. Intercost Workshop on Bioremediation. Sorrento, Italy, pp 119-120

Dönmez G (2002) Bioaccumulation of the reactive textile dyes by *Candida tropicalis* growing in molasses medium. Enzyme Microb Technol 30:363-366

Douglas CJ (1996) Phenylpropanoid metabolism and lignin biosynthesis: from weeds to trees. Trends Plant Sci 1:171-178

Ducros V, Brzozowski AM, Wilson KS, Brown SH, Østergaard P, Schneider P, Yaver DS, Pedersen AH, Davies GJ (1998) Crystal structure of the type-2 Cu depleted laccase from *Coprinus cinereus* at 2.2 Å resolution. Nat Struct Biol 5:310-316

Eggert C, Temp U, Eriksson K-EL (1997) Laccase is essential for lignin degradation by the white-rot fungus *Pycnoporus cinnabarinus*. FEBS Lett 407:89-92

Eggert C, Temp U, Dean JF, Eriksson KE (1996) A fungal metabolite mediates degradation of non-phenolic lignin structures and synthetic lignin by laccase. FEBS Lett 391:144-148

Farmer VC, Henderson MEK, Russell JD (1960) Aromatic-alcohol-oxidase activity in the growth medium of *Polystictus versicolor*. Biochem J 74:257-262

Faure D, Bouillant M-L, Bally R (1995) Comparative study of substrates and inhibitors of *Azospirillum lipoferum* and *Pyricularia oryzae* laccases. Appl Environ Microbiol 61:1144-1146

Field JA, de Jong E, Feijoo-Costa G, de Bont JAM (1993) Screening for ligninolytic fungi applicable to the biodegradation of xenobiotics. Trends Biotechnol 11:44-49

Font X, Caminal G, Gabarrel X, Romero S, Vicent MT (2003) Black liquor detoxification by laccase of *Trametes versicolor* pellets. J Chem Tech Biotechnol 78:548-554

Fu Y, Viraraghavan T (2001) Fungal decolorization of dye wastewaters: a review. Bioresource Technol 79:251-262

Galeno GD, Agosin ET (1990) Screening of white-rot fungi for efficient decolourization of bleach pulp effluents. Biotechnol Lett 12:869-872

Gianfreda L, Xu F, Bollag J-M (1999) Laccases: a useful group of oxidoreductive enzymes. Bioremediation J 3:1-25

Gill PK, Arora DS, Chander M (2002) Biodecolourization of azo and triphenylmethane dyes by *Dichomitus squalens* and *Phlebia* spp. J Ind Microbiol Biotechnol 28:201-203

Givaudan A, Effosse A, Faure D, Potier P, Bouillant ML, Bally R (1993) Polyphenol oxidase in *Azospirillum lipoferum* isolated from rice rhizosphere: evidence for laccase activity in non-motile strains of *Azospirillum lipoferum*. FEMS Microbiol Lett 108:205-210

Glenn JK, Akileswaran L, Gold MH (1986) Mn(II) oxidation is the principal function of the extracellular Mn-peroxidase from *Phanerochaete chrysosporium*. Arch Biochem Biophys 251:688-696

Glenn JK, Morgan MA, Mayfield MB, Kuwahara M, Gold MH (1983) An extracellular $H_2O_2$-requiring enzyme preparation involved in lignin biodegradation by the white rot basidiomycete *Phanerochaete chrysosporium*. Biochem Biophys Res Commun 114:1077-1083

Gramss G, Gunther Th, Fritsche W (1998) Spot tests for oxidative enzymes in ectomycorrhizal, wood- and litter-decaying fungi. Mycol Res 102:67-72

Gramss G, Voigt K, Kirsche B (2000) Oxidoreductases enzymes liberated by plant roots and their effects on soil humic material. Chemosphere 38:1481-1494

Gunther H, Perner B, Gramss G (1998) Activities of phenol oxidizing enzymes of ectomycorrhizal fungi in axenic culture and in symbiosis with Scots pine (*Pinus sylvestris* L.). J Basic Microb 38:197-206

Hakulinen N, Kiiskinen LL, Kruus K, Saloheimo M, Paananen A, Koivula A, Rouvinen J (2002) Crystal structure of a laccase from *Melanocarpus albomyces* with an intact trinuclear copper site. Nat Struct Biol 9:601-605

Jarosz-Wilkolazka A, Kochmanska-Rdest J, Malarczyk E, Wardas W, Leonowicz A (2002) Fungi and their ability to decolourize azo and anthraquinonic dyes. Enzyme Microb Technol 30:566-572

Karam J, Nicell JA (1997) Potential applications of enzymes in waste treatment. J Chem Tech Biotechnol 69:141-153

Keharia H, Madamwar D (2002) Transformation of textile dyes by white-rot fungus *Trametes versicolor*. Appl Biochem Biotechnol 102-103:99-108

Keilin D, Mann T (1939) Laccase, a blue copper-protein oxidase from the latex of *Rhus succedanea*. Nature 143:23-24

Kersten PJ, Kirk TK (1987) Involvement of a new enzyme, glyoxal oxidase, in extracellular $H_2O_2$ production by *Phanerochaete chrysosporium*. J Bacteriol 169:2195-2201

Knapp JS, Newby PS (1995) The microbiological decolorization of an industry effluent containing a diazo-linked chromophore. Water Res 29:1807-1809

Kumar SVS, Phale PS, Durani S, Wangikar PP (2003) Combined sequence and structure analysis of the fungal laccase family. Biotechnol Bioeng 83:386-394

Kuwahara M, Glenn JK, Morgan MA, Gold MH (1984) Separation and characterization of two extracellular $H_2O_2$-dependent oxidases from ligninolytic cultures of *Phanerochaete-chrysosporium*. FEBS Lett 169:247-250

Langergrabe G (2000) Application of constructed wetlands for wastewater treatment. In: Intercost Workshop on Bioremediation. Sorrento, Italy, pp 115-118

Lee SK, George SD, Antholine WE, Hedman B, Hodgson KO, Solomon EI (2002) Nature of the intermediate formed in the reduction of $O_2$ to $H_2O$ at the trinuclear copper cluster active site in native laccase. J Am Chem Soc 124:6180-6193

Leonowicz A, Grzywnowicz K (1981) Quantitative estimation of laccase forms in some white-rot fungi using syringaldazine as a substrate. Enzyme Microb Technol 3:55-58

Leonowicz A, Matuszewska A, Luterek J, Ziegenhagen D, Wojtas-Wasilewska M, Cho N-S, Hofrichter M, Rogalski J (1999) Biodegradation of lignin by white rot fungi. Fungal Genet Biol 27:175-185

Longhinotti E, Pozza F, Furlan L, Sanchez MNM, Klug M, Laranjeira MCM, Favere VT (1998) Adsorption of anionic dyes on the biopolymer chitin. J Braz Chem Soc 9:435-440

Low KS, Lee CK, Tan KK (1995) Biosorption of basic-dyes by water hyacinth roots. Bioresource Technol 52:79-83

Malliga P, Uma L, Subramanian G (1996) Lignolytic activity of the cyanobacterium *Anabaena azollae* ML2 and the value of coir waste as a carrier for BGA biofertilizer. Microbios 86:175-183

Mayer AM, Staples RC (2002) Laccase: new functions for an old enzyme. Phytochemistry 60:551-565

Moldes D, Lorenzo M, Sanroman MA (2004) Different proportions of laccase isoenzymes produced by submerged cultures of *Trametes versicolor* grown on lignocellulosic wastes. Biotechnol Lett 26:327-30

Nagai M, Sato T, Watanabe H, Saito K, Kawata M, Enei H (2002) Purification and characterization of an extracellular laccase from the edible mushroom *Lentinula edodes*, and decolorization of chemically different dyes. Appl Microbiol Biotechnol 60:327-335

Niku-Paavola ML, Raaska L, Itavaara M (1990) Detection of white-rot fungi by a non-toxic stain. Mycol Res 94:27-31

Nyanhongo GS, Gomes J, Gubitz GM, Zvauya R, Read J, Steiner W (2002) Decolorization of textile dyes by laccases from a newly isolated strain of *Trametes modesta*. Water Res 36:1449-1456

Pagga V, Taeger K (1994) Development of a method for adsorption of dyestuffs on activated sludge. Water Res 28:1051-1057

Palmer AE, Lee SK, Solomon EI (2001) Decay of the peroxide intermediate in laccase: reductive cleavage of the O-O bond. J Am Chem Soc 121:7138-7149

Paul EA, Clark FE (1989) *Soil Microbiology and Biochemistry*. Academic Press, New York

Pérez J, Munoz-Dorado J, de la Rubia T, Martínez J (2002) Biodegradation and biological treatments of cellulose, hemicellulose and lignin: an overview. Int Microbiol 5:53-63

Perkins WS, Baughman GL (2002) Color in textile wastewater: influence of dye dispersing agents. AATCC Review 2(8):65-67

Pierce J (1994) Colour in textile effluents – the origins of the problem. J Soc Dyers Colourists 110:131-133

Piontek K, Antorini M, Choinowski T (2002) Crystal structure of a laccase from the fungus *Trametes versicolor* at 1.90-Å resolution containing a full complement of coppers. J Biol Chem 277:37663-37669

Pointing SB, Bucher VVC, Vrijmoed LLP (2000) Dye decolorization by sub-tropical basidiomycetous fungi and the effect of metals on decolorizing ability. World J Microbiol Biotechnol 16:199-205

Prasad MNV, Freitas HMD (2003) Metal hyperaccumulation in plants - biodiversity prospecting for phytoremediation technology. Electronic J Biotechnol 6:285-321

Ramalho PA, Cardoso MH, Cavaco-Paulo A, Ramalho MT (2004) Characterization of azo reduction activity in a novel ascomycete yeast strain. Appl Environ Microbiol 70:2279- 2288

Reddy CA (1995) The potential for white-rot fungi in the treatment of pollutants. Curr Opin Biotechnol 6:320-328

Reid ID (1995) Biodegradation of lignin. Can J Bot 73 (suppl.):S1011-S1018

Rodríguez E, Pickard MA, Vazquez-Duhalt R (1999) Industrial dye decolorization by laccases from ligninolytic fungi. Curr Microbiol 38:27-32

Saloheimo M, Niku-Paavola ML (1991) Heterologous production of a ligninolytic enzyme: expression of the *Phlebia radiata* laccase gene in *Trichoderma reesei*. Bio/Technology 9:987-990

Schneider IAH, Rubio J (1999) Sorption of heavy metal ions by the non living biomass of freshwater macrophytes. Environ Sci Technol 33:2213-2217

Schwitzguébel JP, Aubert S, Grosse W, Laturnus F (2002) Sulphonated aromatic pollutants - Limits of microbial degradability and potential of phytoremediation. Environ Sci Pollut Res 9:62-72

Shaul GM, Holdsworth TJ, Dempsey CR, Dostal KA (1991) Fate of water-soluble azo dyes in the activated-sludge process. Chemosphere 22:107-119

Shimada M, Higuchi T (1983) Biochemical aspects of the secondary metabolism of xenobiotic lignin and veratryl alcohol biosynthesis in *Phanerochaete chrysosporium*. In: Higuchi T, Chang H-M, Kirk TK (eds) Proc of Recent Advances in Lignin Biodegradation Research, Tokyo Univ. Publ., pp. 195-208

Soares GMB, Amorim MTP, Hrdina R, Costa-Ferreira M (2002) Studies on the biotransformation of novel disazo dyes by laccase. Process Biochem 37:581-587

Sterjiades R, Dean JFD, Eriksson KEL (1992) Laccase from sycamore maple (*Acer pseudoplatanus*) polymerizes monolignols. Plant Physiol 99:1162-1168

Stolz A (2001) Basic and applied aspects in the microbial degradation of azo dyes. Appl Microbiol Biotechnol 56:69-80

Susarla S, Medina VF, McCutcheon SC (2002) Phytoremediation: an ecological solution to organic chemical contamination. Ecol Engineering 18:647-658

Suzuki T, Endo K, Ito M, Tsujibo H, Miyamoto K, Inamori Y (2003) A thermostable laccase from *Streptomyces lavendulae* REN-7: purification, characterization, nucleotide sequence, and expression. Biosci Biotechnol Biochem 67:2167-2175

Sweeney EA, Chipman JK, Forsythe SJ (1994) Evidence for direct-acting oxidative genotoxicity by reduction products of azo dyes. Environ Health Persp 102:119-122

Tekere M, Mswaka AY, Zvauya R, Read JS (2001) Growth, dye degradation and ligninolytic activity studies on Zimbabwean white rot fungi. Enzyme Microb Technol 28:420-426

ten Have R, Teunissen PJM (2001) Oxidative mechanisms involved in lignin degradation by white-rot fungi. Chem Rev 101:3397-3413

Thurston CF (1994) The structure and function of fungal laccases. Microbiology 140:19-26

Tien M, Kirk TK (1983) Lignin-degrading enzyme from the hymenomycete *Phanerochaete chrysosporium* Burds. Science 221:661-663

Tuor U, Winterhalter K, Fiechter A (1995) Enzymes of white-rot fungi involved in lignin degradation and ecological determinants for wood decay. J Biotechnol 41:1-17

Waranusantigul P, Pokethitiyook P, Kruatrachue M, Upatham ES (2003) Kinetics of basic dye (methylene blue) biosorption by giant duckweed (*Spirodela polyrrhiza*). Environ Pollution 125:385-392

Wariishi H, Valli K, Gold MH (1992) Manganese(II) oxidation by manganese peroxidase from the basidiomycete *Phanerochaete chrysosporium*: kinetic mechanism and role of chelators. J Biol Chem 267:23688-23695

Weber EJ, Stickney VC (1993) Hydrolysis kinetics of reactive blue 19-vinyl sulfone. Water Res 27:63-67

Wesenberg D, Buchon F, Agathos SN (2002) Degradation of dye-containing textile effluent by the agaric white-rot fungus *Clitocybula dusenii*. Biotechnol Lett 24:989-993

Wesenberg D, Kyriakides I, Agathos SN (2003) White-rot fungi and their enzymes for the treatment of industrial dye effluents. Biotechnol Adv 22: 161-187

Winthrop CA, Hook PB, Biederman JA, Stein OR (2002) Temperature and wetland plant species effects on wastewater treatment and root zone oxidation. J Environ Qual 31:1010-1016

Xu F, Berka RM, Wahleithner JA, Nelson BA, Shuster JR, Brown SH, Palmer AE, Solomon EI (1998) Site-directed mutations in fungal laccase: effect on redox potential, activity and pH profile. Biochem J 334:63-70

Yang Q, Yang M, Pritsch K, Yediler A, Hagn A, Schloter M, Kettrup A (2003) Decolorization of synthetic dyes and production of manganese-dependent peroxidase by new fungal isolates. Biotechnol Lett 25:709-713

Yoshida H (1883) Chemistry of lacquer (Urushi), part I. J Chem Soc (Tokyo) 43:472-486

Youn H-D, Hah YC, Kang S-O (1995) Role of laccase in lignin degradation by white-rot fungi. FEMS Microbiol Lett 132:183-188

Zollinger H (1991) Color Chemistry: Synthesis, Properties and Applications of Organic Dyes and Pigments. VCH Publishers, New York

# Fungal-Based Remediation: Treatment of PCP Contaminated Soil in New Zealand

J.M. Thwaites[1], R.L. Farrell[1], S.D. Duncan[1], R.T. Lamar[2] and R.B. White[2]

[1]Department of Biological Sciences, University of Waikato, Private Bag 3105, Hamilton, NEW ZEALAND, Email: r.farrell@waikato.ac.nz; [2]Earthfax Development Corporation, Logan, UTAH

## 1. Introduction

Contamination of soil, water, and air with toxic chemicals is a serious and on going problem facing the world today. Hazardous compounds, such as polycyclic aromatic hydrocarbons (PAHs), pentachlorophenol (PCP), polychlorinated biphenyls (PCBs), 1,1,1-trichloro-2,2-bis(4-chlorophenyl)ethane (DDT), and trinitrotoluene (TNT) are persistent in the environment and are known to have carcinogenic and mutagenic effects. Removing these pollutants from the environment in an ecologically responsible, safe, rapid, and cost-effective way is a priority for land management agencies. Bioremediation, using microbial organisms, is one way to achieve this target. Extensive laboratory studies have shown the capability of various organisms to remediate contaminated soil and water. More research, however, to determine the applicability and practicability of utilizing these microorganisms in contaminated field sites needs to be achieved. This review and case study presents an evidence of successful bioremediation of PCP and related dioxins using fungal-based technology.

## 2. Fungal-based Remediation

Fungal-based remediation is an *ex situ* form of bioremediation, in which hazardous organics are degraded or detoxified by fungi that are introduced into the contaminated soil via a fungal inoculum (i.e. a lignocellulosic substrate support which is colonized by the fungus). The soil inoculum mixture is then treated in a forced aeration biopile, in which temperature and moisture conditions are monitored and maintained to provide optimum fungal growth and activity.

Fungi are robust organisms that are tolerant to high concentrations of pollutants (Evans and Hedger 2001). *Phanerochaete chrysosporium* was the first reported microorganism to show degradation of an extremely diverse group of environmental pollutants (Bumpus et al. 1985; Eaton 1985). Since then, the majority of research on bioremediation employed the pollutant-degrading abilities of an ecological group of fungi, including *P. chrysosporium*, referred to as white rot fungi. These fungi are saprophytes that obtain their carbon for energy and biomass from the dead organic matter, and include members, such as the common edible mushrooms, *Pleurotus ostreatus* (oyster mushroom), *Lentinulus edodes* (Shitake), and *Agaricus bisporus* (white button mushroom). White rot fungi degrade cellulose, hemicellulose and most importantly the lignin component of the wood cell wall. After degradation, the residual wood is typically fibrous with a whitish yellow to tan discoloration due to the removal of lignin. Most white rot fungi are basidiomycetes, possessing dikaryotic hyphae and clamp connections along the septate hyphae (Jennings and Lysek 1999). These fungi are uniquely equipped as soil remediation agents (Reddy 1995). As filamentous organisms, they have the natural propensity to extend through soil in search of new substrates to exploit, and thus can colonize places that bacteria are unable to reach. They possess the ability to oxidize extremely hydrophobic substrates due to the highly oxidative nature of the enzymes that comprise the extracellular component of their lignin-degrading system. These extracellular enzymes extend the fungus degradative influence beyond the hyphae. The fungus does not utilize the pollutant for grwoth, and thus, the amount of pollutant degraded is not a function of the concentration of fungus within the soil.

Lignin has a heterogeneous aromatic structure with many different types of subunit linkages. Both direct (pollutant oxidation by lignin-degrading enzymes) and indirect (pollutant mineralization during ligninolysis) evidences indicated that the lignin-degrading systems of white rot fungi were involved in pollutant degradation (Kirk and Farrell 1987). This enzymatic degradation, termed "Enzyme Combustion", is highly oxidative, extracellular and non-specific (Kirk and Farrell 1987). In addition to oxidative agents, the fungi possess reductive agents that are also involved in the degradation of aromatic sub-structures of lignin that are produced from its depolymerization. It makes perfect sense then that the aromatic pollutants, such as PCP and PAHs, that are degraded by white rot fungi, closely resemble the aromatic sub-structures that are produced during lignin depolymerization.

Much of the work on fungal-based remediation has been conducted using PCP. PCP was used as a wide spectrum pesticide and wood preservative throughout the world. In New Zealand, the primary use was in the timber industry as a treatment for the interim protection of timber against sapstain fungi and as a preservative in diesel oil. It was estimated that over a period of forty years, 5000 tonnes of PCP was used in New Zealand alone (Yu and Shepard 1997). Significant use of PCP, in New Zealand, ceased in 1988 and the chemical was formally deregistered by the New Zealand Pesticides Board in 1991. It is currently banned in most countries and is listed as a priority pollutant by the United States Environmental Protection Agency. Soil contamination through accidental spillage and inappropriate disposal at many wood treatment facilities plus the relative stability of PCP in the environment means that it continues to be a major problem.

PCP is toxic to organisms because it is an inhibitor of oxidative phosphorylation (Crosby 1981) and it is also hydrophobic with low water solubility; this contributes to its persistence in the environment. The products of the PCP manufacturing process, polychlorinated dibenzo-p-dioxins and polychlorinated dibenzofurans (PCDFs), are also a problem at many sites where PCP was manufactured or used (Eduljee 1999). The major dioxin congeners identified are the hexa-, hepta-, and octachloro congeners (Eduljee 1999) and all are highly persistant and toxic in the environment.

## 2.1 Fungal Degradation of PCP

White rot fungi have been shown to cause a considerable depletion of PCP from soil in the laboratory experiments (Lamar et al. 1990a, b; Okeke et al. 1993; Okeke et al. 1994; Lestan et al. 1996) and under field conditions (Lamar and Dietrich 1990; Lamar et al. 1993; Lamar et al. 1994). In several small-scale field trials (total weight of soil for all treatments < 24 tonnes) significant but less extensive PCP decreases (88-91%) were reported (Lamar and Dietrich 1990). In this study, *Phanerochaete sordida* and *P. chrysosporium* were inoculated into a sandy gravel soil pH 9.67, contaminated with a commercial PCP containing wood preservative (250-400mg PCP/kg soil) and the reductions were evident after 6.5 weeks.

## 2.2 Fate of PCP after Bioremediation

Numerous publications have shown that white rot fungi can efficiently deplete PCP in the contaminated soil, but the question of the fate of the portion of PCP that is converted to non-extractable products remains unanswered (Lamar et al. 1990a,b; Lamar and Dietrich 1990; Lamar et al. 1993). In liquid fungal culture, chlorinated phenols, were shown to be degraded via a series of reactions that remove all the chlorines, after which the aromatic ring is oxidatively cleaved and further degraded to $CO_2$ (Valli and Gold 1991; Joshi and Gold 1993). In soil, mineralization (degradation to $CO_2$ and $H_2O$) was demonstrated to be a minor fate for PCP. Work by Rüttiman-Johnson and Lamar (1997) demonstrated that a large part of PCP present in the contaminated soil is bound to the soil humic materials by the action of extracellular oxidative enzymes. The covalent binding of pollutants to soil fractions is important as it may reduce their bioavailability and therefore their toxicity. The oxidized transformation products of xenobiotics, like PCP (Rüttimann-Johnson and Lamar 1997), PAHs (Eschenback et al. 1998; Kastner et al. 1999), and chlorocatechols (Stott et al. 1983), can be readily incorporated into soil humic materials (a phenomenon referred to as humification) because of their structural similarity to natural aromatic substrates originating mainly from lignin degradation. The underlying mechanism of the humification process involves oxidation of the chlorinated phenols or other aromatics to free radicals or quinone products that subsequently couple directly to humic acids, fulvic acids and/or humin, all of which possess oxygen-containing functional groups (hydroxyls, carboxyls, and quinones) or to naturally-occurring phenols that are also subject to oxidation. If the unpaired electron of a free radical is located at an aromatic carbon which is substituted by a chlorine atom, dehalogenation occurs during the coupling reaction (Hatcher et al. 1993). The occurrence of dehalogenation during oxidative polymerization provides a direct evidence for the formation of covalent bonds between the chlorophenol transformation product and humic acid. Covalent binding is considered the strongest type of bonding, and therefore, together with the dehalogenation process, is a desired reaction for the removal of chlorinated phenols, and other aromatic xenobiotics from the environment.

Phenol oxidases are produced by white rot fungi, are present in terrestrial systems and enriched in fungal-based remediation systems (Stevenson 1994). The ability of oxidases to mediate oxidative coupling reactions has been demonstrated in a number of experiments in which selected xenobiotics were combined with humic monomers or natural humic acids in the presence of a phenoloxidase. Rüttmann-Johnson and Lamar (1996) reported that high molecular weight humic polymers were produced when a reaction mixture containing PCP and fulvic acid (a humic/fulvic acid precursor), in the presence of a surfactant and $H_2O_2$, were mixed with a crude, concentrated supernatant from cultures of *P. chrysosporium*. Pure polyphenoloxidases (manganese-dependent peroxidase, lignin peroxidase, and laccase) also catalyzed the reaction. Humification of oxidized transformation products of 2,4-DCP and other chlorinated phenols in the presence of humic acid precursors (e.g. syringic acid) and laccase was demonstrated by Bollag and colleagues (1980). They showed the production of hybrid polymers in which the aromatic ring of oxidized 2,4-DCP becomes an integral part of the humic polymer. As the complexity and the size of humic polymers increases, the stability of the aromatic ring that was associated with the original xenobiotic also increases, as additional covalent bonds are formed with other aromatic structures through further oxidative polymerization reactions.

While *in vitro* experiments do not provide a direct evidence for covalent binding, they do demonstrate that fungal phenol oxidases are capable of mediating the covalent bonding of xenobiotic transformation products with humic materials. Direct evidence for the covalent nature of bonds formed between oxidized transformation products of chlorinated phenols and humic compounds produced in phenol oxidase-mediated reactions, comes from analyses of the hybrid polymers by MS and NMR spectrometric methods. Hatcher and colleagues (1993) investigated the horseradish peroxidase-mediated bonding of $^{13}C$-labeled 2,4-DCP to natural humic acid. After incubation, the humic acid was analyzed using $^{13}C$ NMR spectroscopy. The NMR spectrum of the humic acid displayed nine major signals that did not appear in the spectrum of the free 2,4-DCP that was just mixed with the humic acid. Evaluation of the signals revealed that the oxidized product of the 2,4-DCP molecules were covalently bound to humic acid through ester, phenolic ether and carbon-carbon linkages.

Covalent bonding of the oxidized transformation products of aromatic xenobiotics, like the *p*-chlorobenzoquinones of chlorophenols, to humic materials, a process that results in the aromatic structures of the original xenobiotics becoming a part of the structure of natural humic polymers, eliminates the bioavailability and thus the toxicity of the original xenobiotics. The decreased bioavailability is a result of the high molecular weight of the polymers which makes them to large to pass through cell membranes. Two phases are observed during the microbial turnover of natural biopolymers in soils. A first phase involving rapid metabolism of parent compounds (e.g. lignin, cellulose, and in contaminated soils, aromatic xenobiotics) is followed by a second phase of slow turnover of derived humic material. The mineralization of formed humic substances decreases to mean turnover rates of 2-7% per year after 250-300 days (Alexander 1977; Führ et al. 1985). With increasing age of the material, the formed molecules become increasingly inert, leading to humic substances with long-term stability. Age analysis of humic substances revealed residence times of several hundred years (Haider 1996). Eschenback and colleagues (1998) investigated the fate and stability of non-extractable residues of $^{14}C$-labeled PAHs in contaminated soils under environmental stress conditions. The work described in their paper is worth examining in detail, because it directly addressed the long-term stability of humified aromatic xenobiotics in soils. The experiments were conducted using non-extractable [$^{14}C$]-PAH residues that were produced in the preceeding long-term bioremediation experiments using the white rot fungus *P. ostreatus* (Eschenback et al. 1995). Soil samples from these experiments bearing non-extractable residues of oxidized transformation products of [$^{14}C$]-naphthalene, [$^{14}C$]-anthracene, [$^{14}C$]-pyrene, or [$^{14}C$]-benzo[a]pyrene were treated by biological, physical, or chemical treatments. The effect of the various treatments was assessed by comparing first, $^{14}CO_2$ mineralized; second, extractable $^{14}C$-activity; and third, $^{14}C$ associated with non-extractable residues in treated and non-treated soils. Biological treatments consisted of inoculating the soils with selected humus-degrading fungi or bacteria, or amending the soils with easily metabolized carbon sources (glucose and starch to initiate a "priming" effect of indigenous soil microbes). Neither, the addition of humus-degrading microbes, or easily metabolized carbon sources, led to an increase in $^{14}CO_2$ or extractables and thus potentially mobile [$^{14}$C]-PAH residues. The transformation activity in the various treated soils (including non-treated soils), as based on mineralization activity, approached, in all cases, similar levels of continuous but very low $^{14}CO_2$ release during the first 100 days of incubation. More importantly, the rate of mineralization (release of $^{14}CO_2$) was comparable to humus turnover rates in soil (2-5 % per year (Saxena and Bartha 1983)). This information, coupled with no change or decreases in the amount of extractable $^{14}$C, indicated that the $^{14}$C, that was released from non-extractable residues was rapidly metabolized. Physical stress factors including frost, rapid temperature variations, and drying and rewetting of soil also did not have any remarkable effect on the stability of the non-extractable $^{14}$C residue fraction. The effect of a chemical change on the stability of the non-extractable [$^{14}$C]-PAH residues was evaluated by using a complexing agent (EDTA) to disrupt the metal-organic complex. EDTA treatment was accomplished by extracting dry soil samples containing non-extractable [$^{14}$C]-PAH residues, with EDTA solutions of varying concentrations. Application of EDTA did result in the release of $^{14}$C-activity into the soil solution from the non-extractable fraction. It was shown that this activity was due to dissolved organic matter-[$^{14}$C]-PAH residue complexes that were released as a result of EDTA complex of metals, thus disrupting metallo-humic complexes and releasing [$^{14}$C]-residues into the soil solution. This type of treatment would be extremely unlikely to occur naturally or even as a result of anthropogenic activities. As discussed above, humic materials in soil are extremely stable over time, their stability increases with age and they are extremely difficult to mobilize due to their high molecular weights and hydrophobicity.

It has been shown in previous studies both in the laboratory and in field situations with *P. chrysosporium* and *P. sordida* that a small percentage of the decrease in the amount of PCP was the result of fungal methylation of PCP to pentachloroanisole (PCA). Both bacteria and fungi have been shown to methylate chlorophenol compounds to their methylated derivatives. Chung and Aust (1995) speculated that methylation of PCP to PCA may be a detoxification mechanism since PCA is not an inhibitor of oxidation phosphorylation and less toxic than PCP to wood rotting fungi, other microbes and fish. The solubility of PCA (~0.2 ppm) is less than that of PCP (2.5 ppm), thus preventing the contamination of groundwater. Lamar and Dietrich (1990) found that only about 9 to 14% of the PCP was converted to PCA. Thus, methylation was not the major route of PCP depletion in the contaminated soil. It was reported that the PCA was accumulated by these fungi during the initial bioremediation period after which it decreases slowly (Lamar and Dietrich 1990; Lamar et al. 1990b and 1993). The use of fungal stains, that do not produce significant amounts of PCA, would be advantageous for bioremediation purposes. Only trace amounts of PCA and 2,3,4,6-tetrachloroanisol (2,3,4,6-TeCA) were formed during the laboratory based remediation of PCP contaminated soils with the fungus *Trametes versicolor* (Tuomela et al. 1999). A part of the $^{14}$C-label was alkali-extractable, indicating that it was bound to humic substances, but this in part was apparently later attacked and mineralized by the fungus. The production of the enzyme laccase by this fungus was thought to enhance the degradation of PCP to other compounds rather than PCA (Tuomela et al. 1999). Lestan and colleagues (1996) also found a negative correlation between manganese peroxidase activity and PCA production by *T. versicolor*, indicating that this enzyme may also be involved in desirable PCP removal from the contaminated sites.

## 2.3 Biodegradation of Dioxins and Furans

White rot fungi have also been shown to degrade PCDDs and PCDFs in aqueous culture (Bumpus et al. 1985; Valli et al. 1992; Takada et al. 1996; Rosenbrock et al. 1997). Rosenbrock et al. (1997) evaluated the mineralization of undifferentiated dibenzo-*p*-dioxins in soil using four species of white rot fungi (*P. chrysosporium*, *Pleurotus* sp., *Dichomitus squalens*, and an unidentified fungus isolated from the site). They found that mineralization of PCDDs was greatly enhanced by inoculation with white rot fungi. Over a 70 day period, the extent of mineralization varied from 30 to 55% depending on the soil and the fungal species. Valli and Gold (1991) also found that both 2,4 dichlorophenol and 2,4,5-trichlorophenol may be degraded by *P. chrysosporium* by a complex pathway, involving oxidative displacement of chloride and O-methylation with the formation of 1,2,4,5-tetrahydroxybenzene before ring fission.

## 2.4 Fungal-based Remediation for Treatment of Contaminated Soil in New Zealand

In a scientifically-controlled experiment, the ability of fungal-based remediation to decrease the concentration of PCP and selected dioxin and furan congeners was evaluated in soil samples collected from a former dip tank wood-treating operation in Whakatane, New Zealand. The study was conducted using two New Zealand strains of white rot fungi, and two other fungi which were isolated from soil at the site (provided from The University of Waikato Fungal Culture Collection). In addition, the degradation was evaluated, for the purpose of comparison, using one strain of a fungus typically used for the remediation of similar compounds in the United States. All fungal strains were grown on locally available lignocellulosic substrates (*Pinus radiata* or *Eucalyptus* sp. wood chips). Technology performance was based on the percent decreases in the concentrations of 1,2,3,4,6,7,8-heptachlorodibenzo furan (HpCDF), 1,2,3,4,6,7,8-heptachlorodibenzo dioxin (HpCDD), and octachlorodibenzo dioxin (OCDD). These contaminants will be collectively referred to as the “analytes.” The following factors were evaluated: fungal species, inoculum application rate, and surfactant addition. The effectiveness of the various treatments was evaluated by determining the concentrations of the analytes immediately after treatment application, after 14, 28 and 56 days of treatment.

A "representative" soil sample was obtained from the site. The soil was air-dried and sieved to pass a 2-mm screen and thoroughly mixed. The soil was then stored dry in a sealed container until use. The concentrations of target chemicals were determined on soil sub-samples using appropriate extraction and analytical techniques. As the regulatory drivers in this soil were the dioxin congeners and these compounds are extremely hydrophobic, we evaluated the effect of amending the contaminated soil with a surfactant to enhance their degradation. The surfactant that was evaluated was emulsified soybean oil (ESO). The ESO was mixed with the water that was used to adjust the moisture content of the soil to provide homogeneous distribution of the surfactant and was applied at a rate of 3% (weight of oil to dry weight of soil). A total of five fungal species were evaluated. The four New Zealand strains (provided by The University of Waikato Fungal Culture Collection) were *Phanerochaete gigantea*, *Resinicium bicolor*, and two fungi isolated from the site soil referred to as the "East side" and "West side" strains based on where the soil they were isolated from was in relation to the former dip tank location. For comparative purposes, we also evaluated a United States strain, *Pleurotus ostreatus*. The latter fungus has a demonstrated ability to degrade PCP and dioxins in soil. Fungal inoculum was prepared by cultivating pure cultures of each of the fungi on sterilized *P. radiata* and/or *Eucalyptus* sp. wood chips. The moisture content of the chips was adjusted to 60% (wet weight basis) and then they were sterilized by autoclaving at 15 psi and 121°C for one hour on two successive days. The chips were inoculated with a mycelial slurry inocula produced from liquid cultures (2% glucose and 2% malt extract) of each fungal species. The inoculated chips were incubated at 30°C until they were thoroughly colonized (about 2 weeks).

The soil treatments were conducted in 8 oz. (272 ml) canning jars with lids modified to allow adequate air exchange. Each jar contained approximately 30 g of the test soil (i.e. wet weight) and the appropriate amount of fungal inoculum and amendments. Three replicates were prepared for each treatment for each sample time, with the exception for day 0. For day 0, a sample was prepared on the side for each treatment from which 2 sub-samples were taken for analysis. The cultures were incubated at 30°C (this would be the optimum biopile temperature) under high relative humidity to prevent moisture loss. Soil moisture content was maintained as needed. Target compound concentrations were evaluated on the following days: 0, 14, 28, and 56.

Soil and soil inoculum mixtures from each experimental unit were air-dried in plastic weigh boats and then ground to a fine powder using a commercial coffee grinder. The ground samples were stored dry in sealed glass containers. To determine the concentrations of PCP, HpCDD, HpCDF, and OCDD, 3 g sub-samples from each sample were extracted with a 50:50 mixture of hexane and acetone with a Dionex Accelerated Solvent Extractor. Sub-samples of the extracts were then analyzed using GC/ECD methods to determine extract concentrations of the analytes. PCP was analyzed as the trimethylsilyl derivative. PCP in extract sub-samples was derivitized using Sylon BTZ (Supelco Co.). GC/ECD analyses of derivatized extracts were performed on a Hewlett-Packard model 5890 gas chromatograph equipped with a $^{63}$Ni electron capture detector, a model 7673A autosampler, and a split-splitless capillary column injection port. Gas flows were: column flow 2 ml/min; total flow 60 ml/min. Operating temperatures were: 220°C (injector) and 300°C (detector); the carrier and makeup gas was nitrogen. The column was a DB-5 fused silica capillary column (30 m by 0.321mm; film thickness 0.25 um). The temperature program was as follows: initial 60°C; hold for 1 min; split off for 0.5 min; ramp A, 10°C/min for 9 min (60 to 150°C); ramp B, 2°C/min for 20 min (150 to 190°C); hold at 190°C for 5 min. GC/ECD analysis of extracts for HpCDD, HpCDF and OCDD were performed on the same instrument using the following conditions: Gas flows were: column flow 2 ml/min; total flow 30 ml/min. Operating temperatures were: 280°C (injector) and 300°C (detector); the carrier and makeup gas was nitrogen. The column was a DB-5 fused silica capillary column (30 m by 0.321mm; film thickness 0.25 um). The temperature program was as follows: initial 185°C; hold for 2 min; split off for 0.5 min; ramp A, 8°C/min for 8 min (85 to 285°C); hold at 285°C for 8 min.

Analyses of variance (ANOVA), using a = 0.05, were performed on the percent difference between concentrations of the analytes on day 0 and day 56. The main effects included in the ANOVA were fungal treatment, inoculum application rate and surfactant addition.

Initial concentrations after treatment applications are given in Table 1. There was significant variation in initial analyte concentrations among the treatments for all four analytes. This is an indication of the heterogeneity of the soil with respect to contaminant concentrations.

Fungal inoculation had a significant effect on the mean percent decreases of all four analytes among fungal inoculation treatments (Table 2). In all cases, inoculation with any of the tested fungi resulted in a significantly greater decrease than no inoculation (control). Among the tested fungi, the greatest percent PCP decrease occurred in soils inoculated with the "East side" strain. There were no significant differences among the fungal treatments in the degradation of HpCDF and HpCDD. Average percent decrease of these compounds was greater than 90% in all fungal treatments. Degradation of OCDD was greatest in soils inoculated with *P. gigantea* (Table 2). The percent OCDD decrease in all other fungal inoculated soils was less, significantly so, in soils inoculated with *R. bicolor*.

**Table 1.** Initial concentrations of PCP, HpCDF, HpCDD, and OCDD immediately after treatment application

| Treatment | PCP (mg/kg) | HpCDF (μg/kg) | HpCDD (μg/kg) | OCDD (μg/kg) |
|---|---|---|---|---|
| Control | 83 | 313 | 135 | 472 |
| *P. ostreatus* | 92 | 262 | 189 | 508 |
| "East side" | 182 | 340 | 351 | 743 |
| "West side" | 115 | 378 | 331 | 1045 |
| *R. bicolor* | 154 | 323 | 307 | 644 |
| *P. gigantea* | 136 | 356 | 380 | 792 |

**Table 2.** Effect of fungal inoculum and control treatments on mean[1] percent decrease of PCP, HpCDF, HpCDD, and OCDD after 56 days of treatment

| Treatment | PCP | HpCDF | HpCDD | OCDD |
|---|---|---|---|---|
| Control | 15.6c | 5.4b | (33.3)b | (22.4)c |
| *P. ostreatus* | 75.2b | 98.5a | 97.8a | 82.1ab |
| "East side" | 90.3a | 95.7a | 95.9a | 69.3ab |
| "West side" | 75.7b | 97.0a | 92.4a | 81.0ab |
| *R. bicolor* | 83.5ab | 95.0a | 91.6a | 68.2b |
| *P.gigantea* | 76.6ab | 93.7a | 91.5a | 86.2a |

[1]Means within columns followed by the same letter are not significantly different

Mean concentrations of all four analytes after 56 days of treatment were significantly less in fungal inoculated treatments compared to control treatments (Table 3). The lowest residual PCP concentration occurred in soils inoculated with the "East side" fungus. There were no significant differences among the fungal treatments in residual concentrations of HpCDF and OCDD. The lowest residual concentration of HpCDD occurred in soil inoculated with *P. ostreatus.* However, as with HpCDF, and OCDD, all the fungal treatments resulted in very extensive decreases in the concetration of HpCDD. The rate of fungal inoculation did not have a significant effect on the average percent decrease of any of the four analytes (Table 4). Application of ESO had no effect

**Table 3.** Mean[1] fungal inoculum treatment concentrations of PCP, HpCDF, HpCDD, and OCDD after 56 days of treatment

| Treatment (mg/kg) | PCP (μg/kg) | HpCDF (μg/kg) | HpCDD (μg/kg) | OCDD (μg/kg) |
|---|---|---|---|---|
| Control | 70c | 263b | 264c | 557b |
| *P. ostreatus* | 28b | 4a | 3a | 98a |
| "East side" | 13a | 12a | 12ab | 210a |
| "West side" | 28b | 15a | 30b | 196a |
| *R. bicolor* | 22ab | 14a | 24b | 188a |
| *P. gigantea* | 32b | 21a | 30b | 95a |

[1]Means followed by the same letter are not significantly different

**Table 4.** Effect of inoculum application rate on mean treatment percent decrease for inocuolum application rate of PCP, HpCDF, HpCDD, and OCDD after 56 days of treatment

| Inoculum application rate (wt inoc/wt soil) | Decrease (%) | | | |
|---|---|---|---|---|
| | PCP | HpCDF | HpCDD | OCDD |
| 10 | 83.7 | 96.1 | 92.6 | 78.9 |
| 20 | 77.1 | 95.9 | 95 | 75.8 |

on the mean inoculum application rate percent decrease of PCP, but significantly decreased the percent degradation of HpCDF, HpCDD, and OCDD (Table 5).

**Table 5.** Effect of ESO application rate on mean percent decrease of PCP, HpCDF, HpCDD, and OCDD after 56 days of surfactant treatment

| ESO addition rate | Decrease (%) | | | |
|---|---|---|---|---|
| | PCP | HpCDF | HpCDD | OCDD |
| 0 | 71.8a | 91.3a | 82.9a | 79.0a |
| 3 | 76.8a | 84.2b | 81.6b | 57.6b |

The treatment combination, that resulted in the greatest overall percent decreases of the four analytes (386.4%), was inoculation with *P. ostreatus* using an inoculum application rate of 10% and augmentation of the soil with 3% ESO (Table 6). The second most effective treatment, with a total percent decrease for the four analytes of 371.7%, was inoculation with the "East side" strain at an inoculum application rate of 10% in the presence of 3% ESO (Table 6). Based on the degradation of PCDD/PCDFs only, the most effective treatments were inoculation with *P. ostreatus* at a rate of 10% with or without ESO and inoculation with the "West side" isolate at a rate of 20% with or without ESO. Because similar results were obtained with or without ESO, it would not be necessary to use it in the field.

## 3. Conclusion

Inoculation of PCP/PCDD/PCDF-contaminated soils with selected isolates of white rot fungi and fungi from New Zealand, grown on locally available radiata pine or eucalyptus pulpwood chips, resulted in the rapid and extensive decreases in the concentrations of the contaminants. In particular, treatment with either of two fungal species isolated from PCP/PCDD/PCDF-contaminated soil from around the former dip tank at the Whakatane site, effectively decreased the concentrations of the PCP, HpCDF, HpCDD, and OCDD. Based on these results, the use of fungal-based remediation of the treatment of New Zealand soils contaminated with PCPs and associated PCDDss/PCDFs has

**Table 6.** Percent decreases in PCP, HpCDF, HpCDD, and OCDD concentrations in fungal inoculation/inoculum application rate/surfactant addition rate treatments

| Treatment | PCP | HpCDF | HpCDD | OCDD | % sum A1[1] | %sum B2[2] |
|---|---|---|---|---|---|---|
| No inoculation/0/0 | 13.6 | 45.6 | (12.4) | 6.7 | 53.5 | 39.9 |
| No inoculation/0/3 | 17.6 | (34.8) | (54.3) | (51.5) | (123.0) | (105.4) |
| *P. ostreatus*/10/0 | 77.3 | 99.4 | 96.4 | 86.0 | 359.1 | 281.8 |
| *P. ostreatus*/10/3 | 87.1 | 99.8 | 99.8 | 99.7 | 386.4 | 299.3 |
| *P. ostreatus*/20/0 | 65.2 | 96.2 | 91.3 | 93.0 | 345.7 | 280.5 |
| *P. ostreatus*/20/3 | 71.2 | 98.3 | 99.8 | 49.5 | 318.8 | 247.6 |
| East side/10/0 | 98.3 | 96.9 | 91.5 | 64.5 | 351.2 | 252.9 |
| East side/10/3 | 99.7 | 98.1 | 97.2 | 76.7 | 371.7 | 272.0 |
| East side/20/0 | 77.2 | 94.2 | 97.1 | 83.9 | 352.4 | 275.2 |
| East side/20/3 | 86.2 | 93.5 | 97.6 | 52.2 | 329.5 | 243.3 |
| West side/10/0 | 75.5 | 94.0 | 89.6 | 85.3 | 344.4 | 268.9 |
| West side/10/3 | 82.5 | 98.2 | 93.1 | 52.3 | 326.1 | 243.6 |
| West side/20/0 | 70.1 | 99.4 | 96.5 | 90.6 | 356.6 | 286.5 |
| West side/20/3 | 74.8 | 96.4 | 90.3 | 95.8 | 357.3 | 282.5 |
| *R. bicolor*/10/0 | 73.8 | 97.3 | 92.7 | 87.6 | 351.4 | 277.6 |
| *R. bicolor*/10/3 | 88.7 | 90.7 | 86.5 | 54.9 | 320.8 | 232.1 |
| *R. bicolor*/20/0 | 92.3 | 95.9 | 91.2 | 74.0 | 353.4 | 261.1 |
| *R. bicolor*/20/3 | 79.4 | 96.2 | 96.2 | 56.2 | 328.0 | 248.6 |
| *P. gigantea*/10/0 | 72.2 | 89.3 | 84.1 | 99.8 | 345.4 | 273.2 |
| *P. gigantea*/10/3 | 77.7 | 96.6 | 95.0 | 73.4 | 342.7 | 265.0 |
| *P. gigantea*/20/0 | 74.9 | 96.1 | 94.2 | 89.8 | 355.0 | 280.1 |
| *P. gigantea*/20/3 | 80.3 | 92.9 | 92.6 | 73.6 | 339.4 | 259.1 |

[1]the sum of the percent decreases of all four analytes
[2]the sum of the percent decreases of HpCDF, HpCDD, and OCDD

excellent potential. Work has been undertaken to demonstrate the effectiveness of fungal-based remediation, using the fungal strain 'Eastside', at pilot-scale and further developmental work is underway to upscale this technology for application at a full scale commercial basis. On the basis of investigations, following conclusions may be drawn:

1. Fungal inoculation greatly stimulated the degradation of PCP, HpCDD, HpCDF, and OCDD in the Whakatane soil.
2. Two New Zealand wood decay basidiomycetes and two unidentified fungi isolated from site soil performed similarly to a US strain (i.e. *P. ostreatus*) that has proven previously to be an effective degrader of PCP and PCDD/PCDFs.
3. Based on PCP-degrading performance alone, the most effective treatment was inoculation with the "East side" strain at a rate of 10% without addition of ESO.

4. Based on PCDD/PCDFs-degrading performance alone, the most effective treatment that included a New Zealand fungal strain was inoculation with the "West side" strain with or without ESO. It would not be worthwhile to use the ESO given the added cost in materials and labor.

Acknowledgements. We thank sincerely for his expertise and contributions to the project Harold H. Burdsall, Jr., PhD, Mycologist Expert of Black Earth Wisconsin, formerly with Forest Products Laboratory, Madison, Wisconsin, USA.

## References

Alexander M (1977) Introduction into Soil Microbiology. John Wiley & Sons, New York

Bollag J-M, Liu SY, Minard RD (1980) Cross-coupling of phenolic humus consituents and 2,4-dichlorophenol. Soil Sci Soc Am J 44:52-56

Bumpus JA, Tien M, Wright D, Aust SD (1985) Oxidation of persistent environmental pollutants by a white rot fungus. Science 228:1434-1436

Chung N, Aust SD (1995) Degradation of pentachlorophenol in soil by *Phanerochaete chrysosporium*. J Hazard Mater 41:177-183

Crosby DG (1981) Environmental chemistry of pentachlorophenol. Pure Appl Chem 53:1051-1080

Eaton DC (1985) Mineralization of polychlorinated biphenyls by *Phanerochaete chrysosporium*: a ligninolytic fungus. Enzyme Microb Tech 7:194-196

Eduljee G (1999) Secondary exposure to dioxins through exposure to PCP and its derivatives. Sci Total Environ 232:193-214

Eschenbach A, Weinberg B, Mahro B (1998) Fate and stability of non-extractable residues of [14C]-PAH in contaminated soils under stress conditions. Environ Sci Technol 32:2585-2590

Eschenbach A, Kastner M, Weinberg R, Mahro B (1995) Microbial PAH degradation in soil material from a contaminated site. In: van den Brink WJ, Bosman R, Arendt F (ed) Contaminated Soil '95, Kluwar Academic Publishers, The Netherlands, pp 377-378

Evans CS, Hedger JN (2001) Degradation of plant cell wall polymers In: Gadd GM (ed) Fungi in Bioremediation, Cambridge University Press, Cambridge, pp 1-26

Führ F, Kloskowski R, Burauel PW (1985) Bedeutung der gebundenen Rickstande In: Pflanzenschutzmittel im Boden (Bundesminister fur Ernahrung, Landwirtschaft und Forsten, Hrsg.), Z. Agrarpolitik Landwirtschaft 198. Paul Parey, Hamburg, pp 106-116

Haider K (1996) Biochemie des Boden. Enke, Stuttgart

Hatcher PG, Bortiatynski JM, Minard RD, Dec J, Bollag J-M (1993) Use of high-resolution 13C NMR to examine the enzymatic covalent binding of 13C-labeled 2,4-dichlorophenol to humic substances. Environ Sci Technol 27:2098-2103

Jennings DH, Lysek G (1999) Fungal Biology: Understanding the fungal lifestyle. BIOS Scientific Publishers Limited, Oxford

Joshi D, Gold MH (1993) Degradation of 2,4,5-trichlorophenol by the lignin-degrading basidiomycete *Phanerochaete chrysosporium*. Appl Env Microbiol 59:1779-1785

Kastner M, Streibich S, Beyrer M, Richnow HH, Fritsche W (1999) Formation of bound residues during microbial degradation of [$^{14}$C]anthracene in soil. Appl Env Microbiol 65:1834-1842

Kirk TK, Farrell RL (1987) Enzymatic combustion: the microbial degradation of lignin. Ann Rev Microbiol 41:465-505

Lamar RT, Davis MW, Dietrich DM, Glaser JA (1994) Treatment of a pentachlorophenol- and creosote- contaminated soil using the lignin-degrading fungus *Phanerochaete sordida*: A field demonstration. Soil Biol Biochem 26:1603-1611

Lamar, R.T, Dietrich, D.M. (1990) *In situ* depletion of pentachlorophenol from contaminated soil by *Phanerochaete* spp. Appl Env Microbiol 56:3093-3100

Lamar RT, Evans JW, Glaser JA (1993) Solid-phase treatment of pentachlorophenol-contaminated soil using lignin-degrading fungi. Environ Sci Technol 27:2566-2571

Lamar RT, Glaser JA, Kirk TK (1990a) Fate of pentachlorophenol (PCP) in sterile soils inoculated with the white-rot basidomycete *Phanerochaete chrysosporium*: mineralization, volatilization and depletion of PCP. Soil Biol Biochem 22:433-440

Lamar RT, Larsen MJ, Kirk TK (1990b) Sensitivity to and degradation of pentachlorophenol by *Phanerochaete* spp. Appl Env Microbiol 56:3519-3526

Lestan D, Lestan M, Chapelle JA, Lamar RT (1996) Biological potential of fungal inocula for bioaugmentation of contaminated soils. J Ind Microbiol 16:2566-2571

Okeke RC, Paterson A, Smith JE, Watson-Craik IA (1994) Relationships between ligninolytic actvities of *Lentinula* spp. and biotransformation of pentachlorophenol in sterile soil. Lett Appl Microbiol 19:284-287

Okeke RC, Smith JE, Paterson A, Watson-Craik IA (1993) Aerobic metabolism of pnetachlorophenol by spent sawdust culture of "Shitake" mushroom (*Lentinulus edodes*) in soil. Biotechnol Lett 15:1077-1080

Reddy CA (1995) The potential for white-rot fungi in the treatment of pollutants. Curr Opin Biotech 6:320-328

Rosenbrock P, Martens R, Buscot F, Zadrazil F, Munch JC (1997) Enhancing the mineralization of [U-$^{14}$C] dibenzo-p-dioxin in three different soils by the addition of organic substrate or inoculation with white rot fungus. Appl Microbiol Biot 48:665-670

Rüttmann-Johnson C, Lamar RT (1996) Polymerization of pentachlorophenol and ferulic acid by fungal extracellular lignin-degrading enzymes. Appl Env Microbiol 62:3890-3893

Rüttimann-Johnson C, Lamar RT (1997) Binding of pentachlorophenol to humic substances in soil by the action of white-rot fungi. Soil Biol Biochem 29:1143-1148

Saxena A, Bartha R (1983) Microbial mineralization of humic acid-3,4-dichloroaniline complexes. Soil Biol Biochem 15:59-62

Stevenson FJ (1994) Humus Chemistry. Genesis, Composition, Reactions, 2$^{nd}$ edn. John Wiley and Sons, New York

Stott DE, Martin JP, Focht DD, Haider K (1983) Biodegradation, stabilization in humus, and incorporation into soil biomass of 2,4-D and chlorocatechol carbons. Soil Sci Soc Am J 47:66-70

Takada S, Nakamura M, Matsueda T, Kondo R, Sakai K (1996) Degradation of polychlorinated dibenzo-p-dioxins and polychlorinated dibenxofurans by the white rot fungus *Phanerochaete sordida* YK-624. Appl Env Microbiol 62:4323-4328

Tuomela M, Lyytikäinen M, Oivanen P, Hatakka A (1999) Mineralization and conversion of pentachlorophenol (PCP) in soil inoculated with the white-rot fungus *Trametes versicolor*. Soil Biol Biochem 31:65-74

Valli K, Gold MH (1991) Degradation of 2,4-dichlorophenol by the lignin-degrading fungus *Phanerochaete chrysosporium*. J Bacteriol 173:345-352
Valli K, Wariishi H, Gold MH (1992) Degradation of 2,7-dichlorodibenzodixin by the lignin-degrading basidomycete *Phanerochaete chrysosporium*. J Bacteriol 174:2131-2137
Yu P, Shepherd J (1997) Pentachlorophenol in New Zealand – Biological treatment option. Australas Biotechnol 7:340-344

# 21

# Biofilms in Porous Media: Mathematical Modeling and Numerical Simulation

Benito M. Chen-Charpentier[1] and Hristo V. Kojouharov[2]

[1]Department of Mathematics, University of Wyoming, Laramie, WY 82071-3036, USA, bchen@uwyo.edu; [2]Department of Mathematics, University of Texas at Arlington, Arlington, TX 76019-0408, USA

## 1. Introduction

The use of microbes for control of organic contaminants in subsurface regions holds significant potential for *in situ* biorestoration strategies. A concept, which appears promising for the successful clean-up of contaminated aquifers, is the creation of biobarriers for containment and remediation of soil and ground water contaminated with organics and heavy metals. Biobarriers are *in situ* barriers that are formed by stimulating growth of biofilm-forming microbes introduced into the subsurface (James et al. 1995). Microbial biomass plugs the free pore space flow paths through porous media, thereby reducing the hydraulic conductivity and mass transport properties (Cunningham et al. 1991). Selective plugging of permeable strata may be used for preventing migration of ground water contaminants from hazardous waste sites. Simple nutritional differences may be used to deliver bacteria to any location in the subsurface environment. Experiments done by Cunningham et al. (1991) and numerical modeling done by Chen and Kojouharov (1999) show that it is possible to substantially reduce the hydraulic conductivity of the porous medium by adequately feeding the biofilm.

While subsurface biobarriers substantially control the movement of contaminants, they do not reduce it to zero as is desirable in practice. Recently, there have been some experiments done by Kolmos et al. (2000), where two different types of bacteria are combined to get better results. One type is a strong biofilm-forming bacteria and the other is a type of bacteria that reacts with the contaminant transforming it into harmless substances. The biofilm-forming bacteria are needed to form the biobarrier, so that the contaminant transport is reduced. That allows for the contaminant-degrading bacteria to establish themselves in the biobarrier and therefore be almost immobile and efficiently destroy the contaminant as it flows by. In real aquifers, the situation is complicated by the natural presence of protozoa. It is known that protozoa, such as ciliates and flagellates, eat bacteria (Berninger et al. 1991). Their predation on biofilms in porous media has been studied by Eisenman et al. (1998). They investigated the predation of bacteria attached to glass beads and determined the grazing rates. In order for the biobarriers to be useful, they need to persist even when grazed by protozoa. The determination of the conditions, under which the biobarriers keep functioning, is of utmost importance for their practical use.

The clean-up of contaminated subsurface regions cannot be effective without a thorough knowledge and understanding of the mechanisms for solute transport, biological and chemical reactions, and natural biodegradation. Mathematical modeling of biofilm systems is a very important tool in biofilm research and applications, and requires an understanding of the transport and accumulation processes of bacteria that occur in aqueous environments. Mathematical models, for flow, transport and biofilm accumulation in porous media, generally lead to strongly coupled systems of nonlinear partial differential equations. The objective is not only to develop accurate mathematical models, but also to develop reliable, accurate and efficient numerical methods for the given models. Without such methods, results of numerical simulations are of doubtful value.

In this paper, we model the water flow, the transport of nutrients and contaminants as well as the growth of biofilm-forming microbes and biodegradation microbes when predation by protozoa exists. In the next two sections, we discuss the physical system and present the corresponding mathematical models. In section four, numerical methods for solving the equations governing the fluid flow and the solute transport in porous media are presented. In section five, we show qualitative results of some single- and dual-species biobarrier simulations, accounting for the effects of protozoan grazing. The purpose and value of the numerical simulations are to guide future multi-species biofilm experiments that could lead to the design of more effective bioremediation strategies. In the last section, we present some conclusions and future research directions.

## 2. The Physical System

Processes governing mass transport, biofilm accumulation, and biotransformation of organic constituents are intrinsically interrelated (Chen and Kojouharov 1999; Larsen et al. 2000). In porous media flows, microbial cells exist in suspension, or get transported by convection and adsorbed firmly to solid surfaces. Some fraction of these adsorbed cells subsequently desorb, returning to suspension through some diffusion-like processes. If environmental conditions are favorable, the adsorbed cells grow on the surface, reproduce, increase the amount of attached biomass, and form an extracellular polymer matrix which binds the cells together. The entire deposit of attached cells and polymer substance, together with captured organic and inorganic particles, is termed biofilm (Allen 1988). As these cells develop into a continuous film, additional cells and particulate matter may attach to and detach from the biofilm surface. This increases the potential for sloughing of biofilm fragments, which may be subsequently entrapped by downstream pore opening. When biofilm thickness bridges across pore channels, accumulation of biofilm is further enhanced by the filtration of particulate matter from suspension. The net biofilm accumulation is, therefore, a result of the biomass substantial contribution from the processes of adsorption, filtration, growth, and attachment (Fig. 1) (Chen et al. 1994).

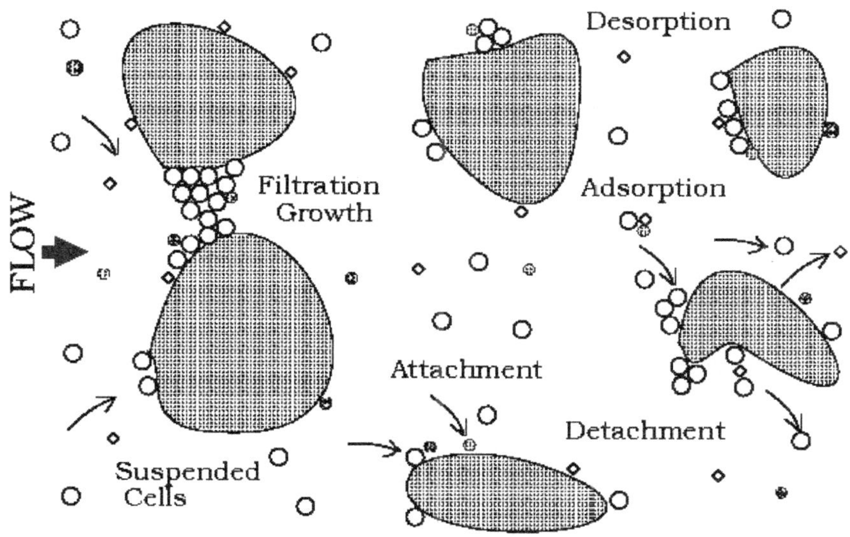


**Fig. 1.** Microbial processes in porous media

The rate of biotransformation of organic constituents is influenced by media mass transport and the fluid dynamic characteristics. As biofilm thickness increases, the effective pore space of the media will decrease, thereby causing a corresponding decrease in media porosity and permeability (Larsen et al. 2000); hydrodynamic dispersivities and molecular diffusivities, for microbes and nutrients, are also affected. Decreased pore velocity results in corresponding reduction of both the convective and dispersive delivery of nutrients and substrate to the growing cells. These, in turn, affect the biomass specific growth rate and thus the rate of biotransformation of organic material. Conversely, decreased pore velocity also reduces the rate of detachment of biofilm cells from the surface. Both biomass growth rate and detachment rate will continue to change unless a steady-state biofilm thickness is achieved.

## 3. The Mathematical Model

In order to model porous media multi-species biofilm interactions in the presence of protozoa, we consider a three-phase mixture consisting of a liquid phase, a solid rock phase and a biofilm phase. Even though the biofilm can be considered to be part of the solid phase, it is simpler to take it as a separate phase. The seven molecular species present in the porous medium are the trichloroethylene (TCE), the *Burkholderia cepatia* microorganism (a TCE-degrading bacteria unable to form a significant biofilm), the *Klebsiella oxytoca* microorganism (a strong biofilm-forming bacteria), organic carbon, biofilm-grazing protozoa, and the water and rock species (Table 1).

**Table 1.** Protozoa-microbes-nutrients interactions and phase mass transfers

| | Water | Protoza | Microbes | | Solutes | | Rock |
|---|---|---|---|---|---|---|---|
| | | | K.Oxytoca | B.Cepatia | TCE | Nutrients | |
| Liquid Phase | √ | √ | √ | √ | √ | √ | |
| Biofilm Phase | | √ | √ | √ | √ | √ | |
| Solid Phase | | | | | | | √ |
| Governing Equations | Flow | Transport and Growth | | | Transport | | |

The fundamental equation for saturated transient ground water flow of constant density, in horizontal direction, can be written in the form (Allen 1988):

$$S_s \frac{\partial h}{\partial t} - \frac{\partial}{\partial x}\left( K \frac{\partial h}{\partial x} \right) = f. \qquad (1)$$

The single fluid-flow equation (1) arises from the mass balance law

$$S_s \frac{\partial h}{\partial t} + \frac{\partial v}{\partial x} = f, \qquad (2)$$

when we substitute for the specific discharge vector $v$ using Darcy's law

$$v = -K \frac{\partial h}{\partial x}. \qquad (3)$$

Here $h$ denotes the hydraulic head, $S_s$ is the specific storage, $K$ is the saturated hydraulic conductivity, and $f$ represents sources or sinks. The specific discharge vector $v$, called superficial or Darcy velocity, represents the speed of the water.

The transport and reaction of nutrients (organic carbon) and contaminants (TCE), and the growth of the two microbial species and the protozoa are governed by a system of partial differential equations (Allen 1988). We assume that the two types of microbes and the protozoa are immobile, as part of the dual-species biofilm structure. Since the rock phase does not change, we assume that the solid rock matrix is stationary and that the diffusion of the two microbial, the protozoan, the nutrient and the contaminant species in the solid phase is negligible. Therefore, we can work only with the liquid and biofilm phases:

$$
\begin{aligned}
\frac{\partial}{\partial t}(\phi^{Bio}\rho_P) &= r_P(\rho_P,\rho_B,\rho_K,\rho_C,\rho_T),\\
\frac{\partial}{\partial t}(\phi^{Bio}\rho_B) &= r_B(\rho_P,\rho_B,\rho_K,\rho_C,\rho_T),\\
\frac{\partial}{\partial t}(\phi^{Bio}\rho_K) &= r_K(\rho_P,\rho_B,\rho_K,\rho_C,\rho_T), \qquad (4)\\
\frac{\partial}{\partial t}(\phi^{L}\rho_C)+\frac{\partial}{\partial x}(v\rho_C)-\frac{\partial}{\partial x}\left(D\frac{\partial\rho_C}{\partial x}\right) &= r_C(\rho_P,\rho_B,\rho_K,\rho_C,\rho_T),\\
\frac{\partial}{\partial t}(\phi^{L}\rho_T)+\frac{\partial}{\partial x}(v\rho_T)-\frac{\partial}{\partial x}\left(D\frac{\partial\rho_T}{\partial x}\right) &= r_T(\rho_P,\rho_B,\rho_K,\rho_C,\rho_T).
\end{aligned}
$$

Here $\rho_i$ $(i = P,B,K,C,T)$ represents the intrinsic mass density of the biofilm-grazing protozoa, the TCE-degrading microbes, the strong biofilm-forming microbes, the nutrients, and the contaminants (TCE), respectively. For a single-fluid flow, the quantity $\phi^L = V_L/(V_L + V_{Bio})$ and the quantity $\phi^{Bio} = V_{Bio}/(V_L + V_{Bio})$, where $V_L$ and $V_{Bio}$ represent the volumes occupied by the liquid and by the biofilm, respectively, are the portions of the void space occupied by the biofilm and the liquid, $D$ is the hydrodynamic dispersion coefficient, and $r_i$ represents the total rate at which species $i$ is produced via reactions and sources.

The protozoan and the two microbial death rates are assumed to be proportional to the size of the corresponding protozoan and microbial populations. The Monod equation is used to describe the kinetics of protozoan transformation of biofilm and of microbial transformations of nutrient and contaminant. The multi-substrate Monod expression $\mu^j$ $(j = B,K)$ is given by:

$$
\mu^j(S_1,S_2,\ldots,S_m) = \mu^j_{max}\prod_{i=1}^{m}\frac{S_i}{K^j_{S_i}+S_i}, \qquad (5)
$$

where $S_i$ is the concentration of the $i$-th transformed species, $\mu_{max}$ is the maximum specific growth rate, and $K_{S_i}$ is the species $S_i$ half saturation constant (Bailey and Ollis 1986).

### 3.1 Single-Species Biobarrier Model

In our first biobarrier model, we assume that there are no protozoa present in the medium and that only the growth and accumulation of the strong biofilm-forming microbes (*K. oxytoca*) in the pore spaces cause changes in the porous media properties. Let $\tilde{X}_B$ and $\tilde{X}_K$ be the current biodegradation and biobarrier-forming microbial concentrations, respectively, then $X_B=\tilde{X}_B/\rho_B$ and $X_K=\tilde{X}_K/\rho_K$ are the corresponding normalized microbial concentrations. It follows that the change in porosity, for small initial biobarrier-forming microbial concentrations (Clement et al. 1996), is given by

$$\phi(X_K) = \phi_0(1 - X_K), \tag{6}$$

where $\phi_0$ is the clean surface porosity. For the saturated hydraulic conductivity $K$, we assume the following form

$$K(X_K) = K_0(1 - X_K)^{n_k}, \tag{7}$$

where $K_0$ is the initial hydraulic conductivity and $n_k$ is an experimentally determined parameter which takes values around 3 (Clement et al. 1996). Furthermore, we assume that direct interactions in the system occur only between the biobarrier-forming microbial and nutrients species, and between the biodegradation microbial and contaminants species.

Invoking all simplifying assumptions to Equations (4) and using normalized concentrations as the unknowns yields the following governing system of differential equations:

$$\begin{aligned}
\frac{\partial X_B}{\partial t} &= \frac{\mu_{max}^B S_T}{K_{S_T}^B + S_T} X_B - k_B B, \\
\frac{\partial X_K}{\partial t} &= \frac{\mu_{max}^K S_C}{K_{S_C}^K + S_C} X_K G(X_K) - k_K X_K, \\
\frac{\partial S_C}{\partial t} + v\frac{\partial S_C}{\partial x} - \frac{\partial}{\partial x}\left(D\frac{\partial S_C}{\partial x}\right) &= -\frac{1}{Y_K}\frac{\mu_{max}^K S_C}{K_{S_C}^K + S_C} X_K G(X_K), \\
\frac{\partial S_T}{\partial t} + v\frac{\partial S_T}{\partial x} - \frac{\partial}{\partial x}\left(D\frac{\partial S_T}{\partial x}\right) &= -\frac{1}{Y_B}\frac{\mu_{max}^B S_T}{K_{S_T}^B + S_T} X_B,
\end{aligned} \tag{8}$$

where $G(X) = \dfrac{1 - X}{1 - X + \gamma}$, with $\gamma$ typically small, is introduced to restrict the growth of biobarrier-forming microbes as the pores are being plugged (Fenchel

1986; Jones and Smith, 2000), $k_B$ and $k_K$ are the first-order microbial decay rates, $Y_B$ and $Y_K$ are the yield rate coefficients (Bailey and Ollis 1986), $h$ is the hydraulic head, $X_K$ is the normalized concentration of biobarrier-forming microbes, $X_B$ is the normalized concentration of biodegradation microbes, $S_C$ is the concentration of the nutrient, and $S_T$ is the concentration of the contaminant.

### 3.2 Biobarrier-Protozoa Model

In our second biobarrier model, we introduce the effects of the protozoa grazing on the biofilm. For simplicity, we assume that the protozoa prays only on the strong biofilm-forming microbes (*K. oxytoca*). Using the assumptions of the single-species biobarrier model in Section 1, we consider a governing system of equations that involves only the protozoa, the *K. oxytoca* microbes $X_K$ and the nutrient species $S_C$. This implies that the changes in porosity $\phi$ and hydraulic conductivity $K$ are given by Equations (6) and (7), respectively. We assume that interactions in the system occur only between the biobarrier-forming microbial, the protozoan and the nutrient species, with no direct interaction between protozoa and nutrients. Let $\tilde{P}$ be the current protozoa concentration, then $P=\tilde{P}/\rho_P$ represents the normalized protozoa concentration.

Invoking all simplifying assumptions to Equations (1) and (4) and using normalized concentrations as the unknowns, gives the final form of the governing system of differential equations:

$$\frac{\partial P}{\partial t} = \frac{\mu_{max}^{P} X_K}{K_{X_K}^{P} + X_K} P - k_P P,$$

$$\frac{\partial X_K}{\partial t} = \frac{\mu_{max}^{K} S_C}{K_{S_C}^{K} + S_C} X_K G(X_K) - \frac{1}{Y_P}\frac{\mu_{max}^{P} X_K}{K_{X_K}^{P} + X_K} P - k_K X_K, \tag{9}$$

$$\frac{\partial S_C}{\partial t} + v\frac{\partial S_C}{\partial x} - \frac{\partial}{\partial x}\left(D\frac{\partial S_C}{\partial x}\right) = -\frac{1}{Y_K}\frac{\mu_{max}^{K} S_C}{K_{S_C}^{K} + S_C} X_K G(X_K), \tag{10}$$

where $k_P$ is the first-order protozoan endogenous decay rate, $K_{X_K}^{P}$ is the half-saturation constant for the protozoa, and $Y_P$ is the protozoan yield rate coefficient (Bailey and Ollis 1986).

### 3.3 Dual-Species Biobarrier Model

In the dual-species biobarrier model, we assume a protozoa-free environment and that the growth and accumulation of both microbial species (*K. oxytoca* and *B. cepatia*) in the pore spaces cause changes in the porous media properties. Using normalized microbial concentrations, the changes in porosity and in saturated hydraulic conductivity, for small initial biobarrier-forming microbial concentrations (Clement et al. 1996), are given by

$$\phi(X_B, X_K) = \phi_0(1 - X_B - X_K), \qquad K(X_B, X_K) = K_0(1 - X_B - X_K)^{n_k} \tag{11}$$

respectively. Furthermore, we assume that direct interactions in the system occur between the strong biofilm-forming microbial and nutrients species, and between the TCE-degrading microbial, the nutrients and the contaminants species.

Incorporating the above simplifying assumptions into Equations (4) and using normalized concentrations as the unknowns, yields the following governing system of differential equations:

$$\frac{\partial X_B}{\partial t} = \mu_{max}^B \left( \frac{S_C}{K_{S_C}^B + S_C} \right) \left( \frac{S_T}{K_{S_T}^B + S_T} \right) X_B G(X_B + X_K) - k_B B \tag{12}$$

$$\frac{\partial X_K}{\partial t} = \frac{\mu_{max}^K S_C}{K_{S_C}^K + S_C} X_K G(X_B + X_K) - k_K X_K$$

$$\begin{aligned} \frac{\partial S_C}{\partial t} + v\frac{\partial S_C}{\partial x} - \frac{\partial}{\partial x}\left( D\frac{\partial S_C}{\partial x} \right) &= -\frac{1}{Y_K} \frac{\mu_{max}^K S_C}{K_{S_C}^K + S_C} X_K G(X_B + X_K) \\ &\quad - \frac{F}{Y_B} \mu_{max}^B \left( \frac{S_C}{K_{S_C}^B + S_C} \right) \left( \frac{S_T}{K_{S_T}^B + S_T} \right) X_B G(X_B + X_K) \end{aligned} \tag{13}$$

$$\begin{aligned} \frac{\partial S_T}{\partial t} + v\frac{\partial S_T}{\partial x} - \frac{\partial}{\partial x}\left( D\frac{\partial S_T}{\partial x} \right) &= \\ &\quad - \frac{1}{Y_B} \mu_{max}^B \left( \frac{S_C}{K_{S_C}^B + S_C} \right) \left( \frac{S_T}{K_{S_T}^B + S_T} \right) X_B G(X_B + X_K) \end{aligned}$$

where $F$ is the ratio of organic carbon to TCE consumed.

## 4. Numerical Solution Techniques

Equations (1)-(4) represent a coupled system of nonlinear, time-dependent ordinary and partial differential equations that are very difficult to solve numerically. One objective of the numerical simulation is to develop time-stepping procedures that are reliable, accurate and computationally stable. Different time-stepping ideas can be applied to solve the governing system of equations (Russell and Wheeler 1983). One such time-stepping approach, that we have adopted in our numerical simulations, is the sequential solution technique. The sequential method first solves implicitly for the Darcy velocity $v$ at the current time-level, by solving the fluid flow equation (1). Then the transport system (4) is solved implicitly for the species concentrations in a decoupled fashion (Ewing and Russell 1982). New values of porosity and permeability are then calculated and the cycle is repeated by calculating the new velocities.

### 4.1 The Fluid Flow Equation

Classical techniques for solving the fluid flow equation (1) include the standard finite difference and Galerkin finite-element methods applied on uniform spatial grids. The resulting linear algebraic systems are symmetric and positive definite, so one can solve for the approximate hydraulic head $\hat{h}$ using a variety of iterative numerical schemes. Having computed $\hat{h}$, one can differentiate numerically to obtain the approximate specific discharge

$$\hat{v} = -K(x,t)\frac{\partial \hat{h}}{\partial x}. \tag{14}$$

A major problem with those approaches is that the approximate Darcy velocity $\hat{v}$ is one order lower in spatial accuracy than the approximate hydraulic head. In groundwater contaminant hydrology, inaccurate velocities are of serious concern, since the hydraulic head appears in the species concentration equations (4) only through its velocity field. To overcome these difficulties, it is more appropriate to choose a numerical method that approximates the velocity field $v$ directly, such as the mixed finite-element method or cell-centered finite differences in space (Allen and Wang 1994).

The mixed finite-element methods, that we have adopted in our numerical simulations, use a different discretization approach than the classical numerical methods. Here, one solves the mass balance and Darcy's laws

$$\begin{aligned} S_s\frac{\partial h}{\partial t}+\frac{\partial v}{\partial x} &= f, \\ v &= -K\frac{\partial h}{\partial x}, \end{aligned} \tag{15}$$

simultaneously. The corresponding mixed finite-element method for solving the groundwater flow equations (15) is as follows: Find a pair $(\hat{v}, \hat{h}) \in \hat{U} \times \hat{Q}$ such that

$$\begin{aligned} \int_\Omega K^{-1}\hat{v}u - \int_\Omega \hat{h}\frac{\partial u}{\partial x} &= 0 \qquad \forall u\hat{U} \\ \int_\Omega qS_s\frac{\partial \hat{h}}{\partial t} + \int_\Omega q\frac{\partial \hat{v}}{\partial x} &= \int_\Omega qf \qquad \forall q\hat{Q} \end{aligned} \tag{16}$$

where $\hat{U}$ and $\hat{Q}$ are finite-dimensional subspaces of given Hilbert spaces $U$ and $Q$, respectively (Allen et al. 1992).

Among the simplest choices for subspaces are the lowest-order Raviart-Thomas spaces (Raviart and Thomas 1977) on uniform grids, where the "hydraulic head" space $\hat{Q}$ consists of piecewise-constant functions and the "velocity space" $\hat{U}$ is the space of functions that are piecewise-linear with respect to the uniform grid on $\Omega$. Adopting a lexicographic ordering of equations and unknowns, the mixed formulation (16) yields a linear system having the following block structure

$$\begin{bmatrix} A & N \\ N^T & \Delta t^{-1}M \end{bmatrix}\begin{bmatrix} V \\ H \end{bmatrix}^{n+1} = \begin{bmatrix} 0 \\ F \end{bmatrix}^{n+1} + \begin{bmatrix} 0 & 0 \\ 0 & \Delta t^{-1}M \end{bmatrix}\begin{bmatrix} V \\ H \end{bmatrix}^{n}. \tag{17}$$

The vector $V$ contains the nodal values of the specific discharge $v$, associated with the cell edges in the grid, and $H$ contains nodal values of the hydraulic head $h$, associated with cell centers (Allen et al. 1992). The block matrix $A$ is symmetric and positive definite, and contains information about the hydraulic conductivity $K$. The matrix $N$ is bidiagonal differencing matrix, $M$ is a diagonal storage matrix, and the vector $F$ contains integrals involving the source-sink term $f$.

Blockwise row reduction of Equation (17) yields the equations

$$\begin{aligned} (\Delta t^{-1}M - N^T A^{-1}N)H^{n+1} &= F^{n+1} + \Delta t^{-1}MH^n, \\ V^{n+1} &= -A^{-1}NH^{n+1}. \end{aligned} \tag{18}$$

The matrix $\Delta t^{-1}M - N^T A^{-1}N$ is symmetric and positive definite; however, solving the system in this form is impractical in large problems, since $\Delta t^{-1}M - N^T A^{-1}N$ is not sparse.

Many efficient iterative schemes have been developed for solving the system (17) and similar systems arising from other mixed methods

(Ewing et al. 1990). One iterative approach, that extends the scheme introduced by (Allen et al. 1992) for steady groundwater flows, uses the matrix splitting

$$\begin{bmatrix} D & N \\ N^T & \Delta t^{-1}M \end{bmatrix} \begin{bmatrix} V \\ H \end{bmatrix}^{n+1,k+1} = \begin{bmatrix} D-A & 0 \\ 0 & 0 \end{bmatrix} \begin{bmatrix} V \\ H \end{bmatrix}^{n+1,k} + \begin{bmatrix} 0 \\ R \end{bmatrix}^n, \tag{19}$$

where the preconditioning matrix $D$ is a diagonal matrix, $D = diag(A)$, and

$$\begin{bmatrix} 0 \\ R \end{bmatrix}^n = \begin{bmatrix} 0 \\ F \end{bmatrix}^{n+1} + \begin{bmatrix} 0 & 0 \\ 0 & \Delta t^{-1}M \end{bmatrix} \begin{bmatrix} V \\ H \end{bmatrix}^n. \tag{20}$$

Computationally, this iterative scheme allows one to solve a pentadiagonal matrix equation for the hydraulic head, instead of solving a system involving the full matrix $\Delta t^{-1}M - N^T A^{-1} N$. The scheme exhibits good convergence properties in the presence of fine spatial grids, and effectively handles variable coefficients $K$ (Allen et al. 1992). Investigation on the parallelism of this approach on distributed-memory computers can be found in the work of Allen and Curran (1992).

### 4.2 The Species Transport Equations

Consider the equations governing transient species transport in porous media (4) in the following form:

$$\frac{\partial c}{\partial t} + v\frac{\partial c}{\partial x} - \frac{\partial}{\partial x}\left(D\frac{\partial c}{\partial x}\right) = r(c). \tag{21}$$

Here, $c$ is the species concentration, $v$ is the velocity, $D$ is the hydrodynamic dispersion tensor, and $r(c)$ represents the nonlinear reaction term.

Classical numerical techniques, such as the standard finite-differences or Galerkin finite-elements, work well for the problems of solute transport that are dominated by the dispersive movement, they suffer from severe oscillations and excessive numerical dispersion when convection, associated with the velocity field $v$, dominates the dispersive effects. Upwind schemes were first used to stabilize convective flows by introducing numerical dispersion (Gray and Pinder 1976). In the streamline diffusion schemes (Johnson and Saranen 1986), the smoothing effects have been limited to the flow direction, in which their influence is much needed. Eulerian-Lagrangian methods have greatly improved time truncation errors, allowing for larger time steps to be taken without significant loss of accuracy in the numerical solution. Many such schemes have been developed, including the modified method of characteristics (Douglas and Russell 1982), the Eulerian-Lagrangian localized adjoin method (Celia et al. 1990), the finite volume Eulerian-Lagrangian localized adjoint method (Healy and Russell, 1993), and the modified method of characteristics incorporating streamline diffusion (Allen and Liu 1995), to name a few. However, still little has been done for numerical solutions of transport problems, in which nonlinear reactions are present. Reactions with unstable equilibria and thresholds can cause small numerical errors to oscillate with increasing amplitude, leading to eventual machine blowup (Liu et al. 1996). Nonlinear reaction terms play a significant role in applications involving bacterial growth and contaminant biodegradation in subsurface regions.

In our numerical simulations, we have adopted a time-splitting algorithm based on a nonstandard finite difference method that efficiently handles the numerically challenging transport equation (21). In the first step, the convection-reaction equation ( $D=0$ ) is solved using a nonstandard method (Kojouharov and Chen 1998, 2004; Chen and Kojouharov 1999; Kojouharov and Welfert 2001). It allows us to follow the transport and track sharp fronts much more accurately than with standard numerical schemes. In the second step of the time-splitting procedure, the diffusion part is computed using standard finite differences or finite elements.

### 4.2.1 Convection-Reaction Equations

We first consider the convection-reaction part of problem (21) with no dispersion ( $D=0$ ), subject to the initial condition $c(x,0)=g(x)$ and periodic boundary conditions. Our goal is to construct an "exact" time-stepping scheme for the equation:

$$\frac{\partial c}{\partial t}+v\frac{\partial c}{\partial x}=r(c). \tag{22}$$

To introduce the concept of "exact" time-stepping schemes, let us consider the following numerical scheme

$$C^n(x)=F(C^{n-1}(x),\Delta t,n), \tag{23}$$

where $\Delta t$ is the time step size and $C^n(x)$ is the numerical solution at time $n\Delta t$. Assume that it has a solution

$$C^n(x)=G(C^0(x),\Delta t,n)=G(g(x),\Delta t,n). \tag{24}$$

The numerical scheme is said to be an "exact" time-stepping scheme, if the relationship $C^n(x)=c(x,n\Delta t)$ holds for arbitrary time step size $\Delta t$ and at every spatial location $x$ (Mitchell and Griffiths, 1980).

**Nonstandard Finite-Difference Method.** As a first case (Kojouharov and Chen 1998), consider $r(c)=\lambda c$. The dimensionless form of the governing equation (22) becomes

$$\frac{\partial c}{\partial t}+v\frac{\partial c}{\partial x}=\lambda c. \tag{25}$$

Using the method of characteristics, the solution to the above equation (25) can be written as

$$c(x,t)=g(s)e^{\lambda t}, \tag{26}$$

with $s = s(x)$ and where $s = \xi(0)$ is the solution at time $t = 0$ of the initial-value problem

$$\frac{d\xi(\tau)}{d\tau} = v(\xi,\tau), \quad \xi(t) = x. \tag{27}$$

Assuming a constant velocity field $v(x,t) = v$, the solution of Problem (27) yields $s = x - vt$. Substitution of $s$ into the expression (26) yields

$$c(x,t) = g(x - vt)e^{\lambda t}. \tag{28}$$

The above expression (28) holds for arbitrary time $t$. Comparison of the analytical solution at time $t$ with the analytical solution at time $t + \Delta t$, where $\Delta t = t^{n+1} - t^n$, gives the following relationship:

$$c(x,t+\Delta t) = c(x - v\Delta t, t)e^{\lambda \Delta t}. \tag{29}$$

Based on it, we construct the "exact" time-stepping scheme

$$\frac{C^{n+1}(x) - C^n(\overline{x}^n)}{\dfrac{e^{\lambda\Delta t} - 1}{\lambda}} = \lambda C^n(\overline{x}^n), \tag{30}$$

where the backtrack point $\overline{x}^n$ has the following expression:

$$\overline{x}^n = x - v\Delta t. \tag{31}$$

The left-hand side of numerical scheme (30) can be viewed as a nonstandard backward difference approximation of the characteristic derivative

$$\frac{Dc}{Dt} = \frac{\partial c}{\partial t} + v\frac{\partial c}{\partial x}.$$

To complete the construction of the nonstandard numerical method, we apply the semi-discrete "exact" time-stepping scheme (30) at each spatial grid point $x_i$, which yields:

$$\frac{C_i^{n+1} - C^n(x_i{}^n)}{\dfrac{e^{\lambda\Delta t} - 1}{\lambda}} = \lambda C^n(x_i{}^n). \tag{32}$$

The backtrack point $\overline{x}_i^n$ is given by $\overline{x}_i^n = \xi(t^n)$ where $\xi$ is the solution of the initial-value problem (27) subject to the condition $\xi(t^{n+1}) = x_i$. For a constant velocity field $v(x,t) = v$, the backtrack point is given by $x_i{}^n = x_i - v\Delta t$.

**Remark**: *For arbitrary velocity fields $v(x,t)$, Problem (27) cannot be integrated exactly and an appropriate approximation of the backtrack point, e.g., $\overline{x}_i^n \approx x_i - v(x_i, t^{n+1})\Delta t^n$ [as in the modified method of characteristics (Douglas and Russell 1982)] must be used. Another possibility is to use the Euler method, the improved Euler method, or a Runge-Kutta method (Arbogast and Wheeler 1995) to solve the initial-value problem (27).*

As a second case (Kojouharov and Chen 1998), consider the logistic-growth reaction term of the form $r(c) = \lambda c(1-c)$. The general solution assumes the form

$$c(x,t) = \frac{g(s)}{e^{-\lambda t} + (1-e^{-\lambda t})g(s)}, \tag{33}$$

where $s = \xi(0)$ is the solution at time $t = 0$ of the initial-value problem (27). Comparison of the solution (33) at time $t$ with the solution at time $t + \Delta t$, at every spatial grid point, yields the nonstandard numerical method:

$$\frac{C_i^{n+1} - C^n(x_i{}^n)}{\frac{e^{\lambda\Delta t} - 1}{\lambda}} = \lambda C^n(x_i{}^n)(1 - C_i^{n+1}). \tag{34}$$

The left-hand side of the numerical scheme (34) represents the same non-standard backward difference approximation of the *characteristic* derivative, as in the linear case (32), while the right-hand side represents a nonlocal modeling of the nonlinear reaction term $r(c) = \lambda c(1-c)$.

As a third case (Chen and Kojouharov 1999), consider the Monod reaction term of the form $r(c) = \frac{\lambda c}{K + c}$, where $\lambda$ and $K$ are constants. The nonstandard finite-difference method is given by the expression

$$\frac{C_i^{n+1} - C^n(x_i{}^n)}{\Delta t} = \lambda - \frac{K}{\Delta t}\ln\left(\frac{C_i^{n+1}}{C^n(x_i{}^n)}\right). \tag{35}$$

As a fourth case (Kojouharov and Chen 2004), consider the following reaction term:

$$r(c) = \frac{\mu}{\sum_{j=0}^{N} a_j c^j}, \tag{36}$$

where $N$ is a positive integer, and $a_j$, $j = 0, \ldots, N$, and $\mu$ are real constants. The nonstandard finite-difference method has the form:

$$\frac{C_i^{n+1} - C^n(x_i{}^n)}{\Delta t} = \frac{\mu}{\sum_{j=0}^{N} a_j \left\{ \frac{\left(C_i^{n+1}\right)^{j+1} - \left(C^n(x_i{}^n)\right)^{j+1}}{(j+1)\left(C_i^{n+1} - C^n(x_i{}^n)\right)} \right\}}. \tag{37}$$

As a fourth case (Kojouharov and Chen, 2004), consider the following reaction term:

$$r(c) = \lambda c + \mu c^N, \tag{38}$$

where $N \neq 1$ is an integer number, and $\lambda$ and $\mu$ are real constants, the nonstandard finite-difference method has the form:

$$\frac{C_i^{n+1} - C^n(x_i^{\ n})}{\dfrac{e^{\lambda \Delta t} - 1}{\lambda}} = \lambda C^n(x_i^{\ n}) + \mu \frac{U(N)\left(C_i^{n+1}\right)^{N-1}\left(C^n(x_i^{\ n})\right)^{N-1} + U(-N)}{\dfrac{\left(C_i^{m+1}\right)^{|N-1|} - \left(e^{\lambda \Delta t} C^n(x_i^{\ n})\right)^{|N-1|}}{\sum_{k=0}^{|N-1|-1} e^{\lambda k \Delta t}\left(C_i^{n+1} - e^{\lambda \Delta t} C^n(x_i^{\ n})\right)}}, \tag{39}$$

where $U(N) = \begin{cases} 0, & N < 0 \\ 1, & N \geq 0 \end{cases}$ is the unit step function.

**Generalized Nonstandard Finite-Difference Method.** Consider the nonlinear reaction term of the form $r(c) = f(c)/f'(c)$, where $f$ is a given function, assumed without loss of generality to be nonnegative (Kojouharov and Welfert 2001). The motivation behind such a choice for $r(c)$ is the fact that the resulting nonlinear differential equation in $c$

$$\frac{\partial c}{\partial t} + v\frac{\partial c}{\partial x} = \frac{f(c)}{f'(c)} \tag{40}$$

reduces to the following linear in $f(c)$ differential equation

$$\frac{\partial f(c)}{\partial t} + v\frac{\partial f(c)}{\partial x} = f(c). \tag{41}$$

For example, in the case of polynomial reactions $r(c) = \prod_{k=1}^{m}(c - \alpha_k)$, with distinct $\alpha_k$'s, solving for the function $f$ yields

$$f(c) = \prod_{k=1}^{m} |c - \alpha_k|^{\gamma_k}, \qquad \gamma_k = \prod_{i=1, i\neq k}^{m} \frac{1}{\alpha_k - \alpha_i}. \tag{42}$$

Comparison of the analytic solution at times $t^n$ and $t^{n+1}$ yields the generalized nonstandard method (Kojouharov and Welfert 2001):

$$\frac{F_i^{n+1} - F^n(\overline{x}_i^n)}{e^{\Delta t} - 1} = F^n(\overline{x}_i^n). \tag{43}$$

Here, the quantity $F_i^{n+1} = f(C_i^{n+1})$ denotes the numerical approximation of $f(c(x_i, t^{n+1}))$ with

$$F_i^0 = f(C_i^0) = f(g(x_i)) \tag{44}$$

and $F^n(\overline{x}_i^n)$ is the numerical solution at $\overline{x}_i^n$. The numerical solution of (40) at time $t^{n+1}$ is then recovered from

$$C_i^{n+1} = f^{-1}\left(F_i^{n+1}\right), \tag{45}$$

where $f^{-1}$ is the inverse function of $f$. Note that because of its special form (42), $f$ is monotonic between two zeros/poles of $r$ so that $C_i^{n+1}$ is well-defined. The determination of $C_i^{n+1}$ from $F_i^{n+1}$ via (45) requires either an explicit expression for $f^{-1}$ or, in general, a numerical procedure (e.g., Newton-Raphson method) for solving $f(C_i^{n+1}) = F_i^{n+1}$.

**Remark**: *The above nonstandard (32),(34),(35),(37),(39) and generalized nonstandard (43) methods represent "exact" time-stepping schemes provided that an exact expression for the backtrack characteristic point $x_i^n$ is given at every spatial grid point $x_i$ and every time $t^n$, i.e., provided the initial-value problem (27) can be integrated exactly.*

### *4.2.2 Convection-Dispersion-Reaction Equations*

We now present a time-splitting method, analyzed in (Dawson and Wheeler 1992), for solving the convection-dispersion-reaction equation (21). The basic idea of a time-splitting approach is to treat processes like convection, dispersion, and reaction on their own in numerical time-stepping, so as to enable an easy use of well prepared, tailored solvers for these different processes.

The solution $c(x,t^{n+1})$ at time $t^{n+1}$ is determined from $c(x,t^n)$ as follows. First the function $c(x,t^n)$ is used as an initial condition $c^{a,r}(x,t^n) = c(x,t^n)$ for the solution of the convection-reaction equation

$$\frac{\partial c^{a,r}}{\partial t} + v\frac{\partial c^{a,r}}{\partial x} = r(c^{a,r}). \tag{46}$$

The solution $c^{a,r}(x,t^{n+1})$ generated from this step is then used as an initial data $c^d(x,t^n) = c^{a,r}(x,t^{n+1})$ for the dispersion equation

$$\frac{\partial c^d}{\partial t} - \frac{\partial}{\partial x}\left(D\frac{\partial c^d}{\partial x}\right) = 0. \tag{47}$$

Finally the new solution at time $t^{n+1}$ is defined by $c(x,t^{n+1}) = c^d(x,t^{n+1})$. For problems with small dispersion, i.e., for convection-dominated transport problems, this splitting approach leads to more accurate representation of the physics of the problem (Dawson and Wheeler 1992).

The numerical implementation of the time-splitting method is as follows. Assume that at time $t^n$ all $\{C_i^n\}_i$ are known. Set $C_i^{n+ar} = C_i^{n+1}$, where $C_i^{n+1}$ is the solution of Equation (46) using the nonstandard finite-difference method introduced in Section 1. In the second time-splitting step, we solve for $\{C_i^{n+1}\}_i$ the following implicit finite-difference scheme:

$$\frac{C_i^{n+1} - C_i^{n+ar}}{\Delta t} - \theta \frac{\partial}{\partial x}\left(D^{n+1}\frac{\partial c}{\partial x}\right)\bigg|_i^{n+1} - (1-\theta)\frac{\partial}{\partial x}\left(D^n \frac{\partial c}{\partial x}\right)\bigg|_i^{n+\frac{1}{2}} = 0, \tag{48}$$

where $\theta \in [0,1]$ is a given parameter and

$$\frac{\partial}{\partial x}\left(D^k \frac{\partial c}{\partial x}\right)\bigg|_j^m \approx \frac{1}{\Delta x}\left(D^k_{j+\frac{1}{2}} \frac{C^m_{j+1} - C^m_j}{\Delta x} - D^k_{j-\frac{1}{2}} \frac{C^m_j - C^m_{j-1}}{\Delta x}\right)$$

is an approximation of the dispersion term (Huyakorn and Pinder 1983) with $D^k_{j\pm\frac{1}{2}} = D\left(\dfrac{x_j + x_{j\pm1}}{2}, t^k\right)$.

## 5. Simulations

In our set of numerical simulations, we assume there are no sources and sinks for the fluid, therefore $f = 0$ in Equation (1). We also assume a piecewise steady-state fluid flow, due to the relatively slow changes in the porous media properties (Cunningham et al. 1991). Also we are modeling very short cores with uniform biofilm distribution, so we can take the velocity to be space-independent (Cunningham et al. 1991). Invoking the above simplifying assumptions to Equations (1) yields the following single-fluid flow equation:

$$-\frac{\partial}{\partial x}\left(K \frac{\partial h}{\partial x}\right) = 0, \tag{49}$$

which we numerically solve for $h$ and the velocity field $v$.

The values of the parameters used in all of the numerical experiments are summarized in Table 2.

### 5.1 Single-Species Biobarrier Model

To validate our mathematical model (8), we first simulate the porous media experiments done by Cunningham et al. (1991) for a $5\,cm$-long reactor packed with $0.70\,mm$, in diameter, sands in the absence of biodegradation microbes.

**Table 2.** Parameter values used in the numerical experiments

| Parameter | Value |
|---|---|
| Initial saturated hydraulic conductivity, $K_0$ | $0.2402\ cm/sec$ |
| Initial porosity, $\phi_0$ | $0.35$ |
| Hydrodynamic dispersion coefficient, $D$ | $5\times10^{-4}\ cm^2/sec$ |
| Protozoa decay coefficient, $k_P$ | $0.000024\ /sec$ |
| K. oxytoca decay coefficient, $k_K$ | $0.0002\ /sec$ |
| B. cepatia decay coefficient, $k_B$ | $0.0001\ /sec$ |
| Maximum specific growth rate, $\mu_{max}^P$ | $0.0000525\ /sec$ |
| K. oxytoca maximum specific growth rate, $\mu_{max}^K$ | $0.0104\ /sec$ |
| B. cepatia maximum specific growth rate, $\mu_{max}^B$ | $0.00527\ /sec$ |
| Protozoa yield coefficient, $Y_P$ | $0.00254\ prot/mic$ |
| K. oxytoca yield coefficient, $Y_K$ | $0.0975\ mic_K/nut$ |
| B. cepatia yield coefficient, $Y_B$ | $0.04875\ mic_B/cont$ |
| Half saturation constant, $K_{X_K}^P$ | $0.05184\ \mu g/ml$ |
| Half saturation constant, $K_{S_C}^K$ | $0.799\ \mu g/ml$ |
| Half saturation constant, $K_{S_C}^B$ | $0.0799\ \mu g/ml$ |
| Half saturation constant, $K_{S_T}^B$ | $0.0799\ \mu g/ml$ |
| Ratio constant, $F$ | $0.5\ nut/cont$ |
| Parameter, $\gamma$ | $0.1$ |
| Parameter, $n_k$ | $3$ |

For ease of calculations, the reactor's length is scaled to 1 and the nutrients' and contaminant's concentrations are scaled by a factor of $1/25$ for graphing purposes.

The initial conditions used in this simulation are:

$$S_C(x,0) = 20\ \frac{\mu g}{ml},\quad S_T(x,0) = 25\ \frac{\mu g}{ml},$$

$$X_K(x,0) = \begin{cases} 0.2, & 0.3 \le x \le 0.4 \\ 0, & \text{otherwise} \end{cases} \tag{50}$$

and the boundary conditions are:

$$h(0,t) = 0.5\ cm, \quad h(1,t) = 0\ cm,$$

$$S_C(0,t) = 20\ \frac{\mu g}{ml}, \tag{51}$$

$$S_T(0,t) = 25\ \frac{\mu g}{ml}, \quad \frac{\partial S_C}{\partial x}(1,t) = \frac{\partial S_T}{\partial x}(1,t) = 0\ \frac{\mu g}{ml \cdot sec}.$$

The boundary and initial conditions considered in the model are in agreement with Cunningham et al. (1991), the reaction parameters are taken from Taylor and Jaffé (1990) and Fenchel (1986), and the parameter $\gamma$ in the function $G$ is taken from Jones and Smith (2000). Numerical simulation results qualitatively agree with published experimental results by the Center for Biofilm Engineering at Montana State University-Bozeman (Cunningham et al. 1991) (see Figs. 2 and 3, above).

For our second simulation, Figures 2 and 3 (below), we introduce the biodegradation microbes, $X_B$, at $t = t^*$:

$$X_B(x,t^*) = \begin{cases} 0.05, & 0.3 \le x \le 0.4 \\ 0, & \text{otherwise} \end{cases},$$

after the biobarrier in the first simulation has stabilized, and observe the effects of the biodegradation microbes on the contaminants. The biodegradation microbe's concentration is scaled by a factor of 1/25 for graphing purposes.

The time scale in Figure 3 (below) is scaled by a factor of 1/100 for graphing purposes.

### 5.2 Biobarrier-Protozoa Model

To validate our mathematical model (9)-(10), we first simulate the same protozoa-free biofilm experiment, as in Section 1, done by Cunningham et al. (1991) [see Figures 4 and 5 (above)]. The nutrients' concentration in this simulation was scaled by a factor of 20 for graphing purposes. The initial (50) and boundary (51) conditions used are the same as in Section 1.

For our second simulation, we introduce the protozoa at time $t = t^*$:

$$P(x,t^*) = \begin{cases} 0.05, & 0.3 \le x \le 0.4 \\ 0, & otherwise \end{cases},$$

after the biofilm in the first simulation has stabilized, and observe the effects of the protozoa on the biofilm performance [see Figs. 4 and 5 (below)].

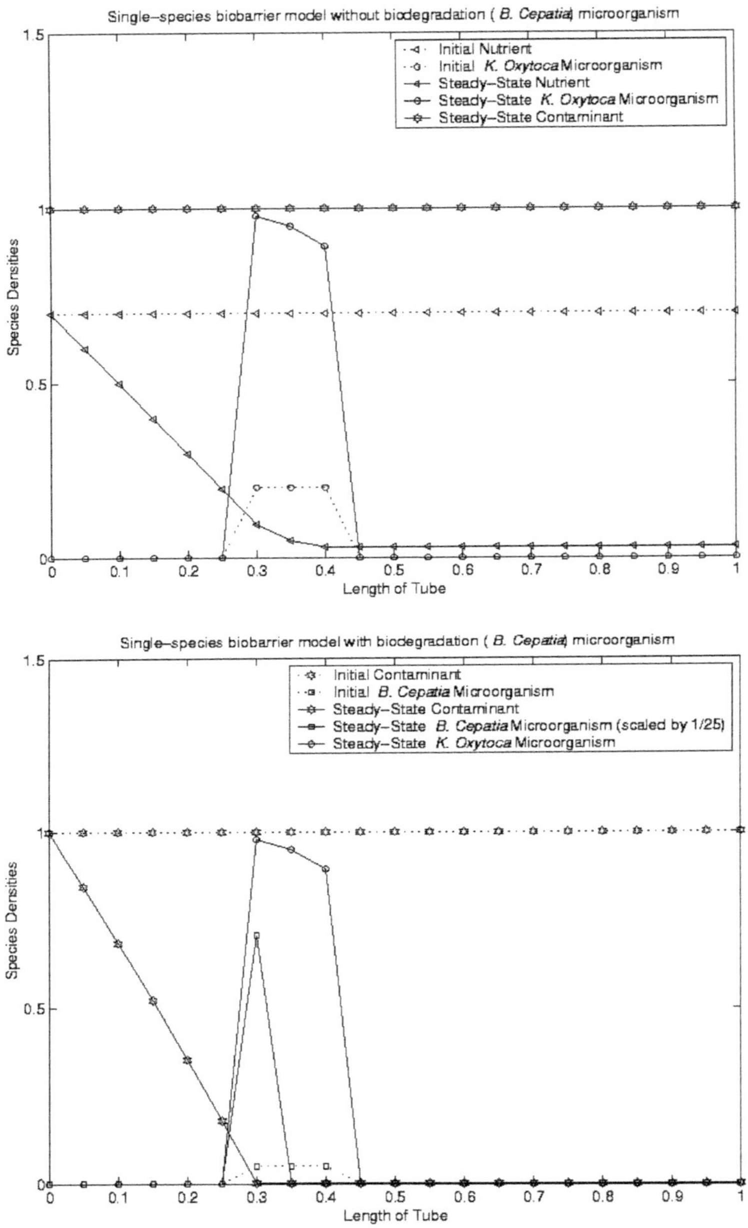


**Fig. 2.** Plots of the initial (dotted lines) and the steady-state (solid lines) normalized concentrations of the biobarrier-forming microbes, $X_K$, the nutrient, $S_C$, and the contaminant, $S_T$ — before (above) and after (below) the introduction of the TCE-degrading microbes into the system

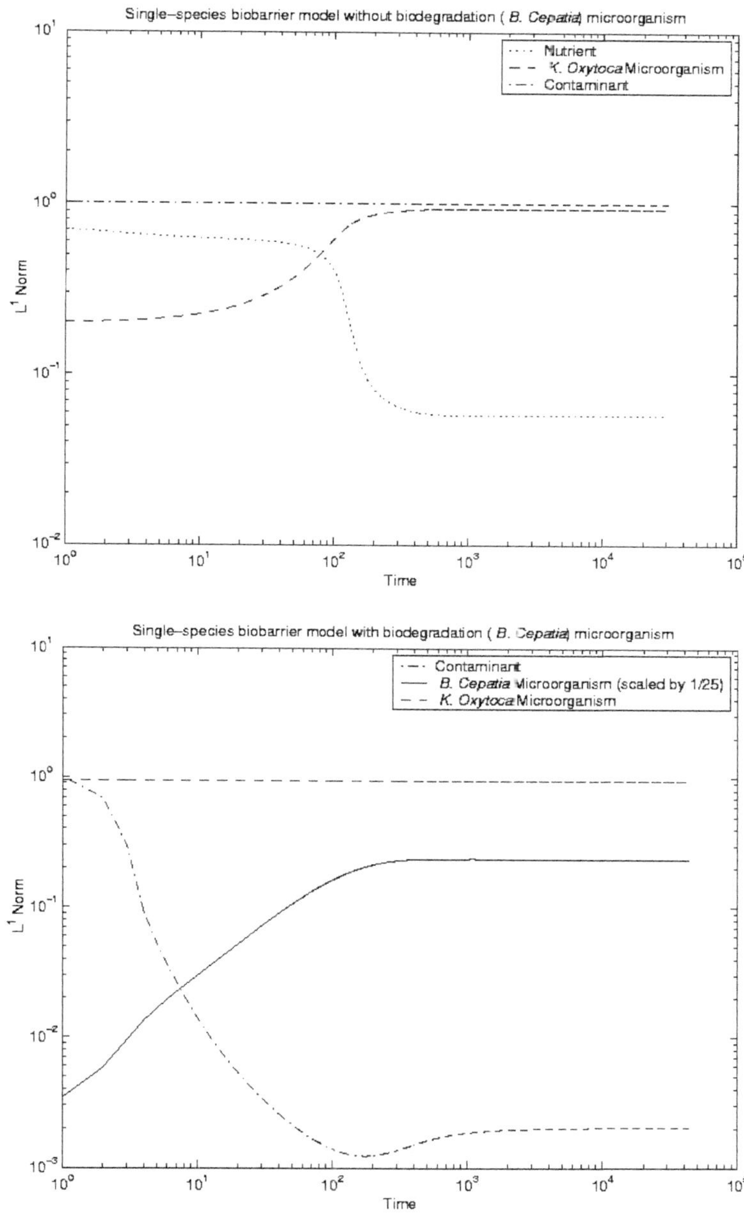


**Fig. 3.** Log-log plots of the $L^1$ norms of normalized concentrations of the biobarrier-forming microbes, $X_K$, the biodegradation microbes, $X_B$, the nutrient, $S_C$, and the contaminant, $S_T$, versus time — before (above) and after (below) the introduction of the biodegradation microbes into the system

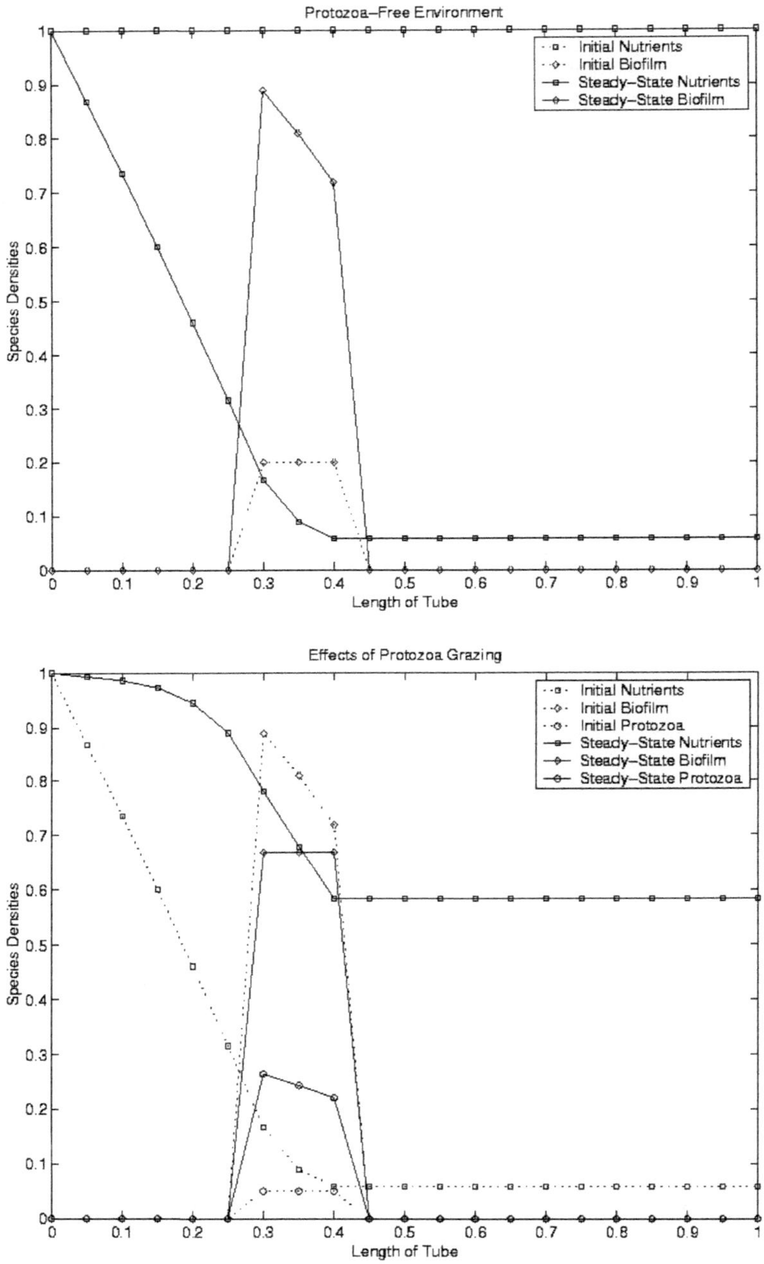


**Fig. 4.** Plots of the initial (dotted lines) and the steady-state (solid lines) normalized concentrations of the biofilm, $X_K$, the protozoa, $P$, and the nutrient, $S_C$ — before (above) and after (below) the introduction of the protozoa into the system

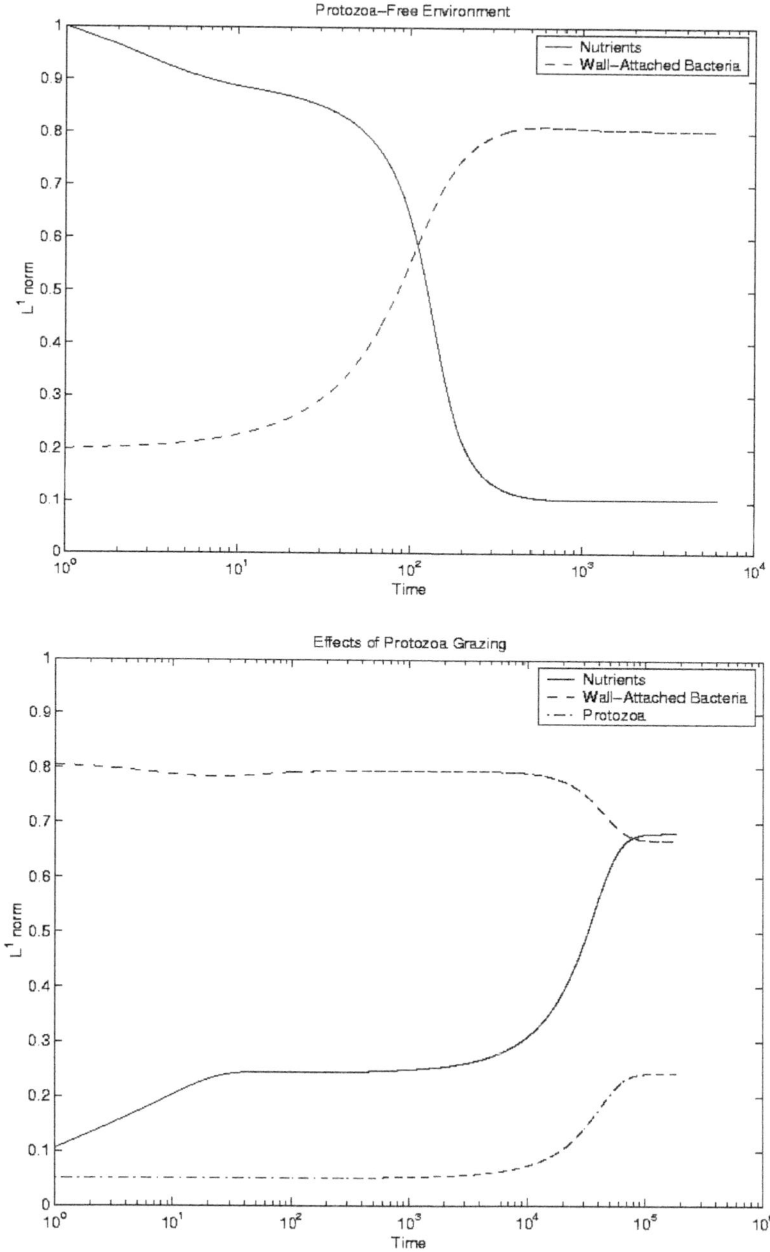


**Fig. 5.** Semi-logarithmic plots of the $L^1$ norms of normalized concentrations of the wall-attached bacteria, $X_K$, the protozoa, $P$, and the nutrient, $S_C$, versus time — before (above) and after (below) the introduction of the protozoa into the system

### 5.3. Dual-Species Biobarrier Model

In this subsection, we simulate the porous media experiments done by Komlos et al. (2002) at the Center for Biofilm Engineering. We use the same parameter values (see Table 2) and boundary conditions (51) as in the single-species biobarrier model.

In the first (high-nutrient-supply) experiment, the initial conditions for $S_T$ and $X_K$ are the same as in (50), and

$$X_B(x,0)=\begin{cases}0.2, & 0.3\le x\le 0.4\\ 0, & \text{otherwise}\end{cases}.$$

However, the initial and the boundary conditions for $S_C$ are ten-times higher than in the single-species experiment, i.e., $S_C(x,0)=S_C(0,t)=200\,\mu g/ml$. In this high substrate experiment, *K. oxytoca*'s population density is almost an order-of-magnitude higher that *B. cepatia*'s population density (Figs. 6 and 7, above).

In the second (low-nutrient-supply) experiment, we use the same initial conditions as (50) and the initial condition for $X_B$ is the same as in the high-substrate dual-species experiment. The low-substrate experiment shows the opposite of what would be expected (Figs. 6 and 7, below). The *B. cepatia*'s population density is almost an order-of-magnitude higher than *K. oxytoca*'s, even though *K. oxytoca* microorganism has a growth rate higher than *B. cepatia* (Table 2). The second numerical experiment confirms what was observed in practice (Camper et al. 1996; Komlos et al. 2002), that slower growing organisms are able to persist at high cell concentrations in some low-nutrient environments. As can be seen from Figures 6 and 7, the biofilm-forming bacteria growth is much slower for low-substrate concentrations which allows the degrading bacteria to grow more and therefore eliminate more contaminants. The dual-species simulation results (Fig. 7), under conditions of low- and high-nutrient supply, qualitatively match actual experiment results done by Komlos et al. (2002).

Flow-rate reduction is greater at the high-substrate concentration since *K. oxytoca*'s growth is the dominant one and there is more biofilm to block the flow, but not enough *B. cepatia* to degrade enough contaminant (Fig. 8, above). TCE degradation potential is larger at low-substrate concentrations, since *K. oxytoca*'s slower growth allows *B. cepatia* to grow more and therefore degrade more TCE (Fig. 8, below). This lower substrate concentration produces a more efficient reacting biobarrier. Long-term numerical simulation results (Fig. 8) of flow-rate reduction and TCE degradation are in agreement with actual experiments presented in (Komlos 2001).

The dual-species biobarrier experiments show that varying the substrate concentration can provide a mechanism to control the fraction of each organism in the dual-species biofilm, and therefore enhance its TCE degradation potential.

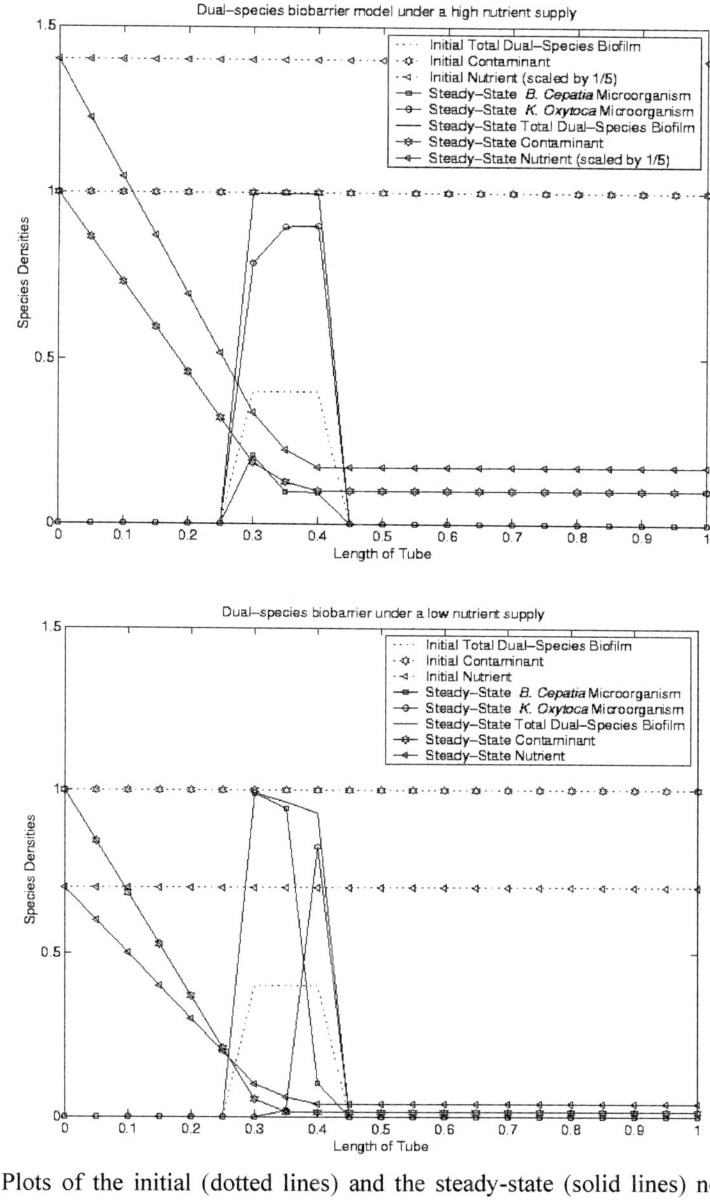


**Fig. 6.** Plots of the initial (dotted lines) and the steady-state (solid lines) normalized concentrations of the total dual-species biofilm, $X_{tot} = X_K + X_B$, the strong biofilm-forming microbes, $X_K$, the TCE-degrading microbes, $X_B$, the nutrient, $S_C$, and the contaminant, $S_T$ — at high nutrient (above) and at low nutrient (below) supply into the system

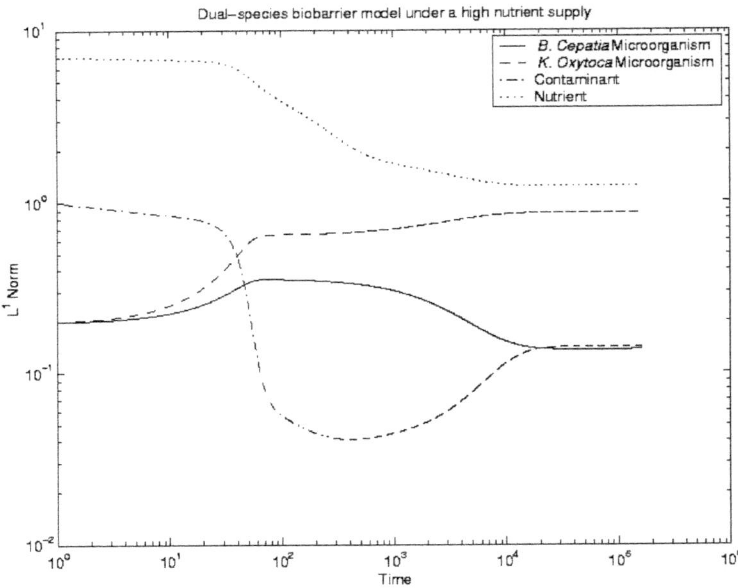


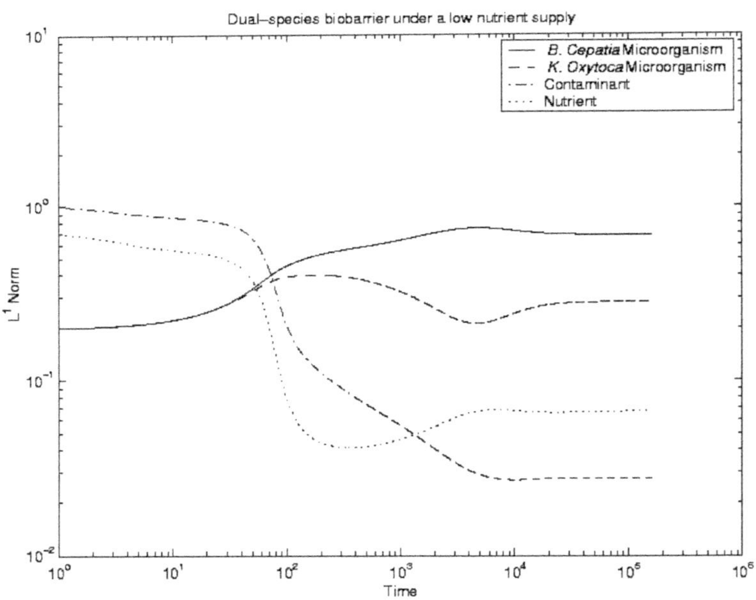


**Fig. 7.** Log-log plots of the $L^1$ norms of normalized concentrations of the strong biofilm-forming microbes, $X_K$, the TCE-degrading microbes, $X_B$, the nutrient, $S_C$, and the contaminant, $S_T$, versus time — at high nutrient (above) and at low nutrient (below) supply into the system

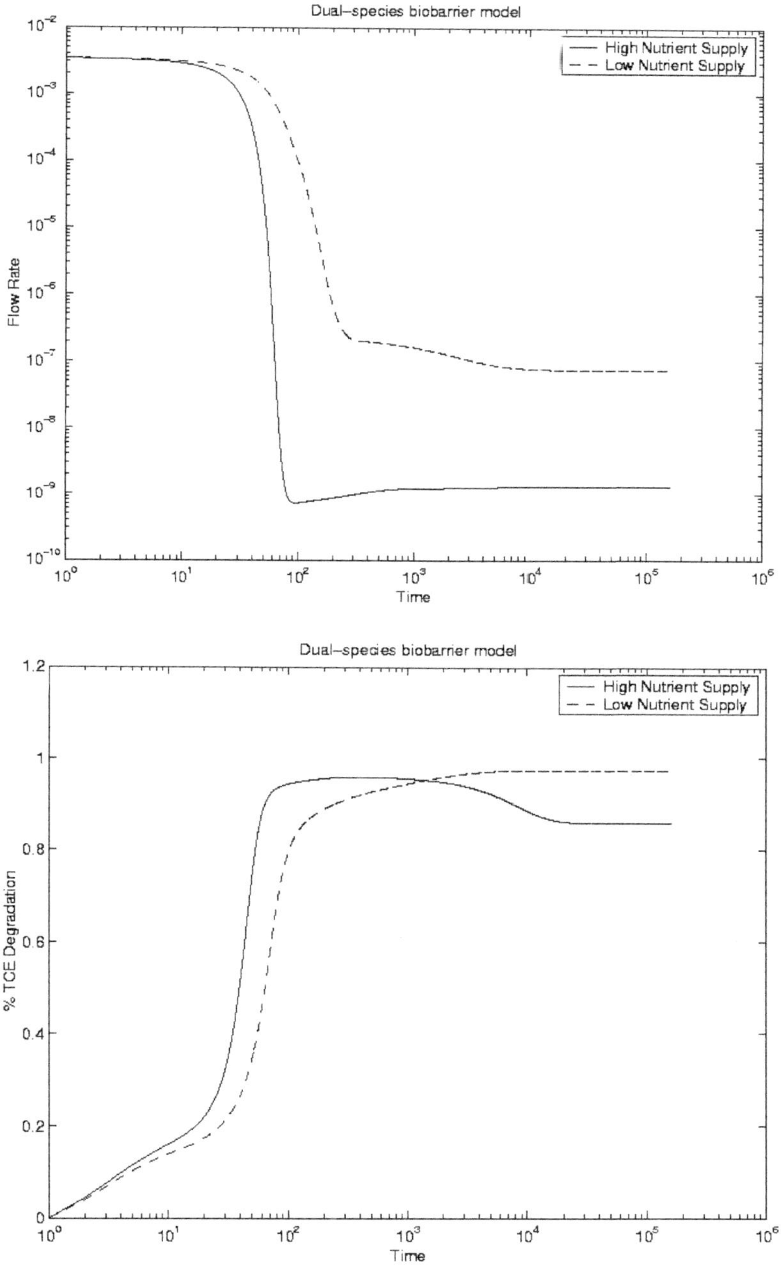


**Fig. 8.** Log-log plot of the flow rate reduction over time (above) and semi-logarithmic plot of the TCE degradation potential (below) at low nutrient (dotted lines) and high nutrient (solid lines) supply into the system

## 6. Conclusions

The motive for this research was to gain a better understanding of the interactions between two microbial species as part of a single biofilm capable of performing multiple functions (bioremediation, biofilm formation, etc.) and also to study the persistence of subsurface biobarriers when grazed by protozoa (ciliates, flagellates, etc.). We have presented a mathematical model for the flow, the transport of nutrients and contaminants, and the growth of two types of microorganisms and protozoan species in porous media. The coupled system of equations was successfully solved using nonstandard numerical methods.

The biobarrier-protozoa simulation results seem to indicate that even though predation tends to destroy the biobarriers, if the amount of nutrients is large enough the barriers will stabilize at a useful size. The colony forming unit ($CFU$) used in the (Komlos 2001; Komlos et al. 2002) is only a qualitative measure of population densities, since results of measurements depend on the viability of the cells that have been sampled from the experiments. Consequently, we are able to make only qualitative comments on the behavior of dual-species biofilms. The use of the two bacteria, *Burkholderia cepatia* and *Klebsiella oxytoca*, looks very promising as shown by the actual experiments (Komlos et al. 2000, 2002; Komlos 2001) and by our numerical simulations (Figs. 6-8).

Some future work will allow the degrading bacteria and the protozoa some movement and will consider more spatial dimensions.

**Nomenclature.** The symbols $L$, $M$, and $T$ denote the dimensions of length, mass, and time (Table 3).

**Table 3.** Parameter-symbol-dimensions relations

| Parameter | Symbol | Dimensions |
|---|---|---|
| Nutrients concentration | $S_C$, $S_T$ | $M/L$ |
| Biofilm concentration | $X_K, X_B$ | $M/L$ |
| Protozoa concentration | $P$ | $M/L$ |
| Specific discharge | $v$ | $L/T$ |
| Dispersion coefficient | $D$ | $L^2/T$ |
| Maximum specific growth rates | $\mu_{max}^K$, $\mu_{max}^B$, $\mu_{max}^P$ | $1/T$ |
| Saturation constants | $K_{X_K}^P$, $K_{S_C}^K$, $K_{S_C}^B$, $K_{S_T}^B$ | $M/L$ |
| Yield coefficients | $Y_K$, $Y_B$, $Y_P$ | $M/M$ |
| Endogenous decay coefficients | $k_K$, $k_B$, $k_P$ | $1/T$ |
| Hydraulic head | $h$ | $L$ |
| Specific storage | $S_s$ | $1/L$ |
| Source/sink term | $f$ | $1/T$ |
| Saturated hydraulic conductivity | $K$ | $L/T$ |
| Porosity | $\phi$ | dimensionless |

Acknowledgements. B. Chen was supported in part by NSF Grant EAR#0083880. H. Kojoukarov was supported in part by NSF Grant DMS #0107439 and UTA Grant REP #14748717. We acknowledge Jeanette Reisenburg's help with editing this paper.

## References

Allen MB (1988) Basic mechanics of oil reservoir flows. In: Brebia CA, Orszag SA (eds) Multiphase Flow in Porous Media, MB Allen III, GA Behie, JA Trangenstein, Lecture Notes in Engineering, Springer-Verlag, New York, NY, 34: 1-81

Allen MB, Curran MC (1992) A multigrid-based solver for the mixed finite-element approximations to groundwater flow. In: Russell TF et al. (eds) Computational Methods in Water Resources IX, Vol. 1: Numerical Methods in Water Resources, Elsevier Applied Science Publishers, London, pp. 579-585

Allen MB, Ewing RE, Lu P (1992) Well conditioned iterative schemes for mixed finite-element models of porous-media flows. SIAM J Sci Stat Comput 13(3):794-814

Allen MB, Liu B (1995) A modified method of characteristics incorporating streamline diffusion. Numer. Methods Partial Differential Equations, 11, pp 155-174

Allen MB, Wang Z (1994) A multigrid-based solver for transient groundwater flows using cell-centered differences. In: Peters A et al. (eds) Computational Methods in Water Resources X, , Kluwer Academic Publishers, the Netherlands, pp 1375-1382

Arbogast T, Wheeler MF (1995) A characteristics-mixed finite element method for advection-dominated transport problems. SIAM J Numer Anal 32:404-424

Bailey JE, Ollis DF (1986) Biochemical Engineering Fundamentals, McGraw-Hill, New York, NY

Berninger U-G, Finlay BJ, Kuuppo-Leinikki P (1991) Protozoan control of bacterial abundances in freshwater. Limnol Oceanogr 36:139-147

Camper AK, Jones WJ, Hayes JT (1996) Effect of growth conditions and substratum composition on the persistence of coliforms in mixed-population biofilms. Appl Environ Microbiol 62:4101-4018

Celia MA, Russell TF, Herrera I, Ewing RE (1990) An Eulerian-Lagrangian localized adjoint method for the advection-diffusion equation. Adv Water Resour 13(4):187-206

Characklis WG, Marshall KC (1990) Biofilms, John Wiley and Sons, Inc., New York

Chen B, Cunningham A, Ewing R, Peralta R, Visser E (1994) Two-dimensional modeling of microscale transport and biotransformation in porous media. Numer Meth PDE 10:65-83

Chen BM, Kojouharov HV (1999) Non-standard numerical methods applied to subsurface biobarrier formation models in porous media. Bull Math Biol 61:779-798

Clement TP, Hooker BS, Skeen RS (1996) Microscopic models for the predicting changes in the saturated porous media properties caused by microbial growth. Ground Water 34(5):934-942

Cunningham AB, Characklis WG, Abedeen F and Crawford D (1991) Influence of the biofilm accumulation on porous media hydrodynamics. Environ Sci Technol 25(7)1305-1311

Dawson CN, Wheeler MF (1992) Time-splitting methods for advection-diffusion-reaction equations arising in contaminant transport, ICIAM 91, (Washington, DC, 1991), SIAM, Philadelphia, PA, pp 71-82

Douglas J Jr, Russell TF (1982) Numerical methods for convection-dominated diffusion problems based on combining the method of characteristics with finite element or finite difference procedures, SIAM J Numer Anal 19:871-885

Eisenmann H, Harms H, Meckenstock R, Meyer EI, Zehnder AJB (1998) Grazing of a *Tetrahymena* sp. on adhered bacteria in percolated columns monitored by *in situ* hybridization with fluorescent oligonucleotide probes. Appl Env Microbio 64(4):1264-1269

Ewing RE, Lazarov RD, Lu P, Vassilevski PS (1990) Preconditioning indefinite systems arising from mixed finite element discretization of second-order elliptic systems. In: Axelsson O, Kolotilina L (eds) Preconditioned Conjugate Gradient Methods, Lecture Notes in Mathematics 1457, Springer-Verlag, Berlin, pp 280-343

Ewing RE, Russell TF (1982) Efficient time-stepping methods for miscible displacement problems in porous media. SIAM J Numer Anal 19:1-66

Fenchel T (1986) The Ecology of Heterotrophic Microflagellates, Advances in Microbial Ecology, Plenum Press, New York, NY, vol 9, pp 57-97

Gray WG, Pinder GF (1976) An analysis of the numerical solution of the transport equation. Water Resour Res 12(3)547-555

Healy RW, Russell TF (1993) A finite-volume Eulerian-Lagrangian localized adjoint method for solution of the advection-dispersion equation. Water Resour Res 29(7):2399-2413

Huyakorn PS, Pinder GF (1983) Computational Methods in Subsurface Flow, Academic Press, New York

James GA, Warwood BK, Cunningham AB, Sturman PJ, Hiebert R, Costerton JW (1995) Evaluation of subsurface biobarrier formation and persistence. In: Erickson LE (ed) Proceedings of the 10th Annual Conference on Hazardous Waste Research, Great Plains/Rocky Mountain HSRC, Manhattan, KS, pp 82-91

Johnson C, Saranen J (1986) Streamline diffusion methods for the incompressible euler and navier-stokes equations. Math Comp 47(175)1-18

Jones D, Smith H (2000) Microbial competition for nutrient and wall sites in plug flow. SIAM J Appl Math 60(5):1576-1600

Kojouharov HV, Chen BM (1998) Non-standard methods for the convective transport equation with nonlinear reactions. Numer Methods Partial Differential Equations 14(4):467-485

Kojouharov HV, Chen BM (2000) Non-standard eulerian-lagrangian methods for advection-diffusion-reaction equations. In: Mickens RE (ed) Applications of the Nonstandard Finite Difference Schemes, World Sci. Publishing, River Edge, NJ, pp 55-108

Kojouharov HV, Chen BM (2004) Nonstandard eulerian-lagrangian methods for multi-dimensional reactive transport problems. Appl Numer Math, to appear

Kojouharov HV, Welfert BD (2001) A new numerical approach for the solution of scalar nonlinear advection-reaction equations. Internat J Appl Sci Comput 8(2):119-126

Kojouharov HV, Welfert BD (2004) Generalized nonstandard numerical methods for nonlinear advection-diffusion-reaction equations, Lecture Notes in Comput Sci, to appear

Komlos J (2001) Effect of co-substrate concentration on dual-species population distribution, permeability reduction and Trichloroethylene (TCE) biodegradation in porous media, Ph.D. Dissertation in Civil Engineering, Montana State University

Kolmos J, Cunningham AB, Sharp RR (2000) Population dynamics in a multi-species biofilm for the creation of a reactive biobarrier. In: Erickson LE (ed) Proceedings of the 1999 Conference on Hazardous Waste Research, Great Plains/Rocky Mountain HSRC, Manhattan, KS, pp 158-166

Komlos J, Cunningham AB, Camper AK, Sharp RR (2002) Effect of substrate concentration on growth rate and population dynamics in a dual-species biofilm, preprint

Larsen RW, Cunningham AB, Characklis WG (2000) Microbial transport and biotransformation processes in porous media, Prepared for Presentation at the American Institute of Chemical Engineers 1990 Summer National Meeting

Liu B, Allen MB, Kojouharov H, Chen B (1996) Finite-element solution of reaction-diffusion equations with advection. In: Aldama AA et al. (eds) Computational Methods in Water Resources XI, Vol. 1: Computational Methods in Subsurface Flow and Transport Problems, Computational Mechanics Publications, Southampton Boston, pp 3-12

Mitchell AR, Griffiths DF (1980) The Finite Difference Method in Partial Differential Equations, John Wiley & Sons, New York

Raviart PA, Thomas JM (1977) A mixed finite element method for second order elliptic problems. In: Galligani I, Magenes E (eds) Mathematical Aspects of Finite Element Methods, Lecture Notes in Mathematics 606, Springer-Verlag, Berlin, pp 292-315

Russell TF, Wheeler MF (1983) Finite element and finite difference methods for continuous flows in porous media. In: Ewing RE (ed) Frontiers in Applied Mathematics, Vol. 1: The Mathematics of Reservoir Simulation, SIAM, Philadelphia, pp 35-106

Taylor SW, Jaffé PR (1990) Subsurface and Biomass Transport in a Porous Medium. Water Resour Res 26:2181-2194

# Index

## G



## H



## I



## K



## L



## M

**U**



**V**

Printing: Krips bv, Meppel
Binding: Stürtz, Würzburg

Printed by Printforce, the Netherlands